ELEMENTARY TRIGONOMETRY

ELEMENTARY TRIGONOMETRY

HENRY SINCLAIR HALL AND
SAMUEL RATCLIFFE KNIGHT

GYAN

Published by

Gyan Publishing House
5, Ansari Road
Daryaganj, New Delhi-110002
Phone: 011-47034999, 9811692060
E-mail: books@gyanbooks.com

Distribution Network
gyanbooks.com
India, USA, Canada, UK, Australia

ISBN: 978-93-7447-254-5 (PB)

First Published , 1906

2nd Impression 2025

Printed at: Gyan Press, Delhi.

ELEMENTARY TRIGONOMETRY
Author: Henry Sinclair Hall and Samuel Ratcliffe Knight

ELEMENTARY TRIGONOMETRY

BY

H. S. HALL, M.A.,

FORMERLY SCHOLAR OF CHRIST'S COLLEGE, CAMBRIDGE.

AND

S. R. KNIGHT, B.A., M.B., CH.B.,

FORMERLY SCHOLAR OF TRINITY COLLEGE, CAMBRIDGE.

FOURTH EDITION, REVISED AND ENLARGED.

London

MACMILLAN AND CO., LIMITED

NEW YORK: THE MACMILLAN COMPANY

1906

PREFACE.

THE distinctive features of the Fourth Edition are:

		PAGE
(1)	Practical Exercises in constructing angles with given ratios, and in finding the trigonometrical ratios of given angles .	7_A
(2)	The Use of Four-Figure Tables of sines, cosines, and tangents	29_A
(3)	Easy Problems requiring Four-Figure Tables	48_A
(4)	Graphs of the Trigonometrical Functions	79_A
(5)	A set of Easy Miscellaneous Examples on Chapters XI and XII	122_A
(6)	The Use of Four-Figure Logarithms and Antilogarithms	163_A
(7)	Solution of triangles with Four-Figure Logarithms	183_A
(8)	Four-Figure Tables of Logarithms, Antilogarithms, Natural and Logarithmic Functions	374

The Tables of Logarithms and Antilogarithms have been taken, with slight modifications, from those published by the Board of Education, South Kensington.

The Four-Figure Tables of Natural and Logarithmic Functions have been reduced from Seven-Figure Tables. For these I am greatly indebted to Mr Frank Castle, who kindly undertook the laborious task of a special compilation for this book.

(9) An easy first course has been mapped out enabling teachers to postpone, if they wish, all but the easier kinds of identities and transformations, so as to reach the more practical parts of the subject as early as possible.

All the special features of earlier editions have been retained, and it is hoped that the present additions will satisfy all modern requirements.

H. S. HALL.

August 1905.

SUGGESTIONS FOR A FIRST COURSE.

In the first eighteen chapters an asterisk has been placed before all articles and sets of examples which may conveniently be omitted from a first course.

For those who wish to postpone the harder identities and transformations, so as to reach practical work with Four-Figure Logarithms at an earlier stage, the following detailed course is recommended.

Chaps. I—III, Arts. 1—30, 32, 33. [Omit Art. 31, Examples III. b.]

Chaps. IV—IX. [Postpone Chaps. XI and XII.]

Chaps. XIII—XV, Arts. 137—170, 182_A—182_F. [Omit Seven-Figure Tables, Arts. 171—182.]

Chaps. XI, XII. [Omit Arts. 127, 136, Examples XI. f. and XII. e.]

Chap. XVI, Arts. 183—187, 197_A—197_D. [Omit Solutions with Seven-Figure Tables, Arts. 188—197.]

Chaps. XVII, XVIII, Arts. 198—218.

From this point the omitted sections must be taken at the discretion of the Teacher.

CONTENTS.

Chapter I. MEASUREMENT OF ANGLES.

Chapter II. TRIGONOMETRICAL RATIOS.

Chapter III. RELATIONS BETWEEN THE RATIOS.

Chapter IV. TRIGONOMETRICAL RATIOS OF CERTAIN ANGLES.

Chapter V. SOLUTION OF RIGHT-ANGLED TRIANGLES.

Chapter VI. EASY PROBLEMS.

Chapter VII. RADIAN OR CIRCULAR MEASURE.

Chapter VIII. RATIOS OF ANGLES OF ANY MAGNITUDE.

Chapter IX. VARIATIONS OF THE FUNCTIONS.

Chapter X. CIRCULAR FUNCTIONS OF ALLIED ANGLES.

Chapter XI. FUNCTIONS OF COMPOUND ANGLES.

Chapter XII. TRANSFORMATION OF PRODUCTS AND SUMS.

Chapter XIII. RELATIONS BETWEEN THE SIDES AND ANGLES OF A TRIANGLE.

Chapter XIV. LOGARITHMS.

Chapter XV. THE USE OF LOGARITHMIC TABLES.

Chapter XVI. SOLUTION OF TRIANGLES WITH LOGARITHMS.

Chapter XVII. HEIGHTS AND DISTANCES.

Chapter XVIII. PROPERTIES OF TRIANGLES AND POLYGONS.

Chapter XIX. GENERAL VALUES AND INVERSE FUNCTIONS.

ELEMENTARY TRIGONOMETRY.

CHAPTER I.

MEASUREMENT OF ANGLES.

1. THE word Trigonometry in its primary sense signifies the measurement of triangles. From an early date the science also included the establishment of the relations which subsist between the sides, angles, and area of a triangle; but now it has a much wider scope and embraces all manner of geometrical and algebraical investigations carried on through the medium of certain quantities called trigonometrical ratios, which will be defined in Chap. II. In every branch of Higher Mathematics, whether Pure or Applied, a knowledge of Trigonometry is of the greatest value.

2. Definition of Angle. Suppose that the straight line OP in the figure is capable of revolving about the point O, and suppose that in this way it has passed successively from the position OA to the positions occupied by OB, OC, OD, ..., then the angle between OA and any position such as OC is measured by the *amount of revolution* which the line OP has undergone in passing from its initial position OA into its final position OC.

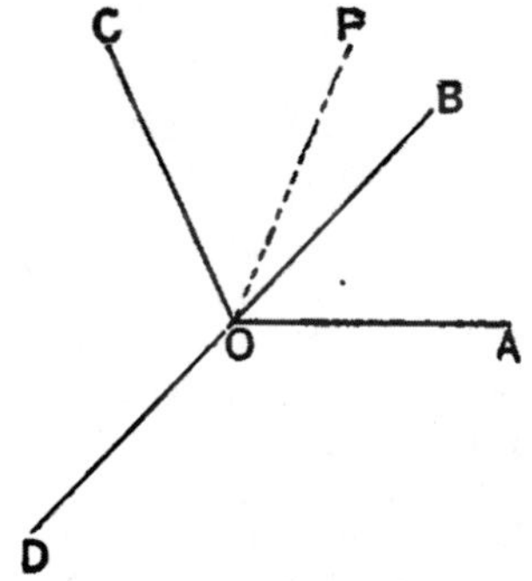

Moreover the line OP may make any number of complete revolutions through the original position OA before taking up its final position.

It will thus be seen that in Trigonometry angles are not restricted as in Geometry, but may be of any magnitude.

The point O is called the *origin*, and OA the *initial line;* the revolving line OP is known as the *generating line* or the *radius vector.*

3. Measurement of Angles. We must first select some *fixed* unit. The natural unit would be a right angle, but as in practice this is inconveniently large, two systems of measurement have been established, in each of which the unit is a certain fraction of a right angle.

4. Sexagesimal Measure. A right angle is divided into 90 equal parts called *degrees*, a degree into 60 equal parts called *minutes*, a minute into 60 equal parts called *seconds.* An angle is measured by stating the number of degrees, minutes, and seconds which it contains.

For shortness, each of these three divisions, degrees, minutes, seconds, is denoted by a symbol; thus the angle which contains 53 degrees 37 minutes 2·53 seconds is expressed symbolically in the form $53^\circ\ 37'\ 2{\cdot}53''$.

5. Centesimal Measure. A right angle is divided into 100 equal parts called *grades*, a grade into 100 equal parts called *minutes*, a minute into 100 equal parts called *seconds.* In this system the angle which contains 53 grades 37 minutes 2·53 seconds is expressed symbolically in the form $53^g\ 37^`\ 2{\cdot}53^{``}$.

It will be noticed that different accents are used to denote sexagesimal and centesimal minutes and seconds; for though they have the same names, a centesimal minute and second are not the same as a sexagesimal minute and second. Thus a right angle contains 90×60 sexagesimal minutes, whereas it contains 100×100 centesimal minutes.

Sexagesimal Measure is sometimes called the English System, and Centesimal Measure the French System.

6. In *numerical* calculations the sexagesimal measure is always used. The centesimal method was proposed at the time of the French Revolution as part of a general system of decimal measurement, but has never been adopted even in France, as it would have made necessary the alteration of Geographical, Nautical, Astronomical, and other tables prepared according to the sexagesimal method. Beyond giving a few examples in transformation from one system to the other which afford exercise in easy Arithmetic, we shall after this rarely allude to centesimal measure.

In *theoretical* work it is convenient to use another method of measurement, where the unit is the angle subtended at the centre of a circle by an arc whose length is equal to the radius. This system is known as **Circular** or **Radian Measure**, and will be fully explained in Chapter VII.

An angle is usually represented by a single letter, different letters $A, B, C, \ldots, \alpha, \beta, \gamma, \ldots, \theta, \phi, \psi, \ldots$, being used to distinguish different angles. For angles estimated in sexagesimal or centesimal measure these letters are used indifferently, but we shall always denote angles in circular measure by letters taken from the Greek alphabet.

7. *If the number of degrees and grades contained in an angle be D and G respectively, to prove that* $\frac{D}{9} = \frac{G}{10}$.

In sexagesimal measure, the given angle when expressed as the fraction of a right angle is denoted by $\frac{D}{90}$. In centesimal measure, the same fraction is denoted by $\frac{G}{100}$;

$$\therefore \frac{D}{90} = \frac{G}{100}; \text{ that is, } \frac{D}{9} = \frac{G}{10}.$$

8. To pass from one system to the other it is advisable first to express the given angle in terms of a right angle.

In centesimal measure any number of grades, minutes, and seconds may be immediately expressed as the decimal of a right angle. Thus

$$23 \text{ grades} = \tfrac{23}{100} \text{ of a right angle} = \cdot 23 \text{ of a right angle};$$

15 minutes $= \frac{15}{100}$ of a grade $= \cdot 15$ of a grade $= \cdot 0015$ of a right angle;

$$\therefore 23^g\ 15' = \cdot 2315 \text{ of a right angle.}$$

Similarly, $15^g\ 7'\ 53{\cdot}4'' = \cdot 1507534$ of a right angle.

Conversely, any decimal of a right angle can be at once expressed in grades, minutes, and seconds. Thus

$$\begin{aligned} \cdot 2173025 \text{ of a right angle} &= 21{\cdot}73025^g \\ &= 21^g\ 73{\cdot}025' \\ &= 21^g\ 73'\ 2{\cdot}5''. \end{aligned}$$

In practice the intermediate steps are omitted.

Example 1. Reduce $2^g\,13^{\backprime}\,4\cdot5^{\backprime\backprime}$ to sexagesimal measure.

This angle $=\cdot0213045$ of a right angle
$=1°\,55'\,2\cdot658''$.

$$
\begin{array}{rl}
\cdot0213045 & \text{of a right angle} \\
90 & \\
\hline
1\cdot917405 & \text{degrees} \\
60 & \\
\hline
55\cdot0443 & \text{minutes} \\
60 & \\
\hline
2\cdot658 & \text{seconds.}
\end{array}
$$

Obs. In the Answers we shall express the angles to the nearest tenth of a second, so that the above result would be written $1°\,55'\,2\cdot7''$.

Example 2. Reduce $12°\,13'\,14\cdot3''$ to centesimal measure.

This angle $=\cdot13578487\ldots$ of a right angle
$=13^g\,57^{\backprime}\,84\cdot9^{\backprime\backprime}$.

$$
\begin{array}{rl}
60\,) & 14\cdot3 \text{ seconds} \\
60\,) & 13\cdot238333\ldots \text{minutes} \\
90\,) & 12\cdot2206388\ldots \text{degrees} \\
 & \cdot13578487\ldots \text{of a right angle.}
\end{array}
$$

EXAMPLES. I.

Express as the decimal of a right angle

1. $67°\,30'$.
2. $11°\,15'$.
3. $37^g\,50^{\backprime}$.
4. $2°\,10'\,12''$.
5. $8°\,0'\,36''$.
6. $2^g\,4^{\backprime}\,4\cdot5^{\backprime\backprime}$.

Reduce to centesimal measure

7. $69°\,13'\,30''$.
8. $19°\,0'\,45''$.
9. $50°\,37'\,5\cdot7''$.
10. $43°\,52'\,38\cdot1''$.
11. $11°\,0'\,38\cdot4''$.
12. $142°\,15'\,45''$.
13. $12'\,9''$.
14. $3'\,26\cdot3''$.

Reduce to sexagesimal measure

15. $56^g\,87^{\backprime}\,50^{\backprime\backprime}$.
16. $39^g\,6^{\backprime}\,25^{\backprime\backprime}$.
17. $40^g\,1^{\backprime}\,25\cdot4^{\backprime\backprime}$.
18. $1^g\,2^{\backprime}\,3^{\backprime\backprime}$.
19. $3^g\,2^{\backprime}\,5^{\backprime\backprime}$.
20. $8^g\,10^{\backprime}\,6\cdot5^{\backprime\backprime}$.
21. $6^{\backprime}\,25^{\backprime\backprime}$.
22. $37^{\backprime}\,5^{\backprime\backprime}$.

23. The sum of two angles is 80^g and their difference is $18°$; find the angles in degrees.

24. The number of degrees in a certain angle added to the number of grades in the angle is 152: what is the angle?

25. If the same angle contains in English measure x minutes, and in French measure y minutes, prove that $50x=27y$.

26. If s and t respectively denote the numbers of sexagesimal and centesimal seconds in any angle, prove that

$$250s=81t.$$

CHAPTER II.

TRIGONOMETRICAL RATIOS.

9. DEFINITION. **Ratio** is the relation which one quantity bears to another of the *same* kind, the comparison being made by considering what multiple, part or parts, one quantity is of the other.

To find what multiple or part A is of B we divide A by B; hence the ratio of A to B may be measured by the fraction $\frac{A}{B}$.

In order to compare two quantities they must be expressed in terms of the same unit. Thus the ratio of 2 yards to 27 inches is measured by the fraction $\frac{2 \times 3 \times 12}{27}$ or $\frac{8}{3}$.

OBS. Since a ratio expresses the *number* of times that one quantity contains another, *every ratio is a numerical quantity.*

10. DEFINITION. If the ratio of any two quantities can be expressed exactly by the ratio of two integers the quantities are said to be **commensurable**; otherwise, they are said to be **incommensurable.** For instance, the quantities $8\frac{4}{5}$ and $5\frac{1}{3}$ are commensurable, while the quantities $\sqrt{2}$ and 3 are incommensurable. But by finding the numerical value of $\sqrt{2}$ we may express the value of the ratio $\sqrt{2} : 3$ by the ratio of two commensurable quantities to any required degree of approximation. Thus to 5 decimal places $\sqrt{2} = 1{\cdot}41421$, and therefore to the same degree of approximation

$$\sqrt{2} : 3 = 1{\cdot}41421 : 3 = 141421 : 300000.$$

Similarly, for the ratio of any two incommensurable quantities.

Trigonometrical Ratios.

11. Let PAQ be any acute angle; in AP one of the boundary lines take a point B and draw BC perpendicular to AQ. Thus a right-angled triangle BAC is formed.

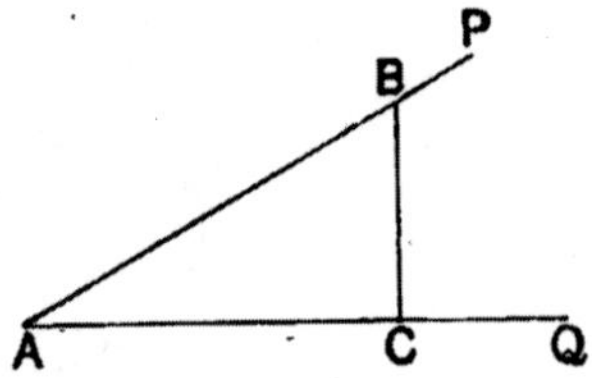

With reference to the angle A the following definitions are employed.

The ratio $\frac{BC}{AB}$ or $\frac{\text{opposite side}}{\text{hypotenuse}}$ is called the **sine of A.**

The ratio $\frac{AC}{AB}$ or $\frac{\text{adjacent side}}{\text{hypotenuse}}$ is called the **cosine of A.**

The ratio $\frac{BC}{AC}$ or $\frac{\text{opposite side}}{\text{adjacent side}}$ is called the **tangent of A.**

The ratio $\frac{AC}{BC}$ or $\frac{\text{adjacent side}}{\text{opposite side}}$ is called the **cotangent of A.**

The ratio $\frac{AB}{AC}$ or $\frac{\text{hypotenuse}}{\text{adjacent side}}$ is called the **secant of A.**

The ratio $\frac{AB}{BC}$ or $\frac{\text{hypotenuse}}{\text{opposite side}}$ is called the **cosecant of A.**

These six ratios are known as the **trigonometrical ratios.** It will be shewn later that as long as the angle remains the same the trigonometrical ratios remain the same. [Art. 19.]

12. Instead of writing in full the words *sine, cosine, tangent, cotangent, secant, cosecant,* abbreviations are adopted. Thus the above definitions may be more conveniently expressed and arranged as follows:

$$\sin A = \frac{BC}{AB}, \qquad \operatorname{cosec} A = \frac{AB}{BC},$$

$$\cos A = \frac{AC}{AB}, \qquad \sec A = \frac{AB}{AC},$$

$$\tan A = \frac{BC}{AC}, \qquad \cot A = \frac{AC}{BC}.$$

In addition to these six ratios, two others, the *versed sine* and *coversed sine* are sometimes used; they are written vers A and covers A and are thus defined:

$$\text{vers } A = 1 - \cos A, \quad \text{covers } A = 1 - \sin A.$$

13. In Chapter VIII. the definitions of the trigonometrical ratios will be extended to the case of angles of any magnitude, but for the present we confine our attention to the consideration of acute angles.

14. Although the verbal form of the definitions of the trigonometrical ratios given in Art. 11 may be helpful to the student at first, he will gain no freedom in their use until he is able to write down from the figure any ratio at sight.

In the adjoining figure, PQR is a right-angled triangle in which $PQ=13$, $PR=5$, $QR=12$.

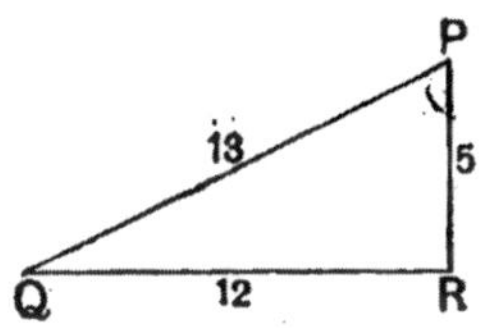

Since PQ is the greatest side, R is the right angle. The trigonometrical ratios of the angles P and Q may be written down at once; for example,

$$\sin Q = \frac{PR}{PQ} = \frac{5}{13}, \qquad \cos Q = \frac{QR}{PQ} = \frac{12}{13},$$

$$\tan P = \frac{QR}{PR} = \frac{12}{5}, \qquad \operatorname{cosec} P = \frac{PQ}{QR} = \frac{13}{12}.$$

15. It is important to observe that *the trigonometrical ratios of an angle are numerical quantities.* Each one of them represents the *ratio of one length to another*, and they must themselves never be regarded as lengths.

16. In every right-angled triangle the hypotenuse is the greatest side; hence from the definitions of Art. 11 it will be seen that those ratios which have the hypotenuse in the *denominator* can never be greater than unity, while those which have the hypotenuse in the *numerator* can never be less than unity. Those ratios which do not involve the hypotenuse are not thus restricted in value, for either of the two sides which subtend the acute angles may be the greater. Hence

the sine and cosine of an angle can never be greater than 1;

the cosecant and secant of an angle can never be less than 1;

the tangent and cotangent may have any numerical value.

Example 1. Draw an angle of 49°, and find by measurement its sine and cosine.

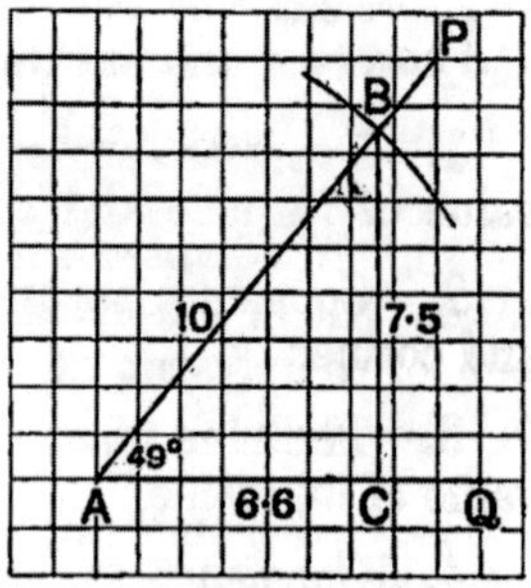

With a protractor make the $\angle PAQ$ equal to 49°. According to the definition we may take any point B on AP, and draw BC perp. to AQ. It will be convenient to use squared paper and to choose B so that $AB=10$ units. Then by measurement $BC=7\cdot5$ units, $AC=6\cdot6$ units.

Hence $\quad \sin 49° = \dfrac{BC}{AB} = \dfrac{7\cdot5}{10} = \cdot75,$

and $\quad \cos 49° = \dfrac{AC}{AB} = \dfrac{6\cdot6}{10} = \cdot66.$

Example 2. Construct an angle whose sine is ·39, and find the approximate value of its cosine.

Since $\cdot39 = \dfrac{3\cdot9}{10}$, we first draw a rt-angled △ whose hypotenuse is 10 units and one of whose sides is 3·9.

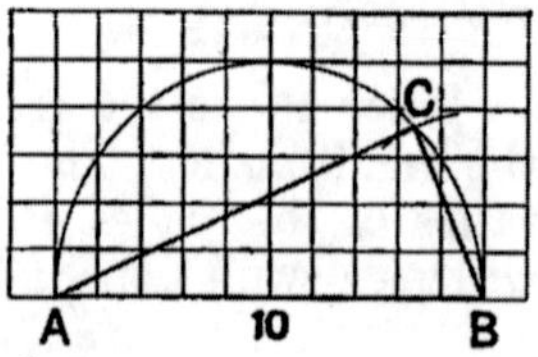

Hence describe a semi-circle of diameter AB, 10 units in length. With centre B and radius 3·9 units draw an arc to cut the semi-circle at C. Then $\angle ACB$ in the semi-circle is 90°.

Hence ABC is a right-angled triangle, and

$$\therefore \sin BAC = \frac{BC}{AB} = \frac{3\cdot9}{10} = \cdot39.$$

Thus BAC is the required angle. Also by measurement $AC=9\cdot2$ units.

$$\therefore \cos BAC = \frac{AC}{AB} = \frac{9\cdot2}{10} = \cdot92.$$

Example 3. Construct an angle whose tangent is 1·2 and find its sine and cosine.

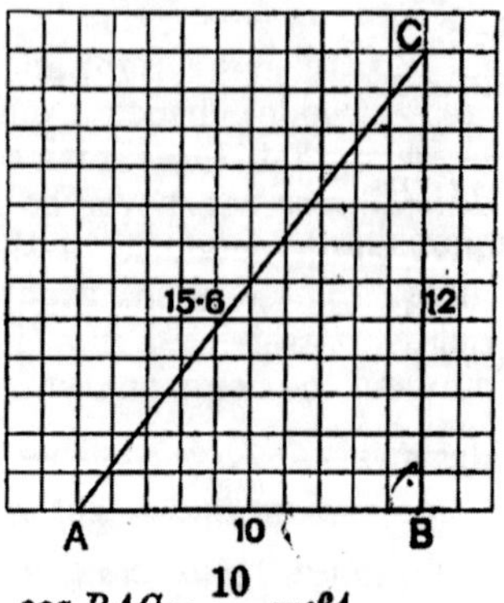

Since $1\cdot2 = \dfrac{12}{10}$, we must construct a right-angled triangle such that the ratio of the sides containing the right angle is 12 : 10. Hence draw $AB=10$ units and erect BC perpendicular to AB, and $=12$ units.

Then $\quad \tan BAC = \dfrac{CB}{AB} = \dfrac{12}{10} = 1\cdot2.$

Hence BAC is the required angle, and by measurement $AC=15\cdot6$.

$$\therefore \sin BAC = \frac{12}{15\cdot6} = \cdot77, \text{ and } \cos BAC = \frac{10}{15\cdot6} = \cdot64.$$

EXAMPLES II. a.

[*Approximate results should be given to two places of decimals.*]

1. Draw an angle of 77°, and find by measurement the value of its sine and cosine.

2. Construct an angle of 39°, and find the value of its sine and cosine.

3. The sine of an angle is ·88; draw the angle and find the value of its cosine.

4. Construct an angle whose cosine is ·34; measure the angle to the nearest degree, and find its sine and tangent.

5. Draw an angle of 42°, and find its tangent and sine.

6. Given $\sec A = 2{\cdot}8$, draw the angle and measure it to the nearest degree.

7. Construct an angle whose sine is ·6; measure the angle to the nearest degree.

8. Construct an angle from each of the following data:

(i) $\tan A = {\cdot}7$; (ii) $\cos B = {\cdot}9$; (iii) $\sin C = {\cdot}71$.

In each case measure the angle to the nearest degree.

Find $\sin A$, $\tan B$, $\cos C$.

9. Construct an angle A such that $\tan A = 1{\cdot}6$. Measure the angle to the nearest degree, and find its sine and cosine.

10. Construct a triangle ABC, right-angled at C, having the hypotenuse 10 cm. in length, and $\tan A = {\cdot}81$. Measure AC and the angle A; and find the values of $\sin A$ and $\cos A$.

11. Find the cosine and cosecant of an angle A whose sine is ·34. Prove that the values approximately satisfy the relation $\sin A \operatorname{cosec} A = 1$.

12. Draw a triangle ABC having $BC = 8$ cm., $\angle ABC = 53°$, $ACB = 72°$. Draw and measure the altitude, and hence find approximately the values of $\tan 53°$, $\cot 72°$.

13. Draw a right-angled triangle ABC from the following data:

$$\tan A = {\cdot}7, \quad \angle C = 90°, \quad b = 2{\cdot}8 \text{ cm.}$$

Measure c and the $\angle A$.

14. Draw the angles whose sines are ·67 and ·94 on the same side of a common arm. Measure their difference in degrees.

17. Let ABC be a right-angled triangle having the right angle at A; then by Geometry,

the sq. on BC

=sum of sqq. on AC and AB,

or, more briefly,

$$BC^2 = AC^2 + AB^2.$$

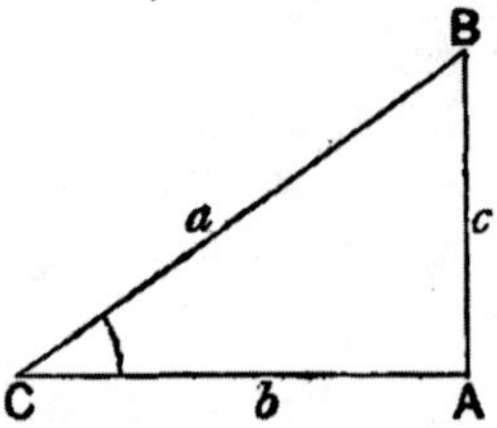

When we use this latter mode of expression it is understood that the sides AB, AC, BC are expressed in terms of some common unit, and the above statement may be regarded as a *numerical relation* connecting the numbers of units of length in the three sides of a right-angled triangle.

It is usual to denote the numbers of units of length in the sides opposite the angles A, B, C by the letters a, b, c respectively. Thus in the above figure we have $a^2 = b^2 + c^2$, so that if the lengths of two sides of a right-angled triangle are known, this equation will give the length of the third side.

Example 1. ABC is a right-angled triangle of which C is the right angle; if $a=3$, $b=4$, find c, and also $\sin A$ and $\cot B$.

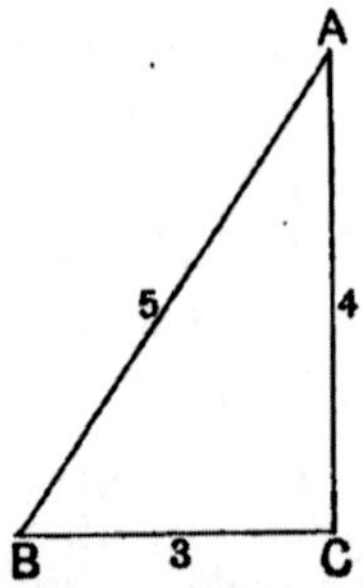

Here $c^2 = a^2 + b^2 = (3)^2 + (4)^2 = 9 + 16 = 25$;

$$\therefore c = 5.$$

Also $$\sin A = \frac{BC}{AB} = \frac{3}{5};$$

$$\cot B = \frac{BC}{AC} = \frac{3}{4}.$$

Example 2. A ladder 17 ft. long is placed with its foot at a distance of 8 ft. from the wall of a house and just reaches a window-sill. Find the height of the window-sill, and the sine and tangent of the angle which the ladder makes with the wall.

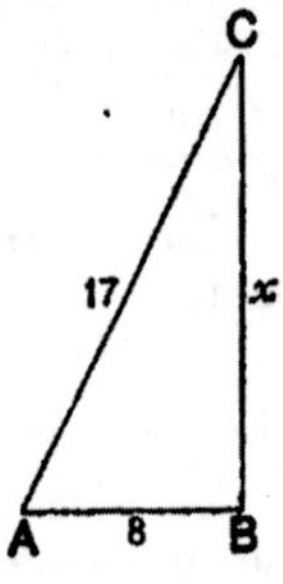

Let AC be the ladder, and BC the wall.

Let x be the number of feet in BC;

then $x^2 = (17)^2 - (8)^2 = (17+8)(17-8) = 25 \times 9$;

$$\therefore x = 5 \times 3 = 15.$$

Also $$\sin C = \frac{AB}{AC} = \frac{8}{17};$$

$$\tan C = \frac{AB}{BC} = \frac{8}{15}.$$

18. The following important proposition depends upon the well-known property of similar triangles. The student who has not read Proportion in Geometry should not fail to notice the result arrived at, even if he is unable at this stage to understand the proof.

19. *To prove that the trigonometrical ratios remain unaltered so long as the angle remains the same.*

Let AOP be any acute angle. In OP take any points B and

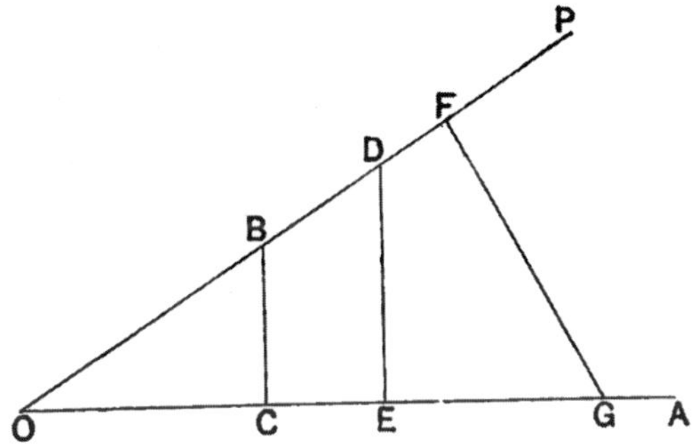

D, and draw BC and DE perpendicular to OA. Also take any point F in OP and draw FG at right angles to OP.

From the triangle BOC, $\sin POA = \frac{BC}{OB}$;

from the triangle DOE, $\sin POA = \frac{DE}{OD}$;

from the triangle FOG, $\sin POA = \frac{FG}{OG}$.

But the triangles BOC, DOE, FOG are equiangular;

$$\therefore \frac{BC}{OB} = \frac{DE}{OD} = \frac{FG}{OG}.$$

Thus the sine of the angle POA is the same whether it is obtained from the triangle BOC, or from the triangle DOE, or from the triangle FOG.

A similar proof holds for each of the other trigonometrical ratios. These ratios are therefore independent of the length of the revolving line and depend only on the magnitude of the angle.

20. If A denote any acute angle, we have proved that all the trigonometrical ratios of A *depend only on the magnitude of the angle A and not upon the lengths of the lines which bound the angle.* It may easily be seen that a change made in the value of A will produce a consequent change in the values of all the trigonometrical ratios of A. This point will be discussed more fully in Chap. IX.

Definition. Any expression which involves a variable quantity x, and whose value is dependent on that of x is called a **function of x.**

Hence the trigonometrical *ratios* may also be defined as trigonometrical *functions*; for the present we shall chiefly employ the term *ratio*, but in a later part of the subject the idea of ratio is gradually lost and the term *function* becomes more appropriate.

21. The use of the principle proved in Art. 19 is well shewn in the following example, where the trigonometrical ratios are employed as a connecting link between the lines and angles.

Example. ABC is a right-angled triangle of which A is the right angle. BD is drawn perpendicular to BC and meets CA produced in D: if $AB=12$, $AC=16$, $BC=20$, find BD and CD.

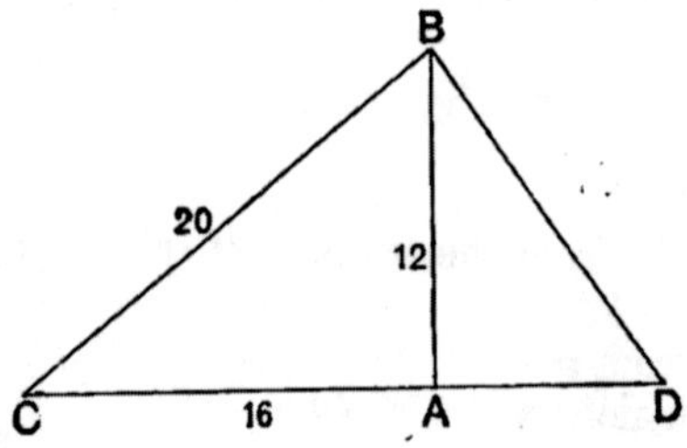

From the right-angled triangle CBD,

$$\frac{BD}{BC}=\tan C;$$

from the right-angled triangle ABC,

$$\frac{AB}{AC}=\tan C;$$

$$\therefore \frac{BD}{BC}=\frac{AB}{AC};$$

$$\therefore \frac{BD}{20}=\frac{12}{16}; \text{ whence } BD=15.$$

Again,
$$\frac{CD}{CB}=\sec C=\frac{BC}{CA};$$

$$\therefore \frac{CD}{20}=\frac{20}{16}; \text{ whence } CD=25.$$

The same results can be obtained by the help of Euc. vi. 8.

EXAMPLES. II. b.

1. The sides AB, BC, CA of a right-angled triangle are 17, 15, 8 respectively; write down the values of $\sin A$, $\sec A$, $\tan B$, $\sec B$.

2. The sides PQ, QR, RP of a right-angled triangle are 13, 5, 12 respectively: write down the values of $\cot P$, $\operatorname{cosec} Q$, $\cos Q$, $\cos P$.

3. ABC is a triangle in which A is a right angle; if $b=15$, $c=20$, find a, $\sin C$, $\cos B$, $\cot C$, $\sec C$.

4. ABC is a triangle in which B is a right angle; if $a=24$, $b=25$, find c, $\sin C$, $\tan A$, $\operatorname{cosec} A$.

5. The sides ED, EF, DF of a right-angled triangle are 35, 37, 12 respectively: write down the values of $\sec E$, $\sec F$, $\cot E$, $\sin F$.

6. The hypotenuse of a right-angled triangle is 15 inches, and one of the sides is 9 inches: find the third side and the sine, cosine and tangent of the angle opposite to it.

7. Find the hypotenuse AB of a right-angled triangle in which $AC=7$, $BC=24$. Write down the sine and cosine of A, and shew that the sum of their squares is equal to 1.

8. A ladder 41 ft. long is placed with its foot at a distance of 9 ft. from the wall of a house and just reaches a window-sill. Find the height of the window-sill, and the sine and cotangent of the angle which the ladder makes with the ground.

9. A ladder is 29 ft. long; how far must its foot be placed from a wall so that the ladder may just reach the top of the wall which is 21 ft. from the ground? Write down all the trigonometrical ratios of the angle between the ladder and the wall.

10. $ABCD$ is a square; C is joined to E, the middle point of AD: find all the trigonometrical ratios of the angle ECD.

11. $ABCD$ is a quadrilateral in which the diagonal AC is at right angles to each of the sides AB, CD: if $AB=15$, $AC=36$, $AD=85$, find $\sin ABC$, $\sec ACB$, $\cos CDA$, $\operatorname{cosec} DAC$.

12. $PQRS$ is a quadrilateral in which the angle PSR is a right angle. If the diagonal PR is at right angles to RQ, and $RP=20$, $RQ=21$, $RS=16$, find $\sin PRS$, $\tan RPS$, $\cos RPQ$, $\operatorname{cosec} PQR$.

CHAPTER III.

RELATIONS BETWEEN THE TRIGONOMETRICAL RATIOS.

22. Reciprocal relations between certain ratios.

(1) Let ABC be a triangle, right-angled at C;

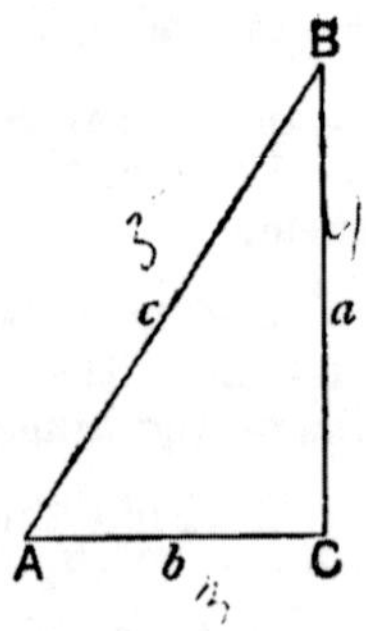

then $$\sin A = \frac{BC}{AB} = \frac{a}{c},$$

and $$\operatorname{cosec} A = \frac{AB}{BC} = \frac{c}{a};$$

$$\therefore \quad \sin A \times \operatorname{cosec} A = \frac{a}{c} \times \frac{c}{a} = 1.$$

Thus $\sin A$ and $\operatorname{cosec} A$ are reciprocals;

$$\therefore \quad \sin A = \frac{1}{\operatorname{cosec} A},$$

and $$\operatorname{cosec} A = \frac{1}{\sin A}.$$

(2) Again,

$$\cos A = \frac{AC}{AB} = \frac{b}{c}, \text{ and } \sec A = \frac{AB}{AC} = \frac{c}{b};$$

$$\therefore \quad \cos A \times \sec A = \frac{b}{c} \times \frac{c}{b} = 1;$$

$$\therefore \quad \cos A = \frac{1}{\sec A}, \text{ and } \sec A = \frac{1}{\cos A}.$$

(3) Also

$$\tan A = \frac{BC}{AC} = \frac{a}{b}, \text{ and } \cot A = \frac{AC}{BC} = \frac{b}{a};$$

$$\therefore \quad \tan A \times \cot A = \frac{a}{b} \times \frac{b}{a} = 1;$$

$$\therefore \quad \tan A = \frac{1}{\cot A}, \text{ and } \cot A = \frac{1}{\tan A}.$$

23. *To express* tan A *and* cot A *in terms of* sin A *and* cos A.

From the adjoining figure we have

$$\tan A = \frac{BC}{AC} = \frac{a}{b} = \frac{a}{c} \div \frac{b}{c}$$

$$= \sin A \div \cos A;$$

$$\therefore \tan A = \frac{\sin A}{\cos A}.$$

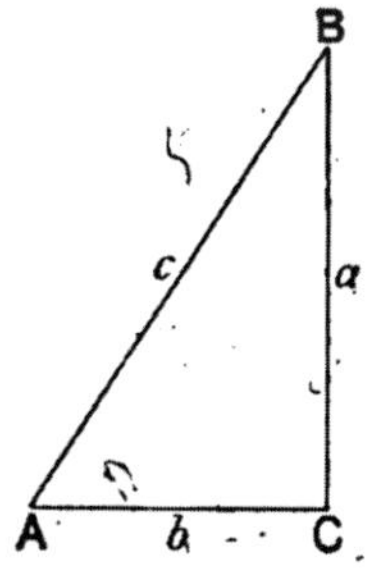

Again, $$\cot A = \frac{AC}{BC} = \frac{b}{a} = \frac{b}{c} \div \frac{a}{c}$$

$$= \cos A \div \sin A;$$

$$\therefore \cot A = \frac{\cos A}{\sin A};$$

which is also evident from the reciprocal relation $\cot A = \frac{1}{\tan A}$.

Example. Prove that cosec A tan A = sec A.

$$\operatorname{cosec} A \tan A = \frac{1}{\sin A} \times \frac{\sin A}{\cos A} = \frac{1}{\cos A}$$

$$= \sec A.$$

24. We frequently meet with expressions which involve the square and other powers of the trigonometrical ratios, such as $(\sin A)^2$, $(\tan A)^3$, ... It is usual to write these in the shorter forms $\sin^2 A$, $\tan^3 A$, ...

Thus $$\tan^2 A = (\tan A)^2 = \left(\frac{\sin A}{\cos A}\right)^2$$

$$= \frac{(\sin A)^2}{(\cos A)^2} = \frac{\sin^2 A}{\cos^2 A}.$$

Example. Shew that $\sin^2 A \sec A \cot^2 A = \cos A$.

$$\sin^2 A \sec A \cot^2 A = \sin^2 A \times \frac{1}{\cos A} \times \left(\frac{\cos A}{\sin A}\right)^2$$

$$= \sin^2 A \times \frac{1}{\cos A} \times \frac{\cos^2 A}{\sin^2 A}$$

$$= \cos A,$$

by cancelling factors common to numerator and denominator.

25. *To prove that* $\sin^2 A + \cos^2 A = 1.$

Let BAC be any acute angle; draw BC perpendicular to

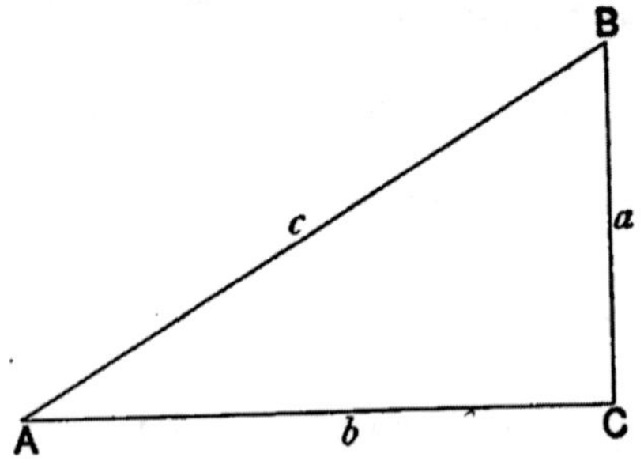

AC, and denote the sides of the right-angled triangle ABC by a, b, c.

By definition, $$\sin A = \frac{BC}{AB} = \frac{a}{c};$$

and $$\cos A = \frac{AC}{AB} = \frac{b}{c};$$

$$\therefore \sin^2 A + \cos^2 A = \frac{a^2}{c^2} + \frac{b^2}{c^2} = \frac{a^2+b^2}{c^2}$$

$$= \frac{c^2}{c^2}$$

$$= 1.$$

COR. $\sin^2 A = 1 - \cos^2 A,\quad \sin A = \sqrt{1-\cos^2 A};$
$\cos^2 A = 1 - \sin^2 A,\quad \cos A = \sqrt{1-\sin^2 A}.$

Example 1. Prove that $\cos^4 A - \sin^4 A = \cos^2 A - \sin^2 A.$

$$\cos^4 A - \sin^4 A = (\cos^2 A + \sin^2 A)(\cos^2 A - \sin^2 A)$$
$$= \cos^2 A - \sin^2 A,$$

since the first factor is equal to 1.

Example 2. Prove that $\cot \alpha \sqrt{1-\cos^2\alpha} = \cos \alpha.$

$$\cot \alpha \sqrt{1-\cos^2\alpha} = \cot\alpha \times \sin\alpha$$
$$= \frac{\cos\alpha}{\sin\alpha} \times \sin\alpha = \cos\alpha.$$

26. *To prove that* $\sec^2 A = 1 + \tan^2 A$.

With the figure of the previous article, we have

$$\sec A = \frac{AB}{AC} = \frac{c}{b};$$

$$\therefore \quad \sec^2 A = \frac{c^2}{b^2} = \frac{b^2 + a^2}{b^2}$$

$$= 1 + \frac{a^2}{b^2}$$

$$= 1 + \tan^2 A.$$

COR. $\sec^2 A - \tan^2 A = 1, \quad \sec A = \sqrt{1 + \tan^2 A},$

$\tan^2 A = \sec^2 A - 1, \quad \tan A = \sqrt{\sec^2 A - 1}.$

Example. Prove that $\cos A \sqrt{\sec^2 A - 1} = \sin A$.

$$\cos A \sqrt{\sec^2 A - 1} = \cos A \times \tan A$$

$$= \cos A \times \frac{\sin A}{\cos A}$$

$$= \sin A.$$

27. *To prove that* $\operatorname{cosec}^2 A = 1 + \cot^2 A$.

With the figure of Art. 25, we have

$$\operatorname{cosec} A = \frac{AB}{BC} = \frac{c}{a};$$

$$\therefore \quad \operatorname{cosec}^2 A = \frac{c^2}{a^2} = \frac{a^2 + b^2}{a^2}$$

$$= 1 + \frac{b^2}{a^2}$$

$$= 1 + \cot^2 A.$$

COR. $\operatorname{cosec}^2 A - \cot^2 A = 1, \quad \operatorname{cosec} A = \sqrt{1 + \cot^2 A},$

$\cot^2 A = \operatorname{cosec}^2 A - 1, \quad \cot A = \sqrt{\operatorname{cosec}^2 A - 1}.$

Example. Prove that $\cot^4 \alpha - 1 = \operatorname{cosec}^4 \alpha - 2 \operatorname{cosec}^2 \alpha$.

$$\cot^4 \alpha - 1 = (\cot^2 \alpha + 1)(\cot^2 \alpha - 1)$$

$$= \operatorname{cosec}^2 \alpha (\operatorname{cosec}^2 \alpha - 1 - 1)$$

$$= \operatorname{cosec}^2 \alpha (\operatorname{cosec}^2 \alpha - 2)$$

$$= \operatorname{cosec}^4 \alpha - 2 \operatorname{cosec}^2 \alpha.$$

28. The formulæ proved in the last three articles are not independent, for they are merely different ways of expressing in trigonometrical symbols the property of a right-angled triangle known as the Theorem of Pythagoras.

29. It will be useful here to collect the formulæ proved in this chapter.

I. $\operatorname{cosec} A \times \sin A = 1, \quad \operatorname{cosec} A = \dfrac{1}{\sin A}, \quad \sin A = \dfrac{1}{\operatorname{cosec} A};$

$$\sec A \times \cos A = 1, \quad \sec A = \frac{1}{\cos A}, \quad \cos A = \frac{1}{\sec A};$$

$$\cot A \times \tan A = 1, \quad \cot A = \frac{1}{\tan A}, \quad \tan A = \frac{1}{\cot A}.$$

II. $$\tan A = \frac{\sin A}{\cos A}, \quad \cot A = \frac{\cos A}{\sin A}.$$

III. $$\sin^2 A + \cos^2 A = 1,$$

$$\sec^2 A = 1 + \tan^2 A,$$

$$\operatorname{cosec}^2 A = 1 + \cot^2 A.$$

Easy Identities.

30. We shall now exemplify the use of these fundamental formulæ in proving *identities*. An identity asserts that two expressions are always equal, and the proof of this equality is called "proving the identity." Some easy illustrations have already been given in this chapter. The general method of procedure is to choose one of the expressions given (usually the more complicated of the two) and to shew by successive transformations that it can be made to assume the form of the other.

Example 1. Prove that $\sin^2 A \cot^2 A + \cos^2 A \tan^2 A = 1$.

Here it will be found convenient to express all the trigonometrical ratios in terms of the sine and cosine.

$$\text{The first side} = \sin^2 A \cdot \frac{\cos^2 A}{\sin^2 A} + \cos^2 A \cdot \frac{\sin^2 A}{\cos^2 A}$$

$$= \cos^2 A + \sin^2 A$$

$$= 1.$$

Example 2. Prove that $\sec^4\theta - \sec^2\theta = \tan^2\theta + \tan^4\theta$.

The form of this identity at once suggests that we should use the secant-tangent formula of Art. 26; hence

$$\begin{aligned}\text{the first side} &= \sec^2\theta\,(\sec^2\theta - 1)\\ &= (1 + \tan^2\theta)\tan^2\theta\\ &= \tan^2\theta + \tan^4\theta.\end{aligned}$$

EXAMPLES. III. a.

Prove the following identities:

1. $\sin A \cot A = \cos A$.
2. $\cos A \tan A = \sin A$.
3. $\cot A \sec A = \operatorname{cosec} A$.
4. $\sin A \sec A = \tan A$.
5. $\cos A \operatorname{cosec} A = \cot A$.
6. $\cot A \sec A \sin A = 1$.
7. $(1 - \cos^2 A)\operatorname{cosec}^2 A = 1$.
8. $(1 - \sin^2 A)\sec^2 A = 1$.
9. $\cot^2\theta\,(1 - \cos^2\theta) = \cos^2\theta$.
10. $(1 - \cos^2\theta)\sec^2\theta = \tan^2\theta$.
11. $\tan \alpha \sqrt{1 - \sin^2\alpha} = \sin\alpha$.
12. $\operatorname{cosec}\alpha \sqrt{1 - \sin^2\alpha} = \cot\alpha$.
13. $(1 + \tan^2 A)\cos^2 A = 1$.
14. $(\sec^2 A - 1)\cot^2 A = 1$.
15. $(1 - \cos^2\theta)(1 + \tan^2\theta) = \tan^2\theta$.
16. $\cos\alpha \operatorname{cosec}\alpha \sqrt{\sec^2\alpha - 1} = 1$.
17. $\sin^2 A\,(1 + \cot^2 A) = 1$.
18. $(\operatorname{cosec}^2 A - 1)\tan^2 A = 1$.
19. $(1 - \cos^2 A)(1 + \cot^2 A) = 1$.
20. $\sin\alpha \sec\alpha \sqrt{\operatorname{cosec}^2\alpha - 1} = 1$.
21. $\cos\alpha \sqrt{\cot^2\alpha + 1} = \sqrt{\operatorname{cosec}^2\alpha - 1}$.
22. $\sin^2\theta \cot^2\theta + \sin^2\theta = 1$.
23. $(1 + \tan^2\theta)(1 - \sin^2\theta) = 1$.
24. $\sin^2\theta \sec^2\theta = \sec^2\theta - 1$.
25. $\operatorname{cosec}^2\theta \tan^2\theta - 1 = \tan^2$

Prove the following identities :

26. $\dfrac{1}{\sec^2 A}+\dfrac{1}{\operatorname{cosec}^2 A}=1.$ **27.** $\dfrac{1}{\cos^2 A}-\dfrac{1}{\cot^2 A}=1.$

28. $\dfrac{\sin A}{\operatorname{cosec} A}+\dfrac{\cos A}{\sec A}=1.$ **29.** $\dfrac{\sec A}{\cos A}-\dfrac{\tan A}{\cot A}=1.$

30. $\sin^4 a-\cos^4 a=2\sin^2 a-1=1-2\cos^2 a.$

31. $\sec^4 a-1=2\tan^2 a+\tan^4 a.$

32. $\operatorname{cosec}^4 a-1=2\cot^2 a+\cot^4 a.$

33. $(\tan a \operatorname{cosec} a)^2-(\sin a \sec a)^2=1.$

34. $(\sec\theta\cot\theta)^2-(\cos\theta\operatorname{cosec}\theta)^2=1.$

35. $\tan^2\theta-\cot^2\theta=\sec^2\theta-\operatorname{cosec}^2\theta.$

***31.** The foregoing examples have required little more than a direct application of the fundamental formulæ; we shall now give some identities offering a greater variety of treatment.

Example 1. Prove that $\sec^2 A+\operatorname{cosec}^2 A=\sec^2 A\operatorname{cosec}^2 A.$

$$\text{The first side } =\frac{1}{\cos^2 A}+\frac{1}{\sin^2 A}=\frac{\sin^2 A+\cos^2 A}{\cos^2 A\sin^2 A}$$

$$=\frac{1}{\cos^2 A\sin^2 A}=\sec^2 A\operatorname{cosec}^2 A.$$

Occasionally it is found convenient to prove the equality of the two expressions by reducing each to the same form.

Example 2. Prove that

$$\sin^2 A\tan A+\cos^2 A\cot A+2\sin A\cos A=\tan A+\cot A.$$

$$\text{The first side } =\sin^2 A\cdot\frac{\sin A}{\cos A}+\cos^2 A\cdot\frac{\cos A}{\sin A}+2\sin A\cos A$$

$$=\frac{\sin^4 A+\cos^4 A+2\sin^2 A\cos^2 A}{\sin A\cos A}$$

$$=\frac{(\sin^2 A+\cos^2 A)^2}{\sin A\cos A}=\frac{1}{\sin A\cos A}.$$

$$\text{The second side} =\frac{\sin A}{\cos A}+\frac{\cos A}{\sin A}=\frac{\sin^2 A+\cos^2 A}{\cos A\sin A}$$

$$=\frac{1}{\sin A\cos A}.$$

$$\text{Thus each side of the identity } =\frac{1}{\sin A\cos A}.$$

Example 3. Prove that $\dfrac{\tan\alpha-\cot\beta}{\tan\beta-\cot\alpha}=\tan\alpha\cot\beta$.

$$\text{The first side}=\frac{\tan\alpha-\cot\beta}{\dfrac{1}{\cot\beta}-\dfrac{1}{\tan\alpha}}=\frac{\tan\alpha-\cot\beta}{\dfrac{\tan\alpha-\cot\beta}{\tan\alpha\cot\beta}}$$

$$=\frac{\tan\alpha-\cot\beta}{1}\times\frac{\tan\alpha\cot\beta}{\tan\alpha-\cot\beta}$$

$$=\tan\alpha\cot\beta.$$

The transformations in the successive steps are usually suggested by the form into which we wish to bring the result. For instance, in this last example we might have proved the identity by substituting for the tangent and cotangent in terms of the sine and cosine. This however is not the best method, for the form in which the right-hand side is given suggests that we should retain $\tan\alpha$ and $\cot\beta$ unchanged throughout the work.

*EXAMPLES. III. b.

Prove the following identities:

1. $\dfrac{\sin\alpha\cot^2\alpha}{\cos\alpha}=\dfrac{1}{\tan\alpha}$.
2. $\dfrac{\sec^2\alpha\cot\alpha}{\operatorname{cosec}^2\alpha}=\tan\alpha$.
3. $1-\operatorname{vers}\theta=\sin\theta\cot\theta$.
4. $\operatorname{vers}\theta\sec\theta=\sec\theta-1$.
5. $\sec\theta-\tan\theta\sin\theta=\cos\theta$.
6. $\tan\theta+\cot\theta=\sec\theta\operatorname{cosec}\theta$.
7. $\sqrt{1+\cot^2 A}\,.\,\sqrt{\sec^2 A-1}\,.\,\sqrt{1-\sin^2 A}=1$.
8. $(\cos\theta+\sin\theta)^2+(\cos\theta-\sin\theta)^2=2$.
9. $(1+\tan\theta)^2+(1-\tan\theta)^2=2\sec^2\theta$.
10. $(\cot\theta-1)^2+(\cot\theta+1)^2=2\operatorname{cosec}^2\theta$.
11. $\sin^2 A\,(1+\cot^2 A)+\cos^2 A\,(1+\tan^2 A)=2$.
12. $\cos^2 A\,(\sec^2 A-\tan^2 A)+\sin^2 A\,(\operatorname{cosec}^2 A-\cot^2 A)=1$.
13. $\cot^2\alpha+\cot^4\alpha=\operatorname{cosec}^4\alpha-\operatorname{cosec}^2\alpha$.
14. $\dfrac{\tan^2\alpha}{1+\tan^2\alpha}\cdot\dfrac{1+\cot^2\alpha}{\cot^2\alpha}=\sin^2\alpha\sec^2\alpha$.
15. $\dfrac{1}{1-\sin\alpha}+\dfrac{1}{1+\sin\alpha}=2\sec^2\alpha$.

Prove the following identities:

16. $\dfrac{\tan a}{\sec a-1}+\dfrac{\tan a}{\sec a+1}=2\operatorname{cosec} a.$

17. $\dfrac{1}{1+\sin^2 a}+\dfrac{1}{1+\operatorname{cosec}^2 a}=1.$

18. $(\sec\theta+\operatorname{cosec}\theta)(\sin\theta+\cos\theta)=\sec\theta\operatorname{cosec}\theta+2.$

19. $(\cos\theta-\sin\theta)(\operatorname{cosec}\theta-\sec\theta)=\sec\theta\operatorname{cosec}\theta-2.$

20. $(1+\cot\theta+\operatorname{cosec}\theta)(1+\cot\theta-\operatorname{cosec}\theta)=2\cot\theta.$

21. $(\sec\theta+\tan\theta-1)(\sec\theta-\tan\theta+1)=2\tan\theta.$

22. $(\sin A+\operatorname{cosec} A)^2+(\cos A+\sec A)^2=\tan^2 A+\cot^2 A+7.$

23. $(\sec^2 A+\tan^2 A)(\operatorname{cosec}^2 A+\cot^2 A)=1+2\sec^2 A\operatorname{cosec}^2 A.$

24. $(1-\sin A+\cos A)^2=2(1-\sin A)(1+\cos A).$

25. $\sin A(1+\tan A)+\cos A(1+\cot A)=\sec A+\operatorname{cosec} A.$

26. $\cos\theta(\tan\theta+2)(2\tan\theta+1)=2\sec\theta+5\sin\theta.$

27. $(\tan\theta+\sec\theta)^2=\dfrac{1+\sin\theta}{1-\sin\theta}.$

28. $\dfrac{2\sin\theta\cos\theta-\cos\theta}{1-\sin\theta+\sin^2\theta-\cos^2\theta}=\cot\theta.$

29. $\cot^2\theta\,.\,\dfrac{\sec\theta-1}{1+\sin\theta}+\sec^2\theta\,.\,\dfrac{\sin\theta-1}{1+\sec\theta}=0.$

[*The following examples contain functions of two angles; in each case the two angles are quite independent of each other.*]

30. $\tan^2 a+\sec^2\beta=\sec^2 a+\tan^2\beta.$

31. $\dfrac{\tan a+\cot\beta}{\cot a+\tan\beta}=\dfrac{\tan a}{\tan\beta}.$

32. $\dfrac{\tan a-\cot\beta}{\cot a-\tan\beta}=-\dfrac{\cot\beta}{\cot a}.$

33. $\cot a\tan\beta(\tan a+\cot\beta)=\cot a+\tan\beta.$

34. $\sin^2 a\cos^2\beta-\cos^2 a\sin^2\beta=\sin^2 a-\sin^2\beta.$

35. $\sec^2 a\tan^2\beta-\tan^2 a\sec^2\beta=\tan^2\beta-\tan^2 a.$

36. $(\sin a\cos\beta+\cos a\sin\beta)^2+(\cos a\cos\beta-\sin a\sin\beta)^2=1.$

32. By means of the relations collected together in Art. 29, all the trigonometrical ratios can be expressed in terms of any one.

Example 1. Express all the trigonometrical ratios of A in terms of $\tan A$.

We have $\cot A = \dfrac{1}{\tan A}$;

$$\sec A = \sqrt{1 + \tan^2 A};$$

$$\cos A = \frac{1}{\sec A} = \frac{1}{\sqrt{1+\tan^2 A}};$$

$$\sin A = \frac{\sin A}{\cos A}\cos A = \tan A \cos A = \frac{\tan A}{\sqrt{1+\tan^2 A}};$$

$$\operatorname{cosec} A = \frac{1}{\sin A} = \frac{\sqrt{1+\tan^2 A}}{\tan A}.$$

Obs. In writing down the ratios we choose the simplest and most natural order. For instance, $\cot A$ is obtained at once by the *reciprocal relation* connecting the tangent and cotangent: $\sec A$ comes immediately from the tangent-secant formula; the remaining three ratios now readily follow.

Example 2. Given $\cos A = \dfrac{5}{13}$, find $\operatorname{cosec} A$ and $\cot A$.

$$\operatorname{cosec} A = \frac{1}{\sin A} = \frac{1}{\sqrt{1-\cos^2 A}}$$

$$= \frac{1}{\sqrt{1-\left(\frac{5}{13}\right)^2}} = \frac{1}{\sqrt{1-\frac{25}{169}}} = \frac{1}{\sqrt{\frac{144}{169}}} = \frac{1}{\frac{12}{13}} = \frac{13}{12}.$$

$$\cot A = \frac{\cos A}{\sin A} = \cos A \times \operatorname{cosec} A$$

$$= \frac{5}{13} \times \frac{13}{12} = \frac{5}{12}.$$

33. *It is always possible to describe a right-angled triangle when two sides are given:* for the third side can be found by Geometry, and the triangle can then be constructed practically. We can thus readily obtain all the trigonometrical ratios when one is given, or express all in terms of any one.

Example 1. Given $\cos A = \frac{5}{13}$, find cosec A and $\cot A$.

Take a right-angled triangle PQR, of which Q is the right angle, having the hypotenuse $PR = 13$ units, and $PQ = 5$ units.

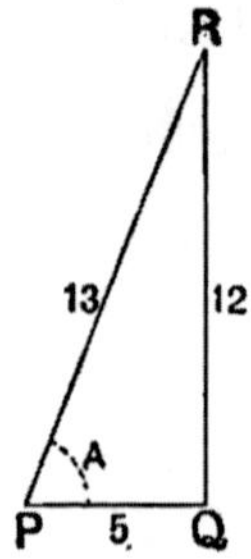

Let $QR = x$ units; then

$$x^2 = (13)^2 - (5)^2 = (13+5)(13-5)$$
$$= 18 \times 8 = 9 \times 2 \times 8;$$
$$\therefore x = 3 \times 4 = 12.$$

Now $$\cos RPQ = \frac{PQ}{PR} = \frac{5}{13},$$

so that $$\angle RPQ = A.$$

Hence $$\operatorname{cosec} A = \frac{PR}{QR} = \frac{13}{12},$$

and $$\cot A = \frac{PQ}{QR} = \frac{5}{12}.$$ [Compare Art. 32, Ex. 2.]

Example 2. Find $\tan A$ and $\cos A$ in terms of cosec A.

Take a triangle PQR right-angled at Q, and having $\angle RPQ = A$. For shortness, denote cosec A by c.

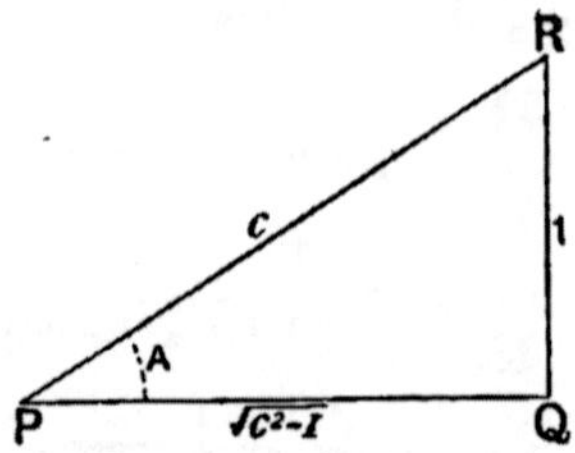

Then $$\operatorname{cosec} A = c = \frac{c}{1};$$

but $$\operatorname{cosec} A = \frac{PR}{QR};$$

$$\therefore \frac{PR}{QR} = \frac{c}{1}.$$

Let QR be taken as the unit of measurement;

then $QR = 1$, and therefore $PR = c$.

Let PQ contain x units; then

$$x^2 = c^2 - 1, \text{ so that } x = \sqrt{c^2 - 1}.$$

Hence $$\tan A = \frac{QR}{PQ} = \frac{1}{\sqrt{c^2-1}} = \frac{1}{\sqrt{\operatorname{cosec}^2 A - 1}},$$

and $$\cos A = \frac{PQ}{PR} = \frac{\sqrt{c^2-1}}{c} = \frac{\sqrt{\operatorname{cosec}^2 A - 1}}{\operatorname{cosec} A}.$$

EXAMPLES. III. c.

1. Given $\sin A = \frac{1}{2}$, find $\sec A$ and $\cot A$.

2. Given $\tan A = \frac{4}{3}$, find $\sin A$ and $\cos A$.

3. Find $\cot\theta$ and $\sin\theta$ when $\sec\theta = 4$.

4. If $\tan\alpha = \frac{1}{2}$, find $\sec\alpha$ and $\operatorname{cosec}\alpha$.

5. Find the sine and cotangent of an angle whose secant is 7.

6. If $25\sin A = 7$, find $\tan A$ and $\sec A$.

7. Express $\sin A$ and $\tan A$ in terms of $\cos A$.

8. Express $\operatorname{cosec}\alpha$ and $\cos\alpha$ in terms of $\cot\alpha$.

9. Find $\sin\theta$ and $\cot\theta$ in terms of $\sec\theta$.

10. Express all the trigonometrical ratios of A in terms of $\sin A$.

11. Given $\sin A - \cos A = 0$, find $\operatorname{cosec} A$.

12. If $\sin A = \frac{m}{n}$, prove that $\sqrt{n^2 - m^2} \, . \tan A = m$.

13. If $p\cot\theta = \sqrt{q^2 - p^2}$, find $\sin\theta$.

14. When $\sec A = \frac{m^2+1}{2m}$, find $\tan A$ and $\sin A$.

15. Given $\tan A = \frac{2pq}{p^2 - q^2}$, find $\cos A$ and $\operatorname{cosec} A$.

16. If $\sec\alpha = \frac{13}{5}$, find the value of $\frac{2\sin\alpha - 3\cos\alpha}{4\sin\alpha - 9\cos\alpha}$.

17. If $\cot\theta = \frac{p}{q}$, find the value of $\frac{p\cos\theta - q\sin\theta}{p\cos\theta + q\sin\theta}$.

CHAPTER IV.

TRIGONOMETRICAL RATIOS OF CERTAIN ANGLES.

34. Trigonometrical Ratios of 45°.

Let BAC be a right-angled isosceles triangle, with the right angle at C; so that $B=A=45°$.

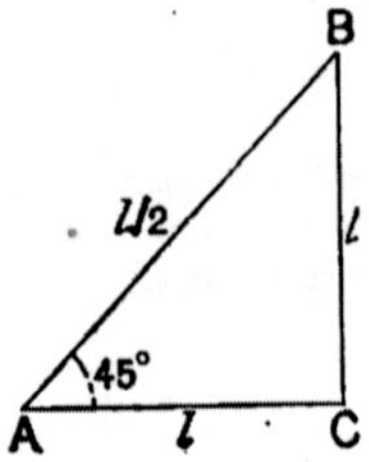

Let each of the equal sides contain l units,

then $$AC=BC=l.$$

Also $$AB^2=l^2+l^2=2l^2;$$

$$\therefore\ AB=l\sqrt{2}.$$

$$\therefore\ \sin 45° = \frac{BC}{AB}=\frac{l}{l\sqrt{2}}=\frac{1}{\sqrt{2}};$$

$$\cos 45°=\frac{AC}{AB}=\frac{l}{l\sqrt{2}}=\frac{1}{\sqrt{2}};$$

$$\tan 45°=\frac{BC}{AC}=\frac{l}{l}=1.$$

The other three ratios are the reciprocals of these; thus

$$\operatorname{cosec} 45°=\sqrt{2},\quad \sec 45°=\sqrt{2},\quad \cot 45°=1;$$

or they may be read off from the figure.

35. Trigonometrical Ratios of 60° and 30°.

Let ABC be an equilateral triangle; thus each of its angles is 60°.

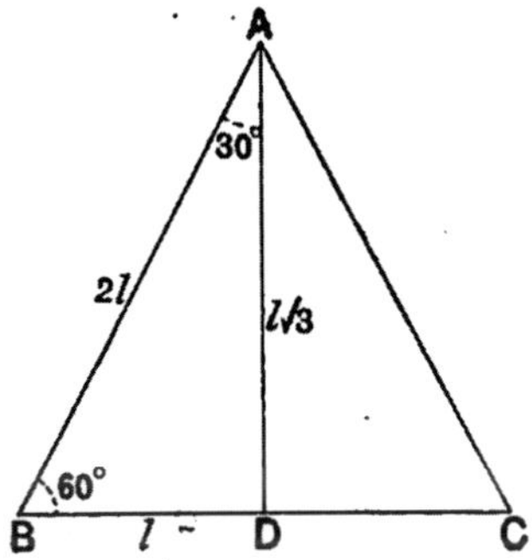

Bisect $\angle BAC$ by AD meeting BC at D; then $\angle BAD = 30°$.

By Euc. I. 4, the triangles ABD, ACD are equal in all respects; therefore $BD = CD$, and the angles at D are right angles.

In the right-angled triangle ADB, let $BD = l$; then

$$AB = BC = 2l;$$

$$\therefore \quad AD^2 = 4l^2 - l^2 = 3l^2;$$

$$\therefore \quad AD = l\sqrt{3}.$$

$$\therefore \quad \sin 60° = \frac{AD}{AB} = \frac{l\sqrt{3}}{2l} = \frac{\sqrt{3}}{2};$$

$$\cos 60° = \frac{BD}{AB} = \frac{l}{2l} = \frac{1}{2};$$

$$\tan 60° = \frac{AD}{BD} = \frac{l\sqrt{3}}{l} = \sqrt{3}.$$

Again,

$$\sin 30° = \frac{BD}{AB} = \frac{l}{2l} = \frac{1}{2};$$

$$\cos 30° = \frac{AD}{AB} = \frac{l\sqrt{3}}{2l} = \frac{\sqrt{3}}{2};$$

$$\tan 30° = \frac{BD}{AD} = \frac{l}{l\sqrt{3}} = \frac{1}{\sqrt{3}}.$$

The other ratios may be read off from the figure.

36. The trigonometrical ratios of 45°, 60°, 30° occur very frequently; it is therefore important that the student should be able to quote readily their numerical values. The exercise which follows will furnish useful practice.

At first it will probably be found safer to make use of the accompanying diagrams than to trust to the memory.

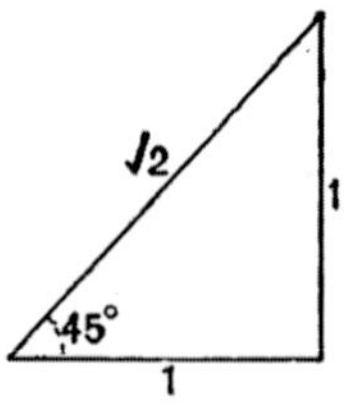

FIG. 1.

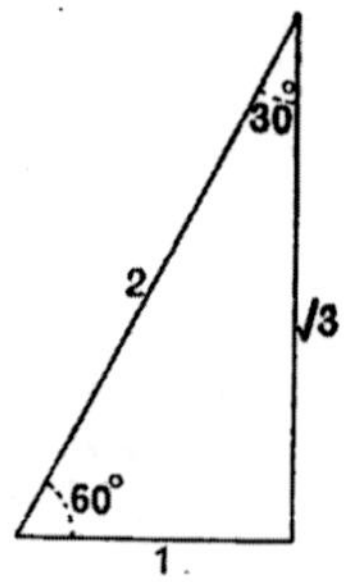

FIG. 2.

The trigonometrical ratios of 45° can be read off from Fig. 1; those of 60° and 30° from Fig. 2.

Example 1. Find the values of $\sec^3 45°$ and $\sin 60° \cot 30° \tan 45°$.

$$\sec^3 45° = (\sec 45°)^3 = (\sqrt{2})^3 = \sqrt{2} \times \sqrt{2} \times \sqrt{2} = 2\sqrt{2}.$$

$$\sin 60° \cot 30° \tan 45° = \frac{\sqrt{3}}{2} \times \sqrt{3} \times 1 = \frac{3}{2}.$$

Example 2. Find the value of

$$2 \cot 45° + \cos^3 60° - 2 \sin^4 60° + \tfrac{3}{4} \tan^2 30°.$$

The value $= (2 \times 1) + \left(\frac{1}{2}\right)^3 - 2\left(\frac{\sqrt{3}}{2}\right)^4 + \frac{3}{4}\left(\frac{1}{\sqrt{3}}\right)^2$

$$= 2 + \frac{1}{8} - 2\left(\frac{3}{4}\right)^2 + \frac{3}{4}\left(\frac{1}{3}\right)$$

$$= 2 + \frac{1}{8} - \frac{9}{8} + \frac{1}{4} = 1\tfrac{1}{4}.$$

EXAMPLES. IV. a.

Find the numerical value of

1. $\tan^2 60° + 2\tan^2 45°$.
2. $\tan^3 45° + 4\cos^3 60°$.
3. $2\,\text{cosec}^2 45° - 3\sec^2 30°$.
4. $\cot 60° \tan 30° + \sec^2 45°$.

5. $2 \sin 30^\circ \cos 30^\circ \cot 60^\circ$.

6. $\tan^2 45^\circ \sin 60^\circ \tan 30^\circ \tan^2 60^\circ$.

7. $\tan^2 60^\circ + 4\cos^2 45^\circ + 3\sec^2 30^\circ$.

8. $\frac{1}{2}\operatorname{cosec}^2 60^\circ + \sec^2 45^\circ - 2\cot^2 60^\circ$.

9. $\tan^2 30^\circ + 2\sin 60^\circ + \tan 45^\circ - \tan 60^\circ + \cos^2 30^\circ$.

10. $\cot^2 45^\circ + \cos 60^\circ - \sin^2 60^\circ - \frac{3}{4}\cot^2 60^\circ$.

11. $3\tan^2 30^\circ + \frac{4}{3}\cos^2 30^\circ - \frac{1}{2}\sec^2 45^\circ - \frac{1}{3}\sin^2 60^\circ$.

12. $\cos 60^\circ - \tan^2 45^\circ + \frac{3}{4}\tan^2 30^\circ + \cos^2 30^\circ - \sin 30^\circ$.

13. $\frac{1}{3}\sin^2 60^\circ - \frac{1}{2}\sec 60^\circ \tan^2 30^\circ + \frac{4}{3}\sin^2 45^\circ \tan^2 60^\circ$.

14. If $\tan^2 45^\circ - \cos^2 60^\circ = x \sin 45^\circ \cos 45^\circ \tan 60^\circ$, find x.

15. Find x from the equation

$$x \sin 30^\circ \cos^2 45^\circ = \frac{\cot^2 30^\circ \sec 60^\circ \tan 45^\circ}{\operatorname{cosec}^2 45^\circ \operatorname{cosec} 30^\circ}.$$

37. DEFINITION. The **complement** of an angle is its *defect from* a right angle.

Two angles are said to be **complementary** when their sum is a right angle.

Thus in every right-angled triangle, each acute angle is the complement of the other. For in the figure of the next article, if B is the right angle, the sum of A and C is 90°.

$$\therefore \quad C = 90^\circ - A, \text{ and } A = 90^\circ - C.$$

Trigonometrical Ratios of Complementary Angles.

38. Let ABC be a right-angled triangle, of which B is the right angle; then the angles at A and C are complementary, so that $C = 90^\circ - A$.

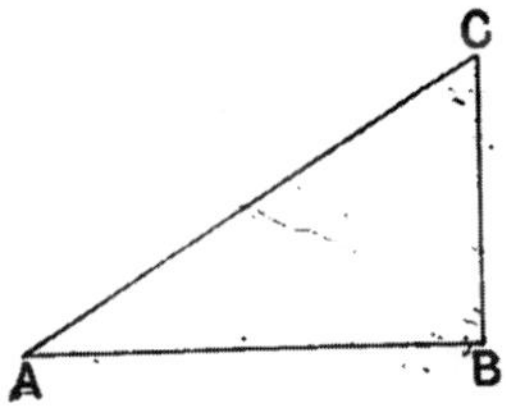

$$\therefore \quad \sin(90^\circ - A) = \sin C = \frac{AB}{AC} = \cos A;$$

$$\text{and } \cos(90^\circ - A) = \cos C = \frac{BC}{AC} = \sin A.$$

Similarly, it may be proved that

$$\left.\begin{aligned} \tan(90^\circ - A) &= \cot A, \\ \cot(90^\circ - A) &= \tan A; \end{aligned}\right\} \text{ and } \left.\begin{aligned} \sec(90^\circ - A) &= \operatorname{cosec} A, \\ \operatorname{cosec}(90^\circ - A) &= \sec A. \end{aligned}\right\}$$

39. If we define the **co-sine**, **co-tangent**, **co-secant**, as the **co-functions** of the angle, the foregoing results may be embodied in a single statement:

each function of an angle is equal to the corresponding co-function of its complement.

As an illustration of this we may refer to Art. 35, from which it will be seen that

$$\sin 60° = \cos 30° = \frac{\sqrt{3}}{2};$$

$$\sin 30° = \cos 60° = \frac{1}{2};$$

$$\tan 60° = \cot 30° = \sqrt{3}.$$

Example 1. Find a value of A when $\cos 2A = \sin 3A$.

Since $\cos 2A = \sin (90° - 2A)$,

the equation becomes $\sin (90° - 2A) = \sin 3A$;

$$\therefore\ 90° - 2A = 3A;$$

whence $A = 18°$.

Thus *one* value of A which satisfies the equation is $A = 18°$. In a later chapter we shall be able to solve the equation more completely, and shew that there are other values of A which satisfy it.

Example 2. Prove that $\sec A \sec (90° - A) = \tan A + \tan (90° - A)$.

Here it will be found easier to begin with the expression on the right side of the identity.

The second side $= \tan A + \cot A$

$$= \frac{\sin A}{\cos A} + \frac{\cos A}{\sin A} = \frac{\sin^2 A + \cos^2 A}{\cos A \sin A}$$

$$= \frac{1}{\cos A \sin A}$$

$$= \sec A \operatorname{cosec} A = \sec A \sec (90° - A).$$

EXAMPLES. IV. b.

Find the complements of the following angles:

1. $67° 30'$.
2. $25° 30''$.
3. $10° 1' 3''$.
4. $45° - A$.
5. $45° + B$.
6. $30° - B$.

7. In a triangle C is 50° and A is the complement of 10°; find B.

8. In a triangle A is the complement of 40°; and B is the complement of 20°; find C.

Find a value of A in each of the following equations:

9. $\sin A = \cos 4A$.
10. $\cos 3A = \sin 7A$.
11. $\tan A = \cot 3A$.
12. $\cot A = \tan A$.
13. $\cot A = \tan 2A$.
14. $\sec 5A = \operatorname{cosec} A$.

Prove the following identities:

15. $\sin(90° - A)\cot(90° - A) = \sin A$.

16. $\sin A \tan(90° - A)\sec(90° - A) = \cot A$.

17. $\cos A \tan A \tan(90° - A)\operatorname{cosec}(90° - A) = 1$.

18. $\sin A \cos(90° - A) + \cos A \sin(90° - A) = 1$.

19. $\cos(90° - A)\operatorname{cosec}(90° - A) = \tan A$.

20. $\operatorname{cosec}^2(90° - A) = 1 + \sin^2 A \operatorname{cosec}^2(90° - A)$.

21. $\sin A \cot A \cot(90° - A)\sec(90° - A) = 1$.

22. $\sec(90° - A) - \cot A \cos(90° - A)\tan(90° - A) = \sin A$.

23. $\tan^2 A \sec^2(90° - A) - \sin^2 A \operatorname{cosec}^2(90° - A) = 1$.

24. $\tan(90° - A) + \cot(90° - A) = \operatorname{cosec} A \operatorname{cosec}(90° - A)$.

25. $\dfrac{\sin(90° - A)}{\sec(90° - A)} \cdot \dfrac{\tan(90° - A)}{\cos A} = \cos A$.

26. $\dfrac{\operatorname{cosec}^2 A \tan^2 A}{\cot(90° - A)} \cdot \dfrac{\cot A}{\sec^2 A} = \sec^2(90° - A) - 1$.

27. $\dfrac{\cot(90° - A)}{\operatorname{cosec}^2 A} \cdot \dfrac{\sec A \cot^3 A}{\sin^2(90° - A)} = \sqrt{\tan^2 A + 1}$.

28. $\dfrac{\cos^2(90° - A)}{\operatorname{vers} A} = 1 + \sin(90° - A)$.

29. $\dfrac{\cot^2 A \sin^2(90° - A)}{\cot A + \cos A} = \tan(90° - A) - \cos A$.

30. If $x \sin(90° - A)\cot(90° - A) = \cos(90° - A)$, find x.

31. Find the value of x which will satisfy

$$\sec A \operatorname{cosec}(90° - A) - x \cot(90° - A) = 1.$$

The Use of Tables.

39_A. Tables have been constructed giving the numerical values of the trigonometrical ratios of all angles between 0° and 90° at suitable intervals. These are called the Tables of **natural sines, cosines, tangents**, ... In *Four-Figure* Tables the interval is usually one-tenth of a degree, or 6 minutes. When the number of minutes in an angle is not an exact multiple of 6, the differences in the trigonometrical ratios corresponding to 1, 2, 3, 4, 5 minutes are given in the *difference columns* printed at the right hand of the Tables [see page 378]. The way in which these differences are used will be explained by examples. We shall assume the properties established in Chap. IX, viz. that as the angle increases from 0° to 90° the sine, cosine, and tangent *increase*, while the cosine, cotangent, and cosecant *decrease*.

Example 1. Find the sine of (i) 35° 18′; (ii) 35° 21′.

The following extract is from the Table of sines on page 378.

	0′	6′	12′	18′	24′	30′	36′	42′	48′	54′	1′	2′	3′	4′	5′
35°	5736	5750	5764	5779	5793	5807	5821	5835	5850	5864	2	5	7	10	12

The above details are sufficient to give the sines of all angles between 35° and 36°, but it must be remembered that the values are fractions, given decimally to four figures, the decimal point being omitted.

(i) Under the column headed 18′ we find the digits 5779. Prefixing the decimal point we have

$$\sin 35^\circ 18' = \cdot 5779.$$

(ii) Here the angle exceeds 35° 18′ by 3′. The Table gives the digit 7 in the column headed 3′. Bearing in mind that all the work is being carried to 4 places of decimals, this digit really represents a difference of 7 *ten-thousandths*, or ·0007. Similarly the differences for 4′, 5′ must be taken as ·0010 and ·0012 respectively.

The work stands as follows:

From the Table $\quad \sin 35^\circ 18' = \cdot 5779$

$\qquad\qquad\qquad\quad$ diff. for $3' = \cdot 0007$

∴, by *addition*, $\quad \sin 35^\circ 21' = \cdot 5786$

The ciphers are usually omitted in writing down the differences.

Example 2. Find the value of cos 26° 28′.

	0′	6′	12′	18′	24′	30′	36′	42′	48′	54′	1′ 2′ 3′	4′ 5′
26°	8988	8980	8973	8965	8957	8949	8942	8934	8926	8918	1 3 4	5 6

Here the angle is *greater* than 26° 24′ by 4′, therefore the cosine will be proportionally *less* than that of 26° 24′.

$$\cos 26° 24' = \cdot 8957$$
$$\text{diff. for } 4' \quad 5$$
$$\therefore, \text{ by } subtraction, \quad \cos 26° 28' = \cdot 8952$$

Example 3. Find θ from the equation $\tan\theta = 1{\cdot}6666$.

We must here look in the Table of tangents (page 383) for the angle whose tangent is nearest in value to 1·6666 and less than it. We shall find

	0′	6′	12′	18′	24′	30′	36′	42′	48′	54′	1′ 2′ 3′	4′ 5′
59°	1·6643	6709	6775	6842	6909	6977	7045	7113	7182	7251	11 23 34	45 56

$$\tan\theta = 1{\cdot}6666$$
$$\tan 59° = 1{\cdot}6643$$
$$\text{by subtraction,} \quad \text{diff.} = 23$$

Thus θ is greater than 59° by a number of minutes corresponding to the difference 23.

From the Table, difference for 2′ = 23,

$$\therefore \theta = 59° 2'.$$

EXAMPLES. IV. c.

Find from the Tables the values of the following Trigonometrical ratios:

1. sin 24° 42′. **2.** sin 24° 47′. **3.** sin 36° 19′.

4. cos 54° 16′. **5.** cos 60° 50′. **6.** cos 48° 20′.

7. tan 45° 18′. **8.** tan 47° 25′. **9.** tan 61° 3′.

Find to the nearest minute the angles given by the following equations:

10. $\sin A = \cdot 8839$. **11.** $\sin A = \cdot 4434$. **12.** $\cos\theta = \cdot 3221$.

13. $\cos\theta = \cdot 8006$. **14.** $\tan A = 1{\cdot}2450$. **15.** $\tan A = \cdot 5711$.

Easy Trigonometrical Equations.

40. We shall now give some examples in trigonometrical equations.

Example 1. Find a value of A which satisfies the equation

$$4\cos A = 3\sec A.$$

By expressing the secant in terms of the cosine, we have

$$4\cos A = \frac{3}{\cos A},$$

$$4\cos^2 A = 3,$$

$$\cos A = \pm\frac{\sqrt{3}}{2}.$$

$$\therefore \cos A = \frac{\sqrt{3}}{2} \quad \text{..............................(1)},$$

or

$$\cos A = -\frac{\sqrt{3}}{2} \quad \text{..............................(2)}.$$

Since $\cos 30° = \frac{\sqrt{3}}{2}$, we see from (1) that $A = 30°$.

The student will be able to understand the meaning of the negative result in (2) after he has read Chap. VIII.

Example 2. Solve $3\sec^2\theta = 8\tan\theta - 2$.

Since $\sec^2\theta = 1 + \tan^2\theta$,

we have $3(1 + \tan^2\theta) = 8\tan\theta - 2$,

or $3\tan^2\theta - 8\tan\theta + 5 = 0$.

This is a quadratic equation in which $\tan\theta$ is the unknown quantity, and it may be solved by any of the rules for solving quadratic equations.

Thus $(\tan\theta - 1)(3\tan\theta - 5) = 0$,

therefore *either* $\tan\theta - 1 = 0$ .... (1),

or $3\tan\theta - 5 = 0$ .... (2).

From (1), $\tan\theta = 1$, so that $\theta = 45°$.

From (2), $\tan\theta = \frac{5}{3} = 1{\cdot}6666 = \tan 59° 2'$ [see Ex. 3, p. 29_B],

$$\therefore \theta = 45°, \text{ or } 59° 2'.$$

41. When an equation involves more than two functions, it will usually be best to express each function in terms of the sine and cosine.

Example. Solve $3\tan\theta + \cot\theta = 5\operatorname{cosec}\theta$.

We have
$$\frac{3\sin\theta}{\cos\theta} + \frac{\cos\theta}{\sin\theta} = \frac{5}{\sin\theta},$$
$$3\sin^2\theta + \cos^2\theta = 5\cos\theta,$$
$$3(1-\cos^2\theta) + \cos^2\theta = 5\cos\theta,$$
$$2\cos^2\theta + 5\cos\theta - 3 = 0,$$
$$(2\cos\theta - 1)(\cos\theta + 3) = 0;$$
therefore *either* $2\cos\theta - 1 = 0$ ....(1),
or $\cos\theta + 3 = 0$ ....(2).

From (1), $\cos\theta = \frac{1}{2}$; so that $\theta = 60°$.

From (2), $\cos\theta = -3$, a result which must be rejected as *impossible*, because the numerical value of the cosine of an angle can never be greater than unity. [Art. 16.]

EXAMPLES. IV. d.

Find a solution of each of the following equations.

(*Tables must be used for Examples marked with an asterisk.*)

1. $2\sin\theta = \operatorname{cosec}\theta$.
2. $\tan\theta = 3\cot\theta$.
3. $\sec\theta = 4\cos\theta$.
4. $\sec\theta - \operatorname{cosec}\theta = 0$.
5. $4\sin\theta = 3\operatorname{cosec}\theta$.
6. $\operatorname{cosec}^2\theta = 4$.
7. $\sqrt{2}\cos\theta = \cot\theta$.
8. $\tan\theta = 2\sin\theta$.
9. $\sec^2\theta = 2\tan^2\theta$.
10. $\operatorname{cosec}^2\theta = 4\cot^2\theta$.
11. $\sec^2\theta = 3\tan^2\theta - 1$.
12. $\sec^2\theta + \tan^2\theta = 7$.
13. $\cot^2\theta + \operatorname{cosec}^2\theta = 3$.
14. $2(\cos^2\theta - \sin^2\theta) = 1$.
15. $2\cos^2\theta + 4\sin^2\theta = 3$.
16. $6\cos^2\theta = 1 + \cos\theta$.
17. $4\sin\theta = 12\sin^2\theta - 1$.
18. $2\sin^2\theta = 3\cos\theta$.

*19. $\tan\theta = 4 - 3\cot\theta$.

20. $\cos^2\theta - \sin^2\theta = 2 - 5\cos\theta$.
21. $\cot\theta + \tan\theta = 2\sec\theta$.
22. $4\operatorname{cosec}\theta + 2\sin\theta = 9$.
23. $\tan\theta - \cot\theta = \operatorname{cosec}\theta$.
24. $2\cos\theta + 2\sqrt{2} = 3\sec\theta$.
25. $2\sin\theta\tan\theta + 1 = \tan\theta + 2\sin\theta$.
26. $6\tan\theta - 5\sqrt{3}\sec\theta + 12\cot\theta = 0$.

*27. $5\tan\theta + 6\cot\theta = 11$.

*28. $\sec^2\theta + \tan^2\theta = 3\tan\theta$.

MISCELLANEOUS EXAMPLES. A.

1. Express as the decimal of a right angle

(1) $25^g\ 37'\ 6{\cdot}4''$; (2) $63°\ 21'\ 36''$.

2. Shew that

$$\sin A \cos A \tan A + \cos A \sin A \cot A = 1.$$

3. A ladder 29 ft. long just reaches a window at a height of 21 ft. from the ground: find the cosine and cosecant of the angle made by the ladder with the ground.

4. If $\operatorname{cosec} A = \frac{17}{15}$, find $\tan A$ and $\sec A$.

5. Shew that $\operatorname{cosec}^2 A - \cot A \cos A \operatorname{cosec} A - 1 = 0$.

---

6. Reduce to sexagesimal measure

(1) $17^g\ 18'\ 75''$; (2) $\cdot 0003$ of a right angle.

7. ABC is a triangle in which B is a right angle; if $c=9$, $a=40$, find b, $\cot A$, $\sec A$, $\sec C$.

8. Which of the following statements are possible and which impossible?

(1) $4 \sin\theta = 1$; (2) $2 \sec\theta = 1$; (3) $7 \tan\theta = 40$.

9. Prove that $\cos\theta \operatorname{vers}\theta (\sec\theta + 1) = \sin^2\theta$.

10. Express $\sec a$ and $\operatorname{cosec} a$ in terms of $\cot a$.

---

11. Find the numerical value of

$$3\tan^2 30° + \frac{1}{4}\sec 60° + 5\cot^2 45° - \frac{2}{3}\sin^2 60°.$$

12. If $\tan a = \frac{m}{n}$, find $\sin a$ and $\sec a$.

13. If m sexagesimal minutes are equivalent to n centesimal minutes, prove that $m = \cdot 54n$.

14. If $\sin A=\frac{4}{5}$, prove that $\tan A+\sec A=3$, when A is an acute angle.

15. Shew that

$$\cot(90^\circ-A)\cot A\cos(90^\circ-A)\tan(90^\circ-A)=\cos A.$$

---

16. PQR is a triangle in which P is a right angle; if $PQ=21$, $PR=20$, find $\tan Q$ and $\operatorname{cosec} Q$.

17. Shew that $(\tan a-\cot a)\sin a\cos a=1-2\cos^2 a$.

18. Find a value of θ which satisfies the equation

$$\sec 6\theta=\operatorname{cosec} 3\theta.$$

19. Prove that

$$\tan^2 60^\circ-2\tan^2 45^\circ=\cot^2 30^\circ-2\sin^2 30^\circ-\frac{3}{4}\operatorname{cosec}^2 45^\circ.$$

20. Solve the equations:

(1) $3\sin\theta=2\cos^2\theta$; (2) $5\tan\theta-\sec^2\theta=3$.

---

21. Prove that $1+2\sec^2 A\tan^2 A-\sec^4 A-\tan^4 A=0$.

22. In the equation

$$6\sin^2\theta-11\sin\theta+4=0,$$

shew that one value of θ is impossible, and find the other value.

23. In a triangle ABC right-angled at C, prove that

$$\tan A+\tan B=\frac{c^2}{ab}.$$

24. If $\cot A=c$, shew that $c+c^{-1}=\sec A\operatorname{cosec} A$.

25. Solve the equation

$$3\sin^2\theta+5\sin\theta=2.$$

CHAPTER V.

SOLUTION OF RIGHT-ANGLED TRIANGLES.

42. EVERY triangle has six *parts*, namely, three sides and three angles. In Trigonometry it is usual to denote the three angles by the capital letters A, B, C, and the lengths of the sides respectively opposite to these angles by the letters a, b, c. It must be understood that a, b, c are *numerical quantities* expressing the number of units of length contained in the three sides.

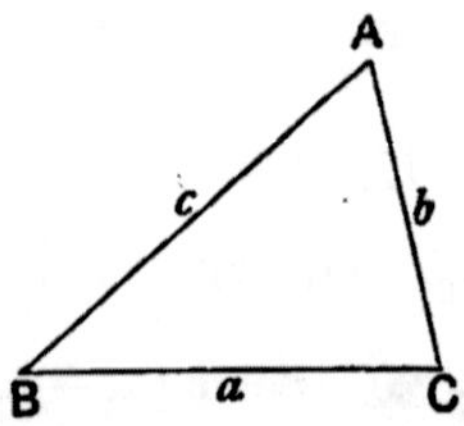

43. We know from Geometry that it is always possible to construct a triangle when any three parts are given, provided that one at least of the parts is a side. Similarly, if the values of suitable parts of a triangle be given, we can by Trigonometry find the remaining parts. The process by which this is effected is called the **Solution of the triangle.**

The *general* solution of triangles will be discussed at a later stage; in this chapter we shall confine our attention to right-angled triangles.

44. From Geometry, we know that when a triangle is right-angled, if any two sides are given the third can be found. Thus in the figure of the next article, where ABC is a triangle right-angled at A, we have $a^2=b^2+c^2$; whence if any two of the three quantities a, b, c are given, the third may be determined.

Again, the two acute angles are complementary, so that if one is given the other is also known.

Hence in the solution of right-angled triangles there are really only two cases to be considered:

I. when *any two sides* are given;

II. when *one side* and *one acute angle* are given.

45. Case I. *To solve a right-angled triangle when two sides are given.*

Let ABC be a right-angled triangle, of which A is the right angle, and suppose that any two sides are given;

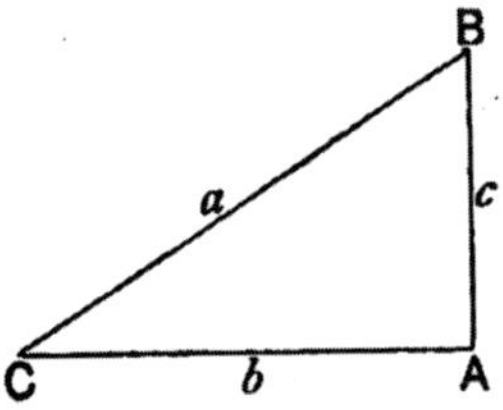

then the third side may be found from the equation

$$a^2=b^2+c^2.$$

Also

$$\cos C=\frac{b}{a}, \text{ and } B=90^\circ-C;$$

whence C and B may be obtained.

Example. Given $B=90^\circ$, $a=20$, $b=40$, solve the triangle.

Here $c^2=b^2-a^2$

$=1600-400=1200;$

$\therefore\ c=20\sqrt{3}.$

Also $\sin A=\frac{a}{b}=\frac{20}{40}=\frac{1}{2};$

$\therefore\ A=30^\circ.$

And $C=90^\circ-A=90^\circ-30^\circ=60^\circ.$

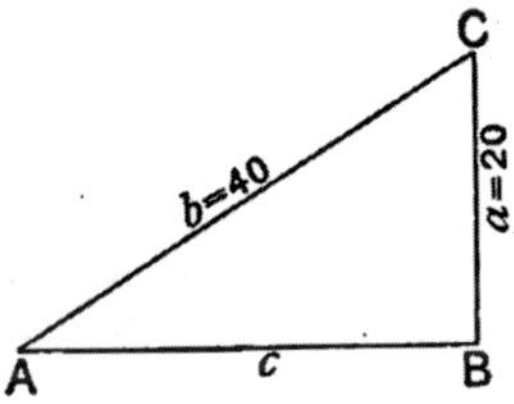

The solution of a trigonometrical problem may often be obtained in more than one way. Here the triangle can be solved without using the geometrical property of a right-angled triangle.

Another solution may be given as follows:

$$\cos C=\frac{a}{b}=\frac{20}{40}=\frac{1}{2};$$

$$\therefore\ C=60^\circ.$$

And $$A=90^\circ-C=90^\circ-60^\circ=30^\circ.$$

Also $$\frac{c}{40}=\cos A=\cos 30^\circ=\frac{\sqrt{3}}{2};$$

$$\therefore\ c=20\sqrt{3}.$$

46. CASE II. *To solve a right-angled triangle when one side and one acute angle are given.*

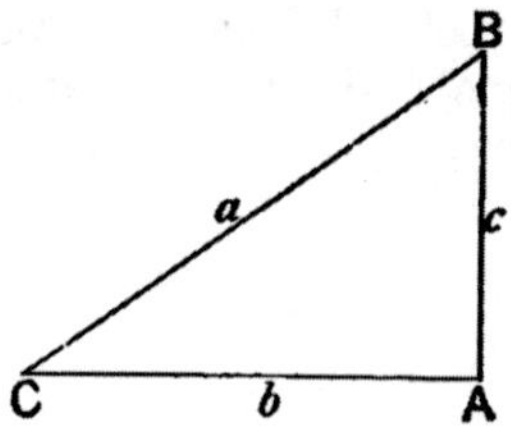

Let ABC be a right-angled triangle of which A is the right angle, and suppose one side b and one acute angle C are given; then

$$B=90^\circ-C, \quad \frac{a}{b}=\sec C, \quad \frac{c}{b}=\tan C;$$

whence B, a, c may be determined.

Example 1. Given $B=90^\circ$, $A=30^\circ$, $c=5$, solve the triangle.

We have $C=90^\circ-A=90^\circ-30^\circ=60^\circ$.

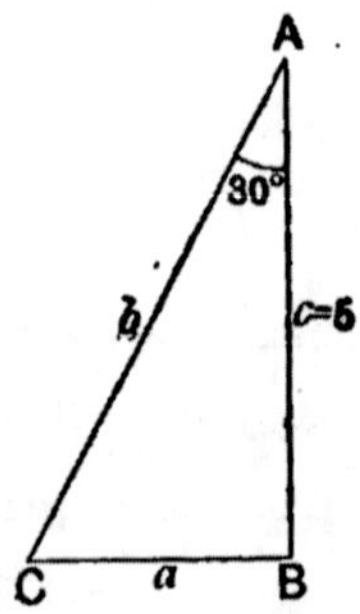

Also $$\frac{a}{5}=\tan 30^\circ;$$

$$\therefore\ a=5\tan 30^\circ=\frac{5}{\sqrt{3}}$$

$$=\frac{5}{\sqrt{3}}\times\frac{\sqrt{3}}{\sqrt{3}}=\frac{5\sqrt{3}}{3}.$$

Again, $$\frac{b}{5}=\sec 30^\circ;$$

$$\therefore\ b=5\sec 30^\circ=5\times\frac{2}{\sqrt{3}}=\frac{10}{\sqrt{3}}=\frac{10\sqrt{3}}{3}.$$

NOTE. The student should observe that in each case we write down a ratio which connects *the side we are finding with that whose value is given*, and a knowledge of the ratios of the given angle enables us to complete the solution.

Example 2. If $C=90^\circ$, $B=25^\circ\,43'$, and $c=100$, solve the triangle by means of Mathematical Tables.

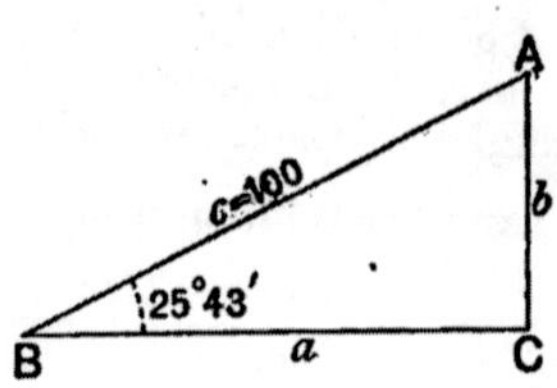

Here $A=90^\circ-B$

$$=90^\circ-25^\circ\,43'=64^\circ\,17'.$$

Now $$\frac{a}{c}=\cos B;$$

that is, $$\frac{a}{100}=\cos 25^\circ\,43';$$

$$\therefore\ a=100\cos 25^\circ\,43'$$

$$=100\times \cdot 9012=90\cdot 12, \text{ from the Tables.}$$

Also $$\frac{b}{a}=\tan B, \text{ or } b=a\tan B;$$

$$\therefore\ b=90\cdot 12\times\tan 25^\circ\,43'=90\cdot 12\times\cdot 4817=43\cdot 41.$$

EXAMPLES. V. a.

(Tables must be used for Examples marked with an asterisk.)

Solve the triangles in which the following parts are given:

1. $A=90°$, $a=4$, $b=2\sqrt{3}$.

2. $c=6$, $b=12$, $B=90°$.

3. $C=90°$, $b=12$, $a=4\sqrt{3}$.

4. $a=60$, $b=30$, $A=90°$.

5. $a=20$, $c=20$, $B=90°$.

6. $a=5\sqrt{3}$, $b=15$, $C=90°$.

7. $b=c=2$, $A=90°$.

8. $2c=b=6\sqrt{3}$, $B=90°$.

9. $C=90°$, $a=9\sqrt{3}$, $A=30°$.

*10. $A=90°$, $B=25°$, $a=4$.

*11. $A=54°$, $c=8$, $C=90°$.

*12. $A=27°$, $C=63°$, $b=6$.

*13. $B=90°$, $C=37°$, $b=100$.

14. $A=30°$, $B=60°$, $b=20\sqrt{3}$.

15. $B=C=45°$, $c=4$.

16. $2B=C=60°$, $a=8$.

17. If $C=90°$, $\cot A=\cdot 07$, $b=49$, find a.

18. If $C=90°$, $A=38° 19'$, $c=50$, find a; given $\sin 38° 19'=\cdot 62$.

*19. If $a=100$, $B=90°$, $C=40° 51'$, find c.

*20. If $b=200$, $A=90°$, $C=78° 12'$, find a to the nearest integer.

21. If $B=90°$, $A=36°$, $c=100$, solve the triangle; given $\tan 36=\cdot 73$, $\sec 36°=1\cdot 24$.

22. If $A=90°$, $c=37$, $a=100$, solve the triangle; given $\sin 21° 43'=\cdot 37$, $\cos 21° 43'=\cdot 93$.

*23. If $A=90°$, $B=39° 24'$, $b=25$, solve the triangle.

*24. If $C=90°$, $a=225$, $b=272$, solve the triangle.

*25. If $C=90°$, $b=22\cdot 75$, $c=25$, solve the triangle.

47. It will be found that all the varieties of the solution of right-angled triangles which can arise are either included in the two cases of Arts. 45 and 46, or in some modification of them. Sometimes the solution of a problem may be obtained by solving *two* right-angled triangles. The two examples we give as illustrations will in various forms be frequently met with in subsequent chapters.

Example 1. In the triangle ABC, the angles A and B are equal to $30°$ and $135°$ respectively, and the side AB is 100 feet; find the length of the perpendicular from C upon AB produced.

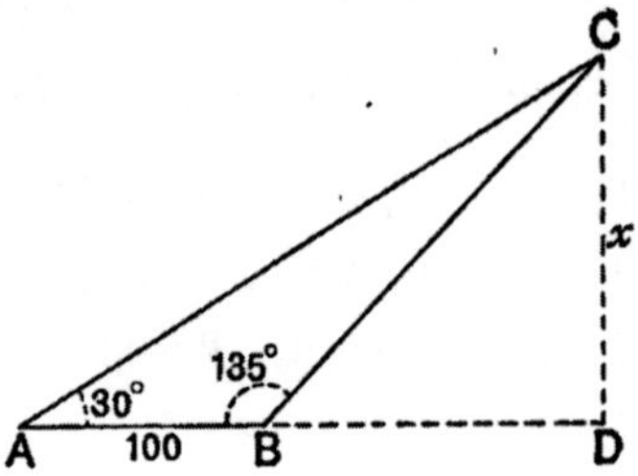

Draw CD perpendicular to AB produced, and let $CD=x$.

Then $\angle CBD = 180° - 135° = 45°$;

$$\therefore BD = CD = x.$$

Now in the right-angled triangle ADC,

$$\frac{CD}{AD} = \tan DAC = \tan 30°;$$

that is,
$$\frac{x}{x+100} = \frac{1}{\sqrt{3}};$$

$$\therefore x\sqrt{3} = x + 100;$$

$$x(\sqrt{3}-1) = 100,$$

$$x = \frac{100}{\sqrt{3}-1} = \frac{100(\sqrt{3}+1)}{3-1} = 50(\sqrt{3}+1);$$

$$\therefore x = 50 \times 2{\cdot}732.$$

Thus the distance required is 136·6 feet.

Example 2. In the triangle ABC, $a=9{\cdot}6$ cm., $c=5{\cdot}4$ cm., $B=37°$. Find the perpendicular from A on BC, and hence find A and C to the nearest degree.

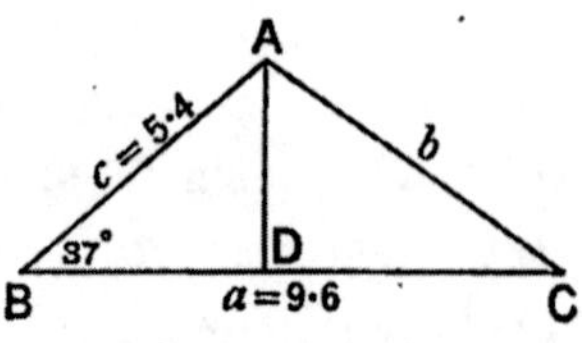

In the right-angled triangle ABD,

$$\frac{BD}{BA} = \cos ABD = \cos 37°;$$

$$\therefore BD = BA\cos 37°$$
$$= 5{\cdot}4 \times {\cdot}7986 \text{ (from the Tables)}$$
$$= 4{\cdot}31 \text{ cm.}$$

Also
$$\frac{AD}{AB} = \sin ABD = \sin 37°;$$

$$\therefore AD = AB\sin 37° = 5{\cdot}4 \times {\cdot}6018 \text{ (from the Tables)}$$
$$= 3{\cdot}25 \text{ cm.}$$

But
$$CD = BC - BD = 9{\cdot}6 - 4{\cdot}31 = 5{\cdot}29 \text{ cm.};$$

$\therefore$ from the right-angled triangle ACD,

$$\tan ACD = \frac{AD}{DC} = \frac{3{\cdot}25}{5{\cdot}29} = {\cdot}6144.$$

But from the Tables,

$$\tan 31° = \cdot 6009, \quad \tan 32° = \cdot 6249;$$

$$\therefore \angle ACD = 32°, \text{ to the nearest degree,}$$

and $$\angle BAC = 180° - 37° - 32° = 111°.$$

Thus $$AD = 3\cdot 25 \text{ cm.}, \quad \angle A = 111°, \quad \angle C = 32°.$$

EXAMPLES. V. b.

(Tables must be used for Examples marked with an asterisk.)

1. ABC is a triangle, and BD is perpendicular to AC produced: find BD, given

$$A = 30°, C = 120°, AC = 20.$$

2. If BD is perpendicular to the base AC of a triangle ABC, find a and c, given

$$A = 30°, C = 45°, BD = 10.$$

3. In the triangle ABC, AD is drawn perpendicular to BC making BD equal to 15 ft.: find the lengths of AB, AC, and AD, given that B and C are equal to 30° and 60° respectively.

4. In a right-angled triangle PQR, find the segments of the hypotenuse PR made by the perpendicular from Q; given

$$QR = 8, \quad \angle QRP = 60°, \quad \angle QPR = 30°.$$

***5.** If PQ is drawn perpendicular to the straight line QRS, find RS, given $PQ = 36$, $\angle RPQ = 35°$, $\angle SPQ = 53°$.

***6.** If PQ is drawn perpendicular to the straight line QRS, find RS, given $PQ = 20$, $\angle PRS = 135°$, $\angle PSR = 25°$.

7. In the triangle ABC, the angles B and C are equal to 45° and 120° respectively; if $a = 40$, find the length of the perpendicular from A on BC produced.

***8.** If CD is drawn perpendicular to the straight line DBA, find DC and BD, given

$$AB = 41\cdot 24, \quad \angle CBD = 45°, \quad \angle CAB = 35° 18'.$$

***9.** In a triangle ABC, $AB = 20$, $BC = 33$, $\angle B = 42°$, find the perpendicular from A on BC, and the angle C.

CHAPTER VI.

EASY PROBLEMS.

48. THE principles explained in the previous chapters may now be applied to the solution of problems in heights and distances. It will be assumed that by the use of suitable instruments the necessary lines and angles can be measured with sufficient accuracy for the purposes required.

After the practice afforded by the examples in the last chapter, the student should be able to write down at once any side of a right-angled triangle in terms of another through the medium of the functions of either acute angle. In the present and subsequent chapters it is of great importance to acquire readiness in this respect.

For instance, from the adjoining figure, we have

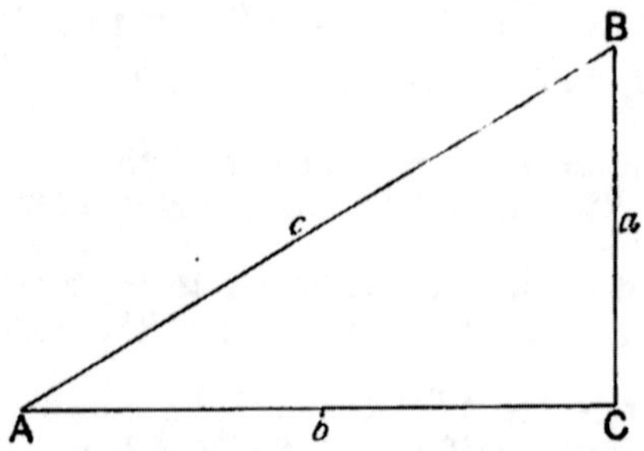

$$a = c \sin A, \quad a = c \cos B, \quad a = b \cot B,$$

$$a = b \tan A, \quad c = a \sec B, \quad b = a \tan B.$$

These relations are not to be committed to memory but in each case should be read off from the figure. There are several other similar relations connecting the parts of the above triangle, and the student should practise himself in obtaining them quickly.

Example. Q, R, T are three points in a straight line, and TP is drawn perpendicular to QT. If $PT = a$, $\angle PQT = \beta$, $\angle PRT = 2\beta$, express the lengths of all the lines of the figure in terms of a and β.

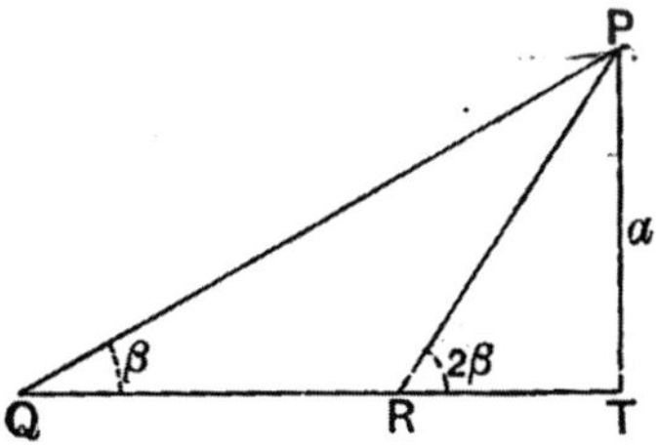

By Euc. I. 32,

$$\angle QPR = \angle PRT - \angle PQR;$$

$$\therefore \angle QPR = 2\beta - \beta = \beta = \angle PQR;$$

$$\therefore QR = PR.$$

In the right-angled triangle PRT,

$$PR = a \operatorname{cosec} 2\beta;$$

$$\therefore QR = a \operatorname{cosec} 2\beta.$$

Also $$TR = a \cot 2\beta.$$

Lastly, in the right-angled triangle PQT,

$$QT = a \cot \beta,$$

$$PQ = a \operatorname{cosec} \beta.$$

49. Angles of elevation and depression. Let OP be a horizontal line in the same vertical plane as an object Q, and let OQ be joined.

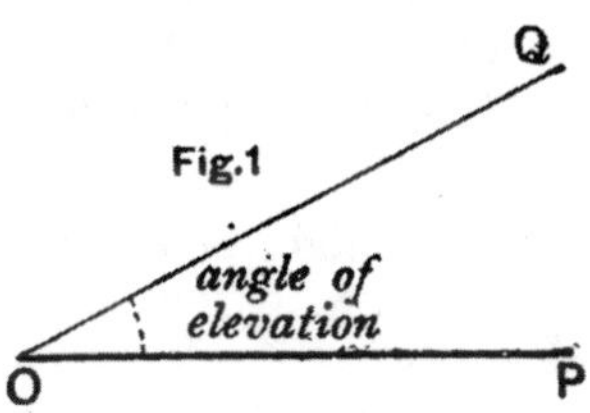

Fig. 1

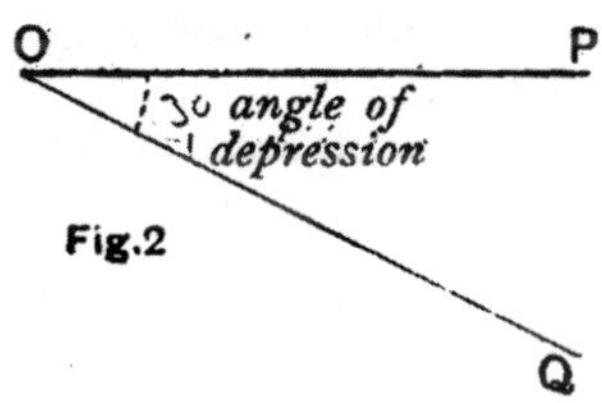

Fig. 2

In Fig. 1, where the object Q is *above* the horizontal line OP, the angle POQ is called the **angle of elevation** of the object Q as seen from the point O.

In Fig. 2, where the object Q is *below* the horizontal line OP, the angle POQ is called the **angle of depression** of the object Q as seen from the point O.

Example I. A flagstaff stands on a horizontal plane, and from a point on the ground at a distance of 30 ft. its angle of elevation is 60°: find its height.

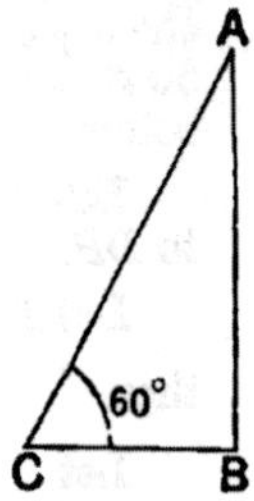

Let AB be the flagstaff, C the point of observation; then

$$AB = BC \tan 60° = 30\sqrt{3}$$
$$= 30 \times 1{\cdot}732 = 51{\cdot}96.$$

Thus the height is 51·96 ft.

EXAMPLES. VI. a.

[*For Examples in the use of Tables see page* 48_A.]

1. The angle of elevation of the top of a chimney at a distance of 300 feet is 30°: find its height.

2. From a ship's masthead 160 feet high the angle of depression of a boat is observed to be 30°: find its distance from the ship.

3. Find the angle of elevation of the sun when the shadow of a pole 6 feet high is $2\sqrt{3}$ feet long.

4. At a distance 86·6 feet from the foot of a tower the angle of elevation of the top is 30°. Find the height of the tower and the observer's distance from the top.

5. A ladder 45 feet long just reaches the top of a wall. If the ladder makes an angle of 60° with the wall, find the height of the wall, and the distance of the foot of the ladder from the wall.

6. Two masts are 60 feet and 40 feet high, and the line joining their tops makes an angle of 33° 41′ with the horizon: find their distance apart, given $\cot 33° 41' = 1{\cdot}5$.

7. Find the distance of the observer from the top of a cliff which is 132 yards high, given that the angle of elevation is 41° 18′, and that $\sin 41° 18' = {\cdot}66$.

8. One chimney is 30 yards higher than another. A person standing at a distance of 100 yards from the lower observes their tops to be in a line inclined at an angle of 27° 2′ to the horizon: find their heights, given $\tan 27° 2' = {\cdot}51$.

Example II. From the foot of a tower the angle of elevation of the top of a column is 60°, and from the top of the tower, which is 50 ft. high, the angle of elevation is 30°: find the height of the column.

Let AB denote the column and CD the tower; draw CE parallel to DB.

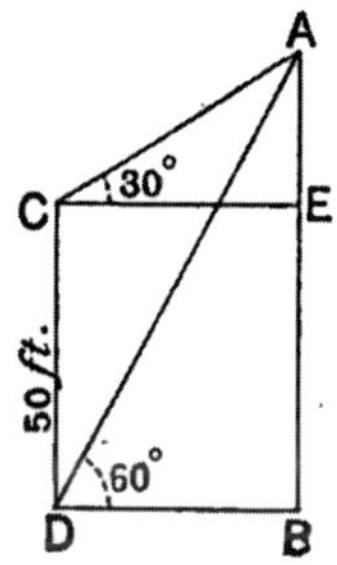

Let $AB=x$;

then $$AE=AB-BE=x-50.$$

Let $$DB=CE=y.$$

From the right-angled triangle ADB,

$$y=x\cot 60°=\frac{x}{\sqrt{3}}.$$

From the right-angled triangle ACE,

$$y=(x-50)\cot 30°=\sqrt{3}\,(x-50).$$

$$\therefore\ \frac{x}{\sqrt{3}}=\sqrt{3}\,(x-50),$$

$$x=3\,(x-50);$$

whence $$x=75.$$

Thus the column is 75 ft. high.

9. The angle of elevation of the top of a tower is 30°; on walking 100 yards nearer the elevation is found to be 60°: find the height of the tower.

10. A flagstaff stands upon the top of a building; at a distance of 40 feet the angles of elevation of the tops of the flagstaff and building are 60° and 30°: find the length of the flagstaff.

11. The angles of elevation of a spire at two places due east of it and 200 feet apart are 45° and 30°: find the height of the spire.

12. From the foot of a post the elevation of the top of a steeple is 45°, and from the top of the post, which is 30 feet high, the elevation is 30°; find the height and distance of the steeple.

13. The height of a hill is 3300 feet above the level of a horizontal plane. From a point A on this plane the angular elevation of the top of the hill is 60°. A balloon rises from A and ascends vertically upwards at a uniform rate; after 5 minutes the angular elevation of the top of the hill to an observer in the balloon is 30°: find the rate of the balloon's ascent in miles per hour.

Example III. From the top of a cliff 150 ft. high the angles of depression of two boats which are due South of the observer are 15° and 75°: find their distance apart, having given

$$\cot 15° = 2 + \sqrt{3} \text{ and } \cot 75° = 2 - \sqrt{3}.$$

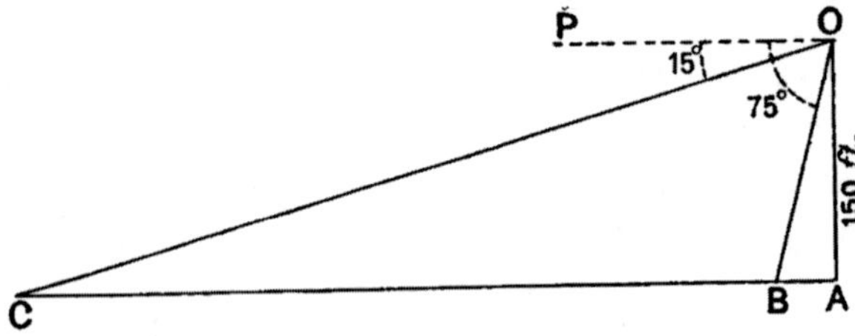

Let OA represent the cliff, B and C the boats. Let OP be a horizontal line through O; then

$$\angle POC = 15° \text{ and } \angle POB = 75°;$$
$$\therefore \angle OCA = 15° \text{ and } \angle OBA = 75°.$$

Let $CB = x$, $AB = y$; then $CA = x + y$.

From the right-angled triangle OBA,

$$y = 150 \cot 75° = 150(2 - \sqrt{3}) = 300 - 150\sqrt{3}.$$

From the right-angled triangle OCA,

$$x + y = 150 \cot 15° = 150(2 + \sqrt{3}) = 300 + 150\sqrt{3}.$$

By subtraction, $x = 300\sqrt{3} = 519{\cdot}6.$

Thus the distance between the boats is 519·6 ft.

14. From the top of a monument 100 feet high, the angles of depression of two objects on the ground due west of the monument are 45° and 30°: find the distance between them.

15. The angles of depression of the top and foot of a tower seen from a monument 96 feet high are 30° and 60°: find the height of the tower.

16. From the top of a cliff 150 feet high the angles of depression of two boats at sea, each due north of the observer, are 30° and 15°: how far are the boats apart?

17. From the top of a hill the angles of depression of two consecutive milestones on a level road running due south from the observer are 45° and 22° respectively. If $\cot 22° = 2{\cdot}475$ find the height of the hill in yards.

18. From the top of a lighthouse 80 yards above the horizon the angles of depression of two rocks due west of the observer are 75° and 15°: find their distance apart, given $\cot 75° = {\cdot}268$ and $\cot 15° = 3{\cdot}732$.

50. Trigonometrical Problems sometimes require a knowledge of the **Points of the Mariner's Compass**, which we shall now explain.

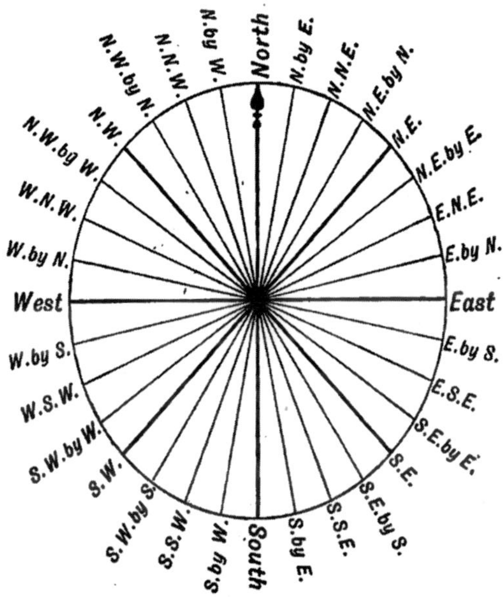

In the above figure, it will be seen that 32 points are taken at equal distances on the circumference of a circle, so that the arc between any two consecutive points subtends at the centre of the circle an angle equal to $\frac{360}{32}^\circ$, that is to $11\frac{1}{4}^\circ$.

The points North, South, East, West are called the **Cardinal Points**, and with reference to them the other *points* receive their names. The student will have no difficulty in learning these if he will carefully notice the arrangement in any one of the principal quadrants.

51. Sometimes a slightly different notation is used; thus N. $11\frac{1}{4}^\circ$ E. means a direction $11\frac{1}{4}^\circ$ east of north, and is therefore the same as N. by E. Again S.W. by S. is 3 *points* from south and may be expressed by S. $33\frac{3}{4}^\circ$ W., or since it is 5 *points* from west it can also be expressed by W. $56\frac{1}{4}^\circ$ S. In each of these cases it will be seen that the angular measurement is made from the direction which is first mentioned.

52. The angle between the directions of any two points is obtained by multiplying $11\frac{1}{4}°$ by the number of intervals between the points. Thus between S. by W. and W.S.W. there are 5 intervals and the angle is $56\frac{1}{4}°$; between N.E. by E. and S.E. there are 7 intervals and the angle is $78\frac{3}{4}°$.

53. If B lies in a certain direction with respect to A, it is said to *bear* in that direction from A; thus Birmingham *bears* N.W. of London, and from Birmingham the *bearing* of London is S.E.

Example 1. From a lighthouse L two ships A and B are observed in directions S.W. and 15° East of South respectively. At the same time B is observed from A in a S.E. direction. If LA is 4 miles find the distance between the ships.

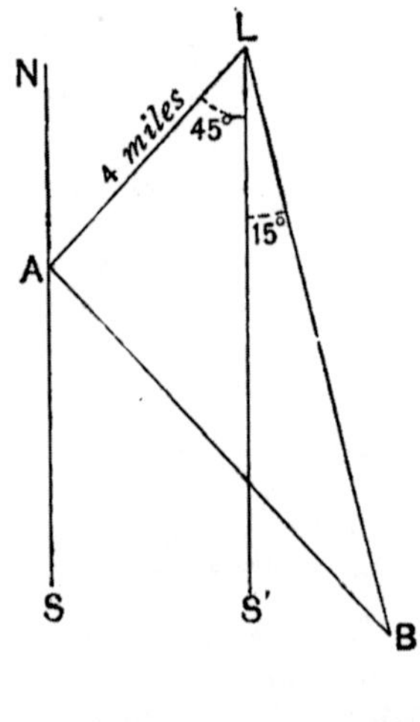

Draw LS' due South; then from the bearings of the two ships,

$$\angle ALS' = 45°, \quad \angle BLS' = 15°,$$

so that $\angle ALB = 60°$.

Through A draw a line NS pointing North and South; then

$$\angle NAL = \angle ALS' = 45°,$$

and $\angle BAS = 45°$, since B bears S.E. from A;

hence $\angle BAL = 180° - 45° - 45° = 90°$.

In the right-angled triangle ABL,

$$AB = AL \tan ALB = 4 \tan 60°$$
$$= 4\sqrt{3} = 6{\cdot}928.$$

Thus the distance between the ships is 6·928 miles.

Example 2. At 9 A.M. a ship which is sailing in a direction E. 37° S. at the rate of 8 miles an hour observes a fort in a direction 53° North of East. At 11 A.M. the fort is observed to bear N. 20° W.: find the distance of the fort from the ship at each observation.

Let A and C be the first and second positions of the ship; B the fort.

Through A draw lines towards the cardinal points of the compass.

From the observations made

$$\angle EAC = 37°, \quad \angle EAB = 53°, \text{ so that } \angle BAC = 90°.$$

Through C draw CN' towards the North; then $\angle BCN' = 20°$, for the bearing of the fort from C is N. 20° W.

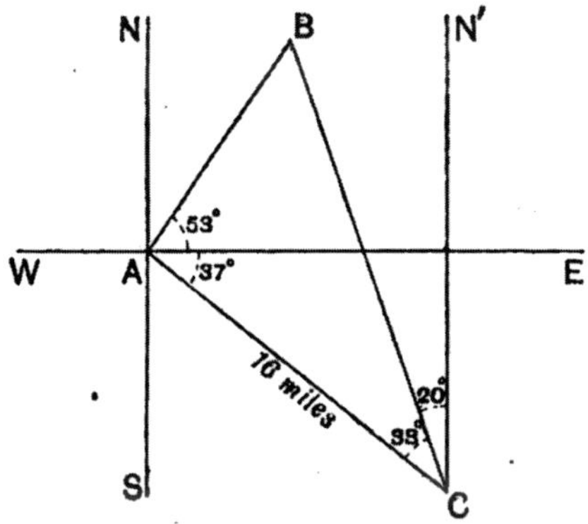

Also $\qquad \angle ACN' = \angle CAS = 90° - 37° = 53°;$

$$\therefore \quad \angle ACB = \angle ACN' - \angle BCN' = 53° - 20° = 33°.$$

In the right-angled triangle ACB,

$$AB = AC \tan ACB = 16 \tan 33° = 16 \times \cdot 6494, \text{ from the Tables;}$$
$$= 10 \cdot 3904.$$

Again

$$BC = AC \sec ACB = 16 \sec 33° = 16 \times 1 \cdot 1924 = 19 \cdot 0784.$$

Thus the distances are 10·39 and 19·08 miles nearly.

EXAMPLES. VI. b.

(*For Examples to be solved by the use of Tables see page* 48_A.)

1. A person walking due E. observes two objects both in the N.E. direction. After walking 800 yards one of the objects is due N. of him, and the other lies N.W.: how far was he from the objects at first?

2. Sailing due E. I observe two ships lying at anchor due S.; after sailing 3 miles the ships bear 60° and 30° S. of W.; how far are they now distant from me?

3. Two vessels leave harbour at noon in directions W. 28° S. and E. 62° S. at the rates 10 and $10\frac{1}{2}$ miles per hour respectively. Find their distance apart at 2 P.M.

4. A lighthouse facing N. sends out a fan-shaped beam extending from N.E. to N.W. A steamer sailing due W. first sees the light when 5 miles away from the lighthouse and continues to see it for $30\sqrt{2}$ minutes. What is the speed of the steamer?

5. A ship sailing due S. observes two lighthouses in a line exactly W. After sailing 10 miles they are respectively N.W. and W.N.W.; find their distances from the position of the ship at the first observation.

6. Two vessels sail from port in directions N. 35° W. and S. 55° W. at the rates of 8 and $8\sqrt{3}$ miles per hour respectively. Find their distance apart at the end of an hour, and the bearing of the second vessel as observed from the first.

7. A vessel sailing S.S.W. is observed at noon to be E.S.E. from a lighthouse 4 miles away. At 1 p.m. the vessel is due S. of the lighthouse: find the rate at which the vessel is sailing. Given $\tan 67\frac{1}{2}° = 2·414$.

8. A, B, C are three places such that from A the bearing of C is N. 10° W., and the bearing of B is N. 50° E.; from B the bearing of C is N. 40° W. If the distance between B and C is 10 miles, find the distances of B and C from A.

9. A ship steaming due E. sights at noon a lighthouse bearing N.E., 15 miles distant; at 1.30 p.m. the lighthouse bears N.W. How many knots per day is the ship making? Given 60 knots = 69 miles.

10. At 10 o'clock forenoon a coaster is observed from a lighthouse to bear 9 miles away to N.E. and to be holding a south-easterly course; at 1 p.m. the bearing of the coaster is 15° S. of E. Find the rate of the coaster's sailing and its distance from the lighthouse at the time of the second observation.

11. The distance between two lighthouses, A and B, is 12 miles and the line joining them bears E. 15° N. At midnight a vessel which is sailing S. 15° E. at the rate of 10 miles per hour is N.E. of A and N.W. of B: find to the nearest minute when the vessel crosses the line joining the lighthouses.

12. From A to B, two stations of a railway, the line runs W.S.W. At A a person observes that two spires, whose distance apart is 1·5 miles, are in the same line which bears N.N.W. At B their bearings are N. $7\frac{1}{2}°$ E. and N. $37\frac{1}{2}°$ E. Find the rate of a train which runs from A to B in 2 minutes.

EXAMPLES. VI. c.

[*In the following Examples Four-Figure Tables will be required.*]

1. At a distance of 83 yards the angle of elevation of the top of a chimney stack is $23° 44'$; find its height to the nearest foot.

2. The shadow cast by a spire at a time when the angular elevation of the sun in $63°$ is 173 feet; find the height of the spire.

3. A balloon held captive by a rope 200 metres long has drifted with the wind till the angle of elevation as observed from the place of ascent is $54°$. How high is the balloon above the ground?

4. A coal seam is inclined at an angle of $23°$ to the horizontal. Find, to the nearest foot, the distance below the level of a man who has walked 500 yards down the seam from the point where it meets the surface.

5. A man standing immediately opposite to a telegraph post on a railway notices that the line joining this post and the next one subtends an angle of $73° 18'$. Assuming that there are 23 telegraph posts to the mile, find his distance from the first post.

6. The middle point of one side of a square is joined to one of the opposite corners of the square; find the size of the two angles formed at this corner.

7. Find without any measurement the angles of an isosceles triangle each of whose equal sides is three times the base.

8. The angles of elevation of a spire at two places due east of it, and 160 feet apart, are $45°$ and $21° 48'$: find the height of the spire to the nearest foot.

9. The angle of elevation of the top of a tower is $27° 12'$, and on walking 100 yards nearer the elevation is found to be $54° 24'$; find the height of the tower to the nearest foot.

10. From the top of a hill the angles of depression of two consecutive milestones on a level road running due east from the observer are $55°$ and $16° 42'$ respectively; find the height of the hill to the nearest yard.

11. A trench is to be dug to measure 15 feet across the top, 9 feet across the bottom, and of uniform depth 8 feet; if one side is inclined at 12° to the vertical, what must be the inclination of the other?

12. Two towers, *A* and *B*, on a level plain subtend an angle of 90° at an observer's eye; he walks directly towards *B* a distance of 630 metres and then finds the angle subtended to be 143° 24′. Find the distance of *A* from each position of the observer.

13. From the roof of a house 30 feet high the angle of elevation of the top of a monument is 42°, and the angle of depression of its foot is 17°. Find its height.

14. From a point on a horizontal plane I find the angle of elevation of the top of a neighbouring hill to be 14°; after walking 700 metres in a straight line towards the hill, I find the elevation to be 35°. Find the height of the hill.

15. At noon a ship which is sailing a straight course due W. at 10 miles an hour observes a lighthouse 32° W. of North. At 1.30 P.M. the lighthouse bears 58° E. of North; find the distance of the lighthouse from the first position of the ship.

16. From two positions, 2 kilometres apart, on a straight road running East and West a house bears 52° W. of N. and 38° E. of N. respectively. Find to the nearest metre how far the house is from the road.

17. A ship sailing due E. observes that a lighthouse known to be 12 miles distant bears N. 34° E. at 3 P.M., and N. 56° W. at 4.10 P.M. How many miles a day is the ship making?

18. A ship which was lying $2\frac{1}{2}$ miles N.W. of a shore battery with an effective range of 4 miles, steers a straight course under cover of darkness until she is due N. of the battery, and just out of range. In what direction does she steam?

19. At 10 A.M. a ship which is sailing E. 41° S. at the rate of 10 miles an hour observes a fort bearing 49° N. of E. At noon the bearing of the fort is N. 15° E.; find the distance of the fort from the ship at each observation.

CHAPTER VII.

RADIAN OR CIRCULAR MEASURE.

54. We shall now return to the system of measuring angles which was briefly referred to in Art. 6. In this system angles are not measured in terms of a submultiple of the right angle, as in the sexagesimal and centesimal methods, but a certain angle known as a *radian* is taken as the standard unit, in terms of which all other angles are measured.

55. Definition. A **radian** is the angle subtended at the centre of any circle by an arc equal in length to the radius of the circle.

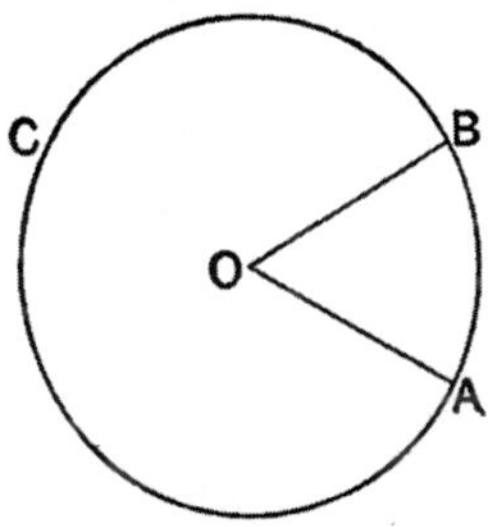

In the above figure, ABC is a circle, and O its centre. If on the circumference we measure an arc AB equal to the radius and join OA, OB, the angle AOB is a radian.

56. In any system of measurement it is essential that the unit should be always the same. In order to shew that a radian, constructed according to the above definition, is of constant magnitude, we must first establish an important property of the circle.

57. *The circumferences of circles are to one another as their radii.*

Take *any* two circles whose radii are r_1 and r_2, and in each circle let a regular polygon of n sides be described.

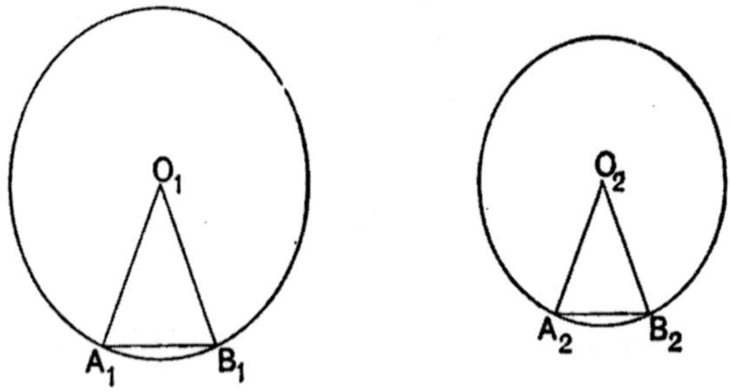

Let A_1B_1 be a side of the first, A_2B_2 a side of the second polygon, and let their lengths be denoted by a_1, a_2. Join their extremities to O_1 and O_2 the centres of the circles. We thus obtain two isosceles triangles whose vertical angles are equal, each being $\frac{1}{n}$ of four right angles.

Hence the triangles are equiangular, and therefore we have by Euc. VI. 4,

$$\frac{A_1B_1}{O_1A_1} = \frac{A_2B_2}{O_2A_2};$$

that is,

$$\frac{a_1}{r_1} = \frac{a_2}{r_2};$$

$$\therefore \frac{na_1}{r_1} = \frac{na_2}{r_2};$$

that is,

$$\frac{p_1}{r_1} = \frac{p_2}{r_2},$$

where p_1 and p_2 are the perimeters of the polygons. This is true whatever be the number of sides in the polygons. By taking n sufficiently large we can make the perimeters of the two polygons differ from the circumferences of the corresponding circles by as small a quantity as we please; so that ultimately

$$\frac{c_1}{r_1} = \frac{c_2}{r_2},$$

where c_1 and c_2 are the circumferences of the two circles.

58. It thus appears that *the ratio of the circumference of a circle to its radius is the same whatever be the size of the circle;* that is,

$$\textit{in all circles } \frac{\textit{circumference}}{\textit{diameter}} \textit{ is a constant quantity.}$$

This constant is incommensurable and is always denoted by the Greek letter π. Though its numerical value cannot be found exactly, it is shewn in a later part of the subject that it can be obtained to any degree of approximation. To ten decimal places its value is 3·1415926536. In many cases $\pi = \frac{22}{7}$, which is true to two decimal places, is a sufficiently close approximation; where greater accuracy is required the value 3·1416 may be used.

59. If c denote the circumference of the circle whose radius is r, we have

$$\frac{\text{circumference}}{\text{diameter}} = \pi;$$

$$\therefore \frac{c}{2r} = \pi,$$

or

$$c = 2\pi r.$$

60. *To prove that all radians are equal.*

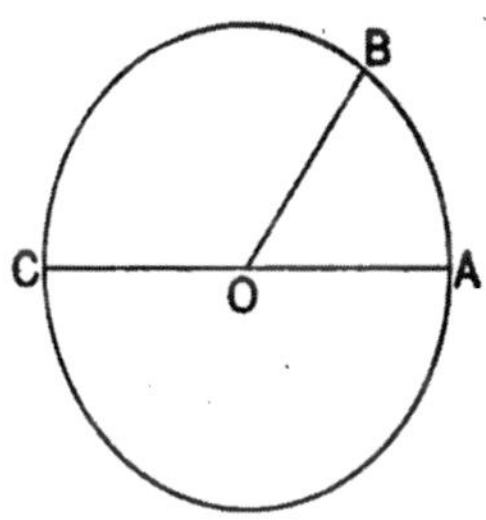

Draw *any* circle; let O be its centre and OA a radius. Let the arc AB be measured equal in length to OA. Join OB; then $\angle AOB$ is a radian. Produce AO to meet the circumference in C. By Euc. VI. 33, angles at the centre of a circle are proportional to the arcs on which they stand; hence

$$\frac{\angle AOB}{\text{two right angles}} = \frac{\text{arc } AB}{\text{arc } ABC}$$

$$= \frac{\text{radius}}{\text{semi-circumference}} = \frac{r}{\pi r} = \frac{1}{\pi},$$

which is constant; that is, *a radian always bears the same ratio to two right angles, and therefore is a constant angle.*

61. Since a radian is constant it is taken as a standard unit, and the *number of radians* contained in any angle is spoken of as its **radian measure** or **circular measure**. [See Art. 71.] In this system, an angle is usually denoted by a *mere number*, the unit being implied. Thus when we speak of an angle 2·5, it is understood that its radian measure is 2·5, or, in other words, that the angle contains $2\frac{1}{2}$ radians.

62. *To find the radian measure of a right angle.*

Let AOC be a right angle at the centre of a circle, and AOB a radian; then

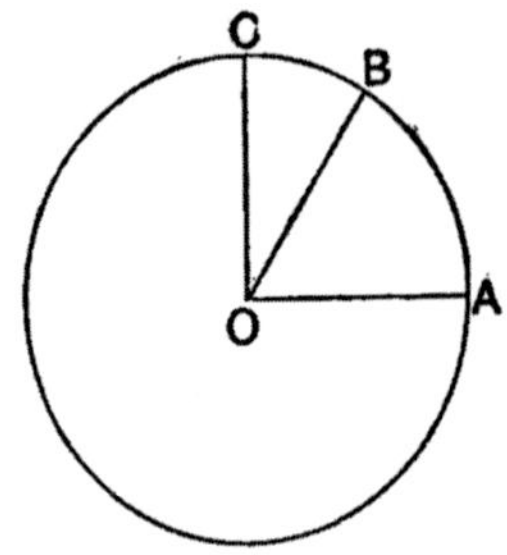

the radian measure of $\angle AOC$

$$=\frac{\angle AOC}{\angle AOB}=\frac{\text{arc } AC}{\text{arc } AB}$$

$$=\frac{\frac{1}{4}\text{(circumference)}}{\text{radius}}=\frac{\frac{1}{4}(2\pi r)}{r}$$

$$=\frac{\pi}{2};$$

that is, a right angle contains $\frac{\pi}{2}$ radians.

63. *To find the number of degrees in a radian.*

From the last article it follows that

$$\pi \textit{ radians}=2 \textit{ right angles}=180 \textit{ degrees}.$$

$$\therefore \text{ a radian}=\frac{180}{\pi} \text{ degrees}.$$

By division we find that $\frac{1}{\pi}=\cdot31831$ nearly;

hence approximately, a radian $=180\times\cdot31831=57\cdot2958$ degrees.

64. The formula

$$\pi \textit{ radians}=180 \textit{ degrees}$$

connecting the sexagesimal and radian measures of an angle, is a useful result which enables us to pass readily from one system to the other.

Example. Express 75° in radian measure, and $\frac{\pi}{54}$ in sexagesimal measure.

(1) Since 180 degrees $= \pi$ radians,

$$75 \text{ degrees} = \frac{75}{180}\pi \text{ radians} = \frac{5\pi}{12} \text{ radians}.$$

Thus the radian measure is $\frac{5\pi}{12}$.

(2) Since π radians $= 180$ degrees,

$$\frac{\pi}{54} \text{ radians} = \frac{180}{54} \text{ degrees}.$$

Thus the angle $= \frac{10}{3}$ degrees $= 3°\ 20'$.

65. It may be well to remind the student that the symbol π always denotes a *number*, viz. 3·14159.... When the symbol stands alone, without reference to any angle, there can be no ambiguity; but even when π is used to denote an angle, it must still be remembered that π is a *number*, namely, the number of radians in two right angles.

NOTE. It is not uncommon for beginners to make statements such as "$\pi = 180$" or "$\frac{\pi}{2} = 90$." Without some modification this mode of expression is quite incorrect. It is true that π *radians* are equal to 180 *degrees*, but the statement '$\pi = 180$' is no more correct than the statement "$20 = 1$" to denote the equivalence of 20 shillings and 1 sovereign.

66. *If the number of degrees and radians in an angle be represented by D and θ respectively, to prove that*

$$\frac{D}{180} = \frac{\theta}{\pi}.$$

In sexagesimal measure, the ratio of the given angle to two right angles is expressed by $\frac{D}{180}$.

In radian measure, the ratio of these same two angles is expressed by $\frac{\theta}{\pi}$.

$$\therefore \frac{D}{180} = \frac{\theta}{\pi}.$$

Example 1. What is the radian measure of 45° 13′ 30″?

If D be the number of degrees in the angle, we have $D = 45{\cdot}225$.

$$\begin{array}{r|l} 60 & 30 \\ \hline 60 & 13{\cdot}5 \\ \hline & {\cdot}225 \end{array}$$

Let θ be the number of radians in the given angle, then

$$\frac{\theta}{\pi} = \frac{45{\cdot}225}{180} = \frac{1{\cdot}005}{4};$$

$$\therefore \theta = \frac{\pi}{4} \times 1{\cdot}005 = \frac{3{\cdot}1416}{4} \times 1{\cdot}005$$

$$= {\cdot}7854 \times 1{\cdot}005 = {\cdot}789327.$$

Thus the radian measure is ·789327.

Example 2. Express in sexagesimal measure the angle whose radian measure is 1·309.

Let D be the number of degrees; then

$$\frac{D}{180} = \frac{1{\cdot}309}{\pi};$$

$$\therefore D = \frac{180 \times 1{\cdot}309}{3{\cdot}1416} = \frac{180 \times 1309 \times 10}{31416}$$

$$= \frac{180 \times 10}{24} = 75.$$

Thus the angle is 75°.

EXAMPLES. VII. a.

[*Unless otherwise stated* $\pi = 3{\cdot}1416$.

It should be noticed that $31416 = 8 \times 3 \times 7 \times 11 \times 17$.]

Express in radian measure as fractions of π:

1. 45°. 2. 30°. 3. 105°. 4. 22° 30′.
5. 18°. 6. 57° 30′. 7. 14° 24′. 8. 78° 45′.

Find numerically the radian measure of the following angles:

9. 25° 50′. 10. 37° 30′. 11. 82° 30′.
12. 68° 45′. 13. 157° 30′. 14. 52° 30′.

Express in sexagesimal measure:

15. $\frac{3\pi}{4}$. **16.** $\frac{7\pi}{45}$. **17.** $\frac{5\pi}{27}$. **18.** $\frac{5\pi}{24}$.

19. ·3927. **20.** ·6720. **21.** ·5201. **22.** 2·8798.

Taking $\pi = \frac{22}{7}$, find the radian measure of:

23. 36° 32′ 24″. **24.** 70° 33′ 36″.

25. 116° 2′ 45·6″. **26.** 171° 41′ 50·4″.

27. Taking $\frac{1}{\pi} = \cdot 31831$, shew that a radian contains 206265 seconds approximately.

28. Shew that a second is approximately equal to ·0000048 of a radian.

67. The angles $\frac{\pi}{4}, \frac{\pi}{3}, \frac{\pi}{6}$ are the equivalents in radian measure of the angles 45°, 60°, 30° respectively.

Hence the results of Arts. 34 and 35 may be written as follows:

$$\sin\frac{\pi}{4} = \frac{1}{\sqrt{2}}, \qquad \cos\frac{\pi}{4} = \frac{1}{\sqrt{2}}, \qquad \tan\frac{\pi}{4} = 1;$$

$$\sin\frac{\pi}{3} = \frac{\sqrt{3}}{2}, \qquad \cos\frac{\pi}{3} = \frac{1}{2}, \qquad \tan\frac{\pi}{3} = \sqrt{3};$$

$$\sin\frac{\pi}{6} = \frac{1}{2}, \qquad \cos\frac{\pi}{6} = \frac{\sqrt{3}}{2}, \qquad \tan\frac{\pi}{6} = \frac{1}{\sqrt{3}}.$$

Example. Find the value of

$$3\tan^2\frac{\pi}{6} + \frac{4}{3}\cos^2\frac{\pi}{6} - \frac{1}{2}\cot^3\frac{\pi}{4} - \frac{2}{3}\sin^2\frac{\pi}{3} + \frac{1}{8}\sec^4\frac{\pi}{3}.$$

$$\text{The value} = 3\left(\frac{1}{\sqrt{3}}\right)^2 + \frac{4}{3}\left(\frac{\sqrt{3}}{2}\right)^2 - \frac{1}{2}(1)^3 - \frac{2}{3}\left(\frac{\sqrt{3}}{2}\right)^2 + \frac{1}{8}(2)^4$$

$$= \left(3 \times \frac{1}{3}\right) + \left(\frac{4}{3} \times \frac{3}{4}\right) - \frac{1}{2} - \left(\frac{2}{3} \times \frac{3}{4}\right) + \left(\frac{1}{8} \times 16\right)$$

$$= 1 + 1 - \frac{1}{2} - \frac{1}{2} + 2 = 3.$$

68. When expressed in radian measure the complement of θ is $\frac{\pi}{2}-\theta$, and corresponding to the formulæ of Art. 38 we now have relations of the form

$$\sin\left(\frac{\pi}{2}-\theta\right)=\cos\theta, \qquad \tan\left(\frac{\pi}{2}-\theta\right)=\cot\theta.$$

Example. Prove that

$$(\cot\theta+\tan\theta)\cot\left(\frac{\pi}{2}-\theta\right)=\operatorname{cosec}^2\left(\frac{\pi}{2}-\theta\right).$$

$$\begin{aligned}\text{The first side} &= (\cot\theta+\tan\theta)\tan\theta\\ &= \cot\theta\tan\theta+\tan^2\theta\\ &= 1+\tan^2\theta=\sec^2\theta\\ &= \operatorname{cosec}^2\left(\frac{\pi}{2}-\theta\right).\end{aligned}$$

69. By means of Euc. I. 32, it is easy to find the number of radians in each angle of a regular polygon.

Example. Express in radians the interior angle of a regular polygon which has n sides.

The sum of the *exterior* angles = 4 right angles. [Euc. I. 32 Cor.]

Let θ be the number of radians in an exterior angle; then

$$n\theta=2\pi, \text{ and therefore } \theta=\frac{2\pi}{n}.$$

But interior angle = two right angles − exterior angle

$$=\pi-\theta=\pi-\frac{2\pi}{n}.$$

Thus each interior angle $=\dfrac{(n-2)\pi}{n}$.

EXAMPLES. VII. b.

Find the numerical value of

1. $\sin\frac{\pi}{3}\cos\frac{\pi}{6}\cot\frac{\pi}{4}$.
2. $\tan\frac{\pi}{6}\cot\frac{\pi}{3}\cos\frac{\pi}{4}$.
3. $\frac{1}{2}\cos\frac{\pi}{3}+2\operatorname{cosec}\frac{\pi}{6}$.
4. $2\sin\frac{\pi}{4}+\frac{1}{2}\sec\frac{\pi}{4}$.

Find the numerical value of

5. $\cot^2\frac{\pi}{6}+4\cos^2\frac{\pi}{4}+3\sec^2\frac{\pi}{6}$.

6. $3\tan^2\frac{\pi}{6}-\frac{1}{3}\sin^2\frac{\pi}{3}-\frac{1}{2}\operatorname{cosec}^2\frac{\pi}{4}+\frac{4}{3}\cos^2\frac{\pi}{6}$.

7. $\left(\sin\frac{\pi}{6}+\cos\frac{\pi}{6}\right)\left(\sin\frac{\pi}{3}-\cos\frac{\pi}{3}\right)\sec\frac{\pi}{3}$.

Prove the following identities :

8. $\sin\theta\sec\left(\frac{\pi}{2}-\theta\right)-\cot\theta\cot\left(\frac{\pi}{2}-\theta\right)=0$.

9. $\sin^2\left(\frac{\pi}{2}-\theta\right)\operatorname{cosec}\theta-\tan^2\left(\frac{\pi}{2}-\theta\right)\sin\theta=0$.

10. $\dfrac{\sin^2\left(\frac{\pi}{2}-\theta\right)}{\operatorname{cosec}\theta}\cdot\dfrac{\sec\theta}{\cot\left(\frac{\pi}{2}-\theta\right)}=\cos^2\theta$.

11. $\tan\theta+\tan\left(\frac{\pi}{2}-\theta\right)=\sec\theta\sec\left(\frac{\pi}{2}-\theta\right)$.

12. $\sec^2\theta+\sec^2\left(\frac{\pi}{2}-\theta\right)=(1+\tan^2\theta)\sec^2\left(\frac{\pi}{2}-\theta\right)$.

13. Find the number of radians in each exterior angle of

(1) a regular octagon, (2) a regular quindecagon.

14. Find the number of radians in each interior angle of

(1) a regular dodecagon, (2) a regular heptagon.

15. Shew that

$$\tan^2\frac{\pi}{3}-\cot^2\frac{\pi}{3}=\frac{\cos^2\frac{\pi}{6}-\cos^2\frac{\pi}{3}}{\cos^2\frac{\pi}{3}\cos^2\frac{\pi}{6}}.$$

16. Shew that the sum of the squares of

$$\sin\theta+\sin\left(\frac{\pi}{2}-\theta\right)\text{ and }\cos\theta-\cos\left(\frac{\pi}{2}-\theta\right)$$

is equal to 2.

70. *To prove that the radian measure of any angle at the centre of a circle is expressed by the fraction* $\dfrac{\textit{subtending arc}}{\textit{radius}}$.

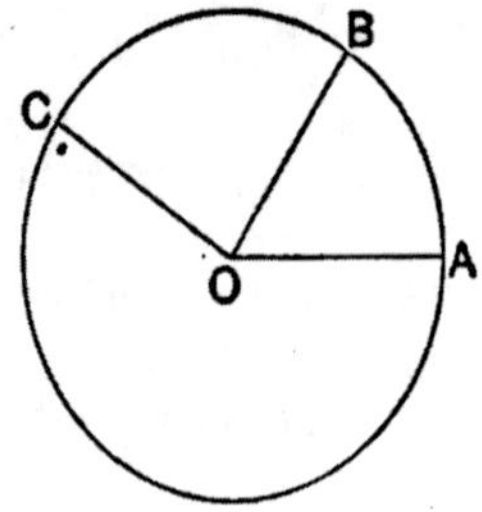

Let AOC be any angle at the centre of a circle, and AOB a radian; then

radian measure of $\angle AOC$

$$=\frac{\angle AOC}{\angle AOB}$$

$$=\frac{\text{arc } AC}{\text{arc } AB}$$

$$=\frac{\text{arc } AC}{\text{radius}},$$

since arc AB=radius;

that is, the radian measure of $\angle AOC=\dfrac{\text{subtending arc}}{\text{radius}}$.

71. If a be the length of the arc which subtends an angle of θ radians at the centre of a circle of radius r, we have seen in the preceding article that

$$\theta=\frac{a}{r}, \text{ and therefore } a=r\theta.$$

The fraction $\dfrac{\text{arc}}{\text{radius}}$ is usually called the *circular measure* of the angle at the centre of the circle subtended by the arc.

The *circular measure* of an angle is therefore equal to its *radian measure*, each denoting the *number of radians* contained in the angle. We have preferred to use the term *radian measure* exclusively, in order to keep prominently in view the unit of measurement, namely the radian.

NOTE. The term *circular measure* is a survival from the times when Mathematicians spoke of the trigonometrical functions of the *arc*. [See page 80.]

Example 1. Find the angle subtended by an arc of 7·5 feet at the centre of a circle whose radius is 5 yards.

Let the angle contain θ radians; then

$$\theta=\frac{\text{arc}}{\text{radius}}=\frac{7\cdot5}{15}=\frac{1}{2}.$$

Thus the angle is half a radian.

Example 2. In running a race at a uniform speed on a circular course, a man in each minute traverses an arc of a circle which subtends $2\frac{6}{7}$ radians at the centre of the course. If each lap is 792 yards, how long does he take to run a mile? $\left[\pi = \frac{22}{7}\right]$.

Let r yards be the radius of the circle; then

$$2\pi r = \text{circumference} = 792;$$

$$\therefore\ r = \frac{792}{2\pi} = \frac{792 \times 7}{2 \times 22} = 126.$$

Let a yards be the length of the arc traversed in each minute; then from the formula $a = r\theta$,

$$a = 126 \times 2\tfrac{6}{7} = \frac{126 \times 20}{7} = 360;$$

that is, the man runs 360 yds. in each minute.

$$\therefore\ \text{the time} = \frac{1760}{360} \text{ or } \frac{44}{9} \text{ minutes.}$$

Thus the time is 4 min. $53\frac{1}{3}$ sec.

Example 3. Find the radius of a globe such that the distance measured along its surface between two places on the same meridian whose latitudes differ by $1°\ 10'$ may be 1 inch, reckoning that $\pi = \frac{22}{7}$.

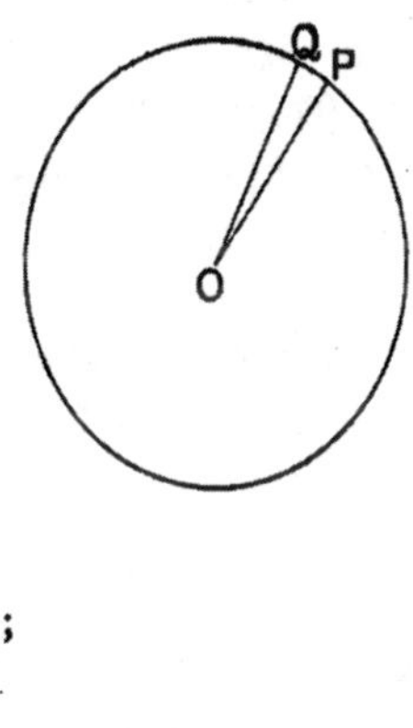

Let the adjoining figure represent a section of the globe through the meridian on which the two places P and Q lie. Let O be the centre, and denote the radius by r inches.

$$\text{Now } \frac{\text{arc } PQ}{\text{radius}} = \text{number of radians in } \angle POQ;$$

but $\text{arc } PQ = 1$ inch, and $\angle POQ = 1°\ 10'$;

$$\therefore\ \frac{1}{r} = \text{number of radians in } 1\tfrac{1}{6}°$$

$$= 1\tfrac{1}{6} \times \frac{\pi}{180} = \frac{7}{6} \times \frac{22}{7} \times \frac{1}{180} = \frac{11}{540};$$

whence
$$r = \frac{540}{11} = 49\tfrac{1}{11}.$$

Thus the radius is $49\frac{1}{11}$ inches.

EXAMPLES. VII. c.

1. Find the radian measure of the angle subtended by an arc of 1·6 yards at the centre of a circle whose radius is 24 feet.

2. An angle whose circular measure is ·73 subtends at the centre of a circle an arc of 219 feet; find the radius of the circle.

3. An angle at the centre of a circle whose radius is 2·5 yards is subtended by an arc of 7·5 feet; what is the angle?

4. What is the length of the arc which subtends an angle of 1·625 radians at the centre of a circle whose radius is 3·6 yards?

5. An arc of 17 yds. 1 ft. 3 in. subtends at the centre of a circle an angle of 1·9 radians; find the radius of the circle in inches.

6. The flywheel of an engine makes 35 revolutions in a second; how long will it take to turn through 5 radians? $\left[\pi = \frac{22}{7}\right]$.

7. The large hand of a clock is 2 ft. 4 in. long; how many inches does its extremity move in 20 minutes? $\left[\pi = \frac{22}{7}\right]$.

8. A horse is tethered to a stake; how long must the rope be in order that, when the horse has moved through 52·36 yards at the extremity of the rope, the angle traced out by the rope may be 75 degrees?

9. Find the length of an arc which subtends 1 minute at the centre of the earth, supposed to be a sphere of diameter 7920 miles.

10. Find the number of seconds in the angle subtended at the centre of a circle of radius 1 mile by an arc $5\frac{1}{2}$ inches long.

11. Two places on the same meridian are 145·2 miles apart; find their difference in latitude, taking $\pi = \frac{22}{7}$, and the earth's diameter as 7920 miles.

12. Find the radius of a globe such that the distance measured along its surface between two places on the same meridian whose latitudes differ by $1\frac{3}{11}^\circ$ may be 1 foot, taking $\pi = \frac{22}{7}$.

MISCELLANEOUS EXAMPLES. B.

1. Express in degrees the angle whose circular measure is ·15708.

2. If $C=90^\circ$, $A=30^\circ$, $c=110$, find b to two decimal places.

3. Find the number of degrees in the unit angle when the angle $\frac{12\pi}{25}$ is represented by $1\frac{3}{5}$.

4. What is the radius of the circle in which an arc of 1 inch subtends an angle of 1′ at the centre?

5. Prove that

(1) $(\sin a+\cos a)(\tan a+\cot a)=\sec a+\operatorname{cosec} a$;

†(2) $(\sqrt{3}+1)(3-\cot 30^\circ)=\tan^3 60^\circ-2\sin 60^\circ$.

6. Find the angle of elevation of the sun when a chimney 60 feet high throws a shadow $20\sqrt{3}$ yards long.

---

7. Prove the identities:

(1) $(\tan\theta+2)(2\tan\theta+1)=5\tan\theta+2\sec^2\theta$;

(2) $1+\dfrac{\cot^2 a}{1+\operatorname{cosec} a}=\operatorname{cosec} a$.

8. One angle of a triangle is 45° and another is $\frac{5\pi}{8}$ radians; express the third angle both in sexagesimal and radian measure.

9. The number of degrees in an angle exceeds 14 times the number of radians in it by 51. Taking $\pi=\frac{22}{7}$, find the sexagesimal measure of the angle.

10. If $B=30^\circ$, $C=90^\circ$, $b=6$, find a, c, and the perpendicular from C on the hypotenuse.

11. Shew that

(1) $\cot\theta+\cot\left(\frac{\pi}{2}-\theta\right)=\operatorname{cosec}\theta\operatorname{cosec}\left(\frac{\pi}{2}-\theta\right)$;

(2) $\operatorname{cosec}^2\theta+\operatorname{cosec}^2\left(\frac{\pi}{2}-\theta\right)=\operatorname{cosec}^2\theta\operatorname{cosec}^2\left(\frac{\pi}{2}-\theta\right)$.

12. The angle of elevation of the top of a pillar is 30°, and on approaching 20 feet nearer it is 60°: find the height of the pillar.

---

13. Shew that $\tan^2 A - \sin^2 A = \sin^4 A \sec^2 A$.

14. In a triangle the angle A is $3x$ degrees, the angle B is x grades, and the angle C is $\frac{\pi x}{300}$ radians: find the number of degrees in each of the angles.

15. Find the numerical value of

$$\sin^3 60^\circ \cot 30^\circ - 2\sec^2 45^\circ + 3\cos 60^\circ \tan 45^\circ - \tan^2 60^\circ.$$

16. Prove the identities:

(1) $(1+\tan A)^2 + (1+\cot A)^2 = (\sec A + \operatorname{cosec} A)^2$;

(2) $(\sec \alpha - 1)^2 - (\tan \alpha - \sin \alpha)^2 = (1-\cos \alpha)^2$.

17. Which of the following statements is possible and which impossible?

(1) $\operatorname{cosec} \theta = \frac{a^2+b^2}{2ab}$; (2) $2\sin\theta = a + \frac{1}{a}$.

18. A balloon leaves the earth at the point A and rises at a uniform pace. At the end of 1·5 minutes an observer stationed at a distance of 660 feet from A finds the angular elevation of the balloon to be 60°; at what rate in miles per hour is the balloon rising?

---

19. Find the number of radians in the angles of a triangle which are in arithmetical progression, the least angle being 36°.

20. Shew that

$$\sin^2\alpha \sec^2\beta + \tan^2\beta \cos^2\alpha = \sin^2\alpha + \tan^2\beta.$$

21. In the triangle ABC if $A = 42^\circ$, $B = 116^\circ\ 33'$, find the perpendicular from C upon AB produced; given

$$c = 55, \quad \tan 42^\circ = \cdot 9, \quad \tan 63^\circ\ 27' = 2.$$

22. Prove the identities :

(1) $\cot a + \dfrac{\sin a}{1+\cos a} = \operatorname{cosec} a$;

(2) $\operatorname{cosec} a(\sec a - 1) - \cot a(1-\cos a) = \tan a - \sin a$.

23. Shew that $\left(\dfrac{1+\cot 60^\circ}{1-\cot 60^\circ}\right)^2 = \dfrac{1+\cos 30^\circ}{1-\cos 30^\circ}$.

24. A man walking N.W. sees a windmill which bears N. 15° W. In half-an-hour he reaches a place which he knows to be W. 15° S. of the windmill and a mile away from it. Find his rate of walking and his distance from the windmill at the first observation.

---

25. Find the number of radians in the complement of $\dfrac{3\pi}{8}$.

26. Solve the equations :

(1) $3\sin\theta + 4\cos^2\theta = 4\frac{1}{2}$; (2) $\tan\theta + \sec 30^\circ = \cot\theta$.

27. If $5\tan a = 4$, find the value of

$$\frac{5\sin a - 3\cos a}{\sin a + 2\cos a}.$$

28. Prove that

$$\frac{1-\sin A\cos A}{\cos A(\sec A - \operatorname{cosec} A)} \times \frac{\sin^2 A - \cos^2 A}{\sin^3 A + \cos^3 A} = \sin A.$$

29. Find the distance of an observer from the top of a cliff which is 195·2 yards high, given that the angle of elevation is 77° 26′, and that sin 77° 26′ = ·976.

30. A horse is tethered to a stake by a rope 27 feet long. If the horse moves along the circumference of a circle always keeping the rope tight, find how far it will have gone when the rope has traced out an angle of 70°. $\left[\pi = \dfrac{22}{7}\right]$.

CHAPTER VIII.

TRIGONOMETRICAL RATIOS OF ANGLES OF ANY MAGNITUDE.

72. In the present chapter we shall find it necessary to take account not only of the magnitude of straight lines, but also of the direction in which they are measured.

Let O be a fixed point in a horizontal line XX', then the position of any other point P in the line, whose distance from O is a given length a, will not be determined unless we know on which side of O the point P lies.

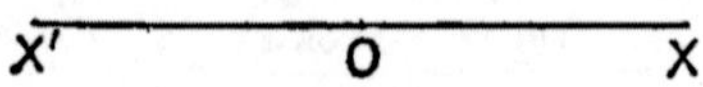

But there will be no ambiguity if it is agreed that distances measured in one direction are positive and distances measured in the opposite direction are negative.

Hence the following **Convention of Signs** is adopted:

lines measured from O to the right are positive,

lines measured from O to the left are negative.

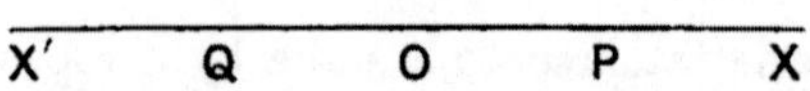

Thus in the above figure, if P and Q are two points on the line XX' at a distance a from O, their positions are indicated by the statements

$$OP = +a, \quad OQ = -a.$$

73. A similar *convention of signs* is used in the case of a plane surface.

Let O be any point in the plane; through O draw two straight lines XX' and YY' in the horizontal and vertical direction respectively, thus dividing the plane into four *quadrants*.

Then it is universally agreed to consider that

(1) *horizontal lines to the right of YY' are positive,*
horizontal lines to the left of YY' are negative;

(2) *vertical lines above XX' are positive,*
vertical lines below XX' are negative.

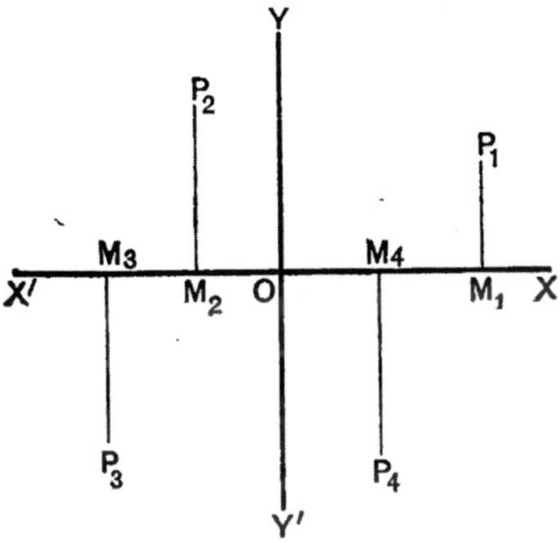

Thus OM_1, OM_4 are positive, OM_2, OM_3 are negative;
M_1P_1, M_2P_2 are positive, M_3P_3, M_4P_4 are negative.

74. Convention of Signs for Angles. In Art. 2 an angle has been defined as the amount of revolution which the radius vector makes in passing from its initial to its final position.

In the adjoining figure the straight line OP may be supposed to have arrived at its present position from the position occupied by OA by revolution about the point O in *either* of the two directions indicated by the arrows. The angle AOP may thus be regarded in two senses according as we suppose the revolution to have been in the same direction as the hands of a clock or in the opposite direction. To distinguish between these cases we adopt the following convention:

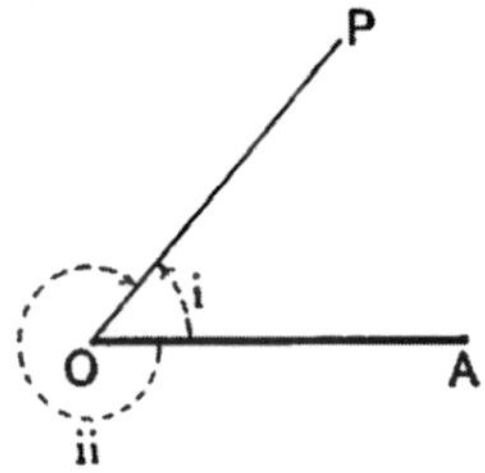

when the revolution of the radius vector is counter-clockwise the angle is positive,

when the revolution is clockwise the angle is negative.

Trigonometrical Ratios of any Angle.

75. Let XX' and YY' be two straight lines intersecting at right angles in O, and let a radius vector starting from OX revolve in either direction till it has traced out an angle A, taking up the position OP.

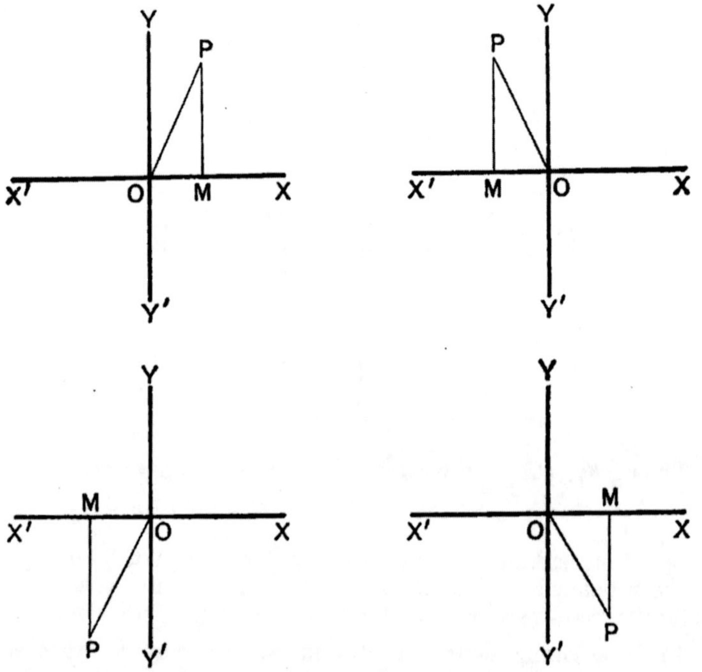

From P draw PM perpendicular to XX'; then in the right-angled triangle OPM, due regard being paid to the signs of the lines,

$$\sin A = \frac{MP}{OP}, \quad \operatorname{cosec} A = \frac{OP}{MP},$$

$$\cos A = \frac{OM}{OP}, \quad \sec A = \frac{OP}{OM},$$

$$\tan A = \frac{MP}{OM}, \quad \cot A = \frac{OM}{MP}.$$

The radius vector OP which only fixes the boundary of the angle is considered to be always positive.

From these definitions it will be seen that any trigonometrical function will be positive or negative according as the fraction which expresses its value has the numerator and denominator of the same sign or of opposite sign.

76. The four diagrams of the last article may be conveniently included in one.

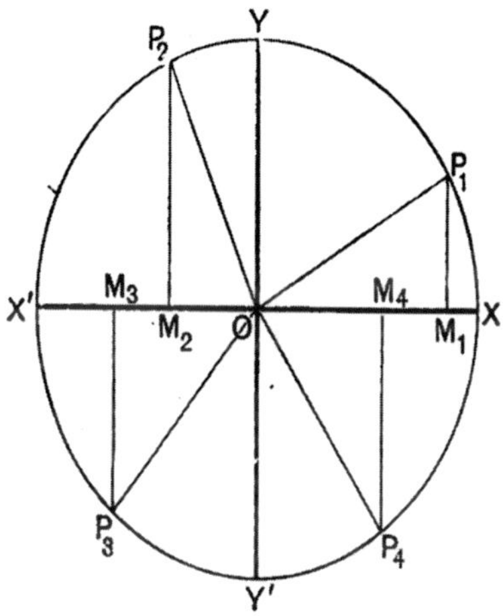

With centre O and fixed radius let a circle be described; then the diameters XX' and YY' divide the circle into four *quadrants* XOY, YOX', $X'OY'$, $Y'OX$, named *first, second, third, fourth* respectively.

Let the positions of the radius vector in the four quadrants be denoted by OP_1, OP_2, OP_3, OP_4, and let perpendiculars P_1M_1, P_2M_2, P_3M_3, P_4M_4 be drawn to XX'; then it will be seen that in the first quadrant all the lines are positive and therefore all the functions of A are positive.

In the second quadrant, OP_2 and M_2P_2 are positive, OM_2 is negative; hence $\sin A$ is positive, $\cos A$ and $\tan A$ are negative.

In the third quadrant, OP_3 is positive, OM_3 and M_3P_3 are negative; hence $\tan A$ is positive, $\sin A$ and $\cos A$ are negative.

In the fourth quadrant, OP_4 and OM_4 are positive, M_4P_4 is negative; hence $\cos A$ is positive, $\sin A$ and $\tan A$ are negative.

77. The following diagrams shew the *signs* of the trigonometrical functions in the four quadrants. It will be sufficient to consider the three principal functions only.

sine

+	+
−	−

cosine

−	+
−	+

tangent

−	+
+	−

The diagram below exhibits the same results in another useful form.

sine positive *cosine negative* *tangent negative*	**all the ratios positive**
tangent positive *sine negative* *cosine negative*	**cosine positive** *sine negative* *tangent negative*

78. When an angle is increased or diminished by any multiple of four right angles, the radius vector is brought back again into the same position after one or more revolutions. There are thus an infinite number of angles which have the same boundary line. Such angles are called **coterminal angles.**

If n is *any* integer, all the angles coterminal with A may be represented by $n \,.\, 360° + A$. Similarly, in radian measure all the angles coterminal with θ may be represented by $2n\pi + \theta$.

From the definitions of Art. 75, we see that the position of the boundary line is alone sufficient to determine the trigonometrical ratios of the angle; hence *all coterminal angles have the same trigonometrical ratios.*

For instance, $\sin (n \,.\, 360° + 45°) = \sin 45° = \dfrac{1}{\sqrt{2}}$;

and $$\cos\left(2n\pi + \frac{\pi}{6}\right) = \cos\frac{\pi}{6} = \frac{\sqrt{3}}{2}.$$

Example. Draw the boundary lines of the angles 780°, $-130°$, $-400°$, and in each case state which of the trigonometrical functions are negative.

(1) Since $780 = (2 \times 360) + 60$, the radius vector has to make two complete revolutions and then turn through 60°. Thus the boundary line is in the first quadrant, so that all the functions are positive.

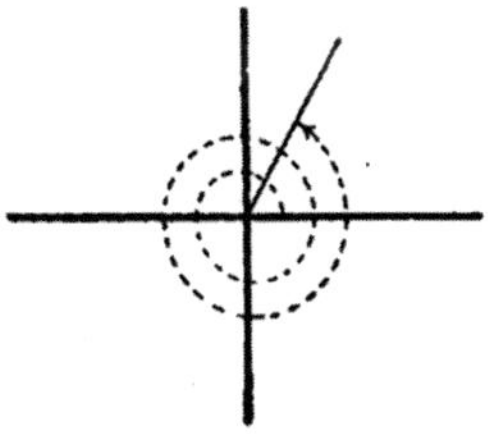

(2) Here the radius vector has to revolve through 130° in the negative direction. The boundary line is thus in the third quadrant, and since *OM* and *MP* are negative, the sine, cosine, cosecant, and secant are negative.

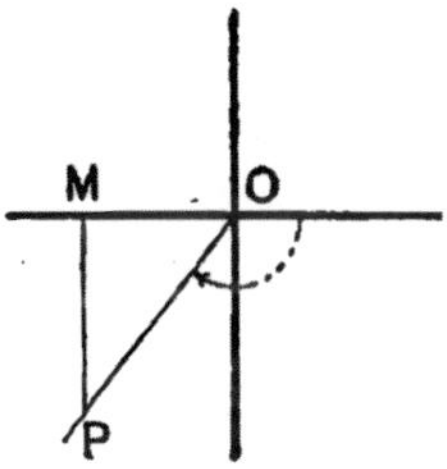

(3) Since $-400 = -(360 + 40)$, the radius vector has to make one complete revolution in the negative direction and then turn through 40°. The boundary line is thus in the fourth quadrant, and since *MP* is negative, the sine, tangent, cosecant, and cotangent are negative.

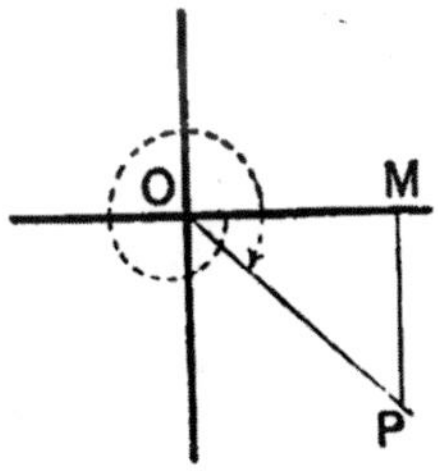

EXAMPLES. VIII. a.

State the quadrant in which the radius vector lies after describing the following angles:

1. 135°. 2. 265°. 3. $-315°$. 4. $-120°$.

5. $\frac{2\pi}{3}$. 6. $\frac{5\pi}{6}$. 7. $\frac{10\pi}{3}$. 8. $-\frac{11\pi}{4}$.

For each of the following angles state which of the three principal trigonometrical functions are positive.

9. 470°. 10. 330°. 11. 575°.

12. $-230°$. 13. $-620°$. 14. $-1200°$.

15. $-\frac{4\pi}{3}$. 16. $\frac{13\pi}{6}$. 17. $-\frac{13\pi}{6}$.

In each of the following cases write down the smallest positive coterminal angle, and the value of the expression.

18. $\sin 420°$. **19.** $\cos 390°$. **20.** $\tan(-315°)$.

21. $\sec 405°$. **22.** $\operatorname{cosec}(-330°)$. **23.** $\operatorname{cosec} 4380°$.

24. $\cot \frac{17\pi}{4}$. **25.** $\sec \frac{25\pi}{3}$. **26.** $\tan\left(-\frac{5\pi}{3}\right)$.

79. Since the definitions of the functions given in Art. 75 are applicable to angles of any magnitude, positive or negative, it follows that all relations derived from these definitions must be true universally. Thus we shall find that the fundamental formulæ given in Art. 29 hold in all cases; that is,

$$\sin A \times \operatorname{cosec} A = 1, \quad \cos A \times \sec A = 1, \quad \tan A \times \cot A = 1;$$

$$\tan A = \frac{\sin A}{\cos A}, \qquad \cot A = \frac{\cos A}{\sin A};$$

$$\sin^2 A + \cos^2 A = 1,$$

$$1 + \tan^2 A = \sec^2 A,$$

$$1 + \cot^2 A = \operatorname{cosec}^2 A.$$

It will be useful practice for the student to test the truth of these formulæ for different positions of the boundary line of the angle A. We shall give one illustration.

80. Let the radius vector revolve from its initial position OX till it has traced out an angle A and come into the position OP indicated in the figure. Draw PM perpendicular to XX'.

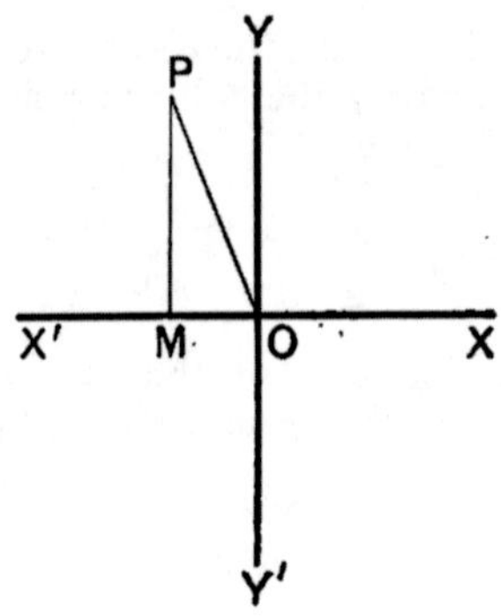

In the right-angled triangle OMP,

$$MP^2 + OM^2 = OP^2 \quad\ldots\ldots\ldots(1).$$

Divide each term by OP^2; thus

$$\left(\frac{MP}{OP}\right)^2 + \left(\frac{OM}{OP}\right)^2 = 1;$$

that is, $$\sin^2 A + \cos^2 A = 1.$$

Divide each term of (1) by OM^2; thus

$$\left(\frac{MP}{OM}\right)^2+1=\left(\frac{OP}{OM}\right)^2;$$

that is, $$\tan^2 A+1=\sec^2 A.$$

Divide each term of (1) by MP^2; thus

$$1+\left(\frac{OM}{MP}\right)^2=\left(\frac{OP}{MP}\right)^2;$$

that is, $$1+\cot^2 A=\operatorname{cosec}^2 A.$$

It thus appears that the truth of these relations depends only on the statement $OP^2=MP^2+OM^2$ in the right-angled triangle OMP, and this will be the case in whatever quadrant OP lies.

NOTE. OM^2 is positive, although the line OM in the figure is negative.

81. In the statement $\cos A=\sqrt{1-\sin^2 A}$, either the positive or the negative sign may be placed before the radical. The sign of the radical hitherto has always been taken positively, because we have restricted ourselves to the consideration of acute angles. It will sometimes be necessary to examine which sign must be taken before the radical in any particular case.

Example 1. Given $\cos 126° 53'=-\frac{3}{5}$, find $\sin 126° 53'$ and $\cot 126° 53'$.

Since $\sin^2 A+\cos^2 A=1$ for angles of any magnitude, we have

$$\sin A=\pm\sqrt{1-\cos^2 A}.$$

Denote $126° 53'$ by A; then the boundary line of A lies in the second quadrant, and therefore $\sin A$ is positive. Hence the sign $+$ must be placed before the radical;

$$\therefore \sin 126° 53'=+\sqrt{1-\frac{9}{25}}=+\sqrt{\frac{16}{25}}=\frac{4}{5};$$

$$\cot 126° 53'=\frac{\cos 126° 53'}{\sin 126° 53'}=\left(-\frac{3}{5}\right)\div\left(\frac{4}{5}\right)=-\frac{3}{4}.$$

The same results may also be obtained by the method used in the following example. The appropriate signs of the lines are shewn in the figure.

Example 2. If $\tan A = -\frac{15}{8}$, find $\sin A$ and $\cos A$.

The boundary line of A will lie either in the second or in the fourth quadrant, as OP or OP'. In either position,

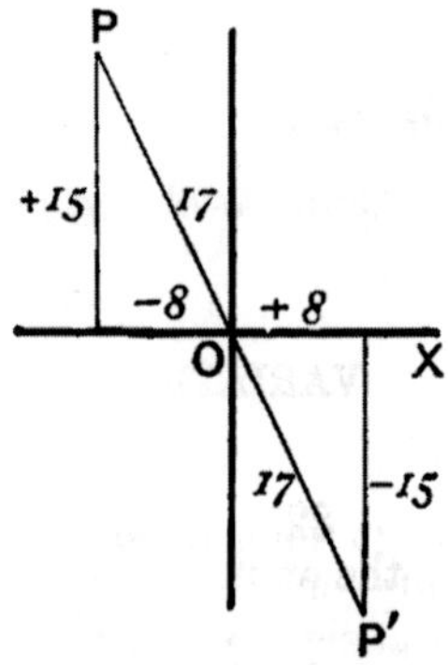

$$\text{the radius vector} = \sqrt{(15)^2 + (8)^2} = \sqrt{289} = 17.$$

Hence $\sin XOP = \frac{15}{17}$, $\cos XOP = -\frac{8}{17}$;

and $\sin XOP' = -\frac{15}{17}$, $\cos XOP' = \frac{8}{17}$.

Thus corresponding to $\tan A$, there are two values of $\sin A$ and two values of $\cos A$. If however it is known in which quadrant the boundary line of A lies, $\sin A$ and $\cos A$ have each a single value.

EXAMPLES. VIII. b.

1. Given $\sin 120° = \frac{\sqrt{3}}{2}$, find $\tan 120°$.

2. Given $\tan 135° = -1$, find $\sin 135°$.

3. Find $\cos 240°$, given that $\tan 240° = \sqrt{3}$.

4. If $A = 202° 37'$ and $\sin A = -\frac{5}{13}$, find $\cos A$ and $\cot A$.

5. If $A = 143° 8'$ and $\operatorname{cosec} A = 1\frac{2}{3}$, find $\sec A$ and $\tan A$.

6. If $A = 216° 52'$ and $\cos A = -\frac{4}{5}$, find $\cot A$ and $\sin A$.

7. Given $\sec \frac{2\pi}{3} = -2$, find $\sin \frac{2\pi}{3}$ and $\cot \frac{2\pi}{3}$.

8. Given $\sin \frac{5\pi}{4} = -\frac{1}{\sqrt{2}}$, find $\tan \frac{5\pi}{4}$ and $\sec \frac{5\pi}{4}$.

9. If $\cos A = \frac{12}{13}$, find $\sin A$ and $\tan A$.

CHAPTER IX.

VARIATIONS OF THE TRIGONOMETRICAL FUNCTIONS.

82. A CAREFUL perusal of the following remarks will render the explanations which follow more easily intelligible.

Consider the fraction $\frac{a}{x}$ in which the numerator a has a *certain fixed value* and the denominator x is a *quantity subject to change;* then it is clear that the smaller x becomes the larger does the value of the fraction $\frac{a}{x}$ become. For instance

$$\frac{a}{\frac{1}{10}}=10a, \quad \frac{a}{\frac{1}{1000}}=1000a, \quad \frac{a}{\frac{1}{10000000}}=10000000a.$$

By making the denominator x sufficiently small the value of the fraction $\frac{a}{x}$ can be made as large as we please; that is, as x approaches to the value 0, the fraction $\frac{a}{x}$ becomes infinitely great.

The symbol ∞ is used to express a quantity infinitely great, or more shortly *infinity*, and the above statement is concisely written

$$\textit{when } x=0, \textit{ the limit of } \frac{a}{x}=\infty.$$

Again, if x is a quantity which gradually increases and finally becomes infinitely large, the fraction $\frac{a}{x}$ becomes infinitely small; that is,

$$\textit{when } x=\infty, \textit{ the limit of } \frac{a}{x}=0.$$

83. DEFINITION. If y is a function of x, and if when x approaches nearer and nearer to the fixed quantity a, the value of y approaches nearer and nearer to the fixed quantity b and can be made to differ from it by as little as we please, then b is called the **limiting value** or the **limit** of y when $x=a$.

84. Trigonometrical Functions of 0°.

Let XOP be an angle traced out by a radius vector OP of fixed length.

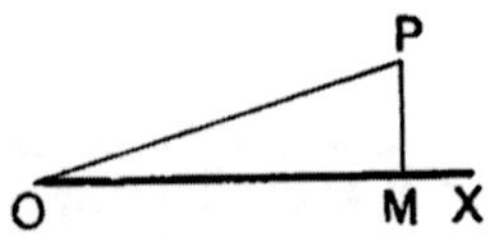

Draw PM perpendicular to OX; then

$$\sin POM = \frac{MP}{OP}.$$

If we suppose the angle POM to be gradually decreasing, MP will also gradually decrease, and if OP ultimately come into coincidence with OM the angle POM vanishes and $MP=0$.

Hence $$\sin 0° = \frac{0}{OP} = 0.$$

Again, $\cos POM = \frac{OM}{OP}$; but when the angle POM vanishes OP becomes coincident with OM.

Hence $$\cos 0° = \frac{OM}{OM} = 1.$$

Also when the angle POM vanishes,

$$\tan 0° = \frac{0}{OM} = 0.$$

And $$\operatorname{cosec} 0° = \frac{1}{\sin 0°} = \frac{1}{0} = \infty;$$

$$\sec 0° = \frac{1}{\cos 0°} = \frac{1}{1} = 1;$$

$$\cot 0° = \frac{1}{\tan 0°} = \frac{1}{0} = \infty.$$

85. Trigonometrical Functions of 90° or $\frac{\pi}{2}$.

Let XOP be an angle traced out by a radius vector of fixed length.

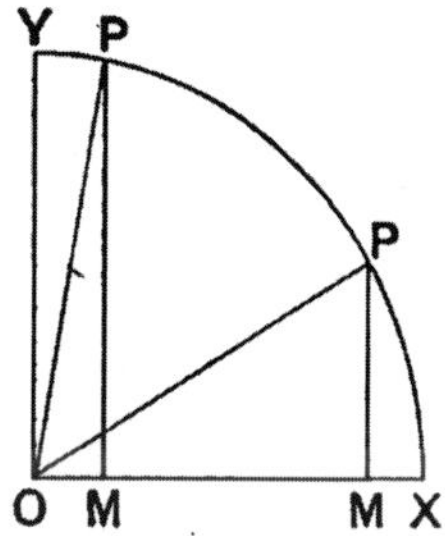

Draw PM perpendicular to OX, and OY perpendicular to OX.

By definition,

$$\sin POM = \frac{MP}{OP}, \quad \cos POM = \frac{OM}{OP}, \quad \tan POM = \frac{MP}{OM}.$$

If we suppose the angle POM to be gradually increasing, MP will gradually increase and OM decrease. When OP comes into coincidence with OY the angle POM becomes equal to 90°, and OM vanishes, while MP becomes equal to OP.

Hence $$\sin 90° = \frac{OP}{OP} = 1;$$

$$\cos 90° = \frac{0}{OP} = 0;$$

$$\tan 90° = \frac{MP}{OM} = \frac{OP}{0} = \infty.$$

And $$\cot 90° = \frac{1}{\tan 90°} = \frac{1}{\infty} = 0;$$

$$\sec 90° = \frac{1}{\cos 90°} = \frac{1}{0} = \infty;$$

$$\operatorname{cosec} 90° = \frac{1}{\sin 90°} = 1.$$

86. *To trace the changes in sign and magnitude of* sin A *as* A *increases from* 0° *to* 360°.

Let XX' and YY' be two straight lines intersecting at right angles in O.

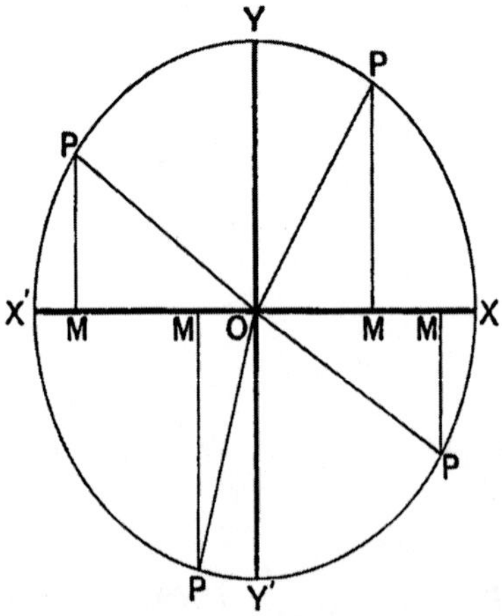

With centre O and any radius OP describe a circle, and suppose the angle A to be traced out by the revolution of OP through the four quadrants starting from OX.

Draw PM perpendicular to OX and let $OP=r$; then

$$\sin A = \frac{MP}{r},$$

and since r does not alter in sign or magnitude, we have only to consider the changes of MP as P moves round the circle.

When $A=0°$, $MP=0$, and $\sin 0° = \frac{0}{r} = 0$.

In the first quadrant, MP is positive and increasing;

$\therefore$ sin A is positive and increasing.

When $A=90°$, $MP=r$, and $\sin 90° = \frac{r}{r} = 1$.

In the second quadrant, MP is positive and decreasing;

$\therefore$ sin A is positive and decreasing.

When $A=180°$, $MP=0$, and $\sin 180° = \frac{0}{r} = 0$.

In the third quadrant, MP is negative and increasing;

$\therefore$ $\sin A$ is negative and increasing.

When $A = 270°$, MP is equal to r, but is negative; hence

$$\sin 270° = -\frac{r}{r} = -1.$$

In the fourth quadrant, MP is negative and decreasing;

$\therefore$ $\sin A$ is negative and decreasing.

When $A = 360°$, $MP = 0$, and $\sin 360° = \frac{0}{r} = 0.$

87. The results of the previous article are concisely shewn in the following diagram:

sin 90° = 1

sin 180° = 0	*sin A positive and decreasing*	*sin A positive and increasing*	*sin 0° = 0*
	sin A negative and increasing	*sin A negative and decreasing*	

sin 270° = −1

88. We leave as an exercise to the student the investigation of the changes in sign and magnitude of $\cos A$ as A increases from 0° to 360°. The following diagram exhibits these changes.

cos 90° = 0

cos 180° = −1	*cos A negative and increasing*	*cos A positive and decreasing*	*cos 0° = 1*
	cos A negative and decreasing	*cos A positive and increasing*	

cos 270° = 0

89. *To trace the changes in sign and magnitude of* tan A *as* A *increases from* 0° *to* 360°.

With the figure of Art. 86, $\tan A = \frac{MP}{OM}$, and its changes will therefore depend on those of MP and OM.

When $A = 0°$, $MP = 0$, $OM = r$; $\therefore \tan 0° = \frac{0}{r} = 0$.

In the first quadrant,

MP is positive and increasing,
OM is positive and decreasing;
$\therefore$ $\tan A$ is positive and increasing.

When $A = 90°$, $MP = r$, $OM = 0$; $\therefore \tan 90° = \frac{r}{0} = \infty$.

In the second quadrant,

MP is positive and decreasing,
OM is negative and increasing;
$\therefore$ $\tan A$ is negative and decreasing.

When $A = 180°$, $MP = 0$; $\therefore \tan 180° = 0$.

In the third quadrant,

MP is negative and increasing,
OM is negative and decreasing;
$\therefore$ $\tan A$ is positive and increasing.

When $A = 270°$, $OM = 0$; $\therefore \tan 270° = \infty$.

In the fourth quadrant,

MP is negative and decreasing,
OM is positive and increasing;
$\therefore$ $\tan A$ is negative and decreasing.

When $A = 360°$, $MP = 0$; $\therefore \tan 360° = 0$.

NOTE. When the numerator of a fraction changes continually from a small positive to a small negative quantity the fraction changes sign by passing through the value 0. When the denominator changes continually from a small positive to a small negative quantity the fraction changes sign by passing through the value ∞. For instance, as A passes through the value 90°, OM changes from a small positive to a small negative quantity, hence $\frac{OM}{OP}$, that is $\cos A$, changes sign by passing through the value 0, while $\frac{PM}{OM}$, that is $\tan A$, changes sign by passing through the value ∞.

90. The results of Art. 89 are shewn in the following diagram:

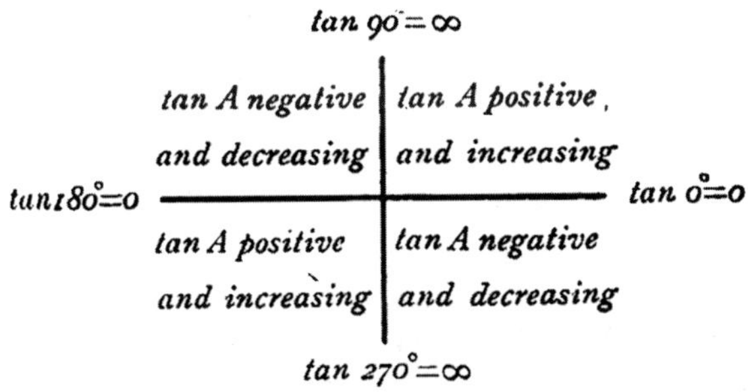

The student will now have no difficulty in tracing the variations in sign and magnitude of the other functions.

91. In Arts. 86 and 89 we have seen that the variations of the trigonometrical functions of the angle XOP depend on the position of P as P moves round the circumference of the circle. On this account the trigonometrical functions of an angle are called **circular functions.** This name is one that we shall use frequently.

EXAMPLES. IX. a.

Trace the changes in sign and magnitude of

1. $\cot A$, between $0°$ and $360°$.
2. $\operatorname{cosec}\theta$, between 0 and π.
3. $\cos\theta$, between π and 2π.
4. $\tan A$, between $-90°$ and $-270°$.
5. $\sec\theta$, between $\frac{\pi}{2}$ and $\frac{3\pi}{2}$.

Find the value of

6. $\cos 0° \sin^2 270° - 2\cos 180° \tan 45°$.
7. $3\sin 0° \sec 180° + 2\operatorname{cosec} 90° - \cos 360°$.
8. $2\sec^2\pi \cos 0 + 3\sin^3\frac{3\pi}{2} - \operatorname{cosec}\frac{\pi}{2}$.
9. $\tan\pi\cos\frac{3\pi}{2} + \sec 2\pi - \operatorname{cosec}\frac{3\pi}{2}$.

Graphs of the Trigonometrical Functions.

91_A. The results of Arts. 86—90 may be illustrated by means of graphs.

Graph of sin x. Put $y = \sin x$, and find the values of y corresponding to values of x differing by 30°.

From a Table of sines we have

x	0°	30°	60°	90°	120°	150°	180°	210°	240°	270°	...
y or $\sin x$	0	·5	·866	1	·866	·5	0	−·5	−·866	−1	...

In the figure on the opposite page let values of x be measured on the horizontal axis OX, each division being taken to represent 10°, and on the vertical axis OY, let 10 divisions be taken as the unit. Then from the series of points given by the above table, we obtain the graph represented by the continuous waving line in the diagram.

From the graph the following points are evident:

(i) The sine of an angle goes through all its changes gradually (without abrupt changes) once as the angle increases through **four** right angles. For this reason $\sin x$ is said to be a **continuous function** whose **period** is 4 right angles. Beyond 360° the graph may be continued indefinitely, the curve already drawn being endlessly repeated.

(ii) The greatest and least values of $\sin x$ are $+1$ and -1. Between these limits the sine of an angle may have any value, positive or negative. The maximum and minimum values are shewn by the ordinates at 90° and 270°.

The graph should be compared with the details given in Art. 86.

Example. Solve the equation $\sin x = \cdot 7$ graphically.

Here $y = \cdot 7$, and every point on the graph whose ordinate is ·7 will furnish a solution of the equation. Hence we have only to note the points in which the graph is cut by a line parallel to the x-axis and at a distance ·7 from it.

Thus $x = 45°, 135°, 405°, 495°, \ldots,$

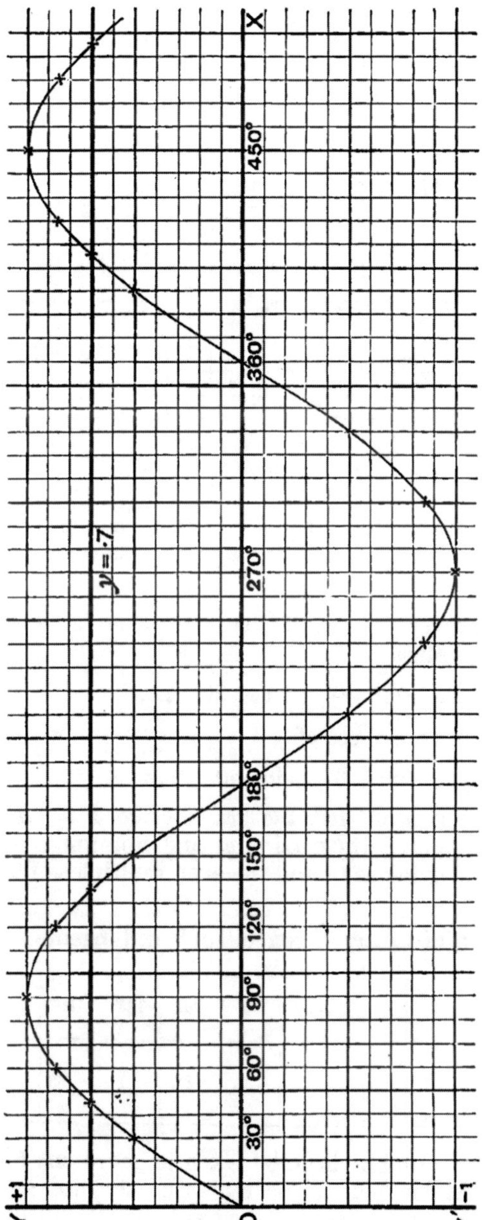

GRAPH OF SIN x.

91B. Graph of cos x. We leave the details of the graph of cos x as an exercise for the student. It should be drawn from the same series of angles, and with the same units as in the graph of sin x. It is given on a small scale in the adjoining figure.

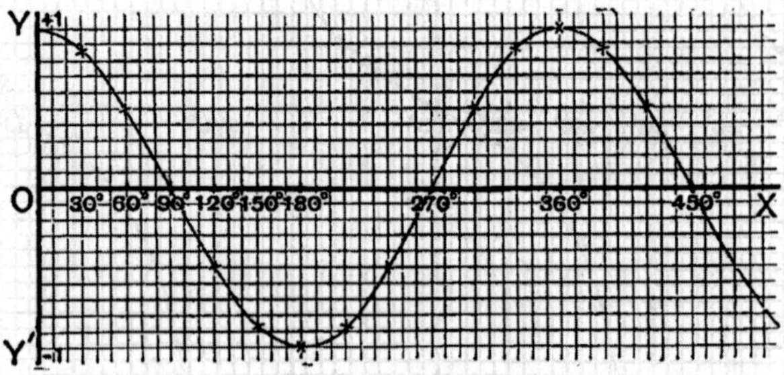

GRAPH OF COS x.

It may be noticed that the graph is the same as if the graph of sin x were moved to the left through a space corresponding to 90°.

It will be proved in Art. 98 that $\cos x = \sin(90° + x)$ so that when x has the values

$$0°, \quad 30°, \quad 60°, \quad 90°, \ldots,$$

the values of cos x are those of

$$\sin 90°, \quad \sin 120°, \quad \sin 150°, \quad \sin 180°, \ldots.$$

As before it is seen that the graph of cos x is **continuous** between 0° and 360°. Its maximum and minimum values are $+1$ and -1, occurring at 0° and the even multiples of 90°.

91C. Graph of tan x. From a Table of tangents we have

x	0°	15°	30°	45°	60°	75°	90°	105°	120°	135°	150°	165°	180°
y or tan x	0	·27	·58	1	1·7	3·7	∞	−3·7	−1·7	−1	−·58	−·27	0

Let each horizontal division be taken to represent 6°, and on the vertical axis let 5 divisions represent the unit; then the graph will be as in the adjoining diagram, and it will be seen to consist of an infinite number of *discontinuous* equal branches.

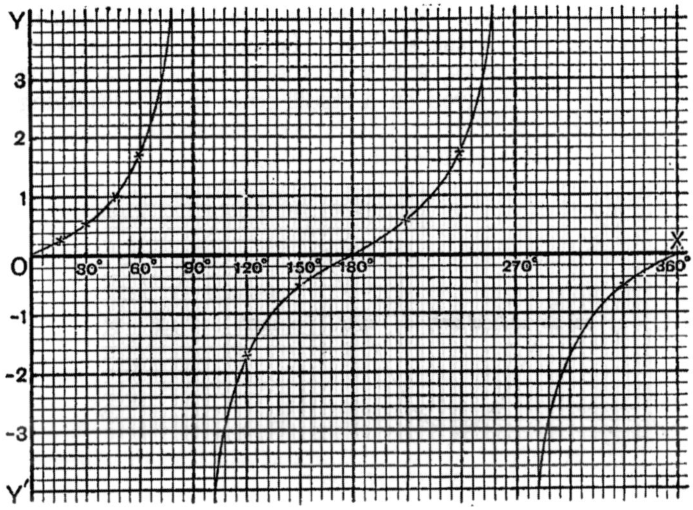

GRAPH OF TAN x.

The following points should be noticed:

(i) The tangent of an angle goes through all its changes once as the angle increases through **two** right angles. As the angle approaches 90°, the tangent increases very rapidly, and the graph is continually approaching nearer to the vertical line through the division marking 90°, but never actually reaching it till $y=\infty$. As the angle passes through 90°, the tangent changes from an infinite positive to an infinite negative value. As the angle increases from 90° to 180°, the numerical values of the tangent are those already traced but in reverse order, and of opposite sign.

(ii) Through every subsequent period of two right angles, the graph is repeated.

(iii) The tangent of an angle may have any numerical value, positive or negative.

NOTE. The above figure was drawn on paper ruled to inches and tenths of an inch and then reduced to half the original size. The student should draw a larger figure for himself.

91D. The graphs of $\cot x$, $\sec x$, and $\operatorname{cosec} x$ may be left as an exercise for the student. He may also consult Arts. 288—291, where the graphs of the trigonometrical functions are discussed in a slightly different manner.

EXAMPLES. IX. b.

1. Draw, with the same axes and units, the graphs of $\sin x$ and $\cos x$ on a scale twice as large as that in Art. 91_A.

[This diagram will be required for Ex. 9.]

2. Draw the graph of $\sin x$ from the following values of x:

$$5^\circ,\ 15^\circ,\ 30^\circ,\ 45^\circ,\ 60^\circ,\ 75^\circ,\ 85^\circ,\ 90^\circ.$$

[Take 1 inch horizontally to represent 25°, and 2 inches vertically to represent unity.]

3. From the figure of Ex. 2 find the value of $\sin 37^\circ$, and the angle whose sine is ·8, to the nearest degree.

4. Find from the Tables the value of $\cos x$ when x has the values

$$0^\circ,\ 10^\circ,\ 20^\circ,\ 30^\circ,\ 40^\circ,\ 50^\circ,\ 60^\circ.$$

Draw a curve on a large scale shewing how $\cos x$ varies as x increases from 0° to 60°.

5. From the figure of Ex. 4, find approximate values of $\cos 25^\circ$ and $\cos 45^\circ$. Verify by means of the Tables.

6. Solve graphically the following equations, giving, to the nearest degree, all the solutions less than 360°.

(i) $15\sin^2\theta - 16\sin\theta + 4 = 0$;

(ii) $\cos^2\theta - 1{\cdot}7\cos\theta + {\cdot}72 = 0$.

7. Draw the graph of $\cot x$, using the following values of x:

$$15^\circ,\ 30^\circ,\ 45^\circ,\ 60^\circ,\ 75^\circ,\ 90^\circ,\ 105^\circ,\ \ldots\ 180^\circ.$$

[Take 1 inch horizontally to represent 15°, and 1 inch vertically to represent unity.]

8. From the figure of Ex. 7 find approximate values of $\cot 48^\circ$ and $\cot 59^\circ$. Verify by means of the Tables.

9. From the graphs in Ex. 1, deduce the graph of

$$\sin x + \cos x.$$

Hence solve the following equations graphically:

(i) $\sin x + \cos x = 0$; (ii) $\sin x + \cos x = 1$.

Note on the old definitions of the Trigonometrical Functions.

Formerly, Mathematicians considered the trigonometrical functions with reference to the *arc* of a given circle, and did not regard them as *ratios* but as the *lengths* of certain straight lines drawn in relation to this arc.

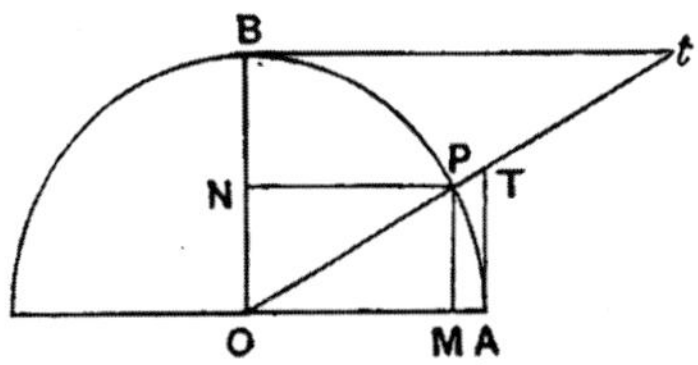

Let OA and OB be two radii of a circle at right angles, and let P be any point on the circumference. Draw PM and PN perpendicular to OA and OB respectively, and let the tangents at A and B meet OP produced in T and t respectively.

The *lines* PM, AT, OT, AM were named respectively the sine, tangent, secant, versed-sine of the arc AP, and PN, Bt, Ot, BN, which are the sine, tangent, secant, versed-sine of the complementary arc BP, were named respectively the cosine, cotangent, cosecant, coversed-sine of the arc AP.

As thus defined each trigonometrical function of the *arc* is equal to the corresponding function of the *angle*, which it subtends at the centre of the circle, multiplied by the radius. Thus

$$\frac{AT}{OA}=\tan POA\text{; that is, } AT=OA\times\tan POA\text{;}$$

and

$$\frac{Ot}{OB}=\sec BOP=\operatorname{cosec} POA\text{; that is, } Ot=OB\times\operatorname{cosec} POA.$$

The values of the functions of the arc therefore depended on the length of the radius of the circle as well as on the angle subtended by the arc at the centre of the circle, so that in Tables of the functions it was necessary to state the magnitude of the radius.

The names of the trigonometrical functions and the abbreviations for them now in use were introduced by different Mathematicians chiefly towards the end of the sixteenth and during the seventeenth century, but were not generally employed until their re-introduction by Euler. The development of the science of Trigonometry may be considered to date from the publication in 1748 of Euler's *Introductio in analysin Infinitorum.*

The reader will find some interesting information regarding the progress of Trigonometry in Ball's *Short History of Mathematics.*

MISCELLANEOUS EXAMPLES. C.

1. Draw the boundary lines of the angles whose tangent is equal to $-\frac{3}{4}$, and find the cosine of these angles.

2. Shew that

$$\cos A\,(2\sec A + \tan A)(\sec A - 2\tan A) = 2\cos A - 3\tan A.$$

3. Given $C=90°$, $b=10{\cdot}5$, $c=21$, solve the triangle.

4. If $\sec A = -\frac{25}{7}$, and A lies between 180° and 270°, find $\cot A$.

5. The latitude of Bombay is 19° N.: find its distance from the equator, taking the diameter of the earth to be 7920 miles.

6. From the top of a cliff 200 ft. high, the angles of depression of two boats due east of the observer are 34° 30′ and 18° 40′: find their distance apart, given

$$\cot 34°\,30' = 1{\cdot}455, \qquad \cot 18°\,40' = 2{\cdot}96.$$

7. If A lies between 180° and 270°, and $3\tan A = 4$, find the value of $2\cot A - 5\cos A + \sin A$.

8. Find, correct to three decimal places, the radius of a circle in which an arc 15 inches long subtends at the centre an angle of 71° 36′ 3·6″.

9. Shew that

$$\frac{\tan^3\theta}{1+\tan^2\theta} + \frac{\cot^3\theta}{1+\cot^2\theta} = \frac{1-2\sin^2\theta\cos^2\theta}{\sin\theta\cos\theta}.$$

10. The angle of elevation of the top of a tower is 68° 11′, and a flagstaff 24 ft. high on the summit of the tower subtends an angle of 2° 10′ at the observer's eye. Find the height of the tower, given

$$\tan 70°\,21' = 2{\cdot}8, \qquad \cot 68°\,11' = {\cdot}4.$$

11. If $\tan A = \frac{1}{2}$, and $\tan B = \frac{1}{3}$, construct the angles A and B on opposite sides of a common arm, and measure the angle $A+B$.

12. Find to the nearest minute the smallest positive angles which satisfy the following equation:

$$12 \tan^2\theta + 7 \tan\theta - 12 = 0.$$

13. Two places on the same meridian have a difference in latitude of 21° 12′. If they are 3·7 inches apart on a globe, find its radius to the nearest tenth of an inch. Also find the number of miles between the two places, assuming the earth's radius to be 4000 miles.

14. A man, walking in a direction 47° N. of E., sees a tower which bears N. 13° E. In 40 minutes he reaches a place which he knows to be E. 13° S. of the tower, and 2 km. from it. Find his rate of walking and his distance from the tower at the first observation.

15. If $\sin(\alpha - \beta) = \cdot 7323$, and $\cos(\alpha + \beta) = \cdot 6218$, find from the Tables the smallest positive values of α and β, to the nearest minute.

16. If $3 \cot \alpha = 2$, find the value of

$$\frac{10 \sin\alpha - 6 \cos\alpha}{4 \sin\alpha + 3 \cos\alpha}.$$

17. At noon a ship sailing W. 16° S., at 12 miles an hour observes a fort in direction S. 31° W. At 1.40 P.M. the fort bears S. 16° E. from the ship; find the distance of the ship from the fort at each observation.

CHAPTER X.

CIRCULAR FUNCTIONS OF CERTAIN ALLIED ANGLES.

92. Circular Functions of 180° – A.

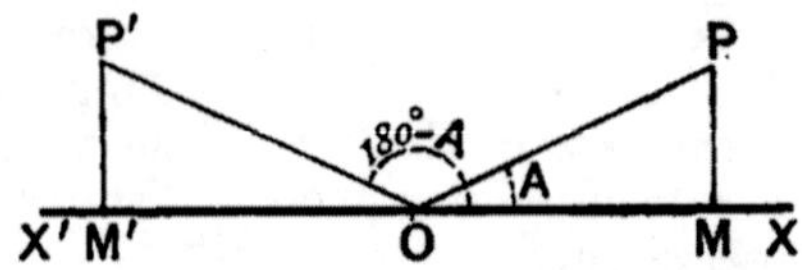

Take any straight line XOX', and let a radius vector starting from OX revolve until it has traced the angle A, taking up the position OP.

Again, let the radius vector starting from OX revolve through 180° into the position OX' and then *back again* through an angle A taking up the final position OP'. Thus XOP' is the angle $180°-A$.

From P and P' draw PM and $P'M'$ perpendicular to XX'; then by Euc. I. 26 the triangles OPM and $OP'M'$ are geometrically equal.

By definition,

$$\sin(180°-A)=\frac{M'P'}{OP'};$$

but $M'P'$ is equal to MP in magnitude and is of the same sign;

$$\therefore \sin(180°-A)=\frac{MP}{OP}=\sin A.$$

Again, $$\cos(180°-A)=\frac{OM'}{OP'};$$

and OM' is equal to OM in magnitude, but is of opposite sign;

$$\therefore \cos(180°-A)=\frac{-OM}{OP}=-\frac{OM}{OP}=-\cos A.$$

Also $$\tan(180°-A)=\frac{M'P'}{OM'}=\frac{MP}{-OM}=-\frac{MP}{OM}=-\tan A.$$

93. In the last article, for the sake of simplicity we have supposed the angle A to be less than a right angle, but all the formulæ of this chapter may be shewn to be true for angles of any magnitude. A general proof of one case is given in Art. 102, and the same method may be applied to all the other cases.

94. If the angles are expressed in radian measure, the formulæ of Art. 92 become

$$\sin(\pi - \theta) = \sin\theta,$$

$$\cos(\pi - \theta) = -\cos\theta,$$

$$\tan(\pi - \theta) = -\tan\theta.$$

Example 1. Find the sine and cosine of 120°.

$$\sin 120° = \sin(180° - 60°) = \sin 60° = \frac{\sqrt{3}}{2}.$$

$$\cos 120° = \cos(180° - 60°) = -\cos 60° = -\frac{1}{2}.$$

Example 2. Find the cosine and cotangent of $\frac{5\pi}{6}$.

$$\cos\frac{5\pi}{6} = \cos\left(\pi - \frac{\pi}{6}\right) = -\cos\frac{\pi}{6} = -\frac{\sqrt{3}}{2}.$$

$$\cot\frac{5\pi}{6} = \cot\left(\pi - \frac{\pi}{6}\right) = -\cot\frac{\pi}{6} = -\sqrt{3}.$$

95. Definition. When the sum of two angles is equal to two right angles each is said to be the **supplement** of the other and the angles are said to be **supplementary**. Thus if A is any angle its supplement is $180° - A$.

96. The results of Art. 92 are so important in a later part of the subject that it is desirable to emphasize them. We therefore repeat them in a verbal form:

the sines of supplementary angles are equal in magnitude and are of the same sign;

the cosines of supplementary angles are equal in magnitude but are of opposite sign;

the tangents of supplementary angles are equal in magnitude but are of opposite sign.

97. Circular Functions of 180°+A.

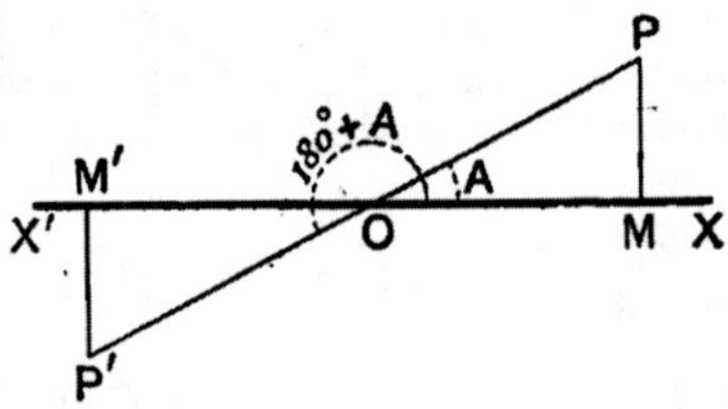

Take any straight line XOX' and let a radius vector starting from OX revolve until it has traced the angle A, taking up the position OP.

Again, let the radius vector starting from OX revolve through 180° into the position OX', and then further through an angle A, taking up the final position OP'. Thus XOP' is the angle $180°+A$.

From P and P' draw PM and $P'M'$ perpendicular to XX'; then OP and OP' are in the same straight line, and by Euc. I. 26 the triangles OPM and $OP'M'$ are geometrically equal.

By definition,

$$\sin(180°+A)=\frac{M'P'}{OP'};$$

and $M'P'$ is equal to MP in magnitude but is of opposite sign;

$$\therefore \sin(180°+A)=\frac{-MP}{OP}=-\frac{MP}{OP}=-\sin A.$$

Again, $$\cos(180°+A)=\frac{OM'}{OP'};$$

and OM' is equal to OM in magnitude but is of opposite sign;

$$\therefore \cos(180°+A)=\frac{-OM}{OP}=-\frac{OM}{OP}=-\cos A.$$

Also $$\tan(180°+A)=\frac{M'P'}{OM'}=\frac{-MP}{-OM}=\frac{MP}{OM}=\tan A.$$

Expressed in radian measure, the above formulæ are written

$$\sin(\pi+\theta)=-\sin\theta,\quad \cos(\pi+\theta)=-\cos\theta,\quad \tan(\pi+\theta)=\tan\theta.$$

In these results we may draw especial attention to the fact that an angle may be increased or diminished by two right angles as often as we please without altering the value of the tangent.

Example. Find the value of cot 210°.

$$\cot 210°=\cot(180°+30°)=\cot 30°=\sqrt{3}.$$

98. Circular Functions of $90°+A$.

Take any straight line XOX', and let a radius vector starting from OX revolve until it has traced the angle A, taking up the position OP.

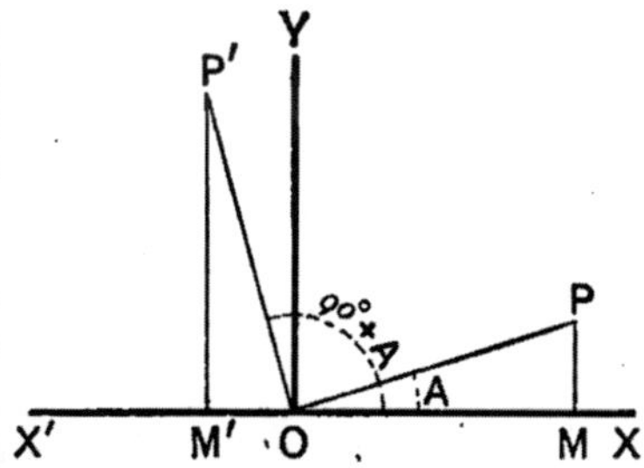

Again, let the radius vector starting from OX revolve through $90°$ into the position OY, and then further through an angle A, taking up the final position OP'. Thus XOP' is the angle $90°+A$.

From P and P' draw PM and $P'M'$ perpendicular to XX'; then $\angle M'P'O = \angle P'OY = A = \angle POM$.

By Euc. I. 26, the triangles OPM and $OP'M'$ are geometrically equal; hence

$M'P'$ is equal to OM in magnitude and is of the same sign,

and OM' is equal to MP in magnitude but is of opposite sign.

By definition,

$$\sin(90°+A)=\frac{M'P'}{OP'}=\frac{OM}{OP}=\cos A;$$

$$\cos(90°+A)=\frac{OM'}{OP'}=\frac{-MP}{OP}=-\frac{MP}{OP}=-\sin A;$$

$$\tan(90°+A)=\frac{M'P'}{OM'}=\frac{OM}{-MP}=-\frac{OM}{MP}=-\cot A.$$

Expressed in radian measure the above formulæ become

$$\sin\left(\frac{\pi}{2}+\theta\right)=\cos\theta,\quad \cos\left(\frac{\pi}{2}+\theta\right)=-\sin\theta,\quad \tan\left(\frac{\pi}{2}+\theta\right)=-\cot\theta.$$

Example 1. **Find the value of sin 120°.**

$$\sin 120°=\sin(90°+30°)=\cos 30°=\frac{\sqrt{3}}{2}.$$

Example 2. **Find the values of $\tan(270^\circ + A)$ and $\cos\left(\frac{3\pi}{2} + \theta\right)$.**

$$\tan(270^\circ + A) = \tan(180^\circ + \overline{90^\circ + A}) = \tan(90^\circ + A) = -\cot A;$$

$$\cos\left(\frac{3\pi}{2} + \theta\right) = \cos\left(\pi + \overline{\frac{\pi}{2} + \theta}\right) = -\cos\left(\frac{\pi}{2} + \theta\right) = \sin\theta.$$

99. Circular Functions of $-A$.

Take any straight line OX and let a radius vector starting from OX revolve until it has traced the angle A, taking up the position OP.

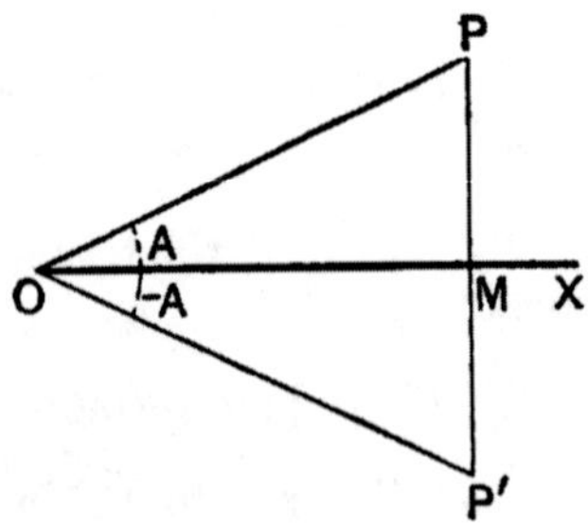

Again, let the radius vector starting from OX revolve in the *opposite* direction until it has traced the angle A, taking up the position OP'. Join PP'; then MP' is equal to MP in magnitude, and the angles at M are right angles. [Euc. I. 4.]

By definition,

$$\sin(-A) = \frac{MP'}{OP'} = \frac{-MP}{OP} = -\sin A;$$

$$\cos(-A) = \frac{OM}{OP'} = \frac{OM}{OP} = \cos A;$$

$$\tan(-A) = \frac{MP'}{OM} = \frac{-MP}{OM} = -\tan A.$$

It is especially worthy of notice that *we may change the sign of an angle without altering the value of its cosine.*

Example. **Find the values of**

$$\text{cosec}(-210^\circ) \text{ and } \cos(A - 270^\circ).$$

$$\text{cosec}(-210^\circ) = -\text{cosec}\,210^\circ = -\text{cosec}(180^\circ + 30^\circ) = \text{cosec}\,30^\circ = 2.$$

$$\cos(A - 270^\circ) = \cos(270^\circ - A) = \cos(180^\circ + \overline{90^\circ - A})$$

$$= -\cos(90^\circ - A) = -\sin A.$$

100. If $f(A)$ denotes a function of A which is unaltered in magnitude and sign when $-A$ is written for A, then $f(A)$ is said to be an **even function** of A. In this case $f(-A)=f(A)$.

If when $-A$ is written for A, the sign of $f(A)$ is changed while the magnitude remains unaltered, $f(A)$ is said to be an **odd function** of A, and in this case $f(-A)=-f(A)$.

From the last article it will be seen that

$\cos A$ *and* $\sec A$ *are even functions of* A,

$\sin A$, $\text{cosec}\, A$, $\tan A$, $\cot A$ *are odd functions of* A.

EXAMPLES. X. a.

Find the numerical value of

1. $\cos 135°$. 2. $\sin 150°$. 3. $\tan 240°$.
4. $\text{cosec}\, 225°$. 5. $\sin(-120°)$. 6. $\cot(-135°)$.
7. $\cot 315°$. 8. $\cos(-240°)$. 9. $\sec(-300°)$.
10. $\tan\frac{3\pi}{4}$. 11. $\sin\frac{4\pi}{3}$. 12. $\sec\frac{2\pi}{3}$.
13. $\text{cosec}\left(-\frac{\pi}{6}\right)$. 14. $\cos\left(-\frac{3\pi}{4}\right)$. 15. $\cot\left(-\frac{5\pi}{6}\right)$.

Express as functions of A:

16. $\cos(270°+A)$. 17. $\cot(270°-A)$. 18. $\sin(A-90°)$.
19. $\sec(A-180°)$. 20. $\sin(270°-A)$. 21. $\cot(A-90°)$.

Express as functions of θ:

22. $\sin\left(\theta-\frac{\pi}{2}\right)$. 23. $\tan(\theta-\pi)$. 24. $\sec\left(\frac{3\pi}{2}-\theta\right)$.

Express in the simplest form:

25. $\tan(180°+A)\sin(90°+A)\sec(90°-A)$.
26. $\cos(90°+A)+\sin(180°-A)-\sin(180°+A)-\sin(-A)$.
27. $\sec(180°+A)\sec(180°-A)+\cot(90°+A)\tan(180°+A)$.

101. In Art. 38 we have established the relations which subsist between the trigonometrical ratios of $90° - A$ and those of A, when A is an acute angle. We shall now give a general proof which is applicable whatever be the magnitude of A.

102. Circular Functions of 90° – A for any value of A.

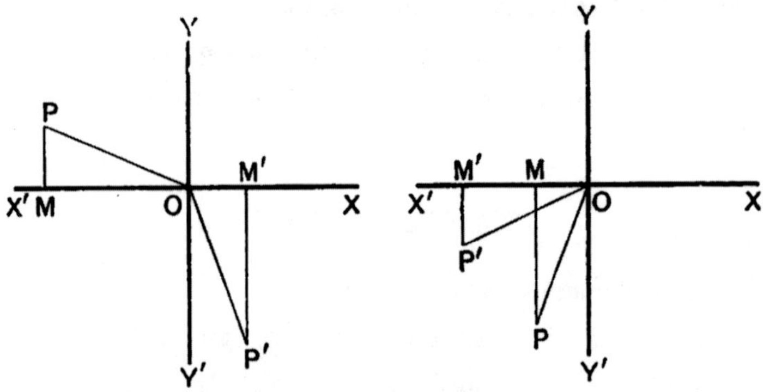

Let a radius vector starting from OX revolve until it has traced the angle A, taking up the position OP in each of the two figures.

Again, let the radius vector starting from OX revolve through $90°$ into the position OY and then *back again* through an angle A, taking up the final position OP' in each of the two figures.

Draw PM and $P'M'$ perpendicular to XX'; then whatever be the value of A, it will be found that $\angle OP'M' = \angle POM$, so that the triangles OMP and $OM'P'$ are geometrically equal, having MP equal to OM', and OM equal to $M'P'$, *in magnitude.*

When P is above XX', P' is to the right of YY',
and when P is below XX', P' is to the left of YY'.

When P' is above XX', P is to the right of YY',
and when P' is below XX', P is to the left of YY'.

Hence MP is equal to OM' in magnitude and is always of the same sign as OM';

and $M'P'$ is equal to OM in magnitude and is always of the same sign as OM.

By definition,

$$\sin(90^\circ - A) = \frac{M'P'}{OP'} = \frac{OM}{OP} = \cos A;$$

$$\cos(90^\circ - A) = \frac{OM'}{OP'} = \frac{MP}{OP} = \sin A;$$

$$\tan(90^\circ - A) = \frac{M'P'}{OM'} = \frac{OM}{MP} = \cot A.$$

A general method similar to the above may be applied to all the other cases of this chapter.

103. Circular Functions of n . 360° + A.

If n is any integer, $n \,.\, 360^\circ$ represents n complete revolutions of the radius vector, and therefore the boundary line of the angle $n \,.\, 360^\circ + A$ is coincident with that of A. The value of each function of the angle $n \,.\, 360^\circ + A$ is thus the same as the value of the corresponding function of A both in magnitude and in sign.

104. Since the functions of all coterminal angles are equal, there is a *recurrence* of the values of the functions each time the boundary line completes its revolution and comes round into its original position. This is otherwise expressed by saying that *the circular functions are periodic*, and 360° is said to be *the amplitude of the period.*

In radian measure, the amplitude of the period is 2π.

NOTE. In the case of the tangent and cotangent the amplitude of the period is half that of the other circular functions, being 180° or π radians. [Art. 97.]

105. Circular Functions of n . 360° − A.

If n is any integer, the boundary line of $n \,.\, 360^\circ - A$ is coincident with that of $-A$. The value of each function of $n \,.\, 360^\circ - A$ is thus the same as the value of the corresponding function of $-A$ both in magnitude and in sign; hence

$$\sin(n \,.\, 360^\circ - A) = \sin(-A) = -\sin A;$$

$$\cos(n \,.\, 360^\circ - A) = \cos(-A) = \cos A;$$

$$\tan(n \,.\, 360^\circ - A) = \tan(-A) = -\tan A.$$

106. We can always express the functions of any angle in terms of the functions of some positive acute angle. In the arrangement of the work it is advisable to follow a uniform plan.

(1) If the angle is negative, use the relations connecting the functions of $-A$ and A. [Art. 99.]

Thus $$\sin(-30^\circ) = -\sin 30^\circ = -\frac{1}{2};$$

$$\cos(-845^\circ) = \cos 845^\circ.$$

(2) If the angle is greater than 360°, by taking off multiples of 360° the angle may be replaced by a coterminal angle less than 360°. [Art. 103.]

Thus $$\tan 735^\circ = \tan(2\times 360^\circ + 15^\circ) = \tan 15^\circ.$$

(3) If the angle is still greater than 180°, use the relations connecting the functions of $180^\circ + A$ and A. [Art. 97.]

Thus $$\cot 585^\circ = \cot(360^\circ + 225^\circ) = \cot 225^\circ$$
$$= \cot(180^\circ + 45^\circ) = \cot 45^\circ = 1.$$

(4) If the angle is still greater than 90°, use the relations connecting the functions of $180^\circ - A$ and A. [Art. 92.]

Thus $$\cos 675^\circ = \cos(360^\circ + 315^\circ) = \cos 315^\circ$$
$$= \cos(180^\circ + 135^\circ) = -\cos 135^\circ$$
$$= -\cos(180^\circ - 45^\circ) = \cos 45^\circ = \frac{1}{\sqrt{2}}.$$

Example. Express $\sin(-1190^\circ)$, $\tan 1000^\circ$, $\cos(-980^\circ)$ as functions of positive acute angles.

$$\sin(-1190^\circ) = -\sin 1190^\circ = -\sin(3\times 360^\circ + 110^\circ) = -\sin 110^\circ$$
$$= -\sin(180^\circ - 70^\circ) = -\sin 70^\circ.$$

$$\tan 1000^\circ = \tan(2\times 360^\circ + 280^\circ) = \tan 280^\circ$$
$$= \tan(180^\circ + 100^\circ) = \tan 100^\circ$$
$$= \tan(180^\circ - 80^\circ) = -\tan 80^\circ.$$

$$\cos(-980^\circ) = \cos 980^\circ = \cos(2\times 360^\circ + 260^\circ) = \cos 260^\circ$$
$$= \cos(180^\circ + 80^\circ) = -\cos 80^\circ.$$

107. From the investigations of this chapter we see that the number of angles which have the same circular function is unlimited. Thus if $\tan\theta=1$, θ may be any one of the angles coterminal with 45° or 225°.

Example. Draw the boundary lines of A when $\sin A=\frac{\sqrt{3}}{2}$; and write down all the angles numerically less than 360° which satisfy the equation.

Since $\sin 60°=\frac{\sqrt{3}}{2}$, if we draw OP making $\angle XOP=60°$, then OP is one position of the boundary line.

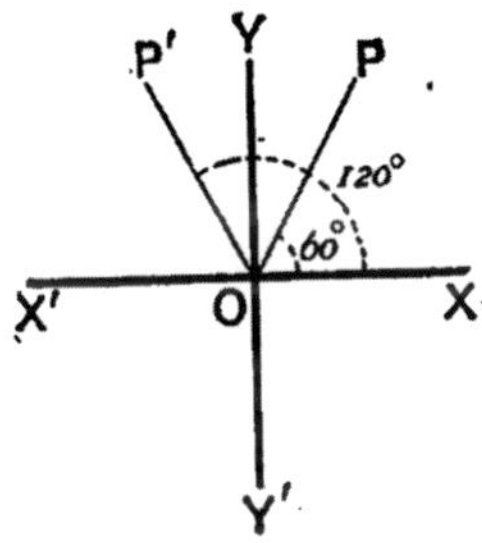

Again, $\sin 60°=\sin(180°-60°)=\sin 120°$, so that another position of the boundary line will be found by making $XOP'=120°$.

There will be no position of the boundary line in the third or fourth quadrant, since in these quadrants the sine is negative.

Thus in one complete revolution OP and OP' are the only two positions of the boundary line of the angle A.

Hence the positive angles are 60° and 120°;

and the negative angles are $-(360°-120°)$ and $-(360°-60°)$; that is, $-240°$ and $-300°$.

EXAMPLES. X. b.

Find the numerical value of

1. $\cos 480°$.
2. $\sin 960°$.
3. $\cos(-780°)$.
4. $\sin(-870°)$.
5. $\sec 900°$.
6. $\tan(-855°)$.
7. $\operatorname{cosec}(-660°)$.
8. $\cot 840°$.
9. $\operatorname{cosec}(-765°)$.
10. $\cos 1125°$.
11. $\cot 990°$.
12. $\sin 855°$.
13. $\sec 1305°$.
14. $\cos 960°$.
15. $\sec(-1575°)$.
16. $\sin\frac{15\pi}{4}$.
17. $\cot\frac{23\pi}{4}$.
18. $\sec\frac{7\pi}{3}$.
19. $\cot\frac{16\pi}{3}$.
20. $\sec\left(\frac{3\pi}{2}+\frac{\pi}{3}\right)$.

Find all the angles numerically less than 360° which satisfy the equations :

21. $\cos\theta=\frac{\sqrt{3}}{2}$. 22. $\sin\theta=-\frac{1}{2}$.

23. $\tan\theta=-\sqrt{3}$. 24. $\cot\theta=-1$.

If A is less than 90°, prove geometrically

25. $\sec(A-180°)=-\sec A$.

26. $\tan(270°+A)=-\cot A$.

27. $\cos(A-90°)=\sin A$.

28. Prove that

$$\tan A+\tan(180°-A)+\cot(90°+A)=\tan(360°-A).$$

29. Shew that

$$\frac{\sin(180°-A)}{\tan(180°+A)}\cdot\frac{\cot(90°-A)}{\tan(90°+A)}\cdot\frac{\cos(360°-A)}{\sin(-A)}=\sin A.$$

Express in the simplest form

30. $\dfrac{\sin(-A)}{\sin(180°+A)}-\dfrac{\tan(90°+A)}{\cot A}+\dfrac{\cos A}{\sin(90°+A)}$.

31. $\dfrac{\operatorname{cosec}(180°-A)}{\sec(180°+A)}\cdot\dfrac{\cos(-A)}{\cos(90°+A)}$.

32. $\dfrac{\cos(90°+A)\sec(-A)\tan(180°-A)}{\sec(360°+A)\sin(180°+A)\cot(90°-A)}$.

33. Prove that $\sin\left(\frac{\pi}{2}+\theta\right)\cos(\pi-\theta)\cot\left(\frac{3\pi}{2}+\theta\right)$

$$=\sin\left(\frac{\pi}{2}-\theta\right)\sin\left(\frac{3\pi}{2}-\theta\right)\cot\left(\frac{\pi}{2}+\theta\right).$$

34. When $a=\frac{11\pi}{4}$, find the numerical value of

$$\sin^2 a-\cos^2 a+2\tan a-\sec^2 a.$$

CHAPTER XI.

FUNCTIONS OF COMPOUND ANGLES.

[*If preferred, Chapters XIII, XIV, XV may be taken before Chapters XI and XII.*]

108. WHEN an angle is made up by the algebraical sum of two or more angles it is called a **compound angle**; thus $A+B$, $A-B$, and $A+B-C$ are compound angles.

109. Hitherto we have only discussed the properties of the functions of single angles, such as A, B, α, θ. In the present chapter we shall prove some fundamental properties relating to the functions of compound angles. We shall begin by finding expressions for the sine, cosine, and tangent of $A+B$ and $A-B$ in terms of the functions of A and B.

It may be useful to caution the student against the prevalent mistake of supposing that a function of $A+B$ is equal to the sum of the corresponding functions of A and B, and a function of $A-B$ to the difference of the corresponding functions.

Thus $\sin(A+B)$ is not equal to $\sin A + \sin B$,

and $\qquad \cos(A-B)$ is not equal to $\cos A - \cos B$.

A numerical instance will illustrate this.

Thus if $A=60°$, $B=30°$, then $A+B=90°$,

so that $$\cos(A+B) = \cos 90° = 0;$$

but $$\cos A + \cos B = \cos 60° + \cos 30° = \frac{1}{2} + \frac{\sqrt{3}}{2}.$$

Hence $\cos(A+B)$ is not equal to $\cos A + \cos B$.

In like manner, $\sin(A+A)$ is not equal to $\sin A + \sin A$;

that is, $\qquad \sin 2A$ is not equal to $2\sin A$.

Similarly $\qquad \tan 3A$ is not equal to $3\tan A$.

110. *To prove the formulæ*

$$\sin(A+B)=\sin A\cos B+\cos A\sin B,$$
$$\cos(A+B)=\cos A\cos B-\sin A\sin B.$$

Let $\angle LOM=A$, and $\angle MON=B$; then $\angle LON=A+B$.

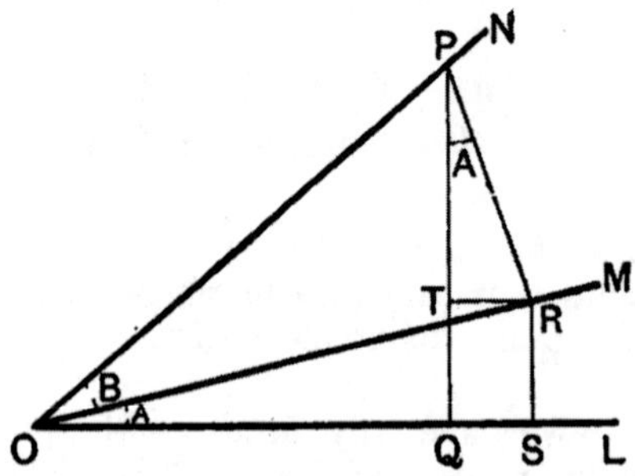

In ON, *the boundary line of the compound angle* $A+B$, take any point P, and draw PQ and PR perpendicular to OL and OM respectively; also draw RS and RT perpendicular to OL and PQ respectively.

By definition,

$$\sin(A+B)=\frac{PQ}{OP}=\frac{RS+PT}{OP}=\frac{RS}{OP}+\frac{PT}{OP}$$
$$=\frac{RS}{OR}\cdot\frac{OR}{OP}+\frac{PT}{PR}\cdot\frac{PR}{OP}$$
$$=\sin A\,.\,\cos B+\cos TPR\,.\,\sin B.$$

But $\angle TPR=90°-\angle TRP=\angle TRO=\angle ROS=A$;

$$\therefore\ \sin(A+B)=\sin A\cos B+\cos A\sin B.$$

Also $$\cos(A+B)=\frac{OQ}{OP}=\frac{OS-TR}{OP}=\frac{OS}{OP}-\frac{TR}{OP}$$
$$=\frac{OS}{OR}\cdot\frac{OR}{OP}-\frac{TR}{PR}\cdot\frac{PR}{OP}$$
$$=\cos A\,.\,\cos B-\sin TPR\,.\,\sin B$$
$$=\cos A\cos B-\sin A\sin B.$$

111. *To prove the formulæ*

$$\sin(A-B)=\sin A\cos B-\cos A\sin B,$$
$$\cos(A-B)=\cos A\cos B+\sin A\sin B.$$

Let $\angle LOM=A$, and $\angle MON=B$; then $\angle LON=A-B$.

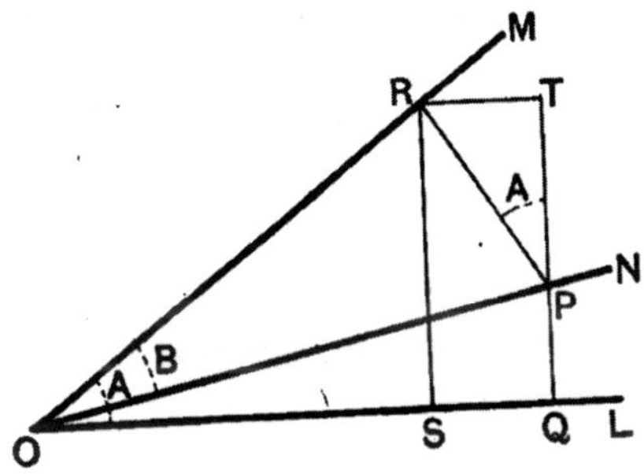

In ON, *the boundary line of the compound angle* $A-B$, take any point P, and draw PQ and PR perpendicular to OL and OM respectively; also draw RS and RT perpendicular to OL and QP respectively.

By definition,

$$\sin(A-B)=\frac{PQ}{OP}=\frac{RS-PT}{OP}=\frac{RS}{OP}-\frac{PT}{OP}$$
$$=\frac{RS}{OR}\cdot\frac{OR}{OP}-\frac{PT}{PR}\cdot\frac{PR}{OP}$$
$$=\sin A\,.\cos B-\cos TPR\,.\sin B.$$

But $\angle TPR=90^\circ-\angle TRP=\angle MRT=\angle MOL=A$;

$$\therefore\ \sin(A-B)=\sin A\cos B-\cos A\sin B.$$

Also $$\cos(A-B)=\frac{OQ}{OP}=\frac{OS+RT}{OP}=\frac{OS}{OP}+\frac{RT}{OP}$$
$$=\frac{OS}{OR}\cdot\frac{OR}{OP}+\frac{RT}{RP}\cdot\frac{RP}{OP}$$
$$=\cos A\,.\cos B+\sin TPR\,.\sin B$$
$$=\cos A\cos B+\sin A\sin B.$$

112. The *expansions* of $\sin(A \pm B)$ and $\cos(A \pm B)$ are frequently called the "Addition Formulæ." We shall sometimes refer to them as the "$A+B$" and "$A-B$" formulæ.

113. In the foregoing geometrical proofs we have supposed that the angles A, B, $A+B$ are all less than a right angle, and that $A-B$ is positive. If the angles are not so restricted some modification of the figures will be required. It is however unnecessary to consider these cases in detail, as in Chap. XXII. we shall shew by the Method of Projections that the Addition Formulæ hold universally. In the meantime the student may assume that they are always true.

Example 1. Find the value of $\cos 75°$.

$$\cos 75° = \cos(45° + 30°) = \cos 45° \cos 30° - \sin 45° \sin 30°$$

$$= \frac{1}{\sqrt{2}} \cdot \frac{\sqrt{3}}{2} - \frac{1}{\sqrt{2}} \cdot \frac{1}{2} = \frac{\sqrt{3}-1}{2\sqrt{2}}.$$

Example 2. If $\sin A = \frac{4}{5}$ and $\sin B = \frac{5}{13}$, find $\sin(A-B)$.

$$\sin(A-B) = \sin A \cos B - \cos A \sin B.$$

But $$\cos A = \sqrt{1 - \sin^2 A} = \sqrt{1 - \frac{16}{25}} = \frac{3}{5};$$

and $$\cos B = \sqrt{1 - \sin^2 B} = \sqrt{1 - \frac{25}{169}} = \frac{12}{13};$$

$$\therefore \sin(A-B) = \frac{4}{5} \cdot \frac{12}{13} - \frac{3}{5} \cdot \frac{5}{13} = \frac{33}{65}.$$

NOTE. Strictly speaking $\cos A = \pm\frac{3}{5}$ and $\cos B = \pm\frac{12}{13}$, so that $\sin(A-B)$ has *four* values. We shall however suppose that in similar cases only the positive value of the square root is taken.

114. *To prove that* $\sin(A+B)\sin(A-B) = \sin^2 A - \sin^2 B$.

The first side

$$= (\sin A \cos B + \cos A \sin B)(\sin A \cos B - \cos A \sin B)$$

$$= \sin^2 A \cos^2 B - \cos^2 A \sin^2 B$$

$$= \sin^2 A\,(1 - \sin^2 B) - (1 - \sin^2 A)\sin^2 B$$

$$= \sin^2 A - \sin^2 B.$$

EXAMPLES. XI. a.

[*The examples printed in more prominent type are important, and should be regarded as standard formulæ.*]

Prove that

1. $\sin(A+45°)=\frac{1}{\sqrt{2}}(\sin A+\cos A)$.

2. $\cos(A+45°)=\frac{1}{\sqrt{2}}(\cos A-\sin A)$.

3. $2\sin(30°-A)=\cos A-\sqrt{3}\sin A$.

4. If $\cos A=\frac{4}{5}$, $\cos B=\frac{3}{5}$, find $\sin(A+B)$ and $\cos(A-B)$.

5. If $\sin A=\frac{3}{5}$, $\cos B=\frac{12}{13}$, find $\cos(A+B)$ and $\sin(A-B)$.

6. If $\sec A=\frac{17}{8}$, $\operatorname{cosec} B=\frac{5}{4}$, find $\sec(A+B)$.

Prove that

7. $\sin 75°=\cos 15°=\frac{\sqrt{3}+1}{2\sqrt{2}}$.

8. $\sin 15°=\cos 75°=\frac{\sqrt{3}-1}{2\sqrt{2}}$.

9. $\frac{\sin(\alpha+\beta)}{\cos\alpha\cos\beta}=\tan\alpha+\tan\beta$.

10. $\frac{\sin(\alpha-\beta)}{\sin\alpha\sin\beta}=\cot\beta-\cot\alpha$.

11. $\frac{\cos(\alpha-\beta)}{\cos\alpha\sin\beta}=\cot\beta+\tan\alpha$.

12. $\cos(A+B)\cos(A-B)=\cos^2 A-\sin^2 B$.

13. $\sin(A+B)\sin(A-B)=\cos^2 B-\cos^2 A$.

14. $\cos(45°-A)-\sin(45°+A)=0$.

15. $\cos(45°+A)+\sin(A-45°)=0$.

16. $\cos(A-B)-\sin(A+B)=(\cos A-\sin A)(\cos B-\sin B)$.

17. $\cos(A+B)+\sin(A-B)=(\cos A+\sin A)(\cos B-\sin B)$.

Prove the following identities:

18. $2\sin(A+45^\circ)\sin(A-45^\circ)=\sin^2 A-\cos^2 A.$

19. $2\cos\left(\frac{\pi}{4}+\alpha\right)\cos\left(\frac{\pi}{4}-\alpha\right)=\cos^2\alpha-\sin^2\alpha.$

20. $2\sin\left(\frac{\pi}{4}+\alpha\right)\cos\left(\frac{\pi}{4}+\beta\right)=\cos(\alpha+\beta)+\sin(\alpha-\beta).$

21. $\frac{\sin(\beta-\gamma)}{\cos\beta\cos\gamma}+\frac{\sin(\gamma-\alpha)}{\cos\gamma\cos\alpha}+\frac{\sin(\alpha-\beta)}{\cos\alpha\cos\beta}=0.$

115. *To expand* $\tan(A+B)$ *in terms of* $\tan A$ *and* $\tan B$.

$$\tan(A+B)=\frac{\sin(A+B)}{\cos(A+B)}=\frac{\sin A\cos B+\cos A\sin B}{\cos A\cos B-\sin A\sin B}.$$

To express this fraction in terms of *tangents*, divide each term of numerator and denominator by $\cos A\cos B$;

$$\therefore\ \tan(A+B)=\frac{\frac{\sin A}{\cos A}+\frac{\sin B}{\cos B}}{1-\frac{\sin A}{\cos A}\cdot\frac{\sin B}{\cos B}};$$

that is,

$$\tan(A+B)=\frac{\tan A+\tan B}{1-\tan A\tan B}.$$

A geometrical proof of this result is given in Chap. XXII.

Similarly, we may prove that

$$\tan(A-B)=\frac{\tan A-\tan B}{1+\tan A\tan B}.$$

Example. Find the value of $\tan 75^\circ$.

$$\tan 75^\circ=\tan(45^\circ+30^\circ)=\frac{\tan 45^\circ+\tan 30^\circ}{1-\tan 45^\circ\tan 30^\circ}$$

$$=\frac{1+\frac{1}{\sqrt{3}}}{1-\frac{1}{\sqrt{3}}}=\frac{\sqrt{3}+1}{\sqrt{3}-1}$$

$$=\frac{(\sqrt{3}+1)(\sqrt{3}+1)}{3-1}=\frac{4+2\sqrt{3}}{2}$$

$$=2+\sqrt{3}.$$

116. *To expand* $\cot(A+B)$ *in terms of* $\cot A$ *and* $\cot B$.

$$\cot(A+B)=\frac{\cos(A+B)}{\sin(A+B)}=\frac{\cos A\cos B-\sin A\sin B}{\sin A\cos B+\cos A\sin B}.$$

To express this fraction in terms of *cotangents*, divide each term of numerator and denominator by $\sin A \sin B$;

$$\therefore\ \cot(A+B)=\frac{\dfrac{\cos A\cos B}{\sin A\sin B}-1}{\dfrac{\cos B}{\sin B}+\dfrac{\cos A}{\sin A}}=\frac{\cot A\cot B-1}{\cot B+\cot A}.$$

Similarly, we may prove that

$$\cot(A-B)=\frac{\cot A\cot B+1}{\cot B-\cot A}.$$

117. *To find the expansion of* $\sin(A+B+C)$.

$$\begin{aligned}\sin(A+B+C)&=\sin\{(A+B)+C\}\\&=\sin(A+B)\cos C+\cos(A+B)\sin C\\&=(\sin A\cos B+\cos A\sin B)\cos C\\&\qquad+(\cos A\cos B-\sin A\sin B)\sin C\\&=\sin A\cos B\cos C+\cos A\sin B\cos C\\&\qquad+\cos A\cos B\sin C-\sin A\sin B\sin C.\end{aligned}$$

118. *To find the expansion of* $\tan(A+B+C)$.

$$\tan(A+B+C)=\tan\{(A+B)+C\}=\frac{\tan(A+B)+\tan C}{1-\tan(A+B)\tan C}$$

$$=\frac{\dfrac{\tan A+\tan B}{1-\tan A\tan B}+\tan C}{1-\dfrac{\tan A+\tan B}{1-\tan A\tan B}\cdot\tan C}$$

$$=\frac{\tan A+\tan B+\tan C-\tan A\tan B\tan C}{1-\tan A\tan B-\tan B\tan C-\tan C\tan A}.$$

COR. If $A+B+C=180°$, then $\tan(A+B+C)=0$; hence the numerator of the above expression must be zero.

$$\therefore\ \tan A+\tan B+\tan C=\tan A\tan B\tan C.$$

EXAMPLES. XI. b.

[*The examples printed in more prominent type are important, and should be regarded as standard formulæ.*]

1. Find $\tan(A+B)$ when $\tan A=\frac{1}{2}$, $\tan B=\frac{1}{3}$.

2. If $\tan A=\frac{4}{3}$, and $B=45°$, find $\tan(A-B)$.

3. If $\cot A=\frac{5}{7}$, $\cot B=\frac{7}{5}$, find $\cot(A+B)$ and $\tan(A-B)$.

4. If $\cot A=\frac{11}{2}$, $\tan B=\frac{7}{24}$, find $\cot(A-B)$ and $\tan(A+B)$.

5. $\mathbf{\tan(45°+A)=\dfrac{1+\tan A}{1-\tan A}}$.

6. $\mathbf{\tan(45°-A)=\dfrac{1-\tan A}{1+\tan A}}$.

7. $\cot\left(\frac{\pi}{4}-\theta\right)=\dfrac{\cot\theta+1}{\cot\theta-1}$.

8. $\cot\left(\frac{\pi}{4}+\theta\right)=\dfrac{\cot\theta-1}{\cot\theta+1}$.

9. $\mathbf{\tan 15°=2-\sqrt{3}}$.

10. $\mathbf{\cot 15°=2+\sqrt{3}}$.

11. Find the expansions of

$$\cos(A+B+C) \text{ and } \sin(A-B+C).$$

12. Express $\tan(A-B-C)$ in terms of $\tan A$, $\tan B$, $\tan C$.

13. Express $\cot(A+B+C)$ in terms of $\cot A$, $\cot B$, $\cot C$.

119. Beginners not unfrequently find a difficulty in the converse use of the $A+B$ and $A-B$ formulæ; that is, they fail to recognise when an expression is merely an expansion belonging to one of the standard forms.

Example 1. Simplify $\cos(\alpha-\beta)\cos(\alpha+\beta)-\sin(\alpha-\beta)\sin(\alpha+\beta)$.

This expression is the expansion of the cosine of the compound angle $(\alpha+\beta)+(\alpha-\beta)$, and is therefore equal to $\cos\{(\alpha+\beta)+(\alpha-\beta)\}$; that is, to $\cos 2\alpha$.

Example 2. Shew that $\dfrac{\tan A+\tan 2A}{1-\tan A\tan 2A}=\tan 3A$.

By Art. 115, the first side is the expansion of $\tan(A+2A)$, and is therefore equal to $\tan 3A$.

Example 3. **Prove that $\cot 2A + \tan A = \text{cosec}\, 2A$.**

$$\text{The first side} = \frac{\cos 2A}{\sin 2A} + \frac{\sin A}{\cos A} = \frac{\cos 2A \cos A + \sin 2A \sin A}{\sin 2A \cos A}$$

$$= \frac{\cos (2A - A)}{\sin 2A \cos A} = \frac{\cos A}{\sin 2A \cos A}$$

$$= \frac{1}{\sin 2A} = \text{cosec}\, 2A.$$

Example 4. **Prove that**

$$\cos 4\theta \cos \theta + \sin 4\theta \sin \theta = \cos 2\theta \cos \theta - \sin 2\theta \sin \theta.$$

$$\text{The first side} = \cos (4\theta - \theta) = \cos 3\theta = \cos (2\theta + \theta)$$

$$= \cos 2\theta \cos \theta - \sin 2\theta \sin \theta.$$

EXAMPLES. XI. c.

Prove the following identities :

1. $\cos (A + B) \cos B + \sin (A + B) \sin B = \cos A.$
2. $\sin 3A \cos A - \cos 3A \sin A = \sin 2A.$
3. $\cos 2a \cos a + \sin 2a \sin a = \cos a.$
4. $\cos (30° + A) \cos (30° - A) - \sin (30° + A) \sin (30° - A) = \frac{1}{2}.$
5. $\sin (60° - A) \cos (30° + A) + \cos (60° - A) \sin (30° + A) = 1.$
6. $\dfrac{\cos 2a}{\sec a} - \dfrac{\sin 2a}{\text{cosec}\, a} = \cos 3a.$
7. $\dfrac{\tan (a - \beta) + \tan \beta}{1 - \tan (a - \beta) \tan \beta} = \tan a.$
8. $\dfrac{\cot (a + \beta) \cot a + 1}{\cot a - \cot (a + \beta)} = \cot \beta.$
9. $\dfrac{\tan 4A - \tan 3A}{1 + \tan 4A \tan 3A} = \tan A.$
10. $\cot \theta - \cot 2\theta = \text{cosec}\, 2\theta.$
11. $1 + \tan 2\theta \tan \theta = \sec 2\theta.$
12. $1 + \cot 2\theta \cot \theta = \text{cosec}\, 2\theta \cot \theta.$
13. $\sin 2\theta \cos \theta + \cos 2\theta \sin \theta = \sin 4\theta \cos \theta - \cos 4\theta \sin \theta.$
14. $\cos 4a \cos a - \sin 4a \sin a = \cos 3a \cos 2a - \sin 3a \sin 2a.$

Functions of Multiple Angles.

120. *To express* $\sin 2A$ *in terms of* $\sin A$ *and* $\cos A$.

$$\sin 2A = \sin (A+A) = \sin A \cos A + \cos A \sin A;$$

that is,
$$\sin 2A = 2 \sin A \cos A.$$

Since A may have any value, this is a perfectly general formula for the sine of an angle in terms of the sine and cosine of the half angle. Thus if $2A$ be replaced by θ, we have

$$\sin \theta = 2 \sin \frac{\theta}{2} \cos \frac{\theta}{2}.$$

Similarly,
$$\sin 4A = 2 \sin 2A \cos 2A$$
$$= 4 \sin A \cos A \cos 2A.$$

121. *To express* $\cos 2A$ *in terms of* $\cos A$ *and* $\sin A$.

$$\cos 2A = \cos (A+A) = \cos A \cos A - \sin A \sin A;$$

that is,
$$\cos 2A = \cos^2 A - \sin^2 A \quad \text{.....................(1)}.$$

There are two other useful forms in which $\cos 2A$ may be expressed, one involving $\cos A$ only, the other $\sin A$ only.

Thus from (1),

$$\cos 2A = \cos^2 A - (1 - \cos^2 A);$$

that is,
$$\cos 2A = 2 \cos^2 A - 1 \quad \text{.....................(2)}.$$

Again, from (1),

$$\cos 2A = (1 - \sin^2 A) - \sin^2 A;$$

that is,
$$\cos 2A = 1 - 2 \sin^2 A \quad \text{.....................(3)}.$$

From formulæ (2) and (3), we obtain by transposition

$$1 + \cos 2A = 2 \cos^2 A \quad \text{.....................(4)},$$

and
$$1 - \cos 2A = 2 \sin^2 A \quad \text{.....................(5)}.$$

By division,
$$\frac{1 - \cos 2A}{1 + \cos 2A} = \tan^2 A \quad \text{.....................(6)}.$$

Example. Express $\cos 4\alpha$ in terms of $\sin \alpha$.

From (3),
$$\cos 4\alpha = 1 - 2 \sin^2 2\alpha = 1 - 2 (4 \sin^2 \alpha \cos^2 \alpha)$$
$$= 1 - 8 \sin^2 \alpha (1 - \sin^2 \alpha)$$
$$= 1 - 8 \sin^2 \alpha + 8 \sin^4 \alpha.$$

122. The six formulæ of the last article deserve special attention. They are universally true so long as one of the angles involved is double of the other. For instance,

$$\cos a = \cos^2 \frac{a}{2} - \sin^2 \frac{a}{2},$$

$$\cos a = 2\cos^2 \frac{a}{2} - 1, \qquad \cos a = 1 - 2\sin^2 \frac{a}{2},$$

$$1 + \cos\theta = 2\cos^2 \frac{\theta}{2}, \qquad 1 - \cos\theta = 2\sin^2 \frac{\theta}{2}.$$

Example. If $\cos\theta = \cdot 28$, find the value of $\tan\frac{\theta}{2}$.

$$\tan^2\frac{\theta}{2} = \frac{1-\cos\theta}{1+\cos\theta} = \frac{1-\cdot 28}{1+\cdot 28} = \frac{\cdot 72}{1\cdot 28} = \frac{72}{128} = \frac{9}{16};$$

$$\therefore \tan\frac{\theta}{2} = \pm\frac{3}{4}.$$

123. *To express* $\tan 2A$ *in terms of* $\tan A$.

$$\tan 2A = \tan(A + A) = \frac{\tan A + \tan A}{1 - \tan A \tan A};$$

that is,

$$\tan 2A = \frac{2\tan A}{1 - \tan^2 A}.$$

124. *To express* $\sin 2A$ *and* $\cos 2A$ *in terms of* $\tan A$.

$$\sin 2A = 2\sin A\cos A = 2\frac{\sin A}{\cos A}\cos^2 A = 2\tan A\cos^2 A;$$

$$\therefore \sin 2A = \frac{2\tan A}{\sec^2 A} = \frac{2\tan A}{1+\tan^2 A}.$$

Again,

$$\cos 2A = \cos^2 A - \sin^2 A = \cos^2 A\,(1 - \tan^2 A);$$

$$\therefore \cos 2A = \frac{1-\tan^2 A}{\sec^2 A} = \frac{1-\tan^2 A}{1+\tan^2 A}.$$

Example. Shew that $\dfrac{1-\tan^2(45^\circ - A)}{1+\tan^2(45^\circ - A)} = \sin 2A$.

The first side $= \cos 2(45^\circ - A) = \cos(90^\circ - 2A) = \sin 2A$.

EXAMPLES. XI. d.

[*The examples printed in more prominent type are important, and should be regarded as standard formulæ.*]

1. If $\cos A = \frac{1}{3}$, find $\cos 2A$.

2. Find $\cos 2A$ when $\sin A = \frac{2}{5}$.

3. If $\sin A = \frac{3}{5}$, find $\sin 2A$.

4. If $\tan \theta = \frac{1}{3}$, find $\tan 2\theta$.

5. If $\tan \theta = \frac{1}{7}$, find $\sin 2\theta$ and $\cos 2\theta$.

6. If $\cos \alpha = \frac{4}{5}$, find $\tan \frac{\alpha}{2}$.

7. Find $\tan A$ when $\cos 2A = \cdot 96$.

Prove the following identities:

8. $\mathbf{\dfrac{\sin 2A}{1+\cos 2A} = \tan A.}$

9. $\mathbf{\dfrac{\sin 2A}{1-\cos 2A} = \cot A.}$

10. $\mathbf{\dfrac{1-\cos A}{\sin A} = \tan \dfrac{A}{2}.}$

11. $\mathbf{\dfrac{1+\cos A}{\sin A} = \cot \dfrac{A}{2}.}$

12. $2 \operatorname{cosec} 2\alpha = \sec \alpha \operatorname{cosec} \alpha$.

13. $\tan \alpha + \cot \alpha = 2 \operatorname{cosec} 2\alpha$.

14. $\cos^4 \alpha - \sin^4 \alpha = \cos 2\alpha$.

15. $\mathbf{\cot \alpha - \tan \alpha = 2 \cot 2\alpha.}$

16. $\mathbf{\cot 2A = \dfrac{\cot^2 A - 1}{2 \cot A}.}$

17. $\dfrac{\cot A - \tan A}{\cot A + \tan A} = \cos 2A$.

18. $\dfrac{1+\cot^2 A}{2 \cot A} = \operatorname{cosec} 2A$.

19. $\dfrac{\cot^2 A + 1}{\cot^2 A - 1} = \sec 2A$.

20. $\dfrac{1+\sec \theta}{\sec \theta} = 2 \cos^2 \dfrac{\theta}{2}$.

21. $\dfrac{\sec \theta - 1}{\sec \theta} = 2 \sin^2 \dfrac{\theta}{2}$.

22. $\dfrac{2-\sec^2 \theta}{\sec^2 \theta} = \cos 2\theta$.

23. $\dfrac{\operatorname{cosec}^2 \theta - 2}{\operatorname{cosec}^2 \theta} = \cos 2\theta$.

Prove the following identities:

24. $\left(\sin\frac{A}{2}+\cos\frac{A}{2}\right)^2=1+\sin A.$

25. $\left(\sin\frac{A}{2}-\cos\frac{A}{2}\right)^2=1-\sin A.$

26. $\frac{\cos 2a}{1+\sin 2a}=\tan(45^\circ-a).$

27. $\frac{\cos 2a}{1-\sin 2a}=\cot(45^\circ-a).$

28. $\sin 8A=8\sin A\cos A\cos 2A\cos 4A.$

29. $\cos 4A=8\cos^4 A-8\cos^2 A+1.$

30. $\sin A=1-2\sin^2\left(45^\circ-\frac{A}{2}\right).$

31. $\cos^2\left(\frac{\pi}{4}-a\right)-\sin^2\left(\frac{\pi}{4}-a\right)=\sin 2a.$

32. $\tan(45^\circ+A)-\tan(45^\circ-A)=2\tan 2A.$

33. $\tan(45^\circ+A)+\tan(45^\circ-A)=2\sec 2A.$

125. Functions of 3A.

$$\begin{aligned}\sin 3A&=\sin(2A+A)=\sin 2A\cos A+\cos 2A\sin A\\&=2\sin A\cos^2 A+(1-2\sin^2 A)\sin A\\&=2\sin A(1-\sin^2 A)+(1-2\sin^2 A)\sin A;\\&=3\sin A-4\sin^3 A.\end{aligned}$$

Similarly it may be proved that

$$\cos 3A=4\cos^3 A-3\cos A.$$

Again, $\tan 3A=\tan(2A+A)=\frac{\tan 2A+\tan A}{1-\tan 2A\tan A};$

by putting $\tan 2A=\frac{2\tan A}{1-\tan^2 A},$

we obtain on reduction

$$\tan 3A=\frac{3\tan A-\tan^3 A}{1-3\tan^2 A}.$$

These formulæ are perfectly general and may be applied to cases of any two angles, one of which is three times the other; thus

$$\cos 6a = 4\cos^3 2a - 3\cos 2a;$$

$$\sin 9A = 3\sin 3A - 4\sin^3 3A.$$

126. *To find the value of* $\sin 18°$.

Let $A = 18°$, then $5A = 90°$, so that $2A = 90° - 3A$.

$$\therefore \ \sin 2A = \sin(90° - 3A) = \cos 3A;$$

$$\therefore \ 2\sin A\cos A = 4\cos^3 A - 3\cos A.$$

Divide by $\cos A$ (which is not equal to zero);

$$\therefore \ 2\sin A = 4\cos^2 A - 3 = 4(1 - \sin^2 A) - 3;$$

$$\therefore \ 4\sin^2 A + 2\sin A - 1 = 0;$$

$$\therefore \ \sin A = \frac{-2 \pm \sqrt{4+16}}{8} = \frac{-1 \pm \sqrt{5}}{4}.$$

Since 18° is an acute angle, we take the positive sign;

$$\therefore \ \sin 18° = \frac{\sqrt{5} - 1}{4}.$$

Example. Find $\cos 18°$ and $\sin 54°$.

$$\cos 18° = \sqrt{1 - \sin^2 18°} = \sqrt{1 - \frac{6 - 2\sqrt{5}}{16}} = \frac{\sqrt{10 + 2\sqrt{5}}}{4}.$$

Since 54° and 36° are complementary, $\sin 54° = \cos 36°$.

Now $\quad \cos 36° = 1 - 2\sin^2 18° = 1 - \dfrac{2(6 - 2\sqrt{5})}{16} = \dfrac{\sqrt{5} + 1}{4};$

$$\therefore \ \sin 54° = \frac{\sqrt{5} + 1}{4}.$$

EXAMPLES. XI. e.

1. If $\cos A = \frac{1}{3}$, find $\cos 3A$.

2. Find $\sin 3A$ when $\sin A = \frac{3}{5}$.

3. Given $\tan A = 3$, find $\tan 3A$.

Prove the following identities:

4. $\dfrac{\sin 3A}{\sin A} - \dfrac{\cos 3A}{\cos A} = 2.$

5. $\cot 3A = \dfrac{\cot^3 A - 3\cot A}{3\cot^2 A - 1}.$

6. $\dfrac{3\cos\alpha + \cos 3\alpha}{3\sin\alpha - \sin 3\alpha} = \cot^3\alpha.$

7. $\dfrac{\sin 3\alpha + \sin^3\alpha}{\cos^3\alpha - \cos 3\alpha} = \cot\alpha.$

8. $\dfrac{\cos^3\alpha - \cos 3\alpha}{\cos\alpha} + \dfrac{\sin^3\alpha + \sin 3\alpha}{\sin\alpha} = 3.$

9. $\sin 18^\circ + \sin 30^\circ = \sin 54^\circ.$

10. $\cos 36^\circ - \sin 18^\circ = \dfrac{1}{2}.$

11. $\cos^2 36^\circ + \sin^2 18^\circ = \dfrac{3}{4}.$

12. $4\sin 18^\circ \cos 36^\circ = 1.$

***127.** The following examples further illustrate the formulæ proved in this chapter.

Example 1. Shew that $\cos^6\alpha + \sin^6\alpha = 1 - \dfrac{3}{4}\sin^2 2\alpha.$

$$\begin{aligned}\text{The first side} &= (\cos^2\alpha + \sin^2\alpha)(\cos^4\alpha + \sin^4\alpha - \cos^2\alpha\sin^2\alpha)\\ &= (\cos^2\alpha + \sin^2\alpha)^2 - 3\cos^2\alpha\sin^2\alpha\\ &= 1 - \frac{3}{4}(4\cos^2\alpha\sin^2\alpha)\\ &= 1 - \frac{3}{4}\sin^2 2\alpha.\end{aligned}$$

Example 2. Prove that $\dfrac{\cos A - \sin A}{\cos A + \sin A} = \sec 2A - \tan 2A.$

$$\text{The right side} = \frac{1}{\cos 2A} - \frac{\sin 2A}{\cos 2A} = \frac{1 - \sin 2A}{\cos 2A},$$

and since $\cos 2A = \cos^2 A - \sin^2 A = (\cos A + \sin A)(\cos A - \sin A)$, this suggests that we should multiply the numerator and denominator of the left side by $\cos A - \sin A$; thus

$$\begin{aligned}\text{the first side} &= \frac{(\cos A - \sin A)(\cos A - \sin A)}{(\cos A + \sin A)(\cos A - \sin A)}\\ &= \frac{\cos^2 A + \sin^2 A - 2\cos A\sin A}{\cos^2 A - \sin^2 A}\\ &= \frac{1 - \sin 2A}{\cos 2A} = \sec 2A - \tan 2A.\end{aligned}$$

Example 3. Shew that $\dfrac{1}{\tan 3A - \tan A} - \dfrac{1}{\cot 3A - \cot A} = \cot 2A.$

The first side $= \dfrac{1}{\dfrac{\sin 3A}{\cos 3A} - \dfrac{\sin A}{\cos A}} - \dfrac{1}{\dfrac{\cos 3A}{\sin 3A} - \dfrac{\cos A}{\sin A}}$

$$= \frac{\cos 3A \cos A}{\sin 3A \cos A - \cos 3A \sin A} - \frac{\sin 3A \sin A}{\cos 3A \sin A - \sin 3A \cos A}$$

$$= \frac{\cos 3A \cos A + \sin 3A \sin A}{\sin 3A \cos A - \cos 3A \sin A}$$

$$= \frac{\cos (3A - A)}{\sin (3A - A)} = \frac{\cos 2A}{\sin 2A} = \cot 2A.$$

NOTE. This example has been given to emphasize the fact that in identities involving the functions of $2A$ and $3A$ it is sometimes best not to substitute their equivalents in terms of functions of A.

*EXAMPLES. XI. f.

Prove the following identities:

1. $\tan 2A - \sec A \sin A = \tan A \sec 2A.$

2. $\tan 2A + \cos A \operatorname{cosec} A = \cot A \sec 2A.$

3. $\dfrac{1 - \cos 2\theta + \sin 2\theta}{1 + \cos 2\theta + \sin 2\theta} = \tan \theta.$

4. $\dfrac{1 + \cos \theta + \cos \dfrac{\theta}{2}}{\sin \theta + \sin \dfrac{\theta}{2}} = \cot \dfrac{\theta}{2}.$

5. $\cos^6 \alpha - \sin^6 \alpha = \cos 2\alpha \left(1 - \dfrac{1}{4} \sin^2 2\alpha\right).$

6. $4 (\cos^6 \theta + \sin^6 \theta) = 1 + 3 \cos^2 2\theta.$

7. $\dfrac{\cos 3\alpha + \sin 3\alpha}{\cos \alpha - \sin \alpha} = 1 + 2 \sin 2\alpha.$

8. $\dfrac{\cos 3\alpha - \sin 3\alpha}{\cos \alpha + \sin \alpha} = 1 - 2 \sin 2\alpha.$

9. $\dfrac{\cos \alpha + \sin \alpha}{\cos \alpha - \sin \alpha} = \tan 2\alpha + \sec 2\alpha.$

Prove the following identities :

10. $\dfrac{\cot a-1}{\cot a+1}=\dfrac{1-\sin 2a}{\cos 2a}$.

11. $\dfrac{1+\sin\theta}{\cos\theta}=\dfrac{1+\tan\dfrac{\theta}{2}}{1-\tan\dfrac{\theta}{2}}$.

12. $\dfrac{\cos\theta}{1-\sin\theta}=\dfrac{\cot\dfrac{\theta}{2}+1}{\cot\dfrac{\theta}{2}-1}$.

13. $\sec A-\tan A=\tan\left(45^\circ-\dfrac{A}{2}\right)$.

14. $\tan A+\sec A=\cot\left(45^\circ-\dfrac{A}{2}\right)$.

15. $\dfrac{1+\sin\theta}{1-\sin\theta}=\tan^2\left(\dfrac{\pi}{4}+\dfrac{\theta}{2}\right)$.

16. $(2\cos A+1)(2\cos A-1)=2\cos 2A+1$.

17. $\dfrac{\sin 2A}{1+\cos 2A}\cdot\dfrac{\cos A}{1+\cos A}=\tan\dfrac{A}{2}$.

18. $\dfrac{\sin 2A}{1-\cos 2A}\cdot\dfrac{1-\cos A}{\cos A}=\tan\dfrac{A}{2}$.

19. $4\sin^3 a\cos 3a+4\cos^3 a\sin 3a=3\sin 4a$.
[*Put* $4\sin^3 a=3\sin a-\sin 3a$ *and* $4\cos^3 a=3\cos a+\cos 3a$.]

20. $\cos^3 a\cos 3a+\sin^3 a\sin 3a=\cos^3 2a$.

21. $4(\cos^3 20^\circ+\cos^3 40^\circ)=3(\cos 20^\circ+\cos 40^\circ)$.

22. $4(\cos^3 10^\circ+\sin^3 20^\circ)=3(\cos 10^\circ+\sin 20^\circ)$.

23. $\tan 3A-\tan 2A-\tan A=\tan 3A\tan 2A\tan A$.
[*Use* $\tan 3A=\tan(2A+A)$.]

24. $\dfrac{\cot\theta}{\cot\theta-\cot 3\theta}+\dfrac{\tan\theta}{\tan\theta-\tan 3\theta}=1$.

25. $\dfrac{1}{\tan 3\theta+\tan\theta}-\dfrac{1}{\cot 3\theta+\cot\theta}=\cot 4\theta$.

[*Some easy miscellaneous Examples on Chapters XI and XII will be found on pages* 122_A, 122_B.]

CHAPTER XII.

TRANSFORMATION OF PRODUCTS AND SUMS.

Transformation of products into sums or differences.

128. In the last chapter we have proved that

$$\sin A \cos B + \cos A \sin B = \sin (A+B),$$

and

$$\sin A \cos B - \cos A \sin B = \sin (A-B).$$

By addition,

$$2 \sin A \cos B = \sin (A+B) + \sin (A-B) \quad \text{.........(1)};$$

by subtraction

$$2 \cos A \sin B = \sin (A+B) - \sin (A-B) \quad \text{.........(2)}.$$

These formulæ enable us to express the product of a sine and cosine as the sum or difference of two sines.

Again,

$$\cos A \cos B - \sin A \sin B = \cos (A+B),$$

and

$$\cos A \cos B + \sin A \sin B = \cos (A-B).$$

By addition,

$$2 \cos A \cos B = \cos (A+B) + \cos (A-B) \quad \text{.........(3)};$$

by subtraction,

$$2 \sin A \sin B = \cos (A-B) - \cos (A+B) \quad \text{.........(4)}.$$

These formulæ enable us to express

(i) the product of two cosines as the sum of two cosines;

(ii) the product of two sines as the difference of two cosines.

129. In each of the four formulæ of the previous article it should be noticed that on the left side we have any two angles A and B, and on the right side the sum and difference of these angles.

For practical purposes the following verbal statements of the results are more useful.

$$2\sin A\cos B = \sin(\textit{sum}) + \sin(\textit{difference});$$
$$2\cos A\sin B = \sin(\textit{sum}) - \sin(\textit{difference});$$
$$2\cos A\cos B = \cos(\textit{sum}) + \cos(\textit{difference});$$
$$2\sin A\sin B = \cos(\textit{difference}) - \cos(\textit{sum}).$$

N.B. In the last of these formulæ, *the difference precedes the sum.*

Example 1. $2\sin 7A\cos 4A = \sin(\textit{sum}) + \sin(\textit{difference})$
$$= \sin 11A + \sin 3A.$$

Example 2. $2\cos 3\theta\sin 6\theta = \sin(3\theta + 6\theta) - \sin(3\theta - 6\theta)$
$$= \sin 9\theta - \sin(-3\theta)$$
$$= \sin 9\theta + \sin 3\theta.$$

Example 3. $\cos\frac{3A}{2}\cos\frac{5A}{2} = \frac{1}{2}\left\{\cos\left(\frac{3A}{2} + \frac{5A}{2}\right) + \cos\left(\frac{3A}{2} - \frac{5A}{2}\right)\right\}$
$$= \frac{1}{2}\{\cos 4A + \cos(-A)\}$$
$$= \frac{1}{2}(\cos 4A + \cos A).$$

Example 4. $2\sin 75^\circ\sin 15^\circ = \cos(75^\circ - 15^\circ) - \cos(75^\circ + 15^\circ)$
$$= \cos 60^\circ - \cos 90^\circ$$
$$= \frac{1}{2} - 0$$
$$= \frac{1}{2}.$$

130. After a little practice the student will be able to omit some of the steps and find the equivalent very rapidly.

Example 1. $2\cos\left(\frac{\pi}{4} + \theta\right)\cos\left(\frac{\pi}{4} - \theta\right) = \cos\frac{\pi}{2} + \cos 2\theta = \cos 2\theta.$

Example 2. $\sin(\alpha - 2\beta)\cos(\alpha + 2\beta) = \frac{1}{2}\{\sin 2\alpha + \sin(-4\beta)\}$
$$= \frac{1}{2}(\sin 2\alpha - \sin 4\beta).$$

EXAMPLES. XII. a.

Express in the form of a sum or difference

1. $2 \sin 3\theta \cos \theta$.
2. $2 \cos 6\theta \sin 3\theta$.
3. $2 \cos 7A \cos 5A$.
4. $2 \sin 3A \sin 2A$.
5. $2 \cos 5\theta \sin 4\theta$.
6. $2 \sin 4\theta \cos 8\theta$.
7. $2 \sin 9\theta \sin 3\theta$.
8. $2 \cos 9\theta \sin 7\theta$.
9. $2 \cos 2\alpha \cos 11\alpha$.
10. $2 \sin 5\alpha \sin 10\alpha$.
11. $\sin 4\alpha \cos 7\alpha$.
12. $\sin 3\alpha \sin \alpha$.
13. $\cos \frac{A}{2} \sin \frac{3A}{2}$.
14. $\sin \frac{5A}{2} \cos \frac{7A}{2}$.
15. $2 \cos \frac{2\theta}{3} \cos \frac{5\theta}{3}$.
16. $\sin \frac{\theta}{4} \sin \frac{3\theta}{4}$.
17. $2 \cos 2\beta \cos (\alpha - \beta)$.
18. $2 \sin 3\alpha \sin (\alpha + \beta)$.
19. $2 \sin (2\theta + \phi) \cos (\theta - 2\phi)$.
20. $2 \cos (3\theta + \phi) \sin (\theta - 2\phi)$.
21. $\cos (60° + \alpha) \sin (60° - \alpha)$.

Transformation of sums or differences into products.

131. Since $\sin (A+B) = \sin A \cos B + \cos A \sin B$,

and $\sin (A-B) = \sin A \cos B - \cos A \sin B$;

by addition,

$$\sin (A+B) + \sin (A-B) = 2 \sin A \cos B \quad\text{.........(1)};$$

by subtraction,

$$\sin (A+B) - \sin (A-B) = 2 \cos A \sin B \quad\text{.........(2)}.$$

Again, $\cos (A+B) = \cos A \cos B - \sin A \sin B$,

and $\cos (A-B) = \cos A \cos B + \sin A \sin B$.

By addition,

$$\cos (A+B) + \cos (A-B) = 2 \cos A \cos B \quad\text{.........(3)};$$

by subtraction,

$$\cos (A+B) - \cos (A-B) = -2 \sin A \sin B$$
$$= 2 \sin A \sin (-B) \quad\text{......(4)}.$$

Let $$A+B=C, \text{ and } A-B=D;$$

then $$A=\frac{C+D}{2}, \text{ and } B=\frac{C-D}{2}.$$

By substituting for A and B in the formulæ (1), (2), (3), (4), we obtain

$$\sin C+\sin D=2\sin\frac{C+D}{2}\cos\frac{C-D}{2},$$

$$\sin C-\sin D=2\cos\frac{C+D}{2}\sin\frac{C-D}{2},$$

$$\cos C+\cos D=2\cos\frac{C+D}{2}\cos\frac{C-D}{2},$$

$$\cos C-\cos D=2\sin\frac{C+D}{2}\sin\frac{D-C}{2}.$$

132. In practice, it is more convenient to quote the formulæ we have just obtained verbally as follows :

sum of two sines = 2 sin (*half-sum*) cos (*half-difference*);

difference of two sines = 2 cos (*half-sum*) sin (*half-difference*);

sum of two cosines = 2 cos (*half-sum*) cos (*half-difference*);

difference of two cosines
= 2 sin (*half-sum*) sin (*half-difference reversed*)

Example 1. $$\sin 14\theta+\sin 6\theta=2\sin\frac{14\theta+6\theta}{2}\cos\frac{14\theta-6\theta}{2}$$
$$=2\sin 10\theta\cos 4\theta.$$

Example 2. $$\sin 9A-\sin 7A=2\cos\frac{9A+7A}{2}\sin\frac{9A-7A}{2}$$
$$=2\cos 8A\sin A.$$

Example 3. $$\cos A+\cos 8A=2\cos\frac{9A}{2}\cos\left(-\frac{7A}{2}\right)$$
$$=2\cos\frac{9A}{2}\cos\frac{7A}{2}.$$

Example 4. $$\cos 70^\circ-\cos 10^\circ=2\sin 40^\circ\sin(-30^\circ)$$
$$=-2\sin 40^\circ\sin 30^\circ=-\sin 40^\circ.$$

EXAMPLES. XII. b.

Express in the form of a product

1. $\sin 8\theta+\sin 4\theta$.
2. $\sin 5\theta-\sin\theta$.
3. $\cos 7\theta+\cos 3\theta$.
4. $\cos 9\theta-\cos 11\theta$.
5. $\sin 7a-\sin 5a$.
6. $\cos 3a+\cos 8a$.
7. $\sin 3a+\sin 13a$.
8. $\cos 5a-\cos a$.
9. $\cos 2A+\cos 9A$.
10. $\sin 3A-\sin 11A$.
11. $\cos 10^\circ-\cos 50^\circ$.
12. $\sin 70^\circ+\sin 50^\circ$.

Prove that

13. $\dfrac{\cos a-\cos 3a}{\sin 3a-\sin a}=\tan 2a$.
14. $\dfrac{\sin 2a+\sin 3a}{\cos 2a-\cos 3a}=\cot\dfrac{a}{2}$.
15. $\dfrac{\cos 4\theta-\cos\theta}{\sin\theta-\sin 4\theta}=\tan\dfrac{5\theta}{2}$.
16. $\dfrac{\cos 2\theta-\cos 12\theta}{\sin 12\theta+\sin 2\theta}=\tan 5\theta$.
17. $\sin(60^\circ+A)-\sin(60^\circ-A)=\sin A$.
18. $\cos(30^\circ-A)+\cos(30^\circ+A)=\sqrt{3}\cos A$.
19. $\cos\left(\dfrac{\pi}{4}+a\right)-\cos\left(\dfrac{\pi}{4}-a\right)=-\sqrt{2}\sin a$.
20. $\dfrac{\cos(2a-3\beta)+\cos 3\beta}{\sin(2a-3\beta)+\sin 3\beta}=\cot a$.
21. $\dfrac{\cos(\theta-3\phi)-\cos(3\theta+\phi)}{\sin(3\theta+\phi)+\sin(\theta-3\phi)}=\tan(\theta+2\phi)$.
22. $\dfrac{\sin(a+\beta)-\sin 4\beta}{\cos(a+\beta)+\cos 4\beta}=\tan\dfrac{a-3\beta}{2}$.

133. The eight formulæ proved in this chapter are of the utmost importance and very little further progress can be made until they have been thoroughly learnt. In the first group, the transformation is from products to sums and differences; in the second group, there is the converse transformation from sums and differences to products.

Many examples admit of solution by applying either of these transformations, but it is absolutely necessary that the student should master all the formulæ and apply them with equal readiness.

134. The following examples should be studied with great care.

Example 1. Prove that

$$\sin 5A + \sin 2A - \sin A = \sin 2A\,(2\cos 3A + 1).$$

The first side $= (\sin 5A - \sin A) + \sin 2A$

$$= 2\cos 3A \sin 2A + \sin 2A$$

$$= \sin 2A\,(2\cos 3A + 1).$$

Example 2. Prove that

$$\cos 2\theta \cos\theta - \sin 4\theta \sin\theta = \cos 3\theta \cos 2\theta.$$

The first side $= \frac{1}{2}(\cos 3\theta + \cos\theta) - \frac{1}{2}(\cos 3\theta - \cos 5\theta)$

$$= \frac{1}{2}(\cos\theta + \cos 5\theta)$$

$$= \cos 3\theta \cos 2\theta.$$

Example 3. Find the value of

$$\cos 20^\circ + \cos 100^\circ + \cos 140^\circ.$$

The expression $= \cos 20^\circ + (\cos 100^\circ + \cos 140^\circ)$

$$= \cos 20^\circ + 2\cos 120^\circ \cos 20^\circ$$

$$= \cos 20^\circ + 2\left(-\frac{1}{2}\right)\cos 20^\circ$$

$$= \cos 20^\circ - \cos 20^\circ = 0.$$

Example 4. Express as the product of three sines

$$\sin(\beta + \gamma - \alpha) + \sin(\gamma + \alpha - \beta) + \sin(\alpha + \beta - \gamma) - \sin(\alpha + \beta + \gamma).$$

The expression $= 2\sin\gamma\cos(\beta - \alpha) + 2\cos(\alpha + \beta)\sin(-\gamma)$

$$= 2\sin\gamma\,\{\cos(\beta - \alpha) - \cos(\alpha + \beta)\}$$

$$= 2\sin\gamma\,(2\sin\beta\sin\alpha)$$

$$= 4\sin\alpha\sin\beta\sin\gamma.$$

Example 5. Express $4\cos\alpha\cos\beta\cos\gamma$ as the sum of four cosines.

The expression $= 2\cos\alpha\,\{\cos(\beta + \gamma) + \cos(\beta - \gamma)\}$

$$= 2\cos\alpha\cos(\beta + \gamma) + 2\cos\alpha\cos(\beta - \gamma)$$

$$= \cos(\alpha + \beta + \gamma) + \cos(\alpha - \beta - \gamma) + \cos(\alpha + \beta - \gamma) + \cos(\alpha - \beta + \gamma)$$

$$= \cos(\alpha + \beta + \gamma) + \cos(\beta + \gamma - \alpha) + \cos(\gamma + \alpha - \beta) + \cos(\alpha + \beta - \gamma).$$

Example 6. Prove that $\sin^2 5x - \sin^2 3x = \sin 8x \sin 2x$.

First solution.

$$\begin{aligned}\sin^2 5x - \sin^2 3x &= (\sin 5x + \sin 3x)(\sin 5x - \sin 3x)\\ &= (2\sin 4x \cos x)(2\cos 4x \sin x)\\ &= (2\sin 4x \cos 4x)(2\sin x \cos x)\\ &= \sin 8x \sin 2x.\end{aligned}$$

Second solution.

$$\begin{aligned}\sin 8x \sin 2x &= \frac{1}{2}(\cos 6x - \cos 10x)\\ &= \frac{1}{2}\{1 - 2\sin^2 3x - (1 - 2\sin^2 5x)\}\\ &= \sin^2 5x - \sin^2 3x.\end{aligned}$$

Third solution.

By using the formula of Art. 114 we have at once

$$\sin^2 5x - \sin^2 3x = \sin(5x + 3x)\sin(5x - 3x) = \sin 8x \sin 2x.$$

EXAMPLES. XII. c.

Prove the following identities :

1. $\cos 3A + \sin 2A - \sin 4A = \cos 3A(1 - 2\sin A)$.

2. $\sin 3\theta - \sin\theta - \sin 5\theta = \sin 3\theta(1 - 2\cos 2\theta)$.

3. $\cos\theta + \cos 2\theta + \cos 5\theta = \cos 2\theta(1 + 2\cos 3\theta)$.

4. $\sin a - \sin 2a + \sin 3a = 4\sin\frac{a}{2}\cos a\cos\frac{3a}{2}$.

5. $\sin 3a + \sin 7a + \sin 10a = 4\sin 5a\cos\frac{7a}{2}\cos\frac{3a}{2}$.

6. $\sin A + 2\sin 3A + \sin 5A = 4\sin 3A\cos^2 A$.

7. $\dfrac{\sin 2a + \sin 5a - \sin a}{\cos 2a + \cos 5a + \cos a} = \tan 2a$.

8. $\dfrac{\sin a + \sin 2a + \sin 4a + \sin 5a}{\cos a + \cos 2a + \cos 4a + \cos 5a} = \tan 3a$.

9. $\dfrac{\cos 7\theta + \cos 3\theta - \cos 5\theta - \cos\theta}{\sin 7\theta - \sin 3\theta - \sin 5\theta + \sin\theta} = \cot 2\theta$.

10. $\cos 3A\sin 2A - \cos 4A\sin A = \cos 2A\sin A$.

Prove the following identities:

11. $\cos 5A \cos 2A - \cos 4A \cos 3A = -\sin 2A \sin A.$

12. $\sin 4\theta \cos \theta - \sin 3\theta \cos 2\theta = \sin \theta \cos 2\theta.$

13. $\cos 5^\circ - \sin 25^\circ = \sin 35^\circ.$

[*Use* $\sin 25^\circ = \cos 65^\circ$.]

14. $\sin 65^\circ + \cos 65^\circ = \sqrt{2} \cos 20^\circ.$

15. $\cos 80^\circ + \cos 40^\circ - \cos 20^\circ = 0.$

16. $\sin 78^\circ - \sin 18^\circ + \cos 132^\circ = 0.$

17. $\sin^2 5A - \sin^2 2A = \sin 7A \sin 3A.$

18. $\cos 2A \cos 5A = \cos^2 \dfrac{7A}{2} - \sin^2 \dfrac{3A}{2}.$

19. $\sin(\alpha+\beta+\gamma) + \sin(\alpha-\beta-\gamma) + \sin(\alpha+\beta-\gamma)$
$+ \sin(\alpha-\beta+\gamma) = 4 \sin\alpha \cos\beta \cos\gamma.$

20. $\cos(\beta+\gamma-\alpha) - \cos(\gamma+\alpha-\beta) + \cos(\alpha+\beta-\gamma)$
$- \cos(\alpha+\beta+\gamma) = 4 \sin\alpha \cos\beta \sin\gamma.$

21. $\sin 2\alpha + \sin 2\beta + \sin 2\gamma - \sin 2(\alpha+\beta+\gamma)$
$= 4 \sin(\beta+\gamma) \sin(\gamma+\alpha) \sin(\alpha+\beta).$

22. $\cos\alpha + \cos\beta + \cos\gamma + \cos(\alpha+\beta+\gamma)$
$= 4 \cos \dfrac{\beta+\gamma}{2} \cos \dfrac{\gamma+\alpha}{2} \cos \dfrac{\alpha+\beta}{2}.$

23. $4 \sin A \sin(60^\circ + A) \sin(60^\circ - A) = \sin 3A.$

24. $4 \cos\theta \cos\left(\dfrac{2\pi}{3}+\theta\right) \cos\left(\dfrac{2\pi}{3}-\theta\right) = \cos 3\theta.$

25. $\cos\theta + \cos\left(\dfrac{2\pi}{3}-\theta\right) + \cos\left(\dfrac{2\pi}{3}+\theta\right) = 0.$

26. $\cos^2 A + \cos^2(60^\circ + A) + \cos^2(60^\circ - A) = \dfrac{3}{2}.$

[*Put* $2\cos^2 A = 1 + \cos 2A$.]

27. $\sin^2 A + \sin^2(120^\circ + A) + \sin^2(120^\circ - A) = \dfrac{3}{2}.$

28. $\cos 20^\circ \cos 40^\circ \cos 80^\circ = \dfrac{1}{8}.$

29. $\sin 20^\circ \sin 40^\circ \sin 80^\circ = \dfrac{1}{8}\sqrt{3}.$

135. Many interesting identities can be established connecting the functions of the three angles A, B, C, which satisfy the relation $A+B+C=180°$. In proving these it will be necessary to keep clearly in view the properties of complementary and supplementary angles. [Arts. 39 and 96.]

From the given relation, the sum of any two of the angles is the supplement of the third; thus

$$\sin(B+C)=\sin A, \qquad \cos(A+B)=-\cos C,$$
$$\tan(C+A)=-\tan B, \qquad \cos B=-\cos(C+A),$$
$$\sin C=\sin(A+B), \qquad \cot A=-\cot(B+C).$$

Again, $\frac{A}{2}+\frac{B}{2}+\frac{C}{2}=90°$, so that each half angle is the complement of the sum of the other two; thus

$$\cos\frac{A+B}{2}=\sin\frac{C}{2}, \quad \sin\frac{C+A}{2}=\cos\frac{B}{2}, \quad \tan\frac{B+C}{2}=\cot\frac{A}{2},$$

$$\cos\frac{C}{2}=\sin\frac{A+B}{2}, \quad \sin\frac{A}{2}=\cos\frac{B+C}{2}, \quad \tan\frac{B}{2}=\cot\frac{C+A}{2}.$$

Example 1. If $A+B+C=180°$, prove that

$$\sin 2A+\sin 2B+\sin 2C=4\sin A\sin B\sin C.$$

The first side $=2\sin(A+B)\cos(A-B)+2\sin C\cos C$

$$=2\sin C\cos(A-B)+2\sin C\cos C$$
$$=2\sin C\{\cos(A-B)+\cos C\}$$
$$=2\sin C\{\cos(A-B)-\cos(A+B)\}$$
$$=2\sin C\times 2\sin A\sin B$$
$$=4\sin A\sin B\sin C.$$

Example 2. If $A+B+C=180°$, prove that

$$\tan A+\tan B+\tan C=\tan A\tan B\tan C.$$

Since $A+B$ is the supplement of C, we have

$$\tan(A+B)=-\tan C;$$

$$\therefore \quad \frac{\tan A+\tan B}{1-\tan A\tan B}=-\tan C;$$

whence by multiplying up and rearranging,

$$\tan A+\tan B+\tan C=\tan A\tan B\tan C.$$

Example 3. If $A+B+C=180°$, prove that

$$\cos A+\cos B+\cos C=1+4\sin\frac{A}{2}\sin\frac{B}{2}\sin\frac{C}{2}.$$

The first side $=2\cos\frac{A+B}{2}\cos\frac{A-B}{2}+\cos C$

$$=2\sin\frac{C}{2}\cos\frac{A-B}{2}+1-2\sin^2\frac{C}{2}$$

$$=1+2\sin\frac{C}{2}\left(\cos\frac{A-B}{2}-\sin\frac{C}{2}\right)$$

$$=1+2\sin\frac{C}{2}\left(\cos\frac{A-B}{2}-\cos\frac{A+B}{2}\right)$$

$$=1+2\sin\frac{C}{2}\left(2\sin\frac{A}{2}\sin\frac{B}{2}\right)$$

$$=1+4\sin\frac{A}{2}\sin\frac{B}{2}\sin\frac{C}{2}.$$

EXAMPLES. XII. d.

If $A+B+C=180°$, prove that

1. $\sin 2A-\sin 2B+\sin 2C=4\cos A\sin B\cos C$.

2. $\sin 2A-\sin 2B-\sin 2C=-4\sin A\cos B\cos C$.

3. $\sin A+\sin B+\sin C=4\cos\frac{A}{2}\cos\frac{B}{2}\cos\frac{C}{2}$.

4. $\sin A+\sin B-\sin C=4\sin\frac{A}{2}\sin\frac{B}{2}\cos\frac{C}{2}$.

5. $\cos A-\cos B+\cos C=4\cos\frac{A}{2}\sin\frac{B}{2}\cos\frac{C}{2}-1$.

6. $\dfrac{\sin B+\sin C-\sin A}{\sin A+\sin B+\sin C}=\tan\frac{B}{2}\tan\frac{C}{2}$.

7. $\tan\frac{B}{2}\tan\frac{C}{2}+\tan\frac{C}{2}\tan\frac{A}{2}+\tan\frac{A}{2}\tan\frac{B}{2}=1$.

[*Use* $\tan\frac{A+B}{2}=\cot\frac{C}{2}$, *and therefore* $\tan\frac{A+B}{2}\tan\frac{C}{2}=1$.]

If $A+B+C=180^\circ$, prove that

8. $\dfrac{1+\cos A-\cos B+\cos C}{1+\cos A+\cos B-\cos C}=\tan\dfrac{B}{2}\cot\dfrac{C}{2}$.

9. $\cos 2A+\cos 2B+\cos 2C+4\cos A\cos B\cos C+1=0$.

10. $\cot B\cot C+\cot C\cot A+\cot A\cot B=1$.

11. $(\cot B+\cot C)(\cot C+\cot A)(\cot A+\cot B)$
$=\operatorname{cosec} A\operatorname{cosec} B\operatorname{cosec} C$.

12. $\cos^2 A+\cos^2 B+\cos^2 C+2\cos A\cos B\cos C=1$.
[*Use* $2\cos^2 A=1+\cos 2A$.]

13. $\sin^2\dfrac{A}{2}+\sin^2\dfrac{B}{2}+\sin^2\dfrac{C}{2}=1-2\sin\dfrac{A}{2}\sin\dfrac{B}{2}\sin\dfrac{C}{2}$.

14. $\cos^2 2A+\cos^2 2B+\cos^2 2C=1+2\cos 2A\cos 2B\cos 2C$.

15. $\dfrac{\cot B+\cot C}{\tan B+\tan C}+\dfrac{\cot C+\cot A}{\tan C+\tan A}+\dfrac{\cot A+\cot B}{\tan A+\tan B}=1$.

16. $\dfrac{\tan A+\tan B+\tan C}{(\sin A+\sin B+\sin C)^2}=\dfrac{\tan\dfrac{A}{2}\tan\dfrac{B}{2}\tan\dfrac{C}{2}}{2\cos A\cos B\cos C}$.

***136.** The following examples further illustrate the formulæ proved in this and the preceding chapter.

Example 1. Prove that $\cot(A+15^\circ)-\tan(A-15^\circ)=\dfrac{4\cos 2A}{2\sin 2A+1}$.

$$\text{The first side}=\frac{\cos(A+15^\circ)}{\sin(A+15^\circ)}-\frac{\sin(A-15^\circ)}{\cos(A-15^\circ)}$$

$$=\frac{\cos(A+15^\circ)\cos(A-15^\circ)-\sin(A+15^\circ)\sin(A-15^\circ)}{\sin(A+15^\circ)\cos(A-15^\circ)}$$

$$=\frac{\cos\{(A+15^\circ)+(A-15^\circ)\}}{\sin(A+15^\circ)\cos(A-15^\circ)}$$

$$=\frac{2\cos 2A}{2\sin(A+15^\circ)\cos(A-15^\circ)}=\frac{2\cos 2A}{\sin 2A+\sin 30^\circ}$$

$$=\frac{4\cos 2A}{2\sin 2A+1}.$$

NOTE. In dealing with expressions which involve numerical angles it is usually advisable to effect some simplification before substituting the known values of the functions of the angles, especially if these contain surds.

Example 2. If $A+B+C=\pi$, prove that

$$\cos\frac{A}{2}+\cos\frac{B}{2}+\cos\frac{C}{2}=4\cos\frac{\pi-A}{4}\cos\frac{\pi-B}{4}\cos\frac{\pi-C}{4}.$$

The second side $=2\cos\frac{\pi-A}{4}\left[\cos\frac{2\pi-(B+C)}{4}+\cos\frac{B-C}{4}\right]$

$$=2\cos\frac{\pi-A}{4}\cos\frac{\pi+A}{4}+2\cos\frac{\pi-A}{4}\cos\frac{B-C}{4}$$

$$=\left(\cos\frac{\pi}{2}+\cos\frac{A}{2}\right)+2\cos\frac{B+C}{4}\cos\frac{B-C}{4}$$

$$=\cos\frac{A}{2}+\cos\frac{B}{2}+\cos\frac{C}{2}.$$

*EXAMPLES. XII. e.

Prove the following identities:

1. $\cos(\alpha+\beta)\sin(\alpha-\beta)+\cos(\beta+\gamma)\sin(\beta-\gamma)$
$+\cos(\gamma+\delta)\sin(\gamma-\delta)+\cos(\delta+\alpha)\sin(\delta-\alpha)=0.$

2. $\dfrac{\sin(\beta-\gamma)}{\sin\beta\sin\gamma}+\dfrac{\sin(\gamma-\alpha)}{\sin\gamma\sin\alpha}+\dfrac{\sin(\alpha-\beta)}{\sin\alpha\sin\beta}=0.$

3. $\dfrac{\sin\alpha+\sin\beta+\sin(\alpha+\beta)}{\sin\alpha+\sin\beta-\sin(\alpha+\beta)}=\cot\dfrac{\alpha}{2}\cot\dfrac{\beta}{2}.$

4. $\sin\alpha\cos(\beta+\gamma)-\sin\beta\cos(\alpha+\gamma)=\cos\gamma\sin(\alpha-\beta).$

5. $\cos\alpha\cos(\beta+\gamma)-\cos\beta\cos(\alpha+\gamma)=\sin\gamma\sin(\alpha-\beta).$

6. $(\cos A-\sin A)(\cos 2A-\sin 2A)=\cos A-\sin 3A.$

7. If $\tan\theta=\dfrac{b}{a}$, prove that $a\cos 2\theta+b\sin 2\theta=a.$

[*See Art.* 124.]

8. Prove that $\sin 2A+\cos 2A=\dfrac{(1+\tan A)^2-2\tan^2 A}{1+\tan^2 A}.$

9. Prove that $\sin 4A=\dfrac{4\tan A(1-\tan^2 A)}{(1+\tan^2 A)^2}.$

10. If $A+B=45°$, prove that
$(1+\tan A)(1+\tan B)=2.$

Prove the following identities:

11. $\cot(15^\circ - A) + \tan(15^\circ + A) = \dfrac{4\cos 2A}{1 - 2\sin 2A}$.

12. $\cot(15^\circ + A) + \tan(15^\circ + A) = \dfrac{4}{\cos 2A + \sqrt{3}\sin 2A}$.

13. $\tan(A + 30^\circ)\tan(A - 30^\circ) = \dfrac{1 - 2\cos 2A}{1 + 2\cos 2A}$.

14. $(2\cos A + 1)(2\cos A - 1)(2\cos 2A - 1) = 2\cos 4A + 1$.

15. $\tan(\beta - \gamma) + \tan(\gamma - a) + \tan(a - \beta)$
$$= \tan(\beta - \gamma)\tan(\gamma - a)\tan(a - \beta).$$

16. $\sin(\beta - \gamma) + \sin(\gamma - a) + \sin(a - \beta)$
$$+ 4\sin\frac{\beta - \gamma}{2}\sin\frac{\gamma - a}{2}\sin\frac{a - \beta}{2} = 0.$$

17. $\cos^2(\beta - \gamma) + \cos^2(\gamma - a) + \cos^2(a - \beta)$
$$= 1 + 2\cos(\beta - \gamma)\cos(\gamma - a)\cos(a - \beta).$$

18. $\cos^2 a + \cos^2\beta - 2\cos a\cos\beta\cos(a + \beta) = \sin^2(a + \beta)$.

19. $\sin^2 a + \sin^2\beta + 2\sin a\sin\beta\cos(a + \beta) = \sin^2(a + \beta)$.

20. $\cos 12^\circ + \cos 60^\circ + \cos 84^\circ = \cos 24^\circ + \cos 48^\circ$.

If $A + B + C = 180^\circ$, shew that

21. $\cos\dfrac{A}{2} + \cos\dfrac{B}{2} + \cos\dfrac{C}{2} = 4\cos\dfrac{B + C}{4}\cos\dfrac{C + A}{4}\cos\dfrac{A + B}{4}$.

22. $\cos\dfrac{A}{2} - \cos\dfrac{B}{2} + \cos\dfrac{C}{2} = 4\cos\dfrac{\pi + A}{4}\cos\dfrac{\pi - B}{4}\cos\dfrac{\pi + C}{4}$.

23. $\sin\dfrac{A}{2} + \sin\dfrac{B}{2} + \sin\dfrac{C}{2} = 1 + 4\sin\dfrac{\pi - A}{4}\sin\dfrac{\pi - B}{4}\sin\dfrac{\pi - C}{4}$.

If $a + \beta + \gamma = \dfrac{\pi}{2}$, shew that

24. $\dfrac{\sin 2a + \sin 2\beta + \sin 2\gamma}{\sin 2a + \sin 2\beta - \sin 2\gamma} = \cot a\cot\beta$.

25. $\tan\beta\tan\gamma + \tan\gamma\tan a + \tan a\tan\beta = 1$.

EXAMPLES. XII. f.

(*Easy Miscellaneous Examples on Chapters XI and XII.*)

Prove the following identities:

1. $\operatorname{cosec}(\alpha+\beta)=\dfrac{\operatorname{cosec}\alpha\operatorname{cosec}\beta}{\cot\alpha+\cot\beta}$.

2. $\operatorname{cosec}(\alpha-\beta)=\dfrac{1+\tan\alpha\tan\beta}{\sec\alpha\sec\beta}$.

3. $\operatorname{cosec}2\theta-\cot 2\theta=\tan\theta$.

4. $2\cos^2(45^\circ-A)=1+\sin 2A$.

5. $2\cos 2x\operatorname{cosec}3x=\operatorname{cosec}x-\operatorname{cosec}3x$.

6. $\cos(A+B)\cos(A-B)+1=\cos^2 A+\cos^2 B$.

7. $1+\cos 4A=2\cos 2A(1-2\sin^2 A)$.

8. Express $\cot 2A$ in terms of $\tan A$, and $\tan 2A$ in terms of $\cot A$.

9. If $\tan\dfrac{A}{2}=t$, prove that

(i) $\sin A+\tan A=\dfrac{4t}{1-t^4}$; (ii) $\sec A+\tan A=\dfrac{(1+t)^2}{1-t^2}$.

10. Express $\dfrac{\sin 3A}{\sin 2A-\sin A}$ in terms of $\cos A$.

11. If $\sin\alpha=\cdot 28$ and $\cos\beta=\cdot 6$, find the value of $\cos(\alpha+\beta)$. Thence from the table of cosines find $\alpha+\beta$ to the nearest minute. Check the result by finding α and β separately from the data.

12. If $\alpha=\beta+\gamma$, shew that

$$\sin(\alpha+\beta+\gamma)+\sin(\alpha+\beta-\gamma)+\sin(\alpha-\beta+\gamma)=4\sin\alpha\cos\beta\cos\gamma.$$

13. Prove that $\cos 57^\circ+\sin 27^\circ=\cos 3^\circ$, and verify the relation by means of the Tables.

14. Express $4\sin 5\alpha\cos 3\alpha\cos 2\alpha$ as the sum of three sines.

15. By means of the Tables find approximately the numerical value of the expression $4\cos 20^\circ\cos 30^\circ\cos 40^\circ$.

[First express the product as the sum of three cosines.]

16. Find the smallest value of θ which satisfies the equation

$$\sin 4\theta \cos \theta = \frac{1}{4} + \sin \frac{5\theta}{2} \cos \frac{5\theta}{2}.$$

17. Prove the identities:

(i) $(x \tan \alpha + y \cot \alpha)(x \cot \alpha + y \tan \alpha) = (x+y)^2 + 4xy \cot^2 2\alpha$;

(ii) $\cos \beta \cos (2\alpha - \beta) = \cos^2 \alpha - \sin^2 (\alpha - \beta)$;

(iii) $\cos \alpha + \cos 3\alpha + \cos 5\alpha + \cos 7\alpha = \frac{1}{2} \sin 8\alpha \operatorname{cosec} \alpha$.

18. If $\cos \theta = \cdot 8$, find the numerical values of $\sin 2\theta$ and $\cos 2\theta$. Check the results by first finding θ from the table of cosines.

19. If $2A + B = 90°$, prove that $\cos A = \sqrt{\frac{1 + \sin B}{2}}$.

20. Prove that

(i) $\frac{1}{\sin 10°} - \frac{\sqrt{3}}{\cos 10°} = 4$; (ii) $\sin 54° = \sin 162° + \sin 30°$.

21. If $B + C = 180°$, prove that

$$2(1 - \sin B \sin C) = \cos^2 B + \cos^2 C.$$

22. If $A + B + C = 180°$, prove that

$$1 - 2 \sin B \sin C \cos A + \cos^2 A = \cos^2 B + \cos^2 C.$$

23. If $A + B + C = 0$, prove that

$$1 + 2 \sin B \sin C \cos A + \cos^2 A = \cos^2 B + \cos^2 C.$$

24. Prove that

$$\sin^2 A - \cos^2 A \cos 2B = \sin^2 B - \cos^2 B \cos 2A.$$

25. Prove that $\tan 50° - \tan 40° = 2 \tan 10°$. Verify the relation by using the table of tangents.

26. If $\cot \theta = \cdot 5$ find $\sin 2\theta$ and $\cos 2\theta$. Check the values as in Ex. 18.

27. By means of the Tables, find the two smallest values of θ which satisfy the equation

$$\cdot 362 \cos \theta + \sin \theta = 1.$$

[From the table of tangents find α such that $\tan \alpha = \cdot 362$; then the equation may be written $\sin (\alpha + \theta) = \cos \alpha = \sin (90° - \alpha)$.]

CHAPTER XIII.

RELATIONS BETWEEN THE SIDES AND ANGLES OF A TRIANGLE.

137. *In any triangle the sides are proportional to the sines of the opposite angles; that is,*

$$\frac{a}{\sin A} = \frac{b}{\sin B} = \frac{c}{\sin C}.$$

(1) Let the triangle ABC be acute-angled.

From A draw AD perpendicular to the opposite side; then

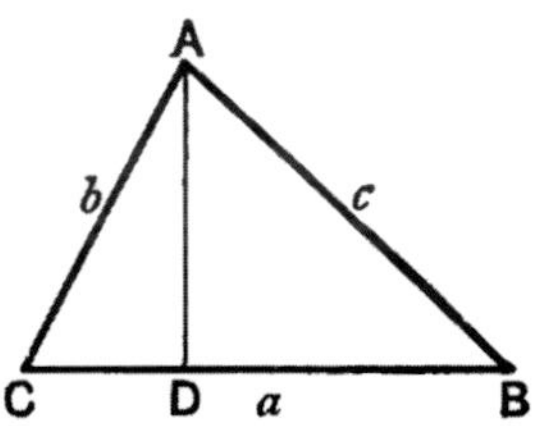

$$AD = AB \sin ABD = c \sin B,$$

and $$AD = AC \sin ACD = b \sin C;$$

$$\therefore \; b \sin C = c \sin B,$$

that is, $$\frac{b}{\sin B} = \frac{c}{\sin C}.$$

(2) Let the triangle ABC have an obtuse angle B.

Draw AD perpendicular to CB produced; then

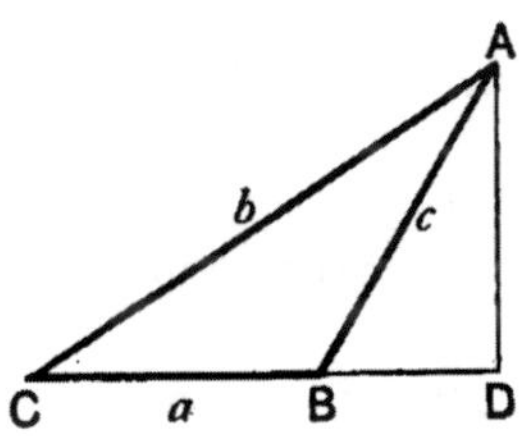

$$AD = AC \sin ACD = b \sin C,$$

and $$AD = AB \sin ABD$$
$$= c \sin (180° - B) = c \sin B;$$
$$\therefore \; b \sin C = c \sin B;$$

that is, $$\frac{b}{\sin B} = \frac{c}{\sin C}.$$

In like manner it may be proved that either of these ratios is equal to $\frac{a}{\sin A}$.

Thus $$\frac{a}{\sin A} = \frac{b}{\sin B} = \frac{c}{\sin C}.$$

138. *To find an expression for one side (c) of a triangle in terms of the other two sides and the included angle (C).*

(1) Let C be an acute angle.

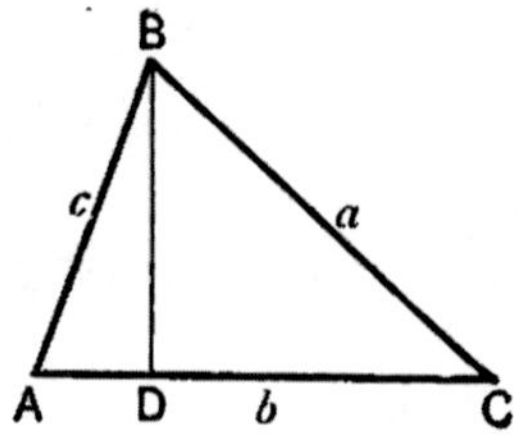

Draw BD perpendicular to AC; then by Euc. II. 13,

$$AB^2 = BC^2 + CA^2 - 2AC \,.\, CD;$$

$$\therefore \; c^2 = a^2 + b^2 - 2b \,.\, a\cos C$$

$$= a^2 + b^2 - 2ab\cos C.$$

(2) Let C be an obtuse angle.

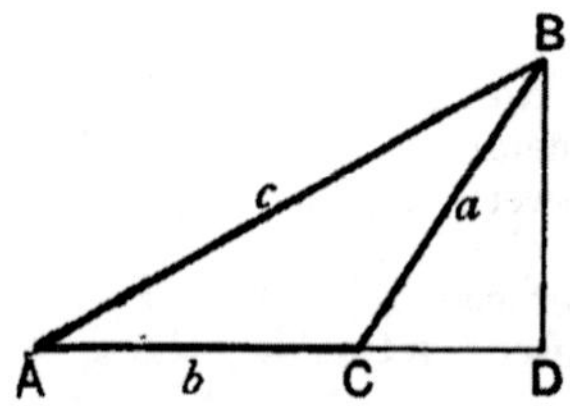

Draw BD perpendicular to AC produced; then by Euc. II. 12,

$$AB^2 = BC^2 + CA^2 + 2AC \,.\, CD;$$

$$\therefore \; c^2 = a^2 + b^2 + 2b \,.\, a\cos BCD$$

$$= a^2 + b^2 + 2ab\cos(180° - C)$$

$$= a^2 + b^2 - 2ab\cos C.$$

Hence in each case, $c^2 = a^2 + b^2 - 2ab\cos C$.

Similarly it may be shewn that

$$a^2 = b^2 + c^2 - 2bc\cos A,$$

and $$b^2 = c^2 + a^2 - 2ca\cos B.$$

139. From the formulæ of the last article, we obtain

$$\cos A = \frac{b^2 + c^2 - a^2}{2bc}; \quad \cos B = \frac{c^2 + a^2 - b^2}{2ca}; \quad \cos C = \frac{a^2 + b^2 - c^2}{2ab}.$$

These results enable us to find the cosines of the angles when the numerical values of the sides are given.

140. *To express one side of a triangle in terms of the adjacent angles and the other two sides.*

(1) Let ABC be an acute-angled triangle.

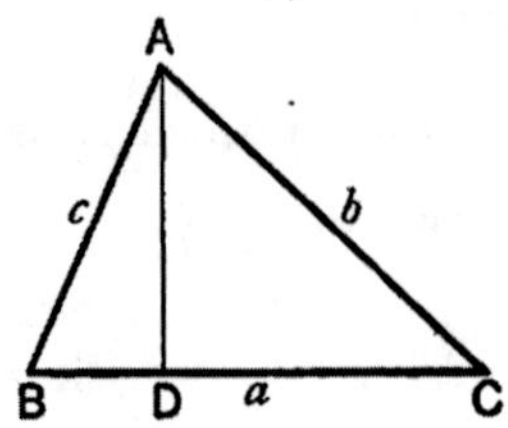

Draw AD perpendicular to BC; then

$$BC = BD + CD$$

$$= AB\cos ABD + AC\cos ACD;$$

that is, $$a = c\cos B + b\cos C.$$

(2) Let the triangle ABC have an obtuse angle C.

Draw AD perpendicular to BC produced; then

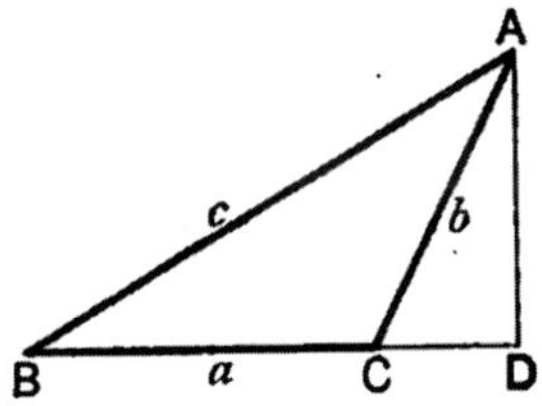

$$BC = BD - CD$$
$$= AB \cos ABD - AC \cos ACD;$$
$$\therefore\ a = c \cos B - b \cos (180° - C)$$
$$= c \cos B + b \cos C.$$

Thus in each case $a = b \cos C + c \cos B$.

Similarly it may be shewn that

$$b = c \cos A + a \cos C, \text{ and } c = a \cos B + b \cos A.$$

NOTE. The formulæ we have proved in this chapter are quite general and may be regarded as the fundamental relations subsisting between the sides and angles of a triangle. The modified forms which they assume in the case of right-angled triangles have already been considered in Chap. V.; it will therefore be unnecessary in the present chapter to make any direct reference to right-angled triangles.

***141.** The sets of formulæ in Arts. 137, 138, and 140 have been established independently of one another; they are however not independent, for from any one set the other two may be derived by the help of the relation $A + B + C = 180°$.

For instance, suppose we have proved as in Art. 137 that

$$\frac{a}{\sin A} = \frac{b}{\sin B} = \frac{c}{\sin C};$$

then since $\sin A = \sin (B + C) = \sin B \cos C + \sin C \cos B$;

$$\therefore\ 1 = \frac{\sin B}{\sin A} \cos C + \frac{\sin C}{\sin A} \cos B;$$

$$\therefore\ 1 = \frac{b}{a} \cos C + \frac{c}{a} \cos B;$$

$$\therefore\ a = b \cos C + c \cos B.$$

Similarly, we may prove that

$$b = c \cos A + a \cos C, \text{ and } c = a \cos B + b \cos A.$$

Multiplying these last three equations by a, b, $-c$ respectively and adding, we have

$$a^2 + b^2 - c^2 = 2ab \cos C;$$
$$\therefore\ c^2 = a^2 + b^2 - 2ab \cos C.$$

Similarly the other relations of Art. 138 may be deduced.

Solution of Triangles.

142. When any three parts of a triangle are given, provided that one at least of these is a side, the relations we have proved enable us to find the numerical values of the unknown parts. For from any equation which connects four quantities three of which are known the fourth may be found. Thus if c, a, B are given, we can find b from the formula

$$b^2=c^2+a^2-2ca\cos B;$$

and if B, C, b are given, we find c from the formula

$$\frac{c}{\sin C}=\frac{b}{\sin B}.$$

We may remark that if the three angles alone are given, the formula

$$\frac{a}{\sin A}=\frac{b}{\sin B}=\frac{c}{\sin C}$$

enables us to find the *ratios* of the sides but not their actual *lengths*, and thus the triangle cannot be completely solved. In such a case there may be an infinite number of equiangular triangles all satisfying the data of the question.

143. CASE I. *To solve a triangle having given the three sides.*

The angles A and B may be found from the formulæ

$$\cos A=\frac{b^2+c^2-a^2}{2bc}, \text{ and } \cos B=\frac{c^2+a^2-b^2}{2ca};$$

then the angle C is known from the equation $C=180°-A-B$.

Example 1. If $a=7$, $b=5$, $c=8$, find the angles A and B, having given that $\cos 38° 11'=\frac{11}{14}$.

$$\cos A=\frac{b^2+c^2-a^2}{2bc}=\frac{5^2+8^2-7^2}{2\times5\times8}=\frac{40}{2\times5\times8}=\frac{1}{2};$$

$$\therefore\ A=60°.$$

$$\cos B=\frac{c^2+a^2-b^2}{2ca}=\frac{8^2+7^2-5^2}{2\times8\times7}=\frac{88}{2\times8\times7}=\frac{11}{14};$$

$$\therefore\ B=38° 11'.$$

Example 2. Find from the Tables the greatest angle of the triangle whose sides are 6, 13, 11.

Let $a=6$, $b=13$, $c=11$. Since the greatest angle is opposite to the greatest side the required angle is B.

And

$$\cos B=\frac{c^2+a^2-b^2}{2ca}=\frac{11^2+6^2-13^2}{2\times 11\times 6}=\frac{-12}{2\times 11\times 6}=-\cdot 0909$$

$$=-\cos 84^\circ\, 47', \text{ from the Tables;}$$

$$\therefore\ B=180^\circ-84^\circ\, 47'=95^\circ\, 13'.$$

Thus the required angle is $95^\circ\, 13'$.

144. CASE II. *To solve a triangle having given two sides and the included angle.*

Let b, c, A be given; then a can be found from the formula

$$a^2=b^2+c^2-2bc\cos A.$$

We may now obtain B from either of the formulæ

$$\cos B=\frac{c^2+a^2-b^2}{2ca}, \text{ or } \sin B=\frac{b\sin A}{a};$$

then C is known from the equation $C=180^\circ-A-B$.

Example. If $a=3$, $b=7$, $C=98^\circ\, 13'$, solve the triangle, with the aid of the Tables.

$$c^2=a^2+b^2-2ab\cos C$$

$$=9+49-2\times 3\times 7\cos 98^\circ\, 13'.$$

Now $\quad \cos 98^\circ\, 13'=-\sin 8^\circ\, 13'$ [Art. 98]

$$=-\cdot 1429, \text{ from the Tables,}$$

$$\therefore\ c^2=58+6\cdot 006=64, \text{ approximately;}$$

$$\therefore\ c=8.$$

$$\cos B=\frac{c^2+a^2-b^2}{2ca}=\frac{64+9-49}{2\times 8\times 3}=\frac{24}{2\times 8\times 3}=\frac{1}{2};$$

$$\therefore\ B=60^\circ.$$

$$\therefore\ B=180^\circ-60^\circ-98^\circ\, 13'=21^\circ\, 47'.$$

145. CASE III. *To solve a triangle having given two angles and a side.*

Let B, C, a be given.

The angle A is found from $A=180°-B-C$; and the sides b and c from

$$b=\frac{a\sin B}{\sin A} \text{ and } c=\frac{a\sin C}{\sin A}.$$

Example. If $A=105°$, $C=60°$, $b=4$, solve the triangle.

$$B=180°-105°-60°=15°.$$

$$\therefore\ c=\frac{b\sin C}{\sin B}=\frac{4\sin 60°}{\sin 15°}=\frac{4\sqrt{3}}{2}\cdot\frac{2\sqrt{2}}{\sqrt{3}-1}=\frac{4\sqrt{6}}{\sqrt{3}-1}$$

$$=\frac{4\sqrt{6}(\sqrt{3}+1)}{3-1}=2\sqrt{6}(\sqrt{3}+1);$$

$$\therefore\ c=6\sqrt{2}+2\sqrt{6}.$$

$$a=\frac{b\sin A}{\sin B}=\frac{4\sin 105°}{\sin 15°}=\frac{4\sin 75°}{\sin 15°}$$

$$=4\times\frac{\sqrt{3}+1}{2\sqrt{2}}\times\frac{2\sqrt{2}}{\sqrt{3}-1}=\frac{4(\sqrt{3}+1)}{\sqrt{3}-1};$$

$$\therefore\ a=4(2+\sqrt{3}).$$

EXAMPLES. XIII. a.

(*Tables must be used for Examples marked with an asterisk.*)

1. If $a=15$, $b=7$, $c=13$, find C.
2. If $a=7$, $b=3$, $c=5$, find A.
3. If $a=5$, $b=5\sqrt{3}$, $c=5$, find the angles.
4. If $a=25$, $b=31$, $c=7\sqrt{2}$, find A.
5. The sides of a triangle are 2, $2\frac{2}{3}$, $3\frac{1}{3}$, find the greatest angle.
6. Solve the triangle when $a=\sqrt{3}+1$, $b=2$, $c=\sqrt{6}$.
7. Solve the triangle when $a=\sqrt{2}$, $b=2$, $c=\sqrt{3}-1$.

*8. If $a=8$, $b=5$, $c=\sqrt{19}$, find C.

*9. If the sides are as 4 : 7 : 5, find the greatest angle.

10. If $a=2$, $b=\sqrt{3}+1$, $C=60°$, find c.

11. Given $a=3$, $c=5$, $B=120°$, find b.

*12. Given $b=7$, $c=6$, $A=75° \, 31'$, find a.

*13. If $b=8$, $c=11$, $A=93° \, 35'$, find a.

*14. If $a=7$, $c=3$, $B=123° \, 12'$, find b.

15. Solve the triangle when $a=2\sqrt{6}$, $c=6-2\sqrt{3}$, $B=75°$.

16. Solve the triangle when $A=72°$, $b=2$, $c=\sqrt{5}+1$.

17. Given $A=75°$, $B=30°$, $b=\sqrt{8}$, solve the triangle.

18. If $B=60°$, $C=15°$, $b=\sqrt{6}$, solve the triangle.

19. If $A=45°$, $B=105°$, $c=\sqrt{2}$, solve the triangle.

20. Given $A=45°$, $B=60°$, shew that $c : a=\sqrt{3}+1 : 2$.

21. If $C=120°$, $c=2\sqrt{3}$, $a=2$, find b.

22. If $B=60°$, $a=3$, $b=3\sqrt{3}$, find c.

23. Given $(a+b+c)(b+c-a)=3bc$, find A.

24. Find the angles of the triangle whose sides are

$$3+\sqrt{3}, \quad 2\sqrt{3}, \quad \sqrt{6}.$$

25. Find the angles of the triangle whose sides are

$$\frac{\sqrt{3}+1}{2\sqrt{2}}, \quad \frac{\sqrt{3}-1}{2\sqrt{2}}, \quad \frac{\sqrt{3}}{2}.$$

26. Two sides of a triangle are $\dfrac{1}{\sqrt{6}-\sqrt{2}}$ and $\dfrac{1}{\sqrt{6}+\sqrt{2}}$, and the included angle is $60°$: solve the triangle.

146. When an angle of a triangle is obtained through the medium of the sine there may be ambiguity, for the sines of supplementary angles are equal in magnitude and are of the same sign, so that there are two angles less than 180° which have the same sine. When an angle is obtained through the medium of the cosine there is no ambiguity, for there is only one angle less than 180° whose cosine is equal to a given quantity.

Thus if $\quad \sin A=\frac{1}{2}$, then $A=30°$ or $150°$;

if $\quad \cos A=\frac{1}{2}$, then $A=60°$.

Example. If $C=60°$, $b=2\sqrt{3}$, $c=3\sqrt{2}$, find A.

From the equation $$\sin B=\frac{b\sin C}{c},$$

we have $$\sin B=\frac{2\sqrt{3}}{3\sqrt{2}}\cdot\frac{\sqrt{3}}{2}=\frac{1}{\sqrt{2}};$$

$$\therefore\ B=45° \text{ or } 135°.$$

The value $B=135°$ is inadmissible, for in this case the sum of B and C would be greater than 180°.

Thus $$A=180°-60°-45°=75°.$$

147. CASE IV. *To solve a triangle having given two sides and an angle opposite to one of them.*

Let a, b, A be given; then B is to be found from the equation

$$\sin B=\frac{b}{a}\sin A.$$

(i) If $a<b\sin A$, then $\frac{b\sin A}{a}>1$, so that $\sin B>1$, which is impossible. Thus there is no solution.

(ii) If $a=b\sin A$, then $\frac{b\sin A}{a}=1$, so that $\sin B=1$, and B has only the value 90°.

(iii) If $a>b\sin A$, then $\frac{b\sin A}{a}<1$, and two values for B may be found from $\sin B=\frac{b\sin A}{a}$. These values are supplementary, so that one angle is acute, the other obtuse.

(1) If $a<b$, then $A<B$, and therefore B may either be acute or obtuse, so that both values are admissible. This is known as **the ambiguous case.**

(2) If $a=b$, then $A=B$; and if $a>b$, then $A>B$; in either case B cannot be obtuse, and therefore only the smaller value of B is admissible.

When B is found, C is determined from $C=180°-A-B$. Finally, c may be found from the equation $c=\frac{a\sin C}{\sin A}$.

From the foregoing investigation it appears that *the only case in which an ambiguous solution can arise is when the smaller of the two given sides is opposite to the given angle.*

148. *To discuss the Ambiguous Case geometrically.*

Let a, b, A be the given parts. Take a line AX unlimited towards X; make $\angle XAC$ equal to A, and AC equal to b. Draw CD perpendicular to AX, then $CD = b \sin A$.

With centre C and radius equal to a describe a circle.

(i) If $a < b \sin A$, the circle will not meet AX; thus no triangle can be constructed with the given parts.

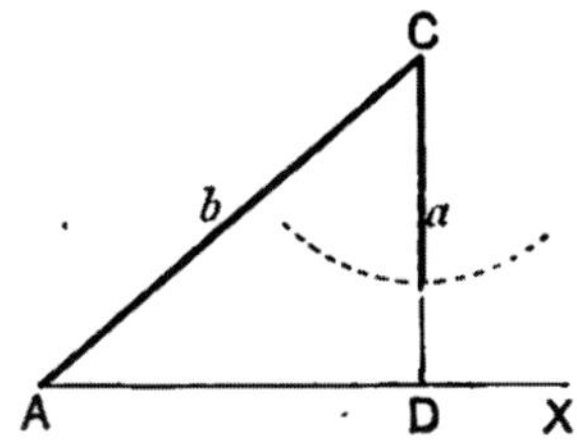

(ii) If $a = b \sin A$, the circle will *touch* AX at D; thus there is a right-angled triangle with the given parts.

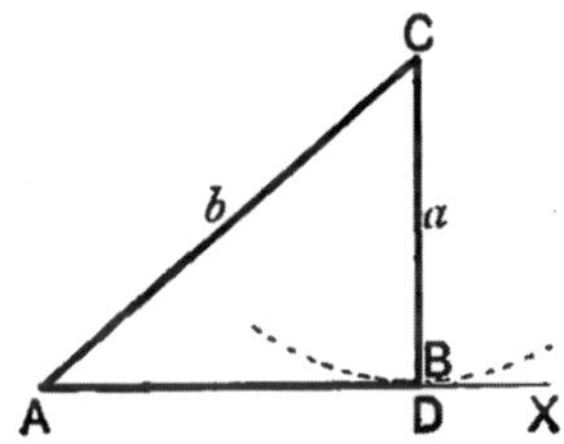

(iii) If $a > b \sin A$, the circle will cut AX in two points B_1, B_2.

(1) These points will be both on the same side of A, when $a < b$, in which case there are two solutions, namely the triangles

$$AB_1C, \quad AB_2C.$$

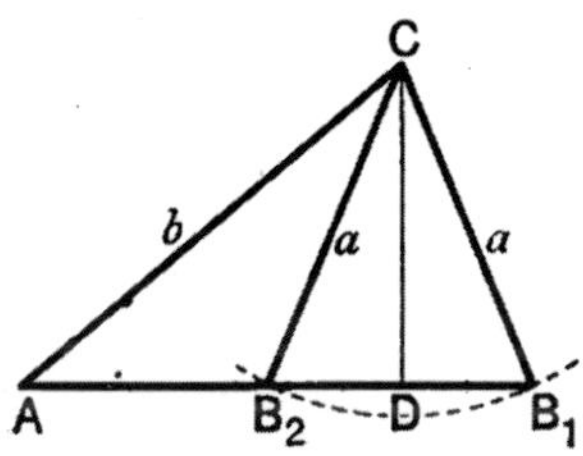

This is *the Ambiguous Case.*

(2) The points B_1, B_2 will be on opposite sides of A when $a > b$.

In this case there is only one solution, for the angle CAB_2 is *the supplement of the given angle,* and thus the triangle AB_2C does not satisfy the data.

(3) If $a = b$, the point B_2 coincides with A, so that there is only one solution.

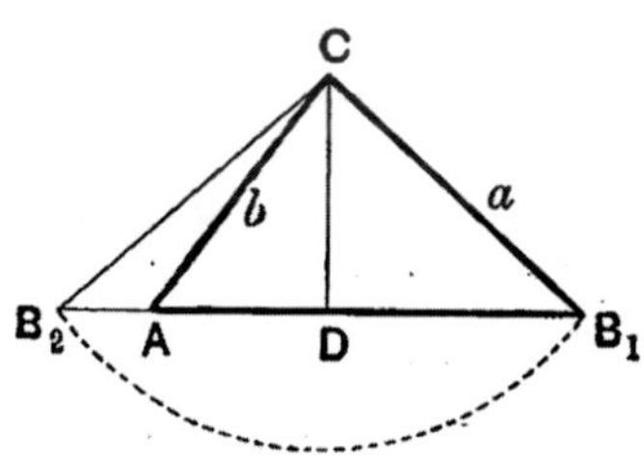

Example. Given $B=45°$, $c=\sqrt{12}$, $b=\sqrt{8}$, solve the triangle.

We have $$\sin C=\frac{c\sin B}{b}=\frac{2\sqrt{3}}{2\sqrt{2}}\cdot\frac{1}{\sqrt{2}}=\frac{\sqrt{3}}{2}.$$

$$\therefore C=60° \text{ or } 120°,$$

and since $b<c$, both these values are admissible. The two triangles which satisfy the data are shewn in the figure.

Denote the sides BC_1, BC_2 by a_1, a_2, and the angles BAC_1, BAC_2 by A_1, A_2 respectively.

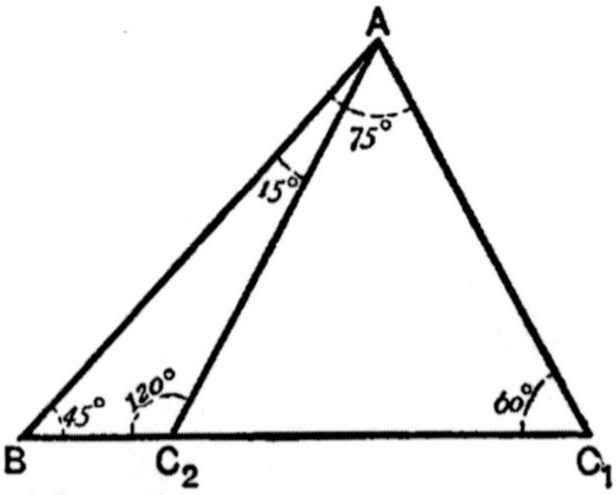

(i) In the $\triangle ABC_1$, $\angle A_1=75°$;

hence $$a_1=\frac{b\sin A_1}{\sin B}=\frac{2\sqrt{2}}{\frac{1}{\sqrt{2}}}\cdot\frac{\sqrt{3}+1}{2\sqrt{2}}=\sqrt{2}(\sqrt{3}+1).$$

(ii) In the $\triangle ABC_2$, $\angle A_2=15°$;

hence $$a_2=\frac{b\sin A_2}{\sin B}=\frac{2\sqrt{2}}{\frac{1}{\sqrt{2}}}\cdot\frac{\sqrt{3}-1}{2\sqrt{2}}=\sqrt{2}(\sqrt{3}-1).$$

Thus the complete solution is $$\begin{cases} C=60°, \text{ or } 120°; \\ A=75°, \text{ or } 15°; \\ a=\sqrt{6}+\sqrt{2}, \text{ or } \sqrt{6}-\sqrt{2}. \end{cases}$$

EXAMPLES. XIII. b.

1. Given $a=1$, $b=\sqrt{3}$, $A=30°$, solve the triangle.

2. Given $b=3\sqrt{2}$, $c=2\sqrt{3}$, $C=45°$, solve the triangle.

3.* If $C=60°$, $a=2$, $c=\sqrt{6}$, solve the triangle.

4. If $A=30°$, $a=2$, $c=5$, solve the triangle.

5. If $B=30°$, $b=\sqrt{6}$, $c=2\sqrt{3}$, solve the triangle.

6. If $B=60°$, $b=3\sqrt{2}$, $c=3+\sqrt{3}$, solve the triangle.

7. If $a=3+\sqrt{3}$, $c=3-\sqrt{3}$, $C=15°$, solve the triangle.

8. If $A=18°$, $a=4$, $b=4+\sqrt{80}$, solve the triangle.

9. If $B=135°$, $a=3\sqrt{2}$, $b=2\sqrt{3}$, solve the triangle.

149. Many relations connecting the sides and angles of a triangle may be proved by means of the formulæ we have established.

Example 1. Prove that $(b-c)\cos\frac{A}{2}=a\sin\frac{B-C}{2}$.

Let $$k=\frac{a}{\sin A}=\frac{b}{\sin B}=\frac{c}{\sin C};$$

then $$a=k\sin A,\quad b=k\sin B,\quad c=k\sin C;$$

$$\therefore\ (b-c)\cos\frac{A}{2}=k(\sin B-\sin C)\cos\frac{A}{2}$$

$$=2k\cos\frac{B+C}{2}\sin\frac{B-C}{2}\cos\frac{A}{2}$$

$$=2k\sin\frac{A}{2}\cos\frac{A}{2}\sin\frac{B-C}{2}$$

$$=k\sin A\sin\frac{B-C}{2}$$

$$=a\sin\frac{B-C}{2}.$$

Example 2. If $a\cos^2\frac{C}{2}+c\cos^2\frac{A}{2}=\frac{3b}{2}$, shew that the sides of the triangle are in A.P.

Since $$2a\cos^2\frac{C}{2}+2c\cos^2\frac{A}{2}=3b,$$

$$\therefore\ a(1+\cos C)+c(1+\cos A)=3b,$$

$$\therefore\ a+c+(a\cos C+c\cos A)=3b,$$

$$\therefore\ a+c+b=3b,$$

$$\therefore\ a+c=2b.$$

Thus the sides a, b, c are in A.P.

Example 3. Prove that

$$(b^2-c^2)\cot A+(c^2-a^2)\cot B+(a^2-b^2)\cot C=0.$$

Let $k=\dfrac{a}{\sin A}=\dfrac{b}{\sin B}=\dfrac{c}{\sin C}$; then

the first side

$$=k^2\left\{(\sin^2 B-\sin^2 C)\frac{\cos A}{\sin A}+\dots\dots+\dots\dots\right\}$$

$$=k^2\left\{\sin(B+C)\sin(B-C)\frac{\cos A}{\sin A}+\dots\dots+\dots\dots\right\}. \text{ [Art. 114].}$$

But $\sin(B+C)=\sin A$, and $\cos A=-\cos(B+C)$;

∴ the first side

$$=-k^2\{\sin(B-C)\cos(B+C)+\dots\dots+\dots\dots\}$$

$$=-\frac{k^2}{2}\{(\sin 2B-\sin 2C)+(\sin 2C-\sin 2A)+(\sin 2A-\sin 2B)\}$$

$$=0.$$

EXAMPLES. XIII. c.

Prove the following identities :

1. $a(\sin B-\sin C)+b(\sin C-\sin A)+c(\sin A-\sin B)=0.$
2. $2(bc\cos A+ca\cos B+ab\cos C)=a^2+b^2+c^2.$
3. $a(b\cos C-c\cos B)=b^2-c^2.$
4. $(b+c)\cos A+(c+a)\cos B+(a+b)\cos C=a+b+c.$
5. $2\left(a\sin^2\dfrac{C}{2}+c\sin^2\dfrac{A}{2}\right)=c+a-b.$
6. $\dfrac{\cos B}{\cos C}=\dfrac{c-b\cos A}{b-c\cos A}.$
7. $\tan A=\dfrac{a\sin C}{b-a\cos C}.$
8. $(b+c)\sin\dfrac{A}{2}=a\cos\dfrac{B-C}{2}.$
9. $\dfrac{a+b}{c}\sin^2\dfrac{C}{2}=\dfrac{\cos A+\cos B}{2}.$
10. $a\sin(B-C)+b\sin(C-A)+c\sin(A-B)=0.$
11. $\dfrac{\sin(A-B)}{\sin(A+B)}=\dfrac{a^2-b^2}{c^2}.$
12. $\dfrac{c\sin(A-B)}{b\sin(C-A)}=\dfrac{a^2-b^2}{c^2-a^2}.$

[*All articles and examples marked with an asterisk may be omitted on the first reading of the subject.*]

***150.** The *ambiguous case* may also be discussed by first finding the third side.

As before, let a, b, A be given, then

$$\cos A = \frac{b^2+c^2-a^2}{2bc};$$

$$\therefore\ c^2 - 2b\cos A \,.\, c + b^2 - a^2 = 0.$$

By solving this quadratic equation in c, we obtain

$$c = b\cos A \pm \sqrt{b^2\cos^2 A + a^2 - b^2}$$
$$= b\cos A \pm \sqrt{a^2 - b^2\sin^2 A}.$$

(i) When $a < b\sin A$, the quantity under the radical is negative, and the values of c are impossible; so that there is no solution.

(ii) When $a = b\sin A$, the quantity under the radical is zero, and $c = b\cos A$. Since $\sin A < 1$, it follows that $a < b$, and therefore $A < B$. Hence the triangle is impossible unless the given angle A is acute, in which case c is positive and there is one solution.

(iii) When $a > b\sin A$, there are three cases to consider.

(1) Suppose $a < b$, then $A < B$, and as before the triangle is impossible unless A is acute. In this case $b\cos A$ is positive.

Also $\sqrt{a^2 - b^2\sin^2 A}$ is real and $< \sqrt{b^2 - b^2\sin^2 A}$;

that is $$\sqrt{a^2 - b^2\sin^2 A} < b\cos A;$$

hence both values of c are real and positive, so that there are two solutions.

(2) Suppose $a > b$, then $\sqrt{a^2 - b^2\sin^2 A} > \sqrt{b^2 - b^2\sin^2 A}$;

that is $$\sqrt{a^2 - b^2\sin^2 A} > b\cos A;$$

hence one value of c is positive and one value is negative, whether A is acute or obtuse, and in each case there is only one solution.

(3) Suppose $a = b$, then $\sqrt{a^2 - b^2\sin^2 A} = b\cos A$;

$$\therefore\ c = 2b\cos A \text{ or } 0;$$

hence there is only one solution when A is acute, and when A is obtuse the triangle is impossible.

Example. If b, c, B are given, and if $b<c$, shew that

$$(a_1-a_2)^2+(a_1+a_2)^2\tan^2 B=4b^2,$$

where a_1, a_2 are the two values of the third side.

From the formula $\cos B=\dfrac{c^2+a^2-b^2}{2ca}$,

we have $a^2-2c\cos B\,.\,a+c^2-b^2=0.$

But the roots of this equation are a_1 and a_2; hence by the theory of quadratic equations

$$a_1+a_2=2c\cos B \text{ and } a_1a_2=c^2-b^2.$$

$$\therefore\ (a_1-a_2)^2=(a_1+a_2)^2-4a_1a_2$$
$$=4c^2\cos^2 B-4(c^2-b^2).$$

$$\therefore\ (a_1-a_2)^2+(a_1+a_2)^2\tan^2 B=4c^2\cos^2 B-4(c^2-b^2)+4c^2\cos^2 B\tan^2 B$$
$$=4c^2(\cos^2 B+\sin^2 B)-4c^2+4b^2$$
$$=4c^2-4c^2+4b^2$$
$$=4b^2.$$

*EXAMPLES. XIII. d.

1. In a triangle in which each base angle is double of the third angle the base is 2 : solve the triangle.

2. If $B=45°$, $C=75°$, and the perpendicular from A on BC is 3, solve the triangle.

3. If $a=2$, $b=4-2\sqrt{3}$, $c=3\sqrt{2}-\sqrt{6}$, solve the triangle.

4. If $A=18°$, $b-a=2$, $ab=4$, find the other angles.

5. Given $B=30°$, $c=150$, $b=50\sqrt{3}$, shew that of the two triangles which satisfy the data one will be isosceles and the other right-angled.

   Find the third side in the greater of these triangles. Would the solution be ambiguous if the data had been $B=30°$, $c=150$, $b=75$?

6. If $A=36°$, $a=4$, and the perpendicular from C upon AB is $\sqrt{5}-1$, find the other angles.

7. If the angles adjacent to the base of a triangle are $22\frac{1}{2}°$ and $112\frac{1}{2}°$, shew that the altitude is half the base.

8. If $a=2b$ and $A=3B$, find the angles and express c in terms of a.

9. The sides of a triangle are $2x+3$, x^2+3x+3, x^2+2x: shew that the greatest angle is 120°.

Shew that in any triangle

10. $(b-a)\cos C+c(\cos B-\cos A)=c\sin\dfrac{A-B}{2}\operatorname{cosec}\dfrac{A+B}{2}$.

11. $a\sin\left(\dfrac{A}{2}+B\right)=(b+c)\sin\dfrac{A}{2}$.

12. $\sin\left(B+\dfrac{C}{2}\right)\cos\dfrac{C}{2}=\dfrac{a+b}{b+c}\cos\dfrac{A}{2}\cos\dfrac{B-C}{2}$.

13. $\dfrac{1+\cos(A-B)\cos C}{1+\cos(A-C)\cos B}=\dfrac{a^2+b^2}{a^2+c^2}$.

14. If $c^4-2(a^2+b^2)c^2+a^4+a^2b^2+b^4=0$, prove that C is 60° or 120°.

15. If a, b, A are given, and if c_1, c_2 are the values of the third side in the ambiguous case, prove that if $c_1>c_2$,

(1) $c_1-c_2=2a\cos B_1$.

(2) $\cos\dfrac{C_1-C_2}{2}=\dfrac{b\sin A}{a}$.

(3) $c_1^2+c_2^2-2c_1c_2\cos 2A=4a^2\cos^2 A$.

(4) $\sin\dfrac{C_1+C_2}{2}\sin\dfrac{C_1-C_2}{2}=\cos A\cos B_1$.

16. If $A=45°$, and c_1, c_2 be the two values of the ambiguous side, shew that

$$\cos B_1CB_2=\frac{2c_1c_2}{c_1^2+c_2^2}.$$

17. If $\cos A+2\cos C : \cos A+2\cos B=\sin B : \sin C$, prove that the triangle is either isosceles or right-angled.

18. If a, b, c are in A. P., shew that

$$\cot\frac{A}{2},\ \cot\frac{B}{2},\ \cot\frac{C}{2}\text{ are also in A. P.}$$

19. Shew that

$$\frac{a^2\sin(B-C)}{\sin B+\sin C}+\frac{b^2\sin(C-A)}{\sin C+\sin A}+\frac{c^2\sin(A-B)}{\sin A+\sin B}=0.$$

MISCELLANEOUS EXAMPLES. D.

1. Prove that (1) $\tan 2\theta \cot \theta - 1 = \sec 2\theta$;

(2) $\sin a - \cot \theta \cos a = -\operatorname{cosec} \theta \cos (a+\theta)$.

2. If $a=48$, $b=35$, $C=60°$, find c.

3. If $\cos a = \frac{8}{17}$ and $\cos \beta = \frac{15}{17}$, find

$\tan (a+\beta)$ and $\operatorname{cosec} (a+\beta)$.

4. If $a = \frac{\pi}{21}$, find the value of $\frac{\sin 23a - \sin 7a}{\sin 2a + \sin 14a}$.

5. Prove that $\sin \theta (\cos 2\theta + \cos 4\theta + \cos 6\theta) = \sin 3\theta \cos 4\theta$.

6. If $b=\sqrt{2}$, $c=\sqrt{3}+1$, $A=45°$, solve the triangle.

7. Prove that

(1) $2 \sin^2 36° = \sqrt{5} \sin 18°$; (2) $4 \sin 36° \cos 18° = \sqrt{5}$.

8. Prove that $\frac{\sin 3a}{\sin a} + \frac{\cos 3a}{\cos a} = 4 \cos 2a$.

9. If $b=c=2$, $a=\sqrt{6}-\sqrt{2}$, solve the triangle.

10. Shew that

(1) $\cos 2a - \cot 3a \sin 2a = \tan a (\sin 2a + \cot 3a \cos 2a)$.

(2) $\cos a + \cos 2a + \cos 3a = 4 \cos a \cos \frac{a}{2} \cos \frac{3a}{2} - 1$.

11. In any triangle, prove that

(1) $b^2 \sin 2C + c^2 \sin 2B = 2bc \sin A$;

(2) $\frac{a^2 \sin (B-C)}{\sin A} + \frac{b^2 \sin (C-A)}{\sin B} + \frac{c^2 \sin (A-B)}{\sin C} = 0$.

12. If A, B, C, D are the angles of a quadrilateral, prove that

$$\frac{\tan A + \tan B + \tan C + \tan D}{\cot A + \cot B + \cot C + \cot D} = \tan A \tan B \tan C \tan D.$$

[*Use* $\tan (A+B) = \tan (360° - C - D)$.]

CHAPTER XIV.

LOGARITHMS.

151. DEFINITION. The **logarithm** of any number to a given **base** is the index of the power to which the base must be raised in order to equal the given number. Thus if $a^x = N$, x is called the logarithm of N to the base a.

Example 1. Since $3^4 = 81$, the logarithm of 81 to base 3 is 4.

Example 2. Since $10^1 = 10$, $10^2 = 100$, $10^3 = 1000$,......
the natural numbers 1, 2, 3,... are respectively the logarithms of 10, 100, 1000,...... to base 10.

Example 3. Find the logarithm of ·008 to base 25.

Let x be the required logarithm; then by definition,

$$25^x = \cdot 008 = \frac{8}{1000} = \frac{1}{125} = \frac{1}{5^3};$$

that is, $$(5^2)^x = 5^{-3}, \text{ or } 5^{2x} = 5^{-3};$$

whence, by equating indices, $2x = -3$, and $x = -1 \cdot 5$.

152. The logarithm of N to base a is usually written $\log_a N$, so that the same meaning is expressed by the two equations

$$a^x = N, \qquad x = \log_a N.$$

From these equations it is evident that $a^{\log_a N} = N$.

Example. Find the value of $\log_{\cdot 01} \cdot 00001$.

Let $\log_{\cdot 01} \cdot 00001 = x$; then $(\cdot 01)^x = \cdot 00001$;

$$\therefore \left(\frac{1}{10^2}\right)^x = \frac{1}{100000}, \text{ or } \frac{1}{10^{2x}} = \frac{1}{10^5}.$$

$$\therefore 2x = 5, \text{ and } x = 2 \cdot 5.$$

153. When it is understood that a particular system of logarithms is in use, the suffix denoting the base is omitted.

Thus in arithmetical calculations in which 10 is the base, we usually write log 2, log 3,...... instead of $\log_{10} 2$, $\log_{10} 3$,......

Logarithms to the base 10 are known as **Common Logarithms**; this system was first introduced in 1615 by Briggs, a contemporary of Napier the inventor of Logarithms.

Before discussing the properties of common logarithms we shall prove some general propositions which are true for all logarithms independently of any particular base.

154. *The logarithm of* 1 *is* 0.

For $a^0=1$ for all values of a; therefore $\log 1=0$, whatever the base may be.

155. *The logarithm of the base itself is* 1.

For $a^1=a$; therefore $\log_a a=1$.

156. *To find the logarithm of a product.*

Let MN be the product; let a be the base of the system, and suppose

$$x=\log_a M, \qquad y=\log_a N;$$

so that

$$a^x=M, \qquad a^y=N.$$

Thus the product $MN=a^x \times a^y=a^{x+y}$;

whence, by definition, $\log_a MN=x+y$

$$=\log_a M+\log_a N.$$

Similarly, $\log_a MNP=\log_a M+\log_a N+\log_a P$;
and so on for any number of factors.

Example. $\log 42=\log(2\times3\times7)=\log 2+\log 3+\log 7$.

157. *To find the logarithm of a fraction.*

Let $\frac{M}{N}$ be the fraction, and suppose

$$x=\log_a M, \qquad y=\log_a N;$$

so that

$$a^x=M, \qquad a^y=N.$$

Thus the fraction $\frac{M}{N}=\frac{a^x}{a^y}=a^{x-y}$;

whence, by definition, $\log_a \frac{M}{N}=x-y$

$$=\log_a M-\log_a N.$$

Example. $\log(2\frac{1}{7}) = \log\frac{15}{7} = \log 15 - \log 7$

$$= \log(3 \times 5) - \log 7 = \log 3 + \log 5 - \log 7.$$

158. *To find the logarithm of a number raised to any power, integral or fractional.*

Let $\log_a(M^p)$ be required, and suppose

$$x = \log_a M, \text{ so that } a^x = M;$$

then $$M^p = (a^x)^p = a^{px};$$

whence, by definition, $$\log_a(M^p) = px;$$

that is, $$\log_a(M^p) = p \log_a M.$$

Similarly, $$\log_a(M^{\frac{1}{r}}) = \frac{1}{r}\log_a M.$$

159. It follows from the results we have proved that

(1) the logarithm of a product is equal to the sum of the logarithms of its factors;

(2) the logarithm of a fraction is equal to the logarithm of the numerator diminished by the logarithm of the denominator;

(3) the logarithm of the pth power of a number is p times the logarithm of the number;

(4) the logarithm of the rth root of a number is $\frac{1}{r}$ of the logarithm of the number.

Thus by the use of logarithms the operations of multiplication and division may be replaced by those of addition and subtraction; the operations of involution and evolution by those of multiplication and division.

Example. Express $\log\frac{a^5\sqrt{b}}{\sqrt[3]{c^2}}$ in terms of $\log a$, $\log b$, $\log c$.

The expression $= \log(a^5\sqrt{b}) - \log\sqrt[3]{c^2}$

$$= \log a^5 + \log\sqrt{b} - \frac{2}{3}\log c$$

$$= 5\log a + \frac{1}{2}\log b - \frac{2}{3}\log c.$$

160. From the equation $10^x = N$, it is evident that common logarithms will not in general be integral, and that they will not always be positive.

For instance, $3154 > 10^3$ and $< 10^4$;

$$\therefore \log 3154 = 3 + \text{a fraction}.$$

Again, $\cdot 06 > 10^{-2}$ and $< 10^{-1}$;

$$\therefore \log \cdot 06 = -2 + \text{a fraction}.$$

161. DEFINITION. The integral part of a logarithm is called the **characteristic**, and the fractional part when expressed as a decimal is called the **mantissa.**

162. The characteristic of the logarithm of any number to base 10 can be written down by inspection, as we shall now shew.

(i) *To determine the characteristic of the logarithm of any number greater than unity.*

It is clear that a number with two digits in its integral part lies between 10^1 and 10^2; a number with three digits in its integral part lies between 10^2 and 10^3; and so on. Hence a number with n digits in its integral part lies between 10^{n-1} and 10^n.

Let N be a number whose integral part contains n digits; then

$$N = 10^{(n-1) + \text{a fraction}};$$

$$\therefore \log N = (n-1) + \text{a fraction}.$$

Hence the characteristic is $n-1$; that is, *the characteristic of the logarithm of a number greater than unity is less by one than the number of digits in its integral part, and is positive.*

Example. The characteristics of

log 314, log 87·263, log 2·78, log 3500

are respectively 2, 1, 0, 3.

(ii) *To determine the characteristic of the logarithm of a number less than unity.*

A decimal with one cipher immediately after the decimal point, such as ·0324, being greater than ·01 and less than ·1, lies between 10^{-2} and 10^{-1}; a number with two ciphers after the decimal point lies between 10^{-3} and 10^{-2}; and so on. Hence a decimal fraction with n ciphers immediately after the decimal point lies between $10^{-(n+1)}$ and 10^{-n}.

Let D be a decimal beginning with n ciphers; then

$$D = 10^{-(n+1)+\text{a fraction}};$$

$$\therefore \log D = -(n+1) + \text{a fraction}.$$

Hence the characteristic is $-(n+1)$; that is, *the characteristic of the logarithm of a number less than one is negative and one more than the number of ciphers immediately after the decimal point.*

Example. The characteristics of

$$\log \cdot 4, \quad \log \cdot 3748, \quad \log \cdot 000135, \quad \log \cdot 08$$

are respectively $\quad -1, \; -1, \; -4, \; -2.$

163. *The mantissæ are the same for the logarithms of all numbers which have the same significant digits.*

For if any two numbers have the same sequence of digits, differing only in the position of the decimal point, one must be equal to the other multiplied or divided by some integral power of 10. Hence their logarithms must *differ by an integer.* In other words, their decimal parts or mantissæ are the same.

Examples. (i) $\log 32700 = \log (3{\cdot}27 \times 10^4) = \log 3{\cdot}27 + \log 10^4$
$= \log 3{\cdot}27 + 4.$

(ii) $\log \cdot 0327 = \log (3{\cdot}27 \times 10^{-2}) = \log 3{\cdot}27 + \log 10^{-2}$
$= \log 3{\cdot}27 - 2.$

(iii) $\log \cdot 000327 = \log (3{\cdot}27 \times 10^{-4}) = \log 3{\cdot}27 + \log 10^{-4}$
$= \log 3{\cdot}27 - 4.$

Thus, log 32700, log ·0327, log ·000327 differ from log 3·27 only in the *integral* part; that is the mantissa is the same in each case.

Note. The characteristics of the logarithms are 4, – 2, – 4 respectively. The foregoing examples shew that by introducing a suitable integral power of 10, all numbers can be expressed in one standard form in which *the decimal point always stands after the first significant digit,* and the characteristics are given by the powers of 10, without using the rules of Art. 162.

164. The logarithms of all integers from 1 to 20000 have been found and tabulated. In Chambers' Mathematical Tables they are given to seven places of decimals, but for many practical purposes sufficient accuracy is secured by using four-figure logarithms. These are available for all numbers from 1 to 9999, and their use is explained on page 163A.

165. Advantages of Common Logarithms. It will now be seen that it is unnecessary to tabulate the characteristics, since they can always be written down by inspection [Art. 162]. Also the Tables need only contain the mantissa of the logarithms of integers [Art. 163].

In order to secure these advantages it is convenient *always to keep the mantissa positive*, and it is usual to write the minus sign over a negative characteristic and not before it, so as to indicate that the characteristic alone is negative. Thus $\bar{4}\cdot30103$, which is the logarithm of $\cdot0002$, is equivalent to $-4+\cdot30103$, and must be distinguished from $-4\cdot30103$, in which both the integer and the decimal are negative.

166. In the course of work we sometimes have to deal with a logarithm which is wholly negative. In such a case an arithmetical artifice is necessary in order to write the logarithm with mantissa positive. Thus a result such as $-3\cdot69897$ may be transformed by subtracting 1 from the integral part and adding 1 to the decimal part. Thus

$$-3\cdot69897 = -3-1+(1-\cdot69897)$$
$$= -4+\cdot30103 = \bar{4}\cdot30103.$$

Example 1. Required the logarithm of $\cdot0002432$.

In the Tables we find that 3859636 is the mantissa of log 2432 (the decimal point as well as the characteristic being omitted); and, by Art. 163, the characteristic of the logarithm of the given number is -4;

$$\therefore \log \cdot0002432 = \bar{4}\cdot3859636.$$

Example 2. Find the cube root of $\cdot0007$, having given

$$\log 7 = \cdot8450980, \quad \log 887904 = 5\cdot9483660.$$

Let x be the required cube root; then

$$\log x = \frac{1}{3}\log(\cdot0007) = \frac{1}{3}(\bar{4}\cdot8450980) = \frac{1}{3}(\bar{6}+2\cdot8450980);$$

that is, $$\log x = \bar{2}\cdot9483660;$$

but $$\log 887904 = 5\cdot9483660;$$

$$\therefore x = \cdot0887904.$$

167. The logarithm of 5 and its powers can easily be obtained from log 2; for

$$\log 5 = \log\frac{10}{2} = \log 10 - \log 2 = 1 - \log 2.$$

EXAMPLES. XIV. a.

1. Find the logarithms respectively

of the numbers	1024,	81,	·125,	·01,	$\cdot\dot{3}$,	100,
to the bases	2,	$\sqrt{3}$,	4,	·001,	$\cdot\dot{1}$,	·01.

2. Find the values of

$$\log_8 16, \quad \log_{81} 243, \quad \log_{\cdot 01} 10, \quad \log_{49} 343\sqrt{7}.$$

3. Find the numbers whose logarithms respectively

to the bases	49,	·25,	·03,	1,	·64,	100,	·1,
are	2,	$\frac{1}{2}$,	-2,	-1,	$-\frac{1}{2}$,	1·5,	-4.

4. Find the respective characteristics

of the logarithms of	325,	1603,	2400,	10000,	19,
to the bases	3,	11,	7,	9,	21.

5. Write down the characteristics of the common logarithms of 3·26, 523·1, ·03, 1·5, ·0002, 3000·1, ·1.

6. The mantissa of log 64439 is ·8091488, write down the logarithms of ·64439, 6443900, ·00064439.

7. The logarithm of 32·5 is 1·5118834, write down the numbers whose logarithms are

$$\cdot 5118834, \quad 2\cdot 5118834, \quad \bar{4}\cdot 5118834.$$

[*When required the following logarithms may be used*

$\log 2 = \cdot 3010300$, $\log 3 = \cdot 4771213$, $\log 7 = \cdot 8450980$.]

Find the value of

8. $\log 768$.

9. $\log 2352$.

10. $\log 35\cdot 28$.

11. $\log \sqrt{6804}$.

12. $\log \sqrt[5]{\cdot 00162}$.

13. $\log \cdot 021\dot{7}$.

14. $\log \cos 60°$.

15. $\log \sin^3 60°$.

16. $\log \sqrt[3]{\sec 45°}$.

Find the numerical value of

17. $2 \log \frac{15}{8} - \log \frac{25}{162} + 3 \log \frac{4}{9}$.

18. Evaluate $16 \log \frac{10}{9} - 4 \log \frac{25}{24} - 7 \log \frac{80}{81}$.

19. Find the seventh root of 7,

given $\log 1{\cdot}320469 = {\cdot}1207283$.

20. Find the cube root of ·00001764,

given $\log 260315 = 5{\cdot}4154995$.

21. Given $\log 3571 = 3{\cdot}5527899$, find the logarithm of

$$3{\cdot}571 \times {\cdot}03571 \times \sqrt[3]{3571}.$$

22. Given $\log 11 = 1{\cdot}0413927$, find the logarithm of

$$({\cdot}00011)^{\frac{1}{3}} \times (1{\cdot}21)^2 \times (13{\cdot}31)^{\frac{4}{3}} \div 12100000.$$

23. Find the number of digits in the integral parts of

$$\left(\frac{21}{20}\right)^{300} \text{ and } \left(\frac{126}{125}\right)^{1000}$$

24. How many positive integers have characteristic 3 when the base is 7?

168. Suppose that we have a table of logarithms of numbers to base a and require to find the logarithms to base b.

Let N be one of the numbers, then $\log_b N$ is required.

Let $y = \log_b N$, so that $b^y = N$.

$$\therefore \log_a (b^y) = \log_a N;$$

that is,

$$y \log_a b = \log_a N;$$

$$\therefore y = \frac{1}{\log_a b} \times \log_a N,$$

or

$$\log_b N = \frac{1}{\log_a b} \times \log_a N \quad \text{.....................(1).}$$

Now since N and b are given, $\log_a N$ and $\log_a b$ are known from the Tables, and thus $\log_b N$ may be found.

Hence it appears that to transform logarithms from base a to base b we have only to multiply them all by $\frac{1}{\log_a b}$; this is a constant quantity and is given by the Tables; it is known as the *modulus*.

If in equation (1) we put a for N, we obtain

$$\log_b a = \frac{1}{\log_a b} \times \log_a a = \frac{1}{\log_a b};$$

$$\therefore \log_b a \times \log_a b = 1.$$

169. In the following examples all necessary logarithms will be given. The use of four-figure Tables will be explained in a future section.

Example 1. Given $\log 2 = \cdot 3010300$ and $\log 4844544 = 6\cdot 6852530$, find the value of $(6\cdot 4)^{\frac{1}{10}} \times (\sqrt[4]{\cdot 256})^3 \div \sqrt{80}$.

Let x be the value of the expression; then

$$\log x = \frac{1}{10}\log\frac{64}{10} + \frac{3}{4}\log\frac{256}{1000} - \frac{1}{2}\log 80$$

$$= \frac{1}{10}(\log 2^6 - 1) + \frac{3}{4}(\log 2^8 - 3) - \frac{1}{2}(\log 2^3 + 1)$$

$$= \left(\frac{6}{10} + 6 - \frac{3}{2}\right)\log 2 - \left(\frac{1}{10} + \frac{9}{4} + \frac{1}{2}\right)$$

$$= \left(5 + \frac{1}{10}\right)\log 2 - 2\tfrac{17}{20}$$

$$= 1\cdot 5051500 + \cdot 0301030 - 2\cdot 85.$$

Thus $\qquad \log x = \bar{2}\cdot 6852530.$

But $\qquad \log 4844544 = 6\cdot 6852530,$

$$\therefore x = \cdot 04844544.$$

Example 2. Find how many ciphers there are between the decimal point and the first significant digit in $(\cdot 0504)^{10}$; having given

$$\log 2 = \cdot 301, \ \log 3 = \cdot 477, \ \log 7 = \cdot 845.$$

Denote the expression by E; then

$$\log E = 10\log\frac{504}{10000}$$

$$= 10(\log 504 - 4)$$

$$= 10\{\log(2^3 \times 3^2 \times 7) - 4\}$$

$$= 10\{3\log 2 + 2\log 3 + \log 7 - 4)$$

$$= 10(2\cdot 702 - 4) = 10(\bar{2}\cdot 702)$$

$$= \overline{20} + 7\cdot 02 = \overline{13}\cdot 02.$$

$$\begin{aligned} 3\log 2 &= \cdot 903 \\ 2\log 3 &= \cdot 954 \\ \log 7 &= \cdot 845 \\ \hline & \ 2\cdot 702 \end{aligned}$$

Thus the number of ciphers is 12. [Art. 162.]

Exponential equations.

170. If in an equation the unknown quantity appears as an exponent, the solution may be effected by the help of logarithms.

Example 1. Solve the equation $8^{5-3x}=12^{4-2x}$, having given

$$\log 2 = \cdot 30103, \text{ and } \log 3 = \cdot 47712.$$

From the given equation, by taking logarithms, we have

$$(5-3x)\log 8 = (4-2x)\log 12;$$

$$\therefore\ 3(5-3x)\log 2 = (4-2x)(2\log 2 + \log 3);$$

$$\therefore\ 15\log 2 - 8\log 2 - 4\log 3 = x(9\log 2 - 4\log 2 - 2\log 3);$$

$$\therefore\ x = \frac{7\log 2 - 4\log 3}{5\log 2 - 2\log 3} = \frac{\cdot 19873}{\cdot 55091}.$$

Thus $x = \cdot 36$ nearly.

$7\log 2 = 2\cdot 10721$
$4\log 3 = 1\cdot 90848$
$\cdot 19873$

$5\log 2 = 1\cdot 50515$
$2\log 3 = \cdot 95424$
$\cdot 55091$

55091) 198730 (36
165273
334570

Example 2. Given $\log 2 = \cdot 30103$, solve the simultaneous equations

$$2^x \,.\, 5^y = 1, \qquad 5^{x+1} \,.\, 2^y = 2.$$

Take logarithms of the given equations;

$$\therefore\ x\log 2 + y\log 5 = 0, \qquad (x+1)\log 5 + y\log 2 = \log 2.$$

For shortness, put $\log 2 = a$, $\log 5 = b$.

Thus $$ax + by = 0,$$

and $$b(x+1) + ay = a, \text{ or } bx + ay = a - b.$$

By eliminating y, $x(a^2 - b^2) = -b(a-b)$,

$$\therefore\ x = -\frac{b}{a+b} = -\frac{\log 5}{\log 2 + \log 5} = -\frac{\log 5}{\log 10} = -\log 5 = -\cdot 69897.$$

And $$y = -\frac{ax}{b} = \frac{a}{b}\log 5 = a = \log 2 = \cdot 30103.$$

EXAMPLES. XIV. b.

[*When required the values of* log 2, log 3, log 7 *given on p.* 145 *may be used.*]

Find the value of

1. $\left(\dfrac{147 \times 375}{126 \times 16}\right)^{\frac{2}{3}}$, given $\log 9{\cdot}076226 = {\cdot}9579053$.

2. $\sqrt[3]{378} \times \sqrt{108} \div (\sqrt[6]{1008} \times \sqrt[3]{486})$,

   given $\log 301824 = 5{\cdot}4797536$.

3. $(1080)^{\frac{1}{2}} \times ({\cdot}24)^{\frac{5}{8}} \times 810$,

   given $\log 2467266 = 6{\cdot}3922160$.

Calculate to two decimal places the values of

4. $\log_{20} 800$. 5. $\log_3 49$. 6. $\log_{125} 4000$.

7. Find how many ciphers there are before the first significant digits in

$$({\cdot}00378)^{\frac{40}{3}} \quad \text{and} \quad ({\cdot}0\dot{2}5\dot{9})^{50}.$$

8. To what base is 3 the logarithm of 11000?

   given $\log 11 = 1{\cdot}0413927$ and $\log 222398 = 5{\cdot}3471309$.

Solve to two decimal places the equations:

9. $2^{x-1} = 5$. 10. $3^{x-4} = 7$. 11. $5^{1-x} = 6^{x-3}$.

12. $5^x = 2^{-y}$ and $5^{2+y} = 2^{2-x}$.

13. $2^x = 3^y$ and $2^{y+1} = 3^{x-1}$.

14. Given $\log 28 = a$, $\log 21 = b$, $\log 25 = c$, find $\log 27$ and $\log 224$ in terms of a, b, c.

15. Given $\log 242 = a$, $\log 80 = b$, $\log 45 = c$, find $\log 36$ and $\log 66$ in terms of a, b, c.

MISCELLANEOUS EXAMPLES. E.

1. Prove that

$\cos(30°+A)\cos(30°-A)-\cos(60°+A)\cos(60°-A)=\frac{1}{2}$.

2. If $A+B+C=180°$, shew that

$$\frac{\sin 2A+\sin 2B+\sin 2C}{\sin A+\sin B+\sin C}=8\sin\frac{A}{2}\sin\frac{B}{2}\sin\frac{C}{2}.$$

3. If $a=2$, $c=\sqrt{2}$, $B=15°$, solve the triangle.

4. Shew that $\cos a+\tan\frac{a}{2}\sin a=\cot\frac{a}{2}\sin a-\cos a$.

5. If $b\cos A=a\cos B$, shew that the triangle is isosceles.

6. Prove that

(1) $\sin\theta(\sin 3\theta+\sin 5\theta+\sin 7\theta+\sin 9\theta)=\sin 6\theta\sin 4\theta$;

(2) $\dfrac{\sin a+\sin 3a+\sin 5a+\sin 7a}{\cos a+\cos 3a+\cos 5a+\cos 7a}=\tan 4a$.

7. Shew that $\dfrac{\cos 3a}{\sin a}+\dfrac{\sin 3a}{\cos a}=2\cot 2a$.

8. If $b=a(\sqrt{3}-1)$, $C=30°$, find A and B.

9. Shew that $\tan 4a=\dfrac{4\tan a-4\tan^3 a}{1-6\tan^2 a+\tan^4 a}$.

10. In a triangle, shew that

(1) $a^2\cos 2B+b^2\cos 2A=a^2+b^2-4ab\sin A\sin B$;

(2) $4\left(bc\cos^2\frac{A}{2}+ca\cos^2\frac{B}{2}+ab\cos^2\frac{C}{2}\right)=(a+b+c)^2$.

11. If $a^4+b^4+c^4=2c^2(a^2+b^2)$, prove that $C=45°$ or $135°$.

[*Solve as a quadratic in* c^2.]

12. If in a triangle $\cos 3A+\cos 3B+\cos 3C=1$, shew that one angle must be $120°$.

CHAPTER XV.

THE USE OF LOGARITHMIC TABLES.

Seven-Figure Tables.

(*The use of Four-Figure Tables will be found explained on page* 163_A.)

171. In a book of Seven-Figure Tables there will usually be found the *mantissæ* of the logarithms of all *integers* from 1 to 100000; the *characteristics* can be written down by inspection and are therefore omitted. [Art. 162.]

The logarithm of any number consisting of not more than 5 significant digits can be obtained directly from these Tables. For instance, suppose the logarithm of 336·34 is required. Opposite to 33634 we find the figures 5267785; this, with the decimal point prefixed, is the mantissa for the logarithms of all numbers whose significant digits are the same as 33634. We have therefore only to prefix the characteristic 2, and we obtain

$$\log 336{\cdot}34 = 2{\cdot}5267785.$$

Similarly, $$\log 33634 = 4{\cdot}5267785,$$

and $$\log {\cdot}0033634 = \bar{3}{\cdot}5267785.$$

172. Suppose now that we required log 33634·392.

Since this number contains more than 5 significant digits it cannot be obtained directly from the tables; but it lies between the two consecutive numbers 33634 and 33635, and therefore its logarithm lies between the logarithms of these two numbers. If we pass from 33634 to 33635, making an increase of 1 in the number, the corresponding increase in the logarithm as obtained from the tables is ·0000129. If now we pass from 33634 to 33634·392, making an increase of ·392 in the number, the increase in the logarithm will be ·392 × ·0000129, provided that the increase in the logarithm is proportional to the increase in the number.

Now it can be proved that *when the increase made is small in comparison with the number, the increase in the logarithm is very nearly proportional to the increase in the number.*

This principle is known as the **Rule of Proportional Parts.**

The application of this rule will be illustrated in the examples which follow.

173. In order to make the explanations more intelligible we give here an Extract from Chambers' *Mathematical Tables.*

No.	0	1	2	3	4	5	6	7	8	9	Diff.
3361	526 4685	4814	4944	5073	5202	5331	5460	5590	5719	5848	
62	5977	6106	6235	6365	6494	6623	6752	6881	7010	7140	129
63	7269	7398	7527	7656	7785	7914	8043	8173	8302	8431	1 13
64	8560	8689	8818	8947	9076	9205	9334	9463	9593	9722	2 26
65	9851	9980	0109	0238	0367	0496	0625	0754	0883	1012	3 39
											4 52
66	527 1141	1270	1399	1528	1657	1786	1915	2044	2173	2302	5 65
67	2431	2560	2689	2818	2947	3076	3205	3334	3463	3592	6 77
68	3721	3850	3979	4108	4237	4366	4494	4623	4752	4881	7 90
69	5010	5139	5268	5397	5526	5655	5783	5912	6041	6170	8 103
70	6299	6428	6557	6686	6814	6943	7072	7201	7330	7459	9 116

174. Suppose that log 33635 is required.

In the third horizontal line we have the logarithms of numbers beginning with 3363. As the next digit is 5 we choose from this line the mantissa which stands under the column 5. We have now only to prefix the characteristic and we obtain $\log 33635 = 4{\cdot}5267914$.

Similarly, $\log 33651 = 4{\cdot}5269980$,

and $\log 33652 = 4{\cdot}5270109$,

the transition in the mantissæ from 526... to 527... being shewn by the bar drawn over 0109. This bar is repeated over each of the subsequent logarithms as far as the end of the line, and in the next line the mantissæ begin with 527.

Example. Find log 33634·392.

$$\begin{array}{rrl} \text{From the Tables,} & \log 33635 = & 4{\cdot}5267914 \\ & \log 33634 = & 4{\cdot}5267785 \\ \hline & \text{difference for } 1 = & {\cdot}0000129 \end{array}$$

Now by the Rule of Proportional Parts, log 33634·392 will be greater than log 33634 by ·392 times the difference for 1; hence to 7 places of decimals, we have

$$\begin{array}{r|l} {\cdot}0000129 & \\ {\cdot}392 & \\ \hline & 258 \\ 11 & 61 \\ 38 & 7 \\ \hline {\cdot}0000050 & 568 \end{array}$$

$$\begin{array}{rl} \log 33634 & = 4{\cdot}5267785 \\ \text{proportional difference for } {\cdot}392 = & \quad {\cdot}0000051 \\ \hline \therefore\ \log 33634{\cdot}392 & = 4{\cdot}5267836 \end{array}$$

In practice, the difference for 1 is usually quoted without the ciphers; if therefore we *treat the difference* 129 *as a whole number*, on multiplying by ·392 we obtain the product 50·568, and we take the digits given by its integral part (51 approximately) as the proportional increase for ·392.

175. The method of calculating the proportional difference for ·392 which we have explained is that which must be adopted when we have nothing given but the logarithms of two consecutive numbers between which lies the number whose logarithm we are seeking.

But when the Tables are used the calculation is facilitated by means of the proportional differences standing in the column to the right. This gives the differences for *tenths* of unity.

The difference for ·392 is obtained as follows.

$${\cdot}392 \times 129 = \left(\frac{3}{10} + \frac{9}{100} + \frac{2}{1000}\right) \times 129 = 39 + 11{\cdot}6 + {\cdot}26 = 50{\cdot}86.$$

The difference for 9 quoted in the margin (really 9 *tenths*) is 116, and therefore the difference for 9 hundredths is 11·6; and similarly the difference for 2 thousandths is ·26.

In practical work, the following arrangement is adopted.

$$\begin{array}{lrr|l}
 & \log 33634 & =4\cdot5267785 & \\
\text{add for} & 3 & 39 & \\
 & 9 & 11 & 6 \\
 & 2 & & 26 \\
\hline
\therefore & \multicolumn{3}{l}{\log 33634\cdot392=4\cdot5267836}
\end{array}$$

176. The following example is solved more concisely as a model for the student. In the column on the left we work from the data of the question; in the column on the right we obtain the logarithm by the use of the Tables independently of the two given logarithms.

Example. Find log 33·656208, having given

$$\log 33656=4\cdot5270625 \text{ and } \log 33657=4\cdot5270754.$$

$$\begin{array}{rr|l}
\log 33\cdot657= & 1\cdot5270754 & \\
\log 33\cdot656= & 1\cdot5270625 & \\
\hline
\text{diff. for } \cdot001= & 129 & \\
 & \cdot208 & \\
\hline
 & 1 & 032 \\
 & 25 & 80 \\
\hline
\text{diff. for } \cdot000208= & 26 & 832 \\
\log 33\cdot656 \quad = & 1\cdot5270625 & \\
\log 33\cdot656208= & 1\cdot5270652 &
\end{array}$$

From the Tables, we have

$$\begin{array}{lrr|l}
 & \log 33\cdot656 & =1\cdot5270625 & \\
\text{add for} & 2 & 26 & \\
 & 0 & 0 & \\
 & 8 & 1 & 03 \\
\end{array}$$

$$\log 33\cdot656208=1\cdot5270652$$

177. The Rule of Proportional Parts also enables us to find the number corresponding to a given logarithm.

Example 1. Find the number whose logarithm is $\bar{2}\cdot5274023$, having given log 3·3683 = ·5274108 and log 3·3682 = ·5273979.

Let x be the required number; then

$$\begin{array}{rr}
\log x= & \bar{2}\cdot5274023 \\
\log \cdot033682= & \bar{2}\cdot5273979 \\
\hline
\text{diff.}= & 44
\end{array}
\qquad
\begin{array}{rr}
\log \cdot033683= & \bar{2}\cdot5274108 \\
\log \cdot033682= & \bar{2}\cdot5273979 \\
\hline
\text{diff. for } \cdot000001= & 129
\end{array}$$

hence x lies between ·033682 and ·033683, and is greater than ·033682 by $\frac{44}{129}\times\cdot000001$, that is by ·00000034.

$$\therefore x=\cdot03368234.$$

$$\begin{array}{r}
129\ )\ 440\ (\ 34 \\
387 \\
\hline
530 \\
516 \\
\hline
\end{array}$$

In working from the Tables, we proceed as follows.

$$\begin{array}{lrr} \log x & = & \bar{2}\cdot5274023 \\ \log \cdot033682 & = & \bar{2}\cdot5273979 \\ \hline & & 44 \\ 3 & & 39 \\ \hline & & 50 \\ 4 & & 52 \\ \hline \end{array}$$

$\therefore\ x = \cdot03368234.$

We are saved the trouble of the division, as the multiples of 129 which occur during the work are given in the approximate forms 39 and 52 in the difference column opposite to the numbers 3 and 4.

Example 2. Find the fifth root of ·0025612, having given
$\log 2\cdot5612 = \cdot4084435$, $\log 3\cdot0317 = \cdot4816862$, $\log 3\cdot0318 = \cdot4817005$.

Let $x = (\cdot0025612)^{\frac{1}{5}}$; then

$$\log x = \frac{1}{5}\log(\cdot0025612) = \frac{1}{5}(\bar{3}\cdot4084435) = \frac{1}{5}(\bar{5} + 2\cdot4084435);$$

$$= \bar{1}\cdot4816887.$$

$$\begin{array}{lrr} \log x & = & \bar{1}\cdot4816887 \\ \log \cdot30317 & = & \bar{1}\cdot4816862 \\ \hline \text{diff.} & = & 25 \end{array} \qquad \begin{array}{lrr} \log \cdot30318 & = & \bar{1}\cdot4817005 \\ \log \cdot30317 & = & \bar{1}\cdot4816862 \\ \hline \text{diff. for } \cdot00001 & = & 143 \end{array}$$

$\therefore$ proportional increase $= \dfrac{25}{143} \times \cdot00001 = \cdot00000175.$

Thus $x = \cdot30317175.$

$$\begin{array}{r} 143\,)\,250\,(\,175 \\ 143 \\ \hline 1070 \\ 1001 \\ \hline 690 \\ 715 \\ \hline \end{array}$$

EXAMPLES. XV. a.

1. Find the value of log 4951634, given that
$\log 49516 = 4\cdot6947456$, $\log 49517 = 4\cdot6947543$.

2. Find log 3·4713026, having given that
$\log 347\cdot13 = 2\cdot5404921$, $\log 34714 = 4\cdot5405047$.

3. Find log 2849614, having given that
$\log 2\cdot8496 = \cdot4547839$, $\log 2\cdot8497 = \cdot4547991$.

4. Find log 57·63325, having given that
$\log 576{\cdot}33 = 2{\cdot}7606712, \quad \log 5763{\cdot}4 = 3{\cdot}7606788.$

5. Given $\log 60814 = 4{\cdot}7840036$, diff. for $1 = 72$, find
$\log 6081465.$

6. Find the number whose logarithm is 4·7461735, given
$\log 55740 = 4{\cdot}7461670, \quad \log 55741 = 4{\cdot}7461748.$

7. Find the number whose logarithm is 2·8283676, given
$\log 6{\cdot}7354 = {\cdot}8283634, \quad \log 67355 = 4{\cdot}8283698.$

8. Find the number whose logarithm is $\bar{2}{\cdot}0288435$, given
$\log 1068{\cdot}6 = 3{\cdot}0288152, \quad \log 1{\cdot}0687 = {\cdot}0288558.$

9. Find the number whose logarithm is $\bar{3}{\cdot}9184377$, given
$\log 8{\cdot}2877 = {\cdot}9184340, \quad \log 8287{\cdot}8 = 3{\cdot}9184392.$

10. Given $\log 253{\cdot}19 = 2{\cdot}4034465$, diff. for $1 = 172$, find the number whose logarithm is $\bar{1}{\cdot}4034508$.

11. Given $\log 2{\cdot}0313 = {\cdot}3077741, \quad \log 2{\cdot}0314 = {\cdot}3077954,$
and $\log 1{\cdot}4271 = {\cdot}1544544,$
find the seventh root of 142·71.

12. Find the eighth root of 13·89492, given
$\log 13894 = 4{\cdot}1428273, \quad \log 138{\cdot}95 = 2{\cdot}1428586.$

13. Find the value of $\sqrt[14]{242447}$, given
$\log 2{\cdot}4244 = {\cdot}3846043$, diff. for $1 = 179$.

14. Find the twentieth root of 2069138, given
$\log 20691 = 4{\cdot}3157815$, diff. for $1 = 210$.

Tables of Natural and Logarithmic Functions.

178. Tables have been constructed giving the values of the trigonometrical functions of all angles between 0° and 90° at intervals of 10″. These are called the Tables of **natural sines, cosines, tangents,...** In the smaller Tables, such as Chambers', the interval is 1′.

The logarithms of the functions have also been calculated. Since many of the trigonometrical functions are less than unity their logarithms are negative, and as the characteristics are not always evident on inspection they cannot be omitted. To avoid the inconvenience of printing the bars over the characteristics, the logarithms are all increased by 10 and are then registered under the name of **tabular logarithmic sines, cosines,...**

The notation used is $L\cos A$, $L\tan\theta$; thus

$$L\sin A = \log\sin A + 10.$$

For instance,

$$L\sin 45^\circ = 10 + \log\sin 45^\circ = 10 + \log\frac{1}{\sqrt{2}}$$

$$= 10 - \frac{1}{2}\log 2 = 9{\cdot}8494850.$$

179. With certain exceptions that need not be here noticed, the rule of proportional parts holds for the natural sines, cosines,... of all angles, and also for their logarithmic sines, cosines,.... In applying this rule it must be remembered that as the angle increases from 0° to 90° the functions sine, tangent, secant increase, while the co-functions cosine, cotangent, cosecant decrease.

Example 1. Find the value of $\sin 29^\circ 37' 42''$.

From the Tables,

$$\begin{aligned} \sin 29^\circ 38' &= {\cdot}4944476 \\ \sin 29^\circ 37' &= {\cdot}4941948 \\ \text{diff. for } 60'' &= \quad\ \ 2528 \end{aligned}$$

$$\therefore \text{ prop}^{\text{l}} \text{ increase for } 42'' = \frac{42}{60} \times 2528 = \quad 1770$$

$$\begin{aligned} \sin 29^\circ 37' &= {\cdot}4941948 \\ \therefore\ \sin 29^\circ 37' 42'' &= {\cdot}4943718. \end{aligned}$$

Example 2. Find the angle whose cosine is ·7280843.

Let A be the required angle; then from the Tables,

$$\begin{aligned} \cos 43^\circ 16' &= {\cdot}7281716 \\ \cos 43^\circ 17' &= {\cdot}7279722 \\ \text{diff. for } 60'' & \quad\ \ 1994 \end{aligned} \qquad \begin{aligned} \cos 43^\circ 16' &= {\cdot}7281716 \\ \cos A &= {\cdot}7280843 \\ \text{prop}^{\text{l}} \text{ part} & \quad\quad\ 873 \end{aligned}$$

But $\cos A$ is *less* than $\cos 43^\circ 16'$; hence A must be *greater* than $43^\circ 16'$ by $\frac{873}{1994} \times 60''$, that is by $26''$ nearly.

Thus the angle is $43^\circ 16' 26''$.

$$\begin{array}{r} 873 \\ 60 \\ \hline 1994\,)\ 52380\ (\ 26 \\ 3988 \\ \hline 12500 \\ 11964 \\ \hline \end{array}$$

180. In order to illustrate the use of the tabular logarithmic functions we give the following extract from the table of logarithmic sines, cosines,... in Chambers' *Mathematical Tables.*

27 Deg.

′	Sine	Diff.	Cosec.	Secant	D.	Cosine	′
0	9·6570468		10·3429532	10·0501191		9·9498809	60
		2478			644		
1	9·6572946		10·3427054	10·0501835		9·9498165	59
		2477			644		
2	9·6575423		10·3424577	10·0502479		9·9497521	58
		2475			645		
3	9·6577898		10·3422102	10·0503124		9·9496876	57
		2473			646		
4	9·6580371		10·3419629	10·0503770		9·9496230	56
56	9·6706576		10·3293424	10·0537968		9·9462032	4
		2382			670		
57	9·6708958		10·3291042	10·0538638		9·9461362	3
		2380			670		
58	9·6711338		10·3288662	10·0539308		9·9460692	2
		2378			671		
59	9·6713716		10·3286284	10·0539979		9·9460021	1
		2377			672		
60	9·6716093		10·3283907	10·0540651		9·9459349	0
′	Cosine	Diff.	Secant	Cosec.	D.	Sine	′

62 Deg.

181. We have quoted here the logarithmic sines, cosecants, secants, and cosines of the angles differing by 1′ between 27° 0′ and 27° 4′, and also between 27° 56′ and 27° 60′. The same extract gives the logarithmic functions of the complements of these angles, namely those between 62° 0′ and 62° 4′, and those between 62° 56′ and 62° 60′.

The column of minutes for 27° is given on the left and increases downwards, the column for 62° is on the right and increases upwards.

The names of the functions printed at the top refer to the angle 27°, the names printed at the foot refer to the angle 62°. Thus

$$L\cos 27^\circ\ 3' = 9{\cdot}9496876, \quad L\operatorname{cosec} 27^\circ\ 58' = 10{\cdot}3288662,$$
$$L\sin 62^\circ\ 2' = 9{\cdot}9460692, \quad L\cos 62^\circ\ 59' = 9{\cdot}6572946.$$

The first *difference column* gives the differences in the logarithms of the sines and cosecants, the second *difference column* gives the differences in the logarithms of the cosines and secants, each difference corresponding to a difference of 1′ in the angle.

Example 1. Find $L\cos 62°\,57'\,12''$.

From the Tables,

$$\begin{aligned} L\cos 62°\,57' &= 9{\cdot}6577898 \\ L\cos 62°\,58' &= 9{\cdot}6575423 \\ \text{diff. for } 60'' &\quad\quad 2475 \end{aligned}$$

$$\therefore \text{ proportional } \textit{decrease} \text{ for } 12'' = \frac{12}{60} \times 2475 = 495.$$

$$\begin{aligned} L\cos 62°\,57' &= 9{\cdot}6577898 \\ \textit{Subtract} \text{ for } 12'' &\quad\quad\quad 495 \\ \therefore\ L\cos 62°\,57'\,12'' &= 9{\cdot}6577403 \end{aligned}$$

Example 2. Given $L\sec 27°\,39' = 10{\cdot}0526648$, diff. for $10'' = 110$, find A when $L\sec A = 10{\cdot}0527253$.

$$\begin{aligned} L\sec A &= 10{\cdot}0527253 \\ L\sec 27°\,39' &= 10{\cdot}0526648 \\ \text{diff.} &\quad\quad\quad 605 \end{aligned}$$

$$\therefore \text{ proportional increase} = \frac{605}{110} \times 10'' = 55''.$$

Thus $\qquad A = 27°\,39'\,55''.$

EXAMPLES. XV. b.

1. Find $\sin 38°\,3'\,35''$, having given that
$\sin 38°\,4' = {\cdot}6165780, \quad \sin 38°\,3' = {\cdot}6163489.$

2. Find $\tan 38°\,24'\,37{\cdot}5''$, having given that
$\tan 38°\,25' = {\cdot}7930640, \quad \tan 38°\,24' = {\cdot}7925902.$

3. Find $\operatorname{cosec} 55°\,21'\,28''$, having given that
$\operatorname{cosec} 55°\,22' = 1{\cdot}2153535, \quad \operatorname{cosec} 55°\,21' = 1{\cdot}2155978.$

4. Find the angle whose secant is $2{\cdot}1809460$, given
$\sec 62°\,43' = 2{\cdot}1815435, \quad \sec 62°\,42' = 2{\cdot}1803139.$

5. Find the angle whose cosine is ${\cdot}8600931$, given
$\cos 30°\,41' = {\cdot}8600007, \quad \cos 30°\,40' = {\cdot}8601491.$

6. Find the angle whose cotangent is ${\cdot}8766003$, given
$\cot 48°\,46' = {\cdot}8764620, \quad \cot 48°\,45' = {\cdot}8769765.$

7. Find $L\sin 44°\,17'\,33''$, given
$L\sin 44°\,18' = 9{\cdot}8441137, \quad L\sin 44°\,17' = 9{\cdot}8439842.$

8. Find $L\cot 36° 26' 16''$, given
$L\cot 36° 27' = 10{\cdot}1315840$, $L\cot 36° 26' = 10{\cdot}1318483$.

9. Find $L\cos 55° 30' 24''$, given
$L\cos 55° 31' = 9{\cdot}7529442$, $L\cos 55° 30' = 9{\cdot}7531280$.

10. Find the angle whose tabular logarithmic sine is 9·8440018, using the data of example 7.

11. Find the angle whose tabular logarithmic cosine is 9·7530075, using the data of example 9.

12. Given $L\tan 24° 50' = 9{\cdot}6653662$, diff. for $1' = 3313$, find $L\tan 24° 50' 52{\cdot}5''$.

13. Given $L\operatorname{cosec} 40° 5' = 10{\cdot}1911808$, diff. for $1' = 1502$, find $L\operatorname{cosec} 40° 4' 17{\cdot}5''$.

182. Considerable practice in the use of logarithmic Tables will be required before the quickness and accuracy necessary in all practical calculations can be attained. Experience shews that mistakes frequently arise from incorrect quotation from the Tables, and from clumsy arrangement. The student is reminded that care in taking out the logarithms from the Tables is of the first importance, and that in the course of the work he should learn to leave out all needless steps, making his solutions as concise as possible consistent with accuracy.

Example 1. Divide 6·6425693 by ·3873007.

From the Tables,

log 6·6425	= ·8223316		
6	40		
9	5	9	
3		20	
log 6·6425693	= ·8223362		
log ·3873007	= $\bar{1}$·5880483		

log ·38730	= $\bar{1}$·5880475	
0		0
7	7	8
log ·3873007	= $\bar{1}$·5880483	

By subtraction, we obtain		1·2342879
From the Tables, log 17·150		= 1·2342641
		238
	9	229
		90
	3	76

Thus the quotient is 17·15093.

Example 2. **The hypotenuse of a right-angled triangle is 3·141024 and one side is 2·593167; find the other side.**

Let c be the hypotenuse, a the given side, and x the side required; then

$$x^2 = c^2 - a^2 = (c+a)(c-a);$$
$$\therefore\ 2 \log x = \log(c+a) + \log(c-a).$$

$c = 3·141024$
$a = 2·593167$
$c+a = 5·734191$
$c-a = ·547858$

From the Tables, log 5·7341	=	·7584653
9		68
1		8
log ·54785	=	$\bar{1}$·7386617
7		55
By addition,		·4971394

Dividing by 2, we have

log x	=	·2485697
log 1·7724	=	·2485617
		80
3		74
		60
2		49

Thus the required side is 1·772432.

EXAMPLES. XV. c.

[*In this Exercise the logarithms are to be taken from Seven-Figure Tables.*]

1. Multiply 300·2618 by ·0078915194.

2. Find the product of 235·6783 and 357·8438.

3. Find the continued product of

   153·2419, 2·8632503, and ·07583646.

4. Divide 1·0304051 by 27·093524.

5. Divide 357·8364 by ·00318973.

6. Find x from the equation

   $$·0178345x = 21·85632.$$

7. Find the value of

   $$3·78956 \times ·0536872 \div ·0072916.$$

8. Find the cube of ·83410039.

9. Find the fifth root of 15063·018.

10. Evaluate $\sqrt[5]{384 \cdot 731}$ and $\sqrt[13]{15 \cdot 7324}$.

11. Find the product of the square root of 1034·3963 and the cube root of 353246.

12. Subtract the square of ·7503269 from the square of 1·035627.

13. Find the value of

$$\frac{(34 \cdot 7326)^{\frac{3}{5}} \times \sqrt[6]{2 \cdot 53894}}{\sqrt[5]{4 \cdot 39682}}.$$

Example 3. Find a third proportional to the cube of ·3172564 and the cube root of 23·32873.

Let x be the required third proportional; then

$$(\cdot 3172564)^3 : (23 \cdot 32873)^{\frac{1}{3}} = (23 \cdot 32873)^{\frac{1}{3}} : x;$$

whence

$$x = (23 \cdot 32873)^{\frac{2}{3}} \div (\cdot 3172564)^3;$$

$$\therefore \log x = \frac{2}{3} \log 23 \cdot 32873 - 3 \log \cdot 3172564.$$

From the Tables,

$$\begin{array}{lr}
\log \cdot 31725 & = \bar{1} \cdot 5014016 \\
\quad 6 & 82 \\
\quad\;\; 4 & 5|5 \\
\hline
 & \bar{1} \cdot 5014103|5 \\
 & 3 \\
\hline
 & \bar{2} \cdot 5042311
\end{array}
\qquad
\begin{array}{lr}
\log 23 \cdot 328 & = 1 \cdot 3678775 \\
\quad 7 & 130 \\
\quad\;\; 3 & 5|6 \\
\hline
 & 1 \cdot 3678910|6 \\
 & 2 \\
\hline
3\,| & 2 \cdot 7357821 \\
\hline
 & \cdot 9119274 \\
 & \bar{2} \cdot 5042311
\end{array}$$

$$\begin{array}{llr}
\text{By subtraction,} & \log x & = 2 \cdot 4076963 \\
 & \log 255 \cdot 67 & = 2 \cdot 4076798 \\
\hline
 & & 165 \\
 & \qquad 9 & 153 \\
\hline
 & & 120 \\
 & \qquad 7 & 119
\end{array}$$

Thus the third proportional is 255·6797.

14. Find a mean proportional between

$$\cdot 0037258169 \text{ and } \cdot 56301078.$$

15. Find a third proportional to the square of ·43607528 and the square root of ·03751786.

16. Find a fourth proportional to

$$56712\cdot 43,\quad 29\cdot 302564,\quad \cdot 33025107.$$

17. Find the geometric mean between

$$(\cdot 035689)^{\frac{2}{5}} \text{ and } (2\cdot 879432)^{\frac{3}{7}}.$$

18. Find a fourth proportional to

$$\sqrt[3]{32\cdot 7812},\quad \sqrt[5]{357\cdot 814},\quad \sqrt[4]{7836\cdot 43}.$$

19. Find the value of

$$\sin 27^\circ\ 13'\ 12'' \times \cos 46^\circ\ 2'\ 15''.$$

20. Find the value of

$$\cot 97^\circ\ 14'\ 16'' \times \sec 112^\circ\ 13'\ 5''.$$

21. Evaluate

$$\sin 20^\circ\ 13'\ 20'' \times \cot 47^\circ\ 53'\ 15'' \times \sec 42^\circ\ 15'\ 30''.$$

22. Find the value of $ab \sin C$, when

$$a = 324\cdot 1368,\quad b = 417\cdot 2431,\quad C = 113^\circ\ 14'\ 16''.$$

23. If $a : b = \sin A : \sin B$, find a, given

$$b = 378\cdot 25,\quad A = 35^\circ\ 15'\ 33'',\quad B = 119^\circ\ 14'\ 18''.$$

24. Find the smallest values of θ which satisfy the equations

(1) $\tan^3\theta = \dfrac{5}{12}$; (2) $3\sin^2\theta + 2\sin\theta = 1$.

25. Find x from the equation

$$x \times \sec 28^\circ\ 17'\ 25'' = \sin 23^\circ\ 18'\ 5'' \times \cot 38^\circ\ 15'\ 13''.$$

26. Find θ from the equation

$$\sin^3\theta = \cos^2\alpha \cot\beta,$$

where $\alpha = 32^\circ\ 47'$ and $\beta = 41^\circ\ 19'$.

Use of Four-Figure Tables.

182$_A$. *To find the logarithm of a given number from the Tables.*

Example 1. Find log 38, log 380, log ·0038.

We first find the number 38 in the left hand column on page 374. Opposite to this we find the digits 5798. This, with the decimal point prefixed, is the mantissa for the logarithms of all numbers whose significant digits are 38. Hence, prefixing the characteristics we have

$$\log 38 = 1\cdot 5798, \quad \log 380 = 2\cdot 5798, \quad \log \cdot 0038 = \bar{3}\cdot 5798.$$

Example 2. Find log 3·86, log ·0386, log 386000.

The same line as before will give the mantissa of the logarithms of all numbers which begin with 38. From this line we choose the mantissa which stands in the column headed 6. This gives ·5866 as the mantissa for all numbers whose significant digits are 386. Hence, prefixing the characteristics, we have

$$\log 3\cdot 86 = \cdot 5866, \quad \log \cdot 0386 = \bar{2}\cdot 5866, \quad \log 386000 = 5\cdot 5866.$$

182$_B$. Similarly the logarithm of any number consisting of not more than 3 significant digits can be obtained directly from the Tables. When the number has 4 significant digits, use is made of the principle that when the difference between two numbers is small compared with either of them, the difference between their logarithms is very nearly proportional to the difference between the numbers. It would be out of place to attempt any demonstration of the principle here. It will be sufficient to point out that differences in the logarithms corresponding to small differences in the numbers have been calculated, and are printed ready for use in the *difference columns* at the right hand of the Tables. The way in which these differences are used is shewn in the following example.

Example. Find (i) log 3·864; (ii) log ·003868.

Here, as before, we can find the mantissa for the sequence of digits 386. This has to be *corrected* by the addition of the figures which stand underneath 4 and 8 respectively in the difference columns

(i)	$\log 3\cdot 86$	$= \cdot 5866$	(ii)	$\log \cdot 00386$	$= \bar{3}\cdot 5866$
	diff. for 4	5		diff. for 8	9
	$\therefore \log 3\cdot 864$	$= \cdot 5871$		$\therefore \log \cdot 003868$	$= \bar{3}\cdot 5875$

Note. After a little practice the necessary 'correction' from the difference columns can be performed mentally.

182$_C$. The number corresponding to a given logarithm is called its **antilogarithm**. Thus in the last example 3·864 and ·003868 are respectively the numbers whose logarithms are ·5871 and $\bar{3}$·5875.

Hence antilog ·5871 = 3·864; antilog $\bar{3}$·5875 = ·003868.

182$_D$. *To find the antilogarithm of a given logarithm.*

In using the Tables of antilogarithms on pages 376, 377, it is important to remember that we are seeking *numbers* corresponding to *given logarithms*. Thus in the left hand column we have the first two digits of the given *mantissæ*, with the decimal point prefixed. The characteristics of the given logarithms will fix the position of the decimal point in the numbers taken from the Tables.

Example 1. Find the antilogarithm of (i) 1·583; (ii) $\bar{2}$·8249.

(i) We first find ·58 in the left hand column and pass along the horizontal line and take the number in the vertical column headed by 3. Thus ·583 is the mantissa of the logarithm of a number whose significant digits are 3828.

Hence antilog 1·583 = 38·28.

(ii) antilog $\bar{2}$·824 = ·06668
diff. for 9   14

∴ antilog $\bar{2}$·8249 = ·06682

Here corresponding to the first 3 digits of the mantissa we find the sequence of digits 6668, and the decimal point is inserted in the position corresponding to the characteristic $\bar{2}$. To the number so found we add 14 from the difference column headed 9, placing it under the two last digits of the given mantissa.

Example 2. Find the product of 72·38 and ·5689.

log 72·3	= 1·8591		
diff. for 8	5		
log ·568	= $\bar{1}$·7543	antilog 1·614	= 41·11
diff. for 9	7	diff. for 6	6
log product	= 1·6146	antilog 1·6146	= 41·17

Thus the required product is 41·17.

Example 3. Find the value of $\dfrac{3{\cdot}274 \times {\cdot}0059}{14{\cdot}83 \times {\cdot}077}$ to four significant digits.

By Art. 157, log *fraction* = log *numerator* − log *denominator*.

Numerator.		**Denominator.**	
log 3·27 =	·5145	log 14·8 =	1·1703
diff. for 4	5	diff. for 3	9
log ·0059 =	$\bar{3}{\cdot}7709$	log ·077 =	$\bar{2}{\cdot}8865$
log *numerator* =	$\bar{2}{\cdot}2859$	log *denominator* =	·0577
	$\bar{2}{\cdot}2859$	antilog $\bar{2}{\cdot}228$ =	·01690
subtract	·0577	diff. for 2	1
log *fraction* =	$\bar{2}{\cdot}2282$	antilog $\bar{2}{\cdot}2282$ =	·01691

Thus $$\frac{3{\cdot}274 \times {\cdot}0059}{14{\cdot}83 \times {\cdot}077} = {\cdot}01691.$$

Example 4. Find the value of $\dfrac{\left(330 \times \frac{1}{49}\right)^4}{\sqrt[3]{22 \times 6{\cdot}9}}$ to the nearest integer.

Denote the expression by x, then

$$\log x = 4\,(\log 330 - \log 49) - \frac{1}{3}\,(\log 22 + \log 6{\cdot}9),$$

log 330 =	2·5185	log 22 =	1·3424
log 49 =	1·6902	log 6·9 =	·8388
	·8283	3	2·1812
	4		·7271
	3·3132		
subtract	·7271		

log x = 2·5861 = antilog 385·6, from the Tables.

∴ x = 386, to the nearest integer.

EXAMPLES. XV. d.

[*Answers to be given to four significant figures.*]

Find by means of the Tables the value of the following products:

1. 2834 × 17·62. **2.** 8·034 × 1893. **3.** ·00567 × ·0297.

4. 3·7 × 8·9 × ·023. **5.** 31·9 × 1·51 × 9·7. **6.** 43 × 8·07 × ·0392.

Find the value of

7. $\dfrac{17{\cdot}3}{294{\cdot}8}$. **8.** $\dfrac{2{\cdot}035}{837{\cdot}6}$. **9.** $\dfrac{{\cdot}2179}{{\cdot}08973}$. **10.** $\dfrac{487}{6398}$.

11. $\dfrac{2\cdot38 \times 3\cdot901}{4\cdot83}$. 12. $\dfrac{14\cdot72 \times 38\cdot05}{387\cdot9}$. 13. $\dfrac{925\cdot9 \times 1\cdot597}{74\cdot03}$.

14. $\dfrac{15\cdot38 \times \cdot0137}{276 \times \cdot0038}$. 15. $\dfrac{2\cdot31 \times \cdot037 \times 1\cdot43}{\cdot0561 \times 3\cdot87 \times \cdot0091}$.

16. $\sqrt{5\cdot1}$. 17. $\sqrt[3]{11}$. 18. $\sqrt[3]{82\cdot56}$. 19. $\sqrt[4]{10\cdot15}$.

20. $(\cdot097)^4$. 21. $(2\cdot301)^5$. 22. $(51\cdot32)^{\frac{2}{3}}$. 23. $(\cdot089)^{\frac{4}{7}}$.

24. $\sqrt{\dfrac{\cdot0137 \times \cdot0296}{873\cdot5}}$. 25. $\dfrac{83 \times \sqrt[3]{92}}{127 \times \sqrt[5]{246}}$.

26. Find the value of $\sqrt{\dfrac{\cdot678 \times 9\cdot01}{\cdot0234}}$ to the nearest integer.

27. Find a mean proportional between 2·87 and 30·08; and a third proportional to ·0238 and 7·805.

28. Find a mean proportional between

$$\sqrt[3]{347\cdot3} \text{ and } \sqrt[5]{256\cdot4}.$$

29. By taking logarithms, and solving for $\log x$ and $\log y$, find values of x and y (to four significant digits) which satisfy the equations

$$x^5y^3 = 5, \quad x^2y^7 = 11.$$

Example 5. Find the value of the expression $\dfrac{\rho\omega\lambda\pi^2\eta^2}{\tau^2}$ when

$$\rho = 7\cdot8, \quad \omega = \cdot66, \quad \tau = \frac{1}{256}, \quad \pi = 3\cdot1416, \quad \lambda = 14, \quad \eta = \cdot025.$$

$$\begin{aligned}
\log\rho &= \log 7\cdot8 &&= \cdot8921 \\
\log\omega &= \log \cdot66 &&= \bar{1}\cdot8195 \\
\log\lambda &= \log 14 &&= 1\cdot1461 \\
\log\pi^2 &= 2\log 3\cdot1416 &&= \cdot9942 \\
\log\eta^2 &= 2\log \cdot025 &&= \bar{4}\cdot7958 \\
\log\frac{1}{\tau^2} &= 2\log 256 &&= 4\cdot8164 \\
& && \overline{4\cdot4641}
\end{aligned}$$

$$\begin{aligned}
\text{antilog } \cdot464 &= 2\cdot911 \\
\text{diff. for } \quad 1 & \quad\quad 1 \\
\text{antilog } \cdot4641 &= \overline{2\cdot912}
\end{aligned}$$

$$\therefore \text{ antilog } 4\cdot4641 = 29120.$$

EXAMPLES. XV. d. (*Continued.*)

30. Find the value of $2\pi\sqrt{\frac{l}{g}}$, when

$$l=2{\cdot}863,\quad g=32{\cdot}19,\quad \pi=3{\cdot}1416.$$

31. When $m=18{\cdot}34$, $v=35{\cdot}28$, find the value of $\frac{1}{2}mv^2$.

32. Calculate the values of

(i) pr^n, where $p=93{\cdot}75$, $r=1{\cdot}03$, $n=4$;

(ii) $\frac{4}{3}\pi r^3$, where $\pi=\frac{355}{113}$, $r=5{\cdot}875$.

33. If $F=\frac{mv^2}{gr}$, find F when

$$m=33{\cdot}47,\quad r=9{\cdot}6,\quad v=60,\quad g=32{\cdot}19.$$

34. Find r from the formula $V=\frac{4}{3}\pi r^3$, given

$$V=537{\cdot}6,\quad \pi=3{\cdot}1416.$$

35. If $s=\frac{1}{2}ft^2$, find f when $s=289{\cdot}3$, $t=3\frac{7}{8}$.

36. If $x^n y=8{\cdot}7\times10^8$, find n when $x=73{\cdot}96$ and $y=27{\cdot}25$.

37. The volume of a sphere of radius r is given by the formula $V=\frac{4}{3}\pi r^3$; find the radius of a sphere whose volume is 33·87 cu. cm.

38. A cubical block of metal, each edge of which is 36·4 cm., is melted down into a sphere. Find the diameter of the sphere as correctly as possible with Four-Figure Tables.

39. If $\frac{v^2}{r}=\frac{g}{289}$, calculate v, having given that

$$r=4000,\quad g=\frac{32{\cdot}2}{5280}.$$

Also shew that the value of $\frac{2\pi r}{v\times60\times60}$, where $\pi=3{\cdot}1416$, is approximately 24.

Four-Figure Tables of Logarithmic Functions.

182_E. The use of Tables of *natural* sines, cosines, and tangents has been explained in Chap. IV. [See Art. 39_A.]

The logarithms of the functions have also been calculated to four places of decimals. Since many of the trigonometrical functions are less than unity their logarithms are negative, and as the characteristics are not always evident on inspection they cannot be omitted. To avoid the inconvenience of printing the bars over the characteristics, the logarithms are increased by 10 and are then registered under the name of **tabular logarithmic functions.**

The notation used is $L\sin A$, $L\tan\theta$; thus $L\sin A$ is equivalent to $\log\sin A + 10$.

The Tables on pages 384--389 give the logarithmic sines, cosines, and tangents of all angles between 0° and 90° at intervals of 6 minutes. For intermediate angles the difference columns are used in the same way as for the natural functions.

182_F. When there is a change in the characteristic in the middle of a horizontal line, the transition is marked by printing a bar over the mantissa.

Thus from the second page of logarithmic tangents we have

	0′	6′	12′	18′	24′	30′	36′	42′	48′	54′
84°	10·9784	9857	9932	$\overline{0008}$	$\overline{0085}$	$\overline{0164}$	$\overline{0244}$	$\overline{0326}$	$\overline{0409}$	$\overline{0494}$
85°	11·0580	0669	0759	0850	0944	1040	1138	1238	1341	1446

Here the transition from **10·9...** to **11·0...** begins at 84° 18′. Thus

$$L\tan 84^\circ\, 18' = 11\cdot0008,$$

and for all subsequent angles in this line the characteristic of the logarithmic tangent is 11 instead of 10.

Obs. As the angle increases from 0° to 90° the logarithmic sine, tangent, and secant *increase*, while the cosine, cotangent, and cosecant *decrease*.

Example 1. Find the value of $L \sin 41° 15'$.

From the Tables,

$$\begin{array}{rr} L \sin 41° 12' = & 9 \cdot 8187 \\ \text{diff. for } \quad 3' & 4 \\ \hline L \sin 41° 15' = & 9 \cdot 8191 \end{array}$$

Example 2. From the equation $L \cos \theta = 9 \cdot 6577$, find θ to the nearest minute.

$$\begin{array}{ll} L \cos \theta = 9 \cdot 6577 & \\ L \cos 63° = 9 \cdot 6570, & \text{from the Tables;} \\ \text{diff.} \qquad\qquad 7, & \text{by subtraction.} \end{array}$$

Hence θ is *less* than 63° by a number of minutes corresponding to a difference 7. Taking the nearest difference in the Table we find that 3′ must be subtracted from 63°. Thus $\theta = 62° 57'$.

182$_G$. The device for securing a positive characteristic mentioned in Art. 182$_E$ is merely a convenience for the purposes of tabulation. In practice it is more expeditious if the 10 is subtracted mentally in copying down the logarithms, as shewn in the following example.

Example. Find from the Tables the value of

$$\frac{\cot 27° 12' \times \sin 34° 17'}{\sec 77° 23'}.$$

Denote the expression by x, then

$$x = \tan 62° 48' \times \sin 34° 17' \times \cos 12° 37'.$$

$$\therefore \log x = \log \tan 62° 48' + \log \sin 34° 17' + \log \cos 12° 37'.$$

$$\begin{array}{lll} \log \tan 62° 48' = \cdot 2891, & \text{subtracting the 10,} & \\ \log \sin 34° 17' = \bar{1} \cdot 7507, & \text{,,} & \text{,,} \\ \log \cos 12° 37' = \bar{1} \cdot 9894, & \text{,,} & \text{,,} \\ \hline \log x \qquad = \cdot 0292 & & \end{array}$$

And $\qquad$ antilog $\cdot 0292 = 1 \cdot 069$.

Thus $\qquad x = 1 \cdot 069$.

EXAMPLES. XV. e.

Find from the Tables the values of

1. $\sin 29° 38'$.
2. $\tan 38° 25'$.
3. $\cos 15° 11'$.
4. $\sec 48° 46'$.
5. $\cot 42° 27'$.
6. $\text{cosec}\, 55° 22'$.

Find to the nearest minute the values of θ from the following equations:

7. $\sin\theta = \cdot 3126$.
8. $\cos\theta = \cdot 1782$.
9. $\tan\theta = \frac{5}{7}$.
10. $\cot\theta = \cdot 7931$.
11. $\cot\theta = 1\cdot 5321$.
12. $\cos\theta = \frac{2}{3}$.

Find the values of

13. $L\sin 44° 17'$.
14. $L\cot 27° 34'$.
15. $L\cos 72° 15'$.

Evaluate

16. $\sin 27° 13' \times \cos 46° 16'$.
17. $\dfrac{\sin 47° 13'}{\tan 22° 27'}$.
18. $\dfrac{\sin 34° 17' \times \tan 82° 6'}{\cos 12° 37'}$.
19. $\dfrac{\cos 28° 14'}{\sec 37° 26'}$.

20. Find the smallest positive angle which satisfies the equation $\tan^7 x = \dfrac{11}{13}$.

21. Find the value of $ab\sin C$ when $a = 32\cdot 73$, $b = 27\cdot 86$, $C = 30° 16'$.

22. Calculate the value of

(i) $na^2 \cot\dfrac{\pi}{n}$, when $a = 2\cdot 0$, $n = 8$;

(ii) $\dfrac{n}{2} r^2 \sin\dfrac{2\pi}{n}$, when $r = 3\cdot 3$, $n = 10$.

23. Given $\tan\phi = \dfrac{2e}{1-e^2}\sin\theta$, find ϕ when $e = \cdot 35$, $\theta = 56° 14'$.

24. The length of the bisector of the angle A of a triangle ABC is $\dfrac{2bc}{b+c}\cos\dfrac{A}{2}$. Assuming this, find its value when

$$b = 32\cdot 78, \quad c = 19\cdot 23, \quad A = 115° 34'.$$

25. Evaluate $\dfrac{2v^2 \sin a}{g(1-e)^2 \cos^2 a}$, when $v = 48$, $a = 23°$, $g = 32\cdot 19$, $e = \cdot 37$.

CHAPTER XVI.

SOLUTION OF TRIANGLES WITH LOGARITHMS.

(The section on Four-Figure Tables, page 183_{A} *may be taken after Art.* 188.)

183. The examples on the solution of triangles in Chap. XIII. furnish a useful exercise on the formulæ connecting the sides and angle of a triangle; but in practical work much of the labour of arithmetical calculation is avoided by the use of logarithms.

We shall now shew how the formulæ of Chap. XIII. may be used or adapted for use in connection with logarithmic Tables.

184. *To find the functions of the half-angles in terms of the sides.*

We have $$2\sin^2\frac{A}{2}=1-\cos A$$

$$=1-\frac{b^2+c^2-a^2}{2bc}$$

$$=\frac{2bc-b^2-c^2+a^2}{2bc}=\frac{a^2-(b^2-2bc+c^2)}{2bc}$$

$$=\frac{a^2-(b-c)^2}{2bc}=\frac{(a+b-c)(a-b+c)}{2bc}.$$

Let $$a+b+c=2s;$$

then $$a+b-c=2s-2c=2(s-c),$$

and $$a-b+c=2s-2b=2(s-b).$$

$$\therefore\ 2\sin^2\frac{A}{2}=\frac{4(s-c)(s-b)}{2bc}=\frac{2(s-b)(s-c)}{bc};$$

$$\therefore\ \sin\frac{A}{2}=\sqrt{\frac{(s-b)(s-c)}{bc}}.$$

Again, $2\cos^2\frac{A}{2}=1+\cos A=1+\frac{b^2+c^2-a^2}{2bc}$

$$=\frac{(b+c)^2-a^2}{2bc}=\frac{(b+c+a)(b+c-a)}{2bc};$$

$$\therefore\quad 2\cos^2\frac{A}{2}=\frac{4s(s-a)}{2bc}=\frac{2s(s-a)}{bc};$$

$$\therefore\quad \cos\frac{A}{2}=\sqrt{\frac{s(s-a)}{bc}}.$$

Also $\tan\frac{A}{2}=\sin\frac{A}{2}\div\cos\frac{A}{2}$

$$=\sqrt{\frac{(s-b)(s-c)}{bc}\times\frac{bc}{s(s-a)}};$$

$$\therefore\quad \tan\frac{A}{2}=\sqrt{\frac{(s-b)(s-c)}{s(s-a)}}.$$

185. Similarly it may be proved that

$$\sin\frac{B}{2}=\sqrt{\frac{(s-c)(s-a)}{ca}},\qquad \sin\frac{C}{2}=\sqrt{\frac{(s-a)(s-b)}{ab}};$$

$$\cos\frac{B}{2}=\sqrt{\frac{s(s-b)}{ca}},\qquad \cos\frac{C}{2}=\sqrt{\frac{s(s-c)}{ab}};$$

$$\tan\frac{B}{2}=\sqrt{\frac{(s-c)(s-a)}{s(s-b)}},\qquad \tan\frac{C}{2}=\sqrt{\frac{(s-a)(s-b)}{s(s-c)}}.$$

In each of these formulæ the positive value of the square root must be taken, for each half angle is less than 90°, so that all its functions are positive.

186. *To find* $\sin A$ *in terms of the sides.*

$$\sin A=2\sin\frac{A}{2}\cos\frac{A}{2}$$

$$=2\sqrt{\frac{(s-b)(s-c)}{bc}\times\frac{s(s-a)}{bc}};$$

$$\therefore\quad \sin A=\frac{2}{bc}\sqrt{s(s-a)(s-b)(s-c)}.$$

We may also obtain this formula in another way which is instructive.

We have

$$\sin^2 A = 1 - \cos^2 A = (1+\cos A)(1-\cos A)$$

$$= \left(1+\frac{b^2+c^2-a^2}{2bc}\right)\left(1-\frac{b^2+c^2-a^2}{2bc}\right)$$

$$= \frac{(b+c)^2-a^2}{2bc} \times \frac{a^2-(b-c)^2}{2bc}$$

$$= \frac{(b+c+a)(b+c-a)(a+b-c)(a-b+c)}{4b^2c^2}$$

$$= \frac{16s(s-a)(s-b)(s-c)}{4b^2c^2};$$

$$\therefore \quad \sin A = \frac{2}{bc}\sqrt{s(s-a)(s-b)(s-c)}.$$

The positive value of the square root must be taken, since the *sine* of an angle of any triangle is always positive.

EXAMPLES. XVI. a.

Prove the following formulæ in any triangle:

1. $b\cos^2\frac{A}{2}+a\cos^2\frac{B}{2}=s.$ 2. $s\tan\frac{B}{2}\tan\frac{C}{2}=s-a.$

3. $\frac{\text{vers } A}{\text{vers } B}=\frac{a(a+c-b)}{b(b+c-a)}.$ 4. $b\sin^2\frac{A}{2}+a\sin^2\frac{B}{2}=s-c.$

5. $(s-a)\tan\frac{A}{2}=(s-b)\tan\frac{B}{2}=(s-c)\tan\frac{C}{2}.$

6. Find the value of $\tan\frac{B}{2}$, when $a=10$, $b=17$, $c=21$.

7. Find $\cot\frac{C}{2}$, when $a=13$, $b=14$, $c=15$.

8. Prove that

$$\frac{1}{a}\cos^2\frac{A}{2}+\frac{1}{b}\cos^2\frac{B}{2}+\frac{1}{c}\cos^2\frac{C}{2}=\frac{s^2}{abc}.$$

9. Prove that

$$\frac{b-c}{a}\cos^2\frac{A}{2}+\frac{c-a}{b}\cos^2\frac{B}{2}+\frac{a-b}{c}\cos^2\frac{C}{2}=0.$$

187. *To solve a triangle when the three sides are given.*

From the formula

$$\tan\frac{A}{2}=\sqrt{\frac{(s-b)(s-c)}{s(s-a)}},$$

$$\log\tan\frac{A}{2}=\frac{1}{2}\{\log(s-b)+\log(s-c)-\log s-\log(s-a)\};$$

whence $\frac{A}{2}$ may be obtained by the help of the Tables.

Similarly B can be found from the formula for $\tan\frac{B}{2}$, and then C from the equation $C=180°-A-B$.

In the above solution, we shall require to look out from the tables *four* logarithms only, namely those of s, $s-a$, $s-b$, $s-c$; whereas if we were to solve from the sine or cosine formulæ we should require *six* logarithms; for

$$\cos\frac{A}{2}=\sqrt{\frac{s(s-a)}{bc}} \quad \text{and} \quad \cos\frac{B}{2}=\sqrt{\frac{s(s-b)}{ca}},$$

so that we should have to look out the logarithms of the *six* quantities s, $s-a$, $s-b$, a, b, c.

If therefore *all* the angles have to be found by the use of the tables it is best to solve from the tangent formulæ; but if *one* angle only is required it is immaterial whether the sine, cosine, or tangent formula is used.

In cases where a solution has to be obtained from certain given logarithms, the choice of formulæ must depend on the data.

NOTE. We shall always find the angles to the nearest second, so that, on account of the multiplication by 2, the half-angles should be found to the nearest tenth of a second.

188. In Art. 178 we have mentioned that 10 is added to each of the logarithmic functions before they are registered as *tabular* logarithms; but this device is introduced only as a convenience for the purposes of tabulation, and in practice it will be found that the work is more expeditious if the tabular logarithms are not used. The 10 should be subtracted mentally in copying down the logarithms. Thus we should write

$$\log\sin 64°\,15'=\bar{1}\cdot9545793, \quad \log\cot 18°\,35'=\cdot4733850,$$

and in the arrangement of the work care must be taken to keep the mantissæ positive.

Example 1. The sides of a triangle are 35, 49, 63; find the greatest angle; given $\log 2 = \cdot 3010300$, $\log 3 = \cdot 4771213$,

$$L \cos 47^\circ 53' = 9\cdot 8264910, \text{ diff. for } 60'' = 1397.$$

Since the *angles* of a triangle depend only on the *ratios* of the sides and not on their actual magnitudes, we may substitute for the sides any lengths proportional to them. Thus in the present case we may take $a = 5$, $b = 7$, $c = 9$; then C is the greatest angle, and

$$\cos\frac{C}{2} = \sqrt{\frac{s(s-c)}{ab}} = \sqrt{\frac{21}{2}\times\frac{3}{2}\times\frac{1}{5\times 7}} = \sqrt{\frac{9}{20}};$$

$$\therefore \log\cos\frac{C}{2} = \frac{1}{2}(2\log 3 - \log 2 - 1).$$

$$\begin{array}{r} 2\log 3 = \cdot 9542426 \\ \underline{1\cdot 3010300} \\ 2\,)\,\underline{\bar{1}\cdot 6532126} \\ \bar{1}\cdot 8266063 \end{array}$$

Thus $\quad \log\cos\frac{C}{2} = \bar{1}\cdot 8266063$

$$\begin{array}{rr} \log\cos 47^\circ 53' = & \underline{\bar{1}\cdot 8264910} \\ \text{diff.} & 1153 \end{array}$$

$$\therefore \text{ proportional } \textit{decrease} = \frac{1153}{1397}\times 60'' = 49\cdot 5'';$$

$$\therefore \frac{C}{2} = 47^\circ 52' 10\cdot 5''.$$

$$\begin{array}{r} 1153 \\ \underline{60} \\ 1397\,)\,69180\,(\,49\cdot 5 \\ \underline{5588} \\ 13300 \\ \underline{12573} \\ 7270 \end{array}$$

Thus the greatest angle is $95^\circ 44' 21''$.

Example 2. If $a = 283$, $b = 317$, $c = 428$, find all the angles.

$$\tan\frac{A}{2} = \sqrt{\frac{(s-b)(s-c)}{s(s-a)}} = \sqrt{\frac{197\times 86}{514\times 231}};$$

$$\therefore \log\tan\frac{A}{2} = \frac{1}{2}(\log 197 + \log 86 - \log 514 - \log 231).$$

$$\begin{array}{r} 283 \\ 317 \\ \underline{428} \\ 2\,)\,\underline{1028} \\ 514 = s \\ 231 = s-a \\ 197 = s-b \\ 86 = s-c \end{array}$$

From the Tables,

$$\begin{array}{rr} \log 197 = & 2\cdot 2944662 \\ \log\ \ 86 = & \underline{1\cdot 9344985} \\ & 4\cdot 2289647 \\ & \underline{5\cdot 0745751} \\ & 2\,)\,\underline{\bar{1}\cdot 1543896} \end{array} \qquad \begin{array}{rr} \log 514 = & 2\cdot 7109631 \\ \log 231 = & \underline{2\cdot 3636120} \\ & 5\cdot 0745751 \end{array}$$

$$\log\tan\frac{A}{2} = \bar{1}\cdot 5771948$$

$$\begin{array}{rr} \log\tan 20^\circ 41' = & \underline{\bar{1}\cdot 5769585} \\ \text{diff.} & 2363 \end{array}$$

But diff. for 60″ is 3822,

$\therefore$ prop^l. increase $= \dfrac{2363}{3822} \times 60'' = 37{\cdot}1''$;

$\therefore \dfrac{A}{2} = 20^\circ\, 41'\, 37{\cdot}1''$ and $A = 41^\circ\, 23'\, 14''$.

```
         2363
           60
3822 ) 141780 ( 37·1
        11466
        -----
         27120
         26754
         -----
          3660
```

Again, $\tan\dfrac{B}{2} = \sqrt{\dfrac{(s-c)(s-a)}{s(s-b)}} = \sqrt{\dfrac{86 \times 231}{514 \times 197}}$;

$\therefore \log\tan\dfrac{B}{2} = \dfrac{1}{2}(\log 86 + \log 231 - \log 514 - \log 197)$.

$$\begin{array}{rl} \log\ 86 = & 1{\cdot}9344985 \\ \log 231 = & 2{\cdot}3636120 \\ \hline & 4{\cdot}2981105 \\ & 5{\cdot}0054293 \\ \hline 2\ ) & \bar{1}{\cdot}2926812 \\ \hline \log\tan\frac{B}{2} = & \bar{1}{\cdot}6463406 \\ \log\tan 23^\circ\, 53' = & \bar{1}{\cdot}6461988 \\ \hline \text{diff.} & 1418 \end{array} \qquad \begin{array}{rl} \log 514 = & 2{\cdot}7109631 \\ \log 197 = & 2{\cdot}2944662 \\ \hline & 5{\cdot}0054293 \end{array}$$

But diff. for 60″ is 3412;

$\therefore$ prop^l. increase $= \dfrac{1418}{3412} \times 60'' = 24{\cdot}9''$.

$\therefore \dfrac{B}{2} = 23^\circ\, 53'\, 24{\cdot}9''$ and $B = 47^\circ\, 46'\, 50''$.

```
         1418
           60
3412 ) 85080 ( 24·9
        6824
        -----
        16840
        13648
        -----
         31920
```

Thus $A = 41^\circ\, 23'\, 14''$, $B = 47^\circ\, 46'\, 50''$, $C = 90^\circ\, 49'\, 56''$.

EXAMPLES. XVI. b.

1. The sides of a triangle are 5, 8, 11; find the greatest angle; given $\log 7 = {\cdot}8450980$,

$$L\sin 56^\circ\, 47' = 9{\cdot}9225205, \quad L\sin 56^\circ\, 48' = 9{\cdot}9226032.$$

2. If $a = 40$, $b = 51$, $c = 43$, find A; given

$$L\tan 24^\circ\, 44'\, 13'' = 9{\cdot}6634464,$$
$$\log 128 = 2{\cdot}1072100, \quad \log 603 = 2{\cdot}7803173.$$

3. The sides a, b, c are as $4 : 5 : 6$, find B; given $\log 2$,

$$L\cos 27^\circ\, 53' = 9{\cdot}9464040, \quad \text{diff. for } 1' = 669.$$

4. Find the greatest angle of the triangle in which the sides are 5, 6, 7; given $\log 6 = \cdot 7781513$,

$$L \cos 39^\circ\ 14' = 9\cdot 8890644, \quad \text{diff. for } 1' = 1032.$$

5. If $a=3$, $b=1\cdot 75$, $c=2\cdot 75$, find C; given $\log 2$,

$$L \tan 32^\circ\ 18' = 9\cdot 8008365, \quad \text{diff. for } 1' = 2796.$$

6. If the sides are 24, 22, 14, find the least angle; given

$$L \tan 17^\circ\ 33' = 9\cdot 500042, \quad \text{diff. for } 1' = 439.$$

7. Find the greatest angle when the sides are 4, 10, 11; given $\log 2$, $\log 3$,

$$L \cos 46^\circ\ 47' = 9\cdot 8355378, \quad \text{diff. for } 1' = 1345.$$

8. If $a : b : c = 15 : 13 : 14$, find the angles; given $\log 2$, $\log 3$, $\log 7$,

$$L \tan 26^\circ\ 33' = 9\cdot 6986847, \quad \text{diff. for } 1' = 3159,$$
$$L \tan 29^\circ\ 44' = 9\cdot 7567587, \quad \text{diff. for } 1' = 2933.$$

9. If $a : b : c = 3 : 4 : 2$, find the angles; given $\log 2$, $\log 3$,

$$L \tan 14^\circ\ 28' = 9\cdot 4116146, \quad \text{diff. for } 10'' = 870,$$
$$L \tan 52^\circ\ 14' = 10\cdot 1108395, \quad \text{diff. for } 10'' = 435.$$

189. *To solve a triangle having given two sides and the included angle.*

Let the given parts be b, c, A, and let

$$k = \frac{\sin B}{b} = \frac{\sin C}{c};$$

then
$$\frac{\sin B - \sin C}{\sin B + \sin C} = \frac{kb - kc}{kb + kc} = \frac{b - c}{b + c};$$

$$\therefore \frac{2 \cos \dfrac{B+C}{2} \sin \dfrac{B-C}{2}}{2 \sin \dfrac{B+C}{2} \cos \dfrac{B-C}{2}} = \frac{b-c}{b+c};$$

$$\therefore \frac{\tan \dfrac{B-C}{2}}{\tan \dfrac{B+C}{2}} = \frac{b-c}{b+c};$$

$$\therefore \tan\frac{B-C}{2}=\frac{b-c}{b+c}\tan\frac{B+C}{2}=\frac{b-c}{b+c}\cot\frac{A}{2},$$

since $\dfrac{B+C}{2}=90^\circ-\dfrac{A}{2}$.

$$\therefore \log\tan\frac{B-C}{2}=\log(b-c)-\log(b+c)+\log\cot\frac{A}{2},$$

from which equation we can find $\dfrac{B-C}{2}$.

Also $\dfrac{B+C}{2}=90^\circ-\dfrac{A}{2}$, and is therefore known.

By addition and subtraction we obtain B and C.

From the equation $\qquad a=\dfrac{b\sin A}{\sin B}$,

$$\log a=\log b+\log\sin A-\log\sin B;$$

whence a may be found.

Example 1. If the sides a and b are in the ratio of 7 to 3, and the included angle C is 60°, find A and B; given

$$\log 2=\cdot 3010300,\quad \log 3=\cdot 4771213,$$
$$L\tan 34^\circ 42'=9\cdot 8403776,\ \text{diff. for } 1'=2699.$$

$$\tan\frac{A-B}{2}=\frac{a-b}{a+b}\cot\frac{C}{2}=\frac{7-3}{7+3}\cot 30^\circ=\frac{4}{10}\sqrt{3};$$

$$\therefore \log\tan\frac{A-B}{2}=2\log 2-1+\frac{1}{2}\log 3;$$

$$\therefore \log\tan\frac{A-B}{2}=\bar{1}\cdot 8406207$$
$$\log\tan 34^\circ 42'=\bar{1}\cdot 8403776$$
$$\text{diff.}\qquad 2431$$

$$\therefore \text{prop}^{l}.\ \text{increase}=\frac{2431}{2699}\times 60''=54'';$$

$$\therefore \frac{A-B}{2}=34^\circ 42' 54''.$$

And $\qquad \dfrac{A+B}{2}=90^\circ-\dfrac{C}{2}=60^\circ.$

By addition, $\qquad A=94^\circ 42' 54''$,

and by subtraction, $\qquad B=25^\circ 17' 6''$.

$$2\log 2=\cdot 6020600$$
$$\tfrac{1}{2}\log 3=\cdot 2385607$$
$$\overline{\cdot 8406207}$$

```
         2431
           60
2699 ) 145860 ( 54
        13495
        -----
        10910
        10796
```

Example 2. If $a=681$, $c=243$, $B=50° 42'$, solve the triangle, by the use of Tables.

$$\tan\frac{A-C}{2}=\frac{a-c}{a+c}\cot\frac{B}{2}=\frac{438}{924}\cot 25° 21';$$

$$\therefore \log\tan\frac{A-C}{2}=\log 438-\log 924 + \log\cot 25° 21'$$

$$\therefore \log\tan\frac{A-C}{2}=\cdot 0002383$$

$$\log\tan 45°=\cdot 0000000$$

$$\text{diff.} \quad 2383$$

$$\begin{aligned} \log 438 &= 2\cdot 6414741 \\ \log\cot 25° 21' &= \cdot 3244362 \\ & 2\cdot 9659103 \\ \log 924 &= 2\cdot 9656720 \\ & \cdot 0002383 \end{aligned}$$

And diff. for 60″ is 2527;

$$\therefore \text{prop}^{l}. \text{ increase}=\frac{2383}{2527}\times 60''=57'';$$

$$\therefore \frac{A-C}{2}=45° 0' 57''.$$

$$\begin{array}{r} 2383 \\ 60 \\ 2527\,)\,142980\,(\,56\cdot 6 \\ 12635 \\ 16630 \\ 15162 \\ 14680 \end{array}$$

Also $\dfrac{A+C}{2}=90°-\dfrac{B}{2}=64° 39'$.

By addition, $A=109° 39' 57''$,

and by subtraction, $C=19° 38' 3''$.

Again, $b=\dfrac{c\sin B}{\sin C}$;

$$\therefore \log b=\log c+\log\sin B-\log\sin C$$

$$=\log 243+\log\sin 50° 42' - \log\sin 19° 38' 3''$$

$$\therefore \log b=2\cdot 7479012$$

$$\log 559\cdot 63 = 2\cdot 7479010$$

$$\therefore b=559\cdot 63.$$

$$\begin{aligned} \log\sin 19° 38' &= \bar{1}\cdot 5263387 \\ \frac{3}{60}\times 3540 &= \phantom{\bar{1}\cdot 5263}177 \\ \log\sin 19° 38' 3'' &= \bar{1}\cdot 5263564 \\ \log 243 &= 2\cdot 3856063 \\ \log\sin 50° 42' &= \bar{1}\cdot 8886513 \\ & 2\cdot 2742576 \\ \log\sin 19° 38' 3'' &= \bar{1}\cdot 5263564 \\ & 2\cdot 7479012 \end{aligned}$$

Thus $A=109° 39' 57''$, $C=19° 38' 3''$, $b=559\cdot 63$.

190. From the formula

$$\tan\frac{B-C}{2}=\frac{b-c}{b+c}\cot\frac{A}{2},$$

it will be seen that if b, c, and $B-C$ are known A can be found; that is, the triangle can be solved when the given parts are two sides and the difference of the angles opposite to them.

EXAMPLES. XVI. c.

1. If $a=9$, $b=6$, $C=60°$, find A and B; given $\log 2$, $\log 3$, $L\tan 19° \, 6'=9{\cdot}5394287$, $L\tan 19° \, 7'=9{\cdot}5398371$.

2. If $a=1$, $c=9$, $B=65°$, find A and C; given $\log 2$,
$L\cot 32° \, 30'=10{\cdot}1958127$,
$L\tan 51° \, 28'=10{\cdot}0988763$, diff. for $1'=2592$.

3. If $17a=7b$, $C=60°$, find A and B; given $\log 2$, $\log 3$,
$L\tan 35° \, 49'=9{\cdot}8583357$, diff. for $10''=2662$.

4. If $b=27$, $c=23$, $A=44° \, 30'$, find B and C; given $\log 2$,
$L\cot 22° \, 15'=10{\cdot}3881591$,
$L\tan 11° \, 3'=9{\cdot}2906713$, diff. for $1'=6711$.

5. If $c=210$, $a=110$, $B=34° \, 42' \, 30''$, find C and A; given $\log 2$,
$L\cot 17° \, 21' \, 15''=10{\cdot}5051500$.

6. Two sides of a triangle are as 5 : 3 and include an angle of $60° \, 30'$: find the other angles; given $\log 2$,
$L\cot 30° \, 15'=10{\cdot}23420$,
$L\tan 23° \, 13'=9{\cdot}63240$, diff. for $1'=35$.

7. If $a=327$, $c=256$, $B=56° \, 28'$, find A and C; given
$\log 7{\cdot}1={\cdot}8512583$, $\log 5{\cdot}83={\cdot}7656686$,
$L\tan 61° \, 46'=10{\cdot}2700705$,
$L\tan 12° \, 46'=9{\cdot}3552267$, diff. for $1'=5859$.

8. If $b=4c$, $A=65°$, find B and C; given $\log 2$, $\log 3$,
$L\tan 57° \, 30'=10{\cdot}1958127$,
$L\tan 43° \, 18'=9{\cdot}9742133$, diff. for $1'=2531$.

9. If $a=23031$, $b=7677$, $C=30° \, 10' \, 5''$, find A and B; given $\log 2$,
$L\tan 15° \, 5'=9{\cdot}4305727$, diff. for $10''=838$,
$L\cot 61° \, 41'=9{\cdot}7314436$, diff. for $10''=504$.

191. *To solve a triangle having given two angles and a side.*

Let the given parts be denoted by B, C, a; then the third angle A is found from the equation $A=180°-B-C$,

and
$$b=\frac{a\sin B}{\sin A};$$

$$\therefore\ \log b=\log a+\log\sin B-\log\sin A:$$

whence b may be found.

Similarly, c may be obtained from the equation
$$\log c=\log a+\log\sin C-\log\sin A.$$

Example. If $b=1000$, $A=45°$, $C=68°\ 17'\ 40''$, find the least side, having given

$$\log 2=\cdot 3010300,\ \log 7\cdot 6986=\cdot 8864118, \text{ diff. for } 1=57,$$
$$L\sin 66°\ 42'=9\cdot 9630538, \text{ diff. for } 1'=544.$$

$B=180°-45°-68°\ 17'\ 40''=66°\ 42'\ 20''$.

The least side $=a=\dfrac{b\sin A}{\sin B}=\dfrac{1000\sin 45°}{\sin 66°\ 42'\ 20''}$;

$$\therefore\ \log a=3+\log\frac{1}{\sqrt{2}}-\log\sin 66°\ 42'\ 20''$$
$$=3-\frac{1}{2}\log 2-\log\sin 66°\ 42'\ 20''$$
$$=3-\cdot 1135869$$

$$\log\sin 66°\,42'=\bar{1}\cdot 9630538$$
$$\frac{20}{60}\times 544=\qquad 181$$
$$\frac{1}{2}\log 2=\ \cdot 1505150$$
$$\cdot 1135869$$

$$\therefore\ \log a=2\cdot 8864131$$
$$\log 769\cdot 86=2\cdot 8864118$$
$$\text{diff.}\qquad 13$$

$$\therefore\ \text{prop}^{l}.\ \text{increase}=\frac{13}{57}=\cdot 22.$$

Thus the least side is $769\cdot 8622$.

EXAMPLES. XVI. d.

1. If $B=60°\ 15'$, $C=54°\ 30'$, $a=100$, find c; given
   $L\sin 54°\ 30'=9\cdot 9106860$, $\log 8\cdot 9646162=\cdot 9525317$,
   $L\sin 65°\ 15'=9\cdot 9581543$.

2. If $A=55°$, $B=65°$, $c=270$, find a; given log 2, log 3,
$\log 25538=4\cdot4071869$, $L \sin 55°=9\cdot9133645$,
$\log 25539=4\cdot4072039$.

3. If $A=45°\ 41'$, $C=62°\ 5'$, $b=100$, find c; given
$\log 9\cdot2788=\cdot96749$, $L \sin 62°\ 5'=9\cdot94627$,
$L \sin 72°\ 14'=9\cdot97878$.

4. If $B=70°\ 30'$, $C=78°\ 10'$, $a=102$, find b and c; given
$\log 1\cdot02=\cdot009$, $\log 1\cdot85=\cdot267$, $\log 1\cdot92=\cdot283$,
$L \sin 70°\ 30'=9\cdot974$, $L \sin 78°\ 10'=9\cdot990$,
$L \sin 31°\ 20'=9\cdot716$.

5. If $a=123$, $B=29°\ 17'$, $C=135°$, find c; given log 2,
$\log\ 123=2\cdot0899051$, $L \sin 15°\ 43'=9\cdot4327777$,
$\log 32110=4\cdot5066403$, $D=135$.

6. If $A=44°$, $C=70°$, $b=1006\cdot62$, find a and c; given
$L \sin 44°=9\cdot8417713$, $\log 100662\ =5\cdot0028656$,
$L \sin 66°=9\cdot9607302$, $\log 103543\ =5\cdot0151212$,
$L \sin 70°=9\cdot9729858$, $\log 7654321=6\cdot8839067$.

7. If $a=1652$, $B=26°\ 30'$, $C=47°\ 15'$, find b and c;
$L \sin 73°\ 45'=9\cdot9822938$, $\log 1\cdot652\ =\cdot2180100$,
$L \sin 26°\ 30'=9\cdot6495274$, $\log 7\cdot6780=\cdot8852481$, $D=57$,
$L \sin 47°\ 15'=9\cdot8658868$, $\log 1\cdot2636=\cdot1016096$, $D=344$.

192. *To solve a triangle when two sides and the angle opposite to one of them are given.*

Let a, b, A be given. Then from $\sin B=\frac{b}{a}\sin A$, we have

$$\log \sin B=\log b-\log a+\log \sin A;$$

whence B may be found;

then C is found from the equation $C=180°-A-B$.

Again, $$c=\frac{a \sin C}{\sin A},$$

$$\therefore\ \log c=\log a+\log \sin C-\log \sin A.$$

If $a<b$, and A is acute the solution is ambiguous and there will be two values of B supplementary to each other, and also two values of C and c. [Art. 147.]

Example. If $b=63$, $c=36$, $C=29° 23' 15''$, find B; given

$$\log 2 = \cdot 3010300, \quad \log 7 = \cdot 8450980.$$

$$L \sin 29° 23' = 9\cdot 6907721, \quad \text{diff. for } 1' = 2243,$$

$$L \sin 59° 10' = 9\cdot 9338222, \quad \text{diff. for } 1' = 755.$$

$$\sin B = \frac{b}{c} \sin C = \frac{63}{36} \sin C$$

$$= \frac{7}{4} \sin 29° 23' 15'';$$

$$\therefore \log \sin B = \log 7 - 2 \log 2 + \log \sin 29° 23' 15'';$$

$$\therefore \log \sin B = \bar{1}\cdot 9338662$$

$$\log \sin 59° 10' = \bar{1}\cdot 9338222$$

$$\text{diff.} \quad 440$$

$$\therefore \text{prop}^{l}. \text{ increase} = \frac{440}{755} \times 60'' = 35'';$$

$$\therefore B = 59° 10' 35''.$$

Also since $c < b$ there is another value of B supplementary to the above, namely $B = 120° 49' 25''$.

$$\log \sin 29° 23' = \bar{1}\cdot 6907721$$

$$\frac{15}{60} \times 2243 = \quad 561$$

$$\log 7 = \cdot 8450980$$

$$\cdot 5359262$$

$$2 \log 2 = \cdot 6020600$$

$$\bar{1}\cdot 9338662$$

$$\begin{array}{r} 440 \\ 60 \\ \hline 755\ )\ 26400\ (\ 35 \\ 2265 \\ \hline 3750 \\ 3775 \\ \hline \end{array}$$

EXAMPLES. XVI. e.

1. If $a=145$, $b=178$, $B=41° 10'$, find A; given

   $\log 178 = 2\cdot 2504200$, $L \sin 41° 10' = 9\cdot 8183919$,
   $\log 145 = 2\cdot 1613680$, $L \sin 32° 25' 35'' = 9\cdot 7293399$.

2. If $A = 26° 26'$, $b = 127$, $a = 85$, find B; given

   $\log 1\cdot 27 = \cdot 1038037$, $L \sin 26° 26' = 9\cdot 6485124$,
   $\log 8\cdot 5 = \cdot 9294189$, $L \sin 41° 41' 28'' = 9\cdot 8228972$.

3. If $c=5$, $b=4$, $C=45°$, find A and B; given

   $\log 2 = \cdot 30103$, $L \sin 34° 26' = 9\cdot 7525750$.

4. If $a = 1405$, $b = 1706$, $A = 40°$, find B; given

   $\log 1\cdot 405 = \cdot 1476763$, $\log 1\cdot 706 = \cdot 2319790$,
   $L \sin 40° = 9\cdot 8080675$, $L \sin 51° 18' = 9\cdot 8923342$,
   diff. for $1' = 1012$.

5. If $B=112° 4'$, $b=573$, $c=394$, find A and C; given

$$\log 573=2{\cdot}7581546, \quad \log 394=2{\cdot}5954962,$$
$$L \sin 39° 35'=9{\cdot}8042757, \quad \text{diff. for } 60''=1527,$$
$$L \cos 22° 4'=9{\cdot}9669614.$$

6. If $b=8{\cdot}4$, $c=12$, $B=37° 36'$, find A; given

$$\log 7={\cdot}8450980, \quad L \sin 37° 36'=9{\cdot}7854332,$$
$$L \sin 60° 39'=9{\cdot}9403381, \quad \text{diff. for } 1'=711.$$

7. Supposing the data for the solution of a triangle to be as in the three following cases, point out whether the solution will be ambiguous or not, and find the third side in the obtuse angled triangle in the ambiguous case:

(i) $A=30°$, $a=125$ feet, $c=250$ feet,
(ii) $A=30°$, $a=200$ feet, $c=250$ feet,
(iii) $A=30°$, $a=200$ feet, $c=125$ feet.

Given log 2,

$$\log 6{\cdot}0389={\cdot}7809578, \quad L \sin 38° 41'=9{\cdot}7958800,$$
$$\log 6{\cdot}0390={\cdot}7809650, \quad L \sin 8° 41'=9{\cdot}1789001.$$

193. Some formulæ which are not primarily suitable for working with logarithms may be adapted to such work by various artifices.

194. *To adapt the formula $c^2=a^2+b^2$ to logarithmic computation.*

We have
$$c^2=a^2\left(1+\frac{b^2}{a^2}\right).$$

Since an angle can always be found whose tangent is equal to a given numerical quantity, we may put $\frac{b}{a}=\tan\theta$, and thus obtain

$$c^2=a^2(1+\tan^2\theta)=a^2\sec^2\theta;$$
$$\therefore \quad c=a\sec\theta;$$
$$\therefore \quad \log c=\log a+\log\sec\theta.$$

The angle θ is called a **subsidiary angle** and is found from the equation

$$\log\tan\theta=\log b-\log a.$$

Thus *any expression which can be put into the form of the sum of two squares can be readily adapted to logarithmic work.*

195. *To adapt the formula* $c^2=a^2+b^2-2ab\cos C$ *to logarithmic computation.*

From the identities

$$\cos C=\cos^2\frac{C}{2}-\sin^2\frac{C}{2}, \text{ and } 1=\cos^2\frac{C}{2}+\sin^2\frac{C}{2},$$

we have

$$c^2=(a^2+b^2)\left(\cos^2\frac{C}{2}+\sin^2\frac{C}{2}\right)-2ab\left(\cos^2\frac{C}{2}-\sin^2\frac{C}{2}\right)$$

$$=(a^2+b^2-2ab)\cos^2\frac{C}{2}+(a^2+b^2+2ab)\sin^2\frac{C}{2}$$

$$=(a-b)^2\cos^2\frac{C}{2}+(a+b)^2\sin^2\frac{C}{2}$$

$$=(a-b)^2\cos^2\frac{C}{2}\left\{1+\left(\frac{a+b}{a-b}\right)^2\tan^2\frac{C}{2}\right\}.$$

Take a subsidiary angle θ, such that

$$\tan\theta=\frac{a+b}{a-b}\tan\frac{C}{2},$$

then $c^2=(a-b)^2\cos^2\frac{C}{2}(1+\tan^2\theta)$

$$=(a-b)^2\cos^2\frac{C}{2}\sec^2\theta;$$

$$\therefore\quad c=(a-b)\cos\frac{C}{2}\sec\theta;$$

$$\therefore\quad \log c=\log(a-b)+\log\cos\frac{C}{2}+\log\sec\theta,$$

where θ is determined from the equation

$$\log\tan\theta=\log(a+b)-\log(a-b)+\log\tan\frac{C}{2}.$$

196. When two sides and the included angle are given, we may solve the triangle by finding the value of the third side first instead of determining the angles first as in Art. 189.

Example. If $a=3$, $c=1$, $B=53°\,7'\,48''$ find b; given
$\log 2=\cdot3010300$, $\log 2\cdot5298=\cdot4030862$, diff. for $1=172$,
$L\cos 26°\,33'\,54''=9\cdot9515452$, $L\tan 26°\,33'\,54''=9\cdot6989700$.

We have $b^2 = c^2 + a^2 - 2ca \cos B$

$$= (a^2+c^2)\left(\cos^2\frac{B}{2} + \sin^2\frac{B}{2}\right) - 2ac\left(\cos^2\frac{B}{2} - \sin^2\frac{B}{2}\right)$$

$$= (a-c)^2 \cos^2\frac{B}{2} + (a+c)^2 \sin^2\frac{B}{2}$$

$$= (a-c)^2 \cos^2\frac{B}{2}\left\{1 + \left(\frac{a+c}{a-c}\right)^2 \tan^2\frac{B}{2}\right\}$$

$$= (a-c)^2 \cos^2\frac{B}{2}(1+\tan^2\theta) \qquad (1),$$

where $\tan\theta = \dfrac{a+c}{a-c}\tan\dfrac{B}{2} = 2\tan 26^\circ\, 33'\, 54''$;

$$\therefore \log\tan\theta = \log 2 + \log\tan 26^\circ\, 33'\, 54''$$
$$= \cdot 3010300 + \bar{1}\cdot 6989700$$
$$= 0;$$

whence $\tan\theta = 1$, and $\theta = 45^\circ$.

From (1), $b = (a-c)\cos\dfrac{B}{2}\sec\theta$

$$= 2\sec 45^\circ \cos\frac{B}{2}$$

$$= 2\sqrt{2}\cos 26^\circ\, 33'\, 54'';$$

$$\therefore \log b = \log 2 + \frac{1}{2}\log 2 + \log\cos 26^\circ\, 33'\, 54''$$

$\log 2 =$	$\cdot 3010300$
$\frac{1}{2}\log 2 =$	$\cdot 1505150$
$\log\cos 26^\circ\, 33'\, 54'' =$	$\bar{1}\cdot 9515452$
	$\cdot 4030902$

$\therefore \log b = \cdot 4030902$

$\log 2\cdot 5298 =$	$\cdot 4030862$
diff.	40

But diff. for 1 is 172;

$$\therefore \text{proportional increase} = \frac{40}{172} = \frac{10}{43} = \cdot 23.$$

Thus the third side is $2\cdot 529823$.

197. The formula $c^2 = a^2 + b^2 - 2ab\cos C$ may also be adapted to logarithmic computation in two other ways by making use of the identities $\cos C = 2\cos^2\dfrac{C}{2} - 1$ and $\cos C = 1 - 2\sin^2\dfrac{C}{2}$.

We shall take the first of these cases, leaving the other as an exercise.

$$c^2 = a^2 + b^2 - 2ab\cos C$$

$$= a^2 + b^2 - 2ab\left(2\cos^2\frac{C}{2} - 1\right)$$

$$= (a+b)^2 - 4ab\cos^2\frac{C}{2}$$

$$= (a+b)^2\left\{1 - \frac{4ab}{(a+b)^2}\cos^2\frac{C}{2}\right\}.$$

Let $$\frac{4ab}{(a+b)^2}\cos^2\frac{C}{2} = \cos^2\theta,$$

then $$c^2 = (a+b)^2(1-\cos^2\theta) = (a+b)^2\sin^2\theta;$$

$$\therefore\ c = (a+b)\sin\theta;$$

$$\therefore\ \log c = \log(a+b) + \log\sin\theta.$$

To determine θ we have the equation

$$\cos\theta = \frac{2\sqrt{ab}}{a+b}\cos\frac{C}{2};$$

$$\therefore\ \log\cos\theta = \log 2 + \frac{1}{2}(\log a + \log b) - \log(a+b) + \log\cos\frac{C}{2}.$$

Since $2\sqrt{ab}$ is never greater than $a+b$ and $\cos\frac{C}{2}$ is positive and less than unity, $\cos\theta$ is positive and less than unity, and thus θ is an acute angle.

EXAMPLES. XVI. f.

1. If $a=8$, $b=7$, $c=9$, find the angles; given $\log 2$, $\log 3$,
$L\tan 24°\ 5' = 9{\cdot}6502809$, diff. for $60'' = 3390$,
$L\tan 36°\ 41' = 9{\cdot}8721123$, diff. for $60'' = 2637$.

2. The difference between the angles at the base of a triangle is 24°, and the sides opposite these angles are 175 and 337: find all the angles; given $\log 2$, $\log 3$,
$L\tan 12° = 9{\cdot}3274745$, $L\cot 56°\ 6'\ 27'' = 9{\cdot}8272293$.

3. One of the sides of a right-angled triangle is two-sevenths of the hypotenuse: find the greater of the two acute angles; given

$\log 2$, $\log 7$, $L\sin 14°\ 11' = 9{\cdot}455921$, $L\sin 14°\ 12' = 9{\cdot}456031$.

4. Find the greatest side when two of the angles are $78°\ 14'$ and $71°\ 24'$ and the side joining them is 2183; given

$\log 2{\cdot}183 = {\cdot}3390537$, $\log 4{\cdot}2274 = {\cdot}6260733$, $D = 103$,
$L\sin 78°\ 14' = 9{\cdot}9907766$, $L\sin 30°\ 22' = 9{\cdot}7037486$.

5. If $b = 2$ ft. 6 in., $c = 2$ ft., $A = 22°\ 20'$, find the other angles; and then shew that the side a is very approximately 1 foot. Given $\log 2$, $\log 3$,

$L\cot 11°\ 10' = 10{\cdot}70465$, $L\sin 49°\ 27'\ 34'' = 9{\cdot}88079$,
$L\sin 22°\ 20' = 9{\cdot}57977$, $L\tan 29°\ 22'\ 26'' = 9{\cdot}75041$.

6. If $a = 1{\cdot}56234$, $b = {\cdot}43766$, $C = 58°\ 42'\ 6''$, find A and B; given $\log 56234 = 4{\cdot}75$,

$\log\cot 29°\ 21' = {\cdot}250015$, $\log\cot 29°\ 22' = {\cdot}249715$.

7. If $a = 9$, $b = 12$, $A = 30°$, find the values of c, having given

$\log 12 = 1{\cdot}07918$, $L\sin 30° = 9{\cdot}69897$,
$\log 9 = {\cdot}95424$, $L\sin 11°\ 48'\ 39'' = 9{\cdot}31108$,
$\log 171 = 2{\cdot}23301$, $L\sin 41°\ 48'\ 39'' = 9{\cdot}82391$,
$\log 368 = 2{\cdot}56635$, $L\sin 108°\ 11'\ 21'' = 9{\cdot}97774$.

8. The sides of a triangle are 9 and 3, and the difference of the angles opposite to them is 90°: find the angles; having given $\log 2$,

$L\tan 26°\ 33' = 9{\cdot}6986847$, $L\tan 26°\ 34' = 9{\cdot}6990006$.

9. Two sides of a triangle are 1404 and 960 respectively, and an angle opposite to one of them is $32°\ 15'$: find the angle contained by the two sides; having given $\log 2$, $\log 3$,

$\log 13 = 1{\cdot}1139434$, $L\operatorname{cosec} 32°\ 15' = 10{\cdot}2727724$,
$L\sin 21°\ 23' = 9{\cdot}5621316$, $L\sin 51°\ 18' = 9{\cdot}8923236$.

10. If $b : c = 11 : 10$ and $A = 35°\ 25'$, use the formula

$$\tan\frac{1}{2}(B - C) = \tan^2\frac{\phi}{2}\cot\frac{A}{2} \text{ to find } B \text{ and } C;$$

given $\log 1{\cdot}1 = {\cdot}041393$, $L\tan 12°\ 18'\ 36'' = 9{\cdot}338891$,
$L\cos 24°\ 37'\ 12'' = 9{\cdot}958607$, $L\cot 17°\ 42'\ 30'' = 10{\cdot}495800$,
$L\tan 8°\ 28'\ 56{\cdot}5'' = 9{\cdot}173582$.

11. If $A=50°$, $b=1071$, $a=873$, find B; given

$$\log 1{\cdot}071=\cdot029789,\quad \log 8{\cdot}73=\cdot941014,$$
$$L\sin 50°=9{\cdot}884254,\quad L\sin 70°\ =9{\cdot}972986,$$
$$L\sin 70°\ 1'=9{\cdot}973032.$$

12. If $a=6$, $b=3$, $C=36°\ 52'\ 12''$, find c without determining A and B; given $\log 2=\cdot30103$, $\log 3=\cdot47712$,

$$\log 40249=4{\cdot}60476,\quad L\sin 18°\ 26'\ 6''=9{\cdot}5,$$
$$L\cot 18°\ 26'\ 6''=10{\cdot}47712.$$

(*In the following Examples the necessary Logarithms must be taken from the Tables.*)

13. Given $a=1000$, $b=840$, $c=1258$, find B.

14. Solve the triangle in which $a=525$, $b=650$, $c=777$.

15. Find the least angle when the sides are proportional to 4, 5, and 6.

16. If $B=90°$, $AC=57{\cdot}321$, $AB=28{\cdot}58$, find A and C.

17. Find the hypotenuse of a right-angled triangle in which the smallest angle is $18°\ 37'\ 29''$ and the side opposite to it is 284 feet.

18. The sides of a triangle are 9 and 7 and the angle between them is $60°$: find the other angles.

19. How long must a ladder be so that when inclined to the ground at an angle of $72°\ 15'$ it may just reach a window 42·37 feet from the ground?

20. If $a=31{\cdot}95$, $b=21{\cdot}96$, $C=35°$, find A and B.

21. Find B, C, a when $b=25{\cdot}12$, $c=13{\cdot}83$, $A=47°\ 15'$.

22. Find the greatest angle of the triangle whose sides are 1837·2, 2385·6, 2173·84.

23. When $a=21{\cdot}352$, $b=45{\cdot}6843$, $c=37{\cdot}2134$, find A, B, and C.

24. If $b=647{\cdot}324$, $c=850{\cdot}273$, $A=103°\ 12'\ 54''$, find the remaining parts.

25. If $b=23{\cdot}2783$, $A=37°\ 57'$, $B=43°\ 13'$, find the remaining sides.

26. Find a and b when $B=72°\ 43'\ 25''$, $C=47°\ 12'\ 17''$, $c=2484{\cdot}3$.

27. If $AB=4517$, $AC=150$, $A=31°\ 30'$, find the remaining parts.

28. Find A, B, and b when

$$a=324{\cdot}68,\quad c=421{\cdot}73,\quad C=35°\ 17'\ 12''.$$

29. Given $a=321{\cdot}7$, $c=435{\cdot}6$, $A=36°\ 18'\ 27''$, find C.

30. If $b=1325$, $c=1665$, $B=52°\ 19'$, solve the obtuse-angled triangle to which the data belong.

31. If $a=3795$, $B=73°\ 15'\ 15''$, $C=42°\ 18'\ 30''$, find the other sides.

32. Find the angles of the two triangles which have $b=17$, $c=12$, and $C=43°\ 12'\ 12''$.

33. Two sides of a triangle are 2·7402 ft. and ·7401 ft. respectively, and contain an angle $59°\ 27'\ 5''$: find the base and altitude of the triangle.

34. The difference between the angles at the base of a triangle is $17°\ 48'$ and the sides subtending these angles are 105·25 ft. and 76·75 ft.: find the angle included by the given sides.

35. From the following data:

(1) $A=43°\ 15'$, $AB=36{\cdot}5$, $BC=20$,
(2) $A=43°\ 15'$, $AB=36{\cdot}5$, $BC=30$,
(3) $A=43°\ 15'$, $AB=36{\cdot}5$, $BC=45$,

point out which solution is impossible and which ambiguous. Find the third side for the triangle the solution of which is neither impossible nor ambiguous.

36. In any triangle prove that $c=(a-b)\sec\theta$, where

$$\tan\theta=\frac{2\sqrt{ab}}{a-b}\sin\frac{C}{2}.$$

If $a=17{\cdot}32$, $b=13{\cdot}47$, $C=47°\ 13'$, find c without finding A and B.

37. If $\tan\phi=\dfrac{a+b}{a-b}\tan\dfrac{C}{2}$, prove that $c=(a-b)\cos\dfrac{C}{2}\sec\phi$.

If $a=27{\cdot}3$, $b=16{\cdot}8$, $C=45°\ 12'$, find ϕ, and thence find c.

Solution of Triangles with Four-Figure Logarithms.

197_A. *To solve a triangle when the three sides are given.*

[For general explanation of method see Art. 187.]

Example. If $a=283$, $b=317$, $c=428$, find all the angles.

$$\tan\frac{A}{2}=\sqrt{\frac{(s-b)(s-c)}{s(s-a)}}=\sqrt{\frac{197\times 86}{514\times 231}};$$

$$\therefore\ \log\tan\frac{A}{2}=\frac{1}{2}(\log 197+\log 86-\log 514-\log 231).$$

$$\begin{array}{r} 283 \\ 317 \\ 428 \\ \hline 2\,)\,1028 \\ \hline 514=s \\ 231=s-a \\ 197=s-b \\ 86=s-c \end{array}$$

From the Tables,

$$\begin{array}{rr} \log 197=2{\cdot}2945 & \log 514=2{\cdot}7110 \\ \log\ 86=1{\cdot}9345 & \log 231=2{\cdot}3636 \\ \hline 4{\cdot}2290 & 5{\cdot}0746 \\ 5{\cdot}0746 & \\ \hline 2\,)\,\bar{1}{\cdot}1544 & \\ \hline \log\tan\frac{A}{2}=\bar{1}{\cdot}5772 & \end{array}$$

$$\log\tan 20^\circ 42'=\bar{1}{\cdot}5773;\ \therefore\ \frac{A}{2}=20^\circ 42',\text{ approx., and } A=41^\circ 24'.$$

Again, $$\tan\frac{B}{2}=\sqrt{\frac{(s-c)(s-a)}{s(s-b)}}=\sqrt{\frac{86\times 231}{514\times 197}};$$

$$\begin{array}{rr} \log\ 86=1{\cdot}9345 & \log 514=2{\cdot}7110 \\ \log 231=2{\cdot}3636 & \log 197=2{\cdot}2945 \\ \hline 4{\cdot}2981 & 5{\cdot}0055 \\ 5{\cdot}0055 & \\ \hline 2\,)\,\bar{1}{\cdot}2926 & \\ \hline \log\tan\frac{B}{2}=\bar{1}{\cdot}6463 & \\ \log\tan 23^\circ 48'=\bar{1}{\cdot}6445 & \\ \hline 18 & \\ \text{diff. for } 5'=\quad 17 & \end{array}$$

$$\therefore\ \frac{B}{2}=23^\circ 53',\text{ approximately, and } B=47^\circ 46'.$$

Thus $A=41^\circ 24'$, $B=47^\circ 46'$, $C=90^\circ 50'$.

This solution may be compared with that of Ex. 2, Art. 188.

[EXAMPLES XVI. g. Nos. 1—10 may be taken here.]

NOTE. Any small error in obtaining the *half* angle is doubled when we multiply by 2. Thus a half angle obtained to the nearest minute will not necessarily give a final result to the same degree of accuracy.

If greater accuracy is required we may proceed as follows:

$\log\tan\frac{A}{2}=\bar{1}\cdot5772$	$\log\tan 20^\circ 42'=\bar{1}\cdot5773$
$\log\tan 20^\circ 36'=\bar{1}\cdot5750$	$\log\tan 20^\circ 36'=\bar{1}\cdot5750$
diff. 22	diff. for 6′ 23

$\therefore\ \frac{A}{2}$ is greater than $20^\circ 36'$ by $\frac{22}{23}\times 6'$ or $5'\cdot 7$;

$\therefore\ \frac{A}{2}=20^\circ 41'\cdot 7$, and $A=41^\circ 23'$.

```
23)132(5·7
   115
   ---
    170
    161
    ---
      9
```

197B. *To solve a triangle having given two sides and the included angle.*

[For general explanation of method see Art. 189.]

Example. If $a=681$, $c=243$, $B=50^\circ 42'$, solve the triangle.

$$\tan\frac{A-C}{2}=\frac{a-c}{a+c}\cot\frac{B}{2}=\frac{438}{924}\cot 25^\circ 21'=\frac{438}{924}\tan 64^\circ 39';$$

$$\therefore\ \log\tan\frac{A-C}{2}=\log 438-\log 924+\log\tan 64^\circ 39';$$

$$\therefore\ \log\tan\frac{A-C}{2}=\cdot 0003;$$

$\log 438=2\cdot6415$
$\log\tan 64^\circ 39'=\cdot3245$
$2\cdot9660$
$\log 924=2\cdot9657$
$\cdot0003$

$$\therefore\ \frac{A-C}{2}=45^\circ 1'.$$

Also $$\frac{A+C}{2}=90^\circ-\frac{B}{2}=64^\circ 39'.$$

By addition $A=109^\circ 40'$,

and by subtraction $C=19^\circ 38'$.

Again, $$b=\frac{c\sin B}{\sin C}=\frac{243\sin 50^\circ 42'}{\sin 19^\circ 38'};$$

$$\therefore\ \log b=\log 243+\log\sin 50^\circ 42'-\log\sin 19^\circ 38';$$

$$\therefore\ \log b=2\cdot7480;$$

whence $b=559\cdot8$.

$\log 243=2\cdot3856$
$\log\sin 50^\circ 42'=\bar{1}\cdot8887$
$2\cdot2743$
$\log\sin 19^\circ 38'=\bar{1}\cdot5263$
$2\cdot7480$

Thus $A=109^\circ 40'$, $C=19^\circ 38'$, $b=559\cdot8$.

This solution may be compared with that on p. 172.

[EXAMPLES XVI. g. Nos. 11—20 may be taken here.]

197_C. *To solve a triangle having given two angles and a side.*

[For general explanation of method see Art. 191.]

Example. If $b=1000$, $A=45°$, $C=68° 18'$, find the least side.

$$B=180°-45°-68° 18'=66° 42';$$

$$\text{the least side}=a=\frac{b\sin A}{\sin B}=\frac{1000\sin 45°}{\sin 66° 42'}.$$

$$\log a=\log 1000+\log\sin 45°-\log\sin 66° 42'$$
$$=2\cdot 8864;$$
whence $a=769\cdot 8$.

$$\log 1000=3$$
$$\log\sin 45°=\bar{1}\cdot 8495$$
$$2\cdot 8495$$
$$\log\sin 66° 42'=\bar{1}\cdot 9631$$
$$2\cdot 8864$$

[EXAMPLES XVI. g. Nos. 21—30 may be taken here.]

197_D. *To solve a triangle when two sides and the angle opposite to one of them are given.*

[For general explanation of method see Art. 192.]

Example. If $b=63$, $c=36$, $C=29° 23'$, solve the triangle.

$$\sin B=\frac{b}{c}\sin C=\frac{63}{36}\sin C$$
$$=\frac{7}{4}\sin 29° 23'$$
$$\log\sin B=\bar{1}\cdot 9338;$$
$$\therefore B=59° 10'.$$

$$\log 7=\cdot 8451$$
$$\log\sin 29° 23'=\bar{1}\cdot 6908$$
$$\cdot 5359$$
$$\log 4=\cdot 6021$$
$$\bar{1}\cdot 9338$$

Also since $c<b$, there is another value of B supplementary to the above, viz. $120° 50'$. [See Art. 148.]

Thus we may say $B_1=59° 10'$, $B_2=120° 50'$;

$$\therefore A_1=180°-59° 10'-29° 23'=91° 27',$$
$$A_2=180°-120° 50'-29° 23'=29° 47'.$$

$$\therefore a_1=\frac{c\sin A_1}{\sin C}=\frac{36\sin 91° 27'}{\sin 29° 23'}$$
$$=\frac{36\sin 88° 33'}{\sin 29° 23'}$$
$$=1\cdot 8654.$$

$$\log 36=1\cdot 5563$$
$$\log\sin 88° 33'=\bar{1}\cdot 9999$$
$$1\cdot 5562$$
$$\log\sin 29° 23'=\bar{1}\cdot 6908$$
$$1\cdot 8654$$

$$\therefore a_1=73\cdot 35, \text{ from the Tables.}$$

Again $a_2 = \frac{c \sin A_2}{\sin C}$

$= \frac{36 \sin 29° 47'}{\sin 29° 23'}$

$= 1·5616.$

$\log 36$	$= 1·5563$
$\log \sin 29° 47'$	$= \bar{1}·6961$
	$1·2524$
$\log \sin 29° 23'$	$= \bar{1}·6908$
	$1·5616$

$\therefore\ a_2 = 36·44$, from the Tables.

Thus finally $\begin{cases} B = 59° 10', \text{ or } 120° 50', \\ A = 91° 27', \text{ or } 29° 47', \\ a = 73·35, \text{ or } 36·44. \end{cases}$

[EXAMPLES XVI. g. Nos. 31—37 may be taken here.]

EXAMPLES. XVI. g.

(Given the three sides.)

1. If $a = 25·3$, $b = 11·7$, $c = 9·0$, find A, using $\sin \frac{A}{2}$.

2. If $a = 68·75$, $b = 93·25$, $c = 63$, find B, using $\cos \frac{B}{2}$.

3. Find the smallest angle of a triangle whose sides are 11·24, 13·65, 9·03, using the sine formula.

4. If $a = 15$, $b = 22$, $c = 9$, find all the angles.

5. Find all the angles, given $a = 31·54$, $b = 46·5$, $c = 63·4$.

6. Find all the angles, given $a = 33·4$, $b = 71·6$, $c = 60·24$.

7. Find the largest angle of a triangle whose sides are 14·75, 6·84, 10·37, using the cosine formula.

8. If the sides are 21·5, 13·7, 29·5, find the smallest angle, using the sine formula.

9. If $a : b : c = 5 : 7 : 8$, find all the angles.

10. If the sides of a triangle are as 1·3 : 1·4 : 1·5, find all the angles.

(Given two sides and the included angle.)

11. If $b = 32·8$, $c = 15·0$, $A = 107° 26'$, find B and C.

12. If $a = 96·7$, $b = 135·4$, $C = 123° 42'$, find A and B.

13. If $a = ·012$, $c = ·576$, $B = 60° 30'$, find A and C.

14. Two sides of a triangle are 27·3 and 16·8, and they contain an angle of 45° 7′, find the remaining angles and side.

15. Given $A=37°$, $b=82·9$, $c=25·1$, solve the triangle.

16. If $b=27·0$, $c=33·48$, $A=60°$, find the other angles.

17. If $13a=29b$, and $C=82° 14'$, find A and B.

18. If one side of a triangle is double another and makes with it an angle of 52° 47′, find the remaining angles.

19. If $87b=131c$, and $A=18° 16'$, find B and C.

20. If $a=17·6$, $b=24·03$, $C=121° 38'$, solve the triangle.

(*Given two angles and a side.*)

21. If $A=40°$, $C=70°$, $b=100$, find a.

22. If $B=42°$, $C=107°$, $a=85·2$, find b.

23. If $A=49° 11'$, $B=21° 15'$, $c=5·23$, find a.

24. If $A=65° 27'$, $C=71° 35'$, $b=873$, find c.

25. If $A=60°$, $B=79° 20'$, $c=60$, find a.

26. The base of a triangle is 3·57 inches, and the angles adjacent to it are 51° 51′ and 87° 43′, find the remaining sides.

27. Given $B=65° 47'$, $C=52° 39'$, $a=125·7$, find the greatest side.

28. Solve the triangle when $A=72° 19'$, $B=83° 17'$, $c=92·93$.

29. In a triangle the side adjacent to two angles of 49° 30′ and 70° 30′ is $4\frac{3}{8}$ inches; find the other sides.

30. If $B:C=4:7$, $C=7A$, and $b=89·36$, find a and c.

(*Given two sides and angle not included.*)

31. If $a=73$, $b=62$, $A=82° 14'$, find B.

32. If $b=41·62$, $c=63·45$, $B=27° 15'$, find C.

33. If $a=17·28$, $b=23·97$, $B=5° 13'$, find A and c.

34. If $a=94·2$, $b=141·3$, $A=40°$, solve the triangle.

35. If $A=20° 41'$, $b=137$, $a=115$, solve the triangle.

36. If $b=1325$, $c=1665$, $B=52° 19'$, solve the obtuse-angled triangle to which the data belong.

37. Find A, B, and b when $a=324{\cdot}7$, $c=421{\cdot}7$, $C=35°$.

(*Miscellaneous.*)

38. If $a=31{\cdot}95$, $b=21{\cdot}96$, $C=35°$, find A and B.

39. Given $a=1000$, $b=840$, $c=1258$, find B, using $\sin\frac{B}{2}$.

40. Find the angles of the two triangles which have $b=17$, $c=12$, and $C=43° 12'$.

41. Solve the triangle in which $a=525$, $b=650$, $c=777$.

42. Find the least angle when the sides are proportional to 4, 5, and 6, using the sine formula.

43. Find B, C, and a, when $b=25{\cdot}12$, $c=13{\cdot}83$, $A=47° 15'$.

44. If $AB=4517$, $AC=150$, $A=31° 30'$, find the remaining parts.

45. Given $a=321{\cdot}7$, $c=435{\cdot}6$, $A=36° 18'$, find C.

46. In any triangle prove that $c=(a-b)\sec\theta$, where

$$\tan\theta=\frac{2\sqrt{ab}}{a-b}\sin\frac{C}{2}.$$

If $a=17{\cdot}32$, $b=13{\cdot}47$, $C=47° 12'$, find c without finding A and B.

47. If $\tan\phi=\frac{a+b}{a-b}\tan\frac{C}{2}$, prove that $c=(a-b)\cos\frac{C}{2}\sec\phi$.

If $a=27{\cdot}3$, $b=16{\cdot}8$, $C=45° 12'$, find ϕ, and thence find c.

CHAPTER XVII.

HEIGHTS AND DISTANCES.

198. Some easy cases of heights and distances depending only on the solution of right-angled triangles have been already dealt with in Chap. VI. The problems in the present chapter are of a more general character, and require for their solution some geometrical skill as well as a ready use of trigonometrical formulæ.

Measurements in one plane.

199. *To find the height and distance of an inaccessible object on a horizontal plane.*

Let A be the position of the observer, CP the object; from P draw PC perpendicular to the horizontal plane; then it is required to find PC and AC.

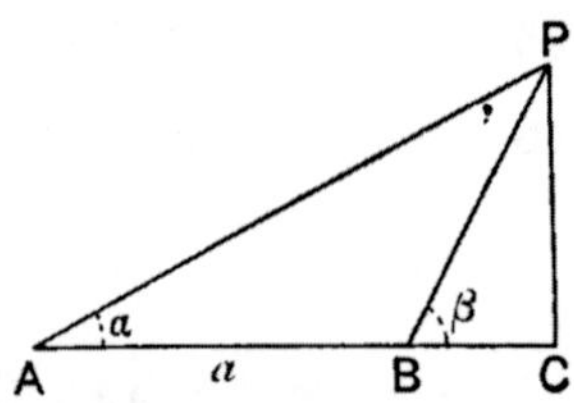

At A observe the angle of elevation PAC. Measure a base line AB in a direct line from A towards the object, and at B observe the angle of elevation PBC.

Let $\qquad \angle PAC=\alpha, \quad \angle PBC=\beta, \quad AB=a.$

From $\triangle PBC, \qquad PC=PB\sin\beta.$

From $\triangle PAB,$

$$PB=\frac{AB\sin PAB}{\sin APB}=\frac{a\sin\alpha}{\sin(\beta-\alpha)};$$

$$\therefore \quad PC=a\sin\alpha\sin\beta\operatorname{cosec}(\beta-\alpha).$$

Also $\qquad AC=PC\cot\alpha=a\cos\alpha\sin\beta\operatorname{cosec}(\beta-\alpha).$

Each of the above expressions is adapted to logarithmic work; thus if $PC=x$, we have

$$\log x=\log a+\log\sin\alpha+\log\sin\beta+\log\operatorname{cosec}(\beta-\alpha).$$

NOTE. Unless the contrary is stated, it will be supposed that the observer's height is disregarded, and that the angles of elevation are measured from the ground.

Example I. A person walking along a straight road observes that at two consecutive milestones the angles of elevation of a hill in front of him are 30° and 75°: find the height of the hill.

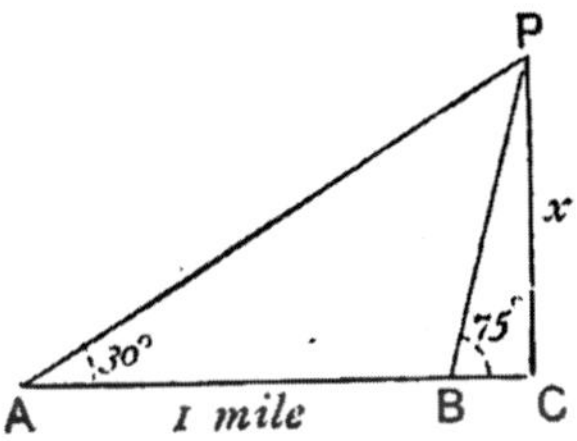

In the adjoining figure,

$\angle PAC = 30°$, $\angle PBC = 75°$, $AB = 1$ mile;

$$\angle APB = 75° - 30° = 45°.$$

Let x be the height in yards; then

$$x = PB \sin 75°;$$

but $PB = \dfrac{AB \sin PAB}{\sin APB} = \dfrac{1760 \sin 30°}{\sin 45°}$;

$$\therefore\ x = \frac{1760 \sin 30° \sin 75°}{\sin 45°}$$

$$= 1760 \times \frac{1}{2} \times \sqrt{2} \times \frac{\sqrt{3}+1}{2\sqrt{2}}$$

$$= 440(\sqrt{3}+1).$$

If we take $\sqrt{3} = 1{\cdot}732$ and reduce to feet, we find that the height is 3606·24 ft.

EXAMPLES. XVII. a.

1. From the top of a cliff 200 ft. above the sea-level the angles of depression of two boats in the same vertical plane as the observer are 45° and 30°: find their distance apart.

2. A person observes the elevation of a mountain top to be 15°, and after walking a mile directly towards it on level ground the elevation is 75°: find the height of the mountain in feet.

3. From a ship at sea the angle subtended by two forts A and B is 30°. The ship sails 4 miles towards A and the angle is then 48°: prove that the distance of B at the second observation is 6·472 miles.

4. From the top of a tower h ft. high the angles of depression of two objects on the horizontal plane and in a line passing through the foot of the tower are $45° - A$ and $45° + A$. Shew that the distance between them is $2h \tan 2A$.

5. An observer finds that the angular elevation of a tower is A. On advancing a feet towards the tower the elevation is 45° and on advancing b feet nearer the elevation is $90° - A$: find the height of the tower.

6. A person observes that two objects A and B bear due N. and N. 30° W. respectively. On walking a mile in the direction N.W. he finds that the bearings of A and B are N.E. and due E. respectively: find the distance between A and B.

7. A tower stands at the foot of a hill whose inclination to the horizon is 9°; from a point 40 ft. up the hill the tower subtends an angle of 54°: find its height.

8. At a point on a level plane a tower subtends an angle α and a flagstaff c ft. in length at the top of the tower subtends an angle β: shew that the height of the tower is

$$c \sin \alpha \operatorname{cosec} \beta \cos (\alpha + \beta).$$

Example II. **The upper three-fourths of a ship's mast subtends at a point on the deck an angle whose tangent is ·6; find the tangent of the angle subtended by the whole mast at the same point.**

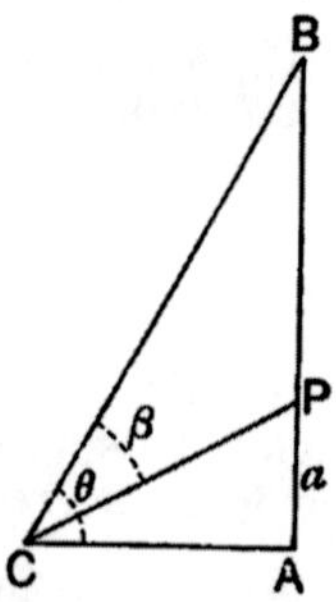

Let C be the point of observation, and let APB be the mast, AP being the lower fourth of it.

Let $AB = 4a$, so that $AP = a$;
also let $AC = b$, $\angle ACB = \theta$, $\angle BCP = \beta$,
so that $\tan \beta = \cdot 6$.

From $\triangle PCA$, $\tan (\theta - \beta) = \dfrac{a}{b}$;

from $\triangle BCA$, $\tan \theta = \dfrac{4a}{b}$;

$$\therefore \tan \theta = 4 \tan (\theta - \beta) = \frac{4(\tan \theta - \tan \beta)}{1 + \tan \theta \tan \beta};$$

$$\therefore \tan \theta = \frac{4\left(\tan \theta - \frac{3}{5}\right)}{1 + \frac{3}{5} \tan \theta} = \frac{4(5 \tan \theta - 3)}{5 + 3 \tan \theta}.$$

On reduction, $\tan^2 \theta - 5 \tan \theta + 4 = 0$;

whence $\tan \theta = 1$ or 4.

NOTE. The student should observe that in examples of this class we make use of right-angled triangles in which the horizontal base line forms one side.

Example III. A tower BCD surmounted by a spire DE stands on a horizontal plane. From the extremity A of a horizontal line BA, it is found that BC and DE subtend equal angles. If $BC=9$ ft., $CD=72$ ft., and $DE=36$ ft., find BA.

Let $$\angle BAC = \angle DAE = \theta,$$
$$\angle DAB = \alpha, \quad AB = x \text{ ft.}$$

Now $BC=9$ ft., $BD=81$ ft., $BE=117$ ft.

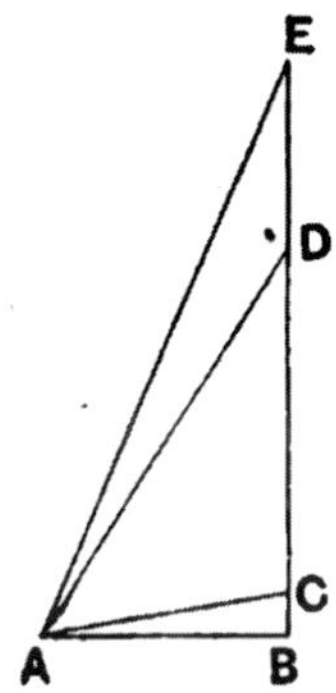

$$\therefore \tan(\alpha+\theta) = \frac{BE}{AB} = \frac{117}{x};$$
$$\tan\alpha = \frac{BD}{AB} = \frac{81}{x};$$
$$\tan\theta = \frac{BC}{AB} = \frac{9}{x}.$$

But $$\tan(\alpha+\theta) = \frac{\tan\alpha+\tan\theta}{1-\tan\alpha\tan\theta};$$

$$\therefore \frac{117}{x} = \frac{\frac{81}{x}+\frac{9}{x}}{1-\frac{81}{x}\cdot\frac{9}{x}} = \frac{90}{x}\cdot\frac{x^2}{x^2-81\times 9}.$$

$$117x^2 - 81\times 9\times 117 = 90x^2;$$
$$\therefore 27x^2 = 81\times 9\times 117;$$
$$\therefore x^2 = 81\times 39;$$
$$\therefore x = 9\sqrt{39}.$$

But $\sqrt{39}=6{\cdot}245$ nearly; $\therefore x=56{\cdot}205$ nearly.
Thus $AB=56{\cdot}2$ ft. nearly.

9. A flagstaff 20 ft. long standing on a wall 10 ft. high subtends an angle whose tangent is ·5 at a point on the ground: find the tangent of the angle subtended by the wall at this point.

10. A statue standing on the top of a pillar 25 feet high subtends an angle whose tangent is ·125 at a point 60 feet from the foot of the pillar: find the height of the statue.

11. A tower BCD surmounted by a spire DE stands on a horizontal plane. From the extremity A of a horizontal line BA it is found that BC and DE subtend equal angles.

If $$BC=9 \text{ ft.},\quad CD=280 \text{ ft.},\quad \text{and } DE=35 \text{ ft.},$$
prove that $BA=180$ ft. nearly.

12. On the bank of a river there is a column 192 ft. high supporting a statue 24 ft. high. At a point on the opposite bank directly facing the column the statue subtends the same angle as a man 6 ft. high standing at the base of the column: find the breadth of the river.

13. A monument $ABCDE$ stands on level ground. At a point P on the ground the portions AB, AC, AD subtend angles α, β, γ respectively. Supposing that $AB=a$, $AC=b$, $AD=c$, $AP=x$, and $\alpha+\beta+\gamma=180°$, shew that $(a+b+c)\,x^2=abc$.

Example IV. The altitude of a rock is observed to be 47°; after walking 1000 ft. towards it up a slope inclined at 32° to the horizon the altitude is 77°. Find the vertical height of the rock above the first point of observation, given $\sin 47° = \cdot 731$.

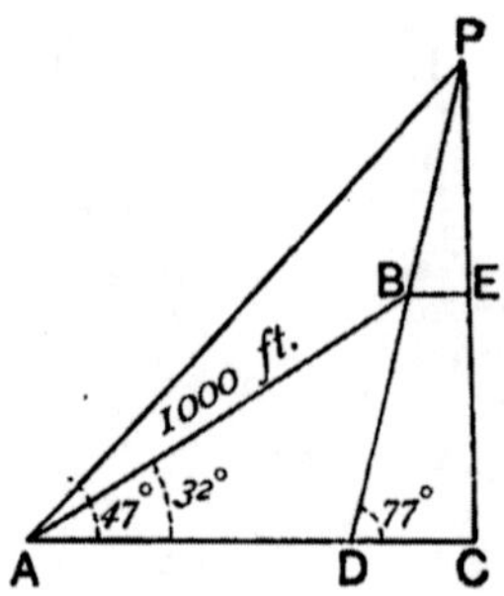

Let P be the top of the rock, A and B the points of observation; then in the figure $\angle PAC=47°$, $\angle BAC=32°$,

$\angle PDC = \angle PBE = 77°$, $AB=1000$ ft.

Let x ft. be the height; then

$$x = PA \sin PAC = PA \sin 47°.$$

We have therefore to find PA in terms of AB.

In $\triangle PAB$, $\angle PAB = 47° - 32° = 15°$;

$\angle APB = 77° - 47° = 30°$;

$\therefore \angle ABP = 135°$;

$$\therefore PA = \frac{AB \sin ABP}{\sin APB}$$

$$= \frac{1000 \sin 135°}{\sin 30°}$$

$$= 1000\sqrt{2};$$

$$\therefore x = PA \sin 47° = 1000\sqrt{2} \times \cdot 731$$

$$= 731\sqrt{2}.$$

If we take $\sqrt{2} = 1\cdot 414$, we find that the height is 1034 ft. nearly.

14. From a point on the horizontal plane, the elevation of the top of a hill is 45°. After walking 500 yards towards its summit up a slope inclined at an angle of 15° to the horizon the elevation is 75°: find the height of the hill in feet.

15. From a station B at the base of a mountain its summit A is seen at an elevation of $60°$; after walking one mile towards the summit up a plane making an angle of $30°$ with the horizon to another station C, the angle BCA is observed to be $135°$: find the height of the mountain in feet.

16. The elevation of the summit of a hill from a station A is α. After walking c feet towards the summit up a slope inclined at an angle β to the horizon the elevation is γ: shew that the height of the hill is $c \sin \alpha \sin (\gamma - \beta) \operatorname{cosec} (\gamma - \alpha)$ feet.

17. From a point A an observer finds that the elevation of Ben Nevis is $60°$; he then walks 800 ft. on a level plane towards the foot, and then 800 ft. further up a slope of $30°$ and finds the elevation to be $75°$: shew that the height of Ben Nevis above A is 4478 ft. approximately.

200. In many of the problems which follow, the solution depends upon the knowledge of some geometrical proposition.

Example I. A tower stands on a horizontal plane. From a mound 14 ft. above the plane and at a horizontal distance of 48 ft. from the tower an observer notices a loophole, and finds that the portions of the tower above and below the loophole subtend equal angles. If the height of the loophole is 30 ft., find the height of the tower.

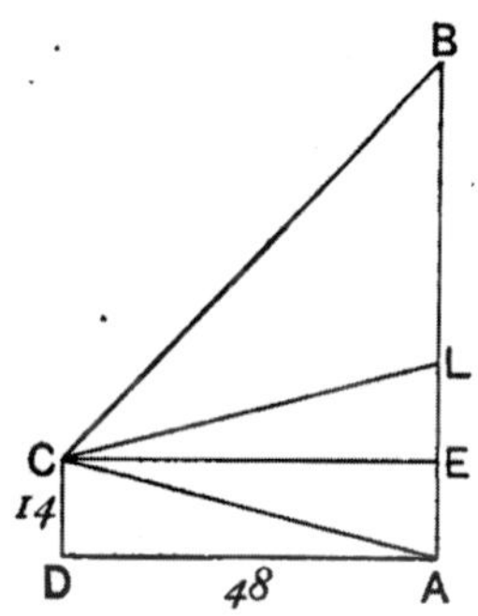

Let AB be the tower, C the point of observation, L the loophole. Draw CD vertical and CE horizontal. Let $AB = x$. We have

$$CD = 14, \ AD = EC = 48, \ BE = x - 14.$$

From $\triangle ADC$, $AC^2 = (14)^2 + (48)^2 = 2500$;

$$\therefore AC = 50.$$

From $\triangle CEB$, $CB^2 = (x - 14)^2 + (48)^2$

$$= x^2 - 28x + 2500.$$

Now $\angle BCL = \angle ACL$;

hence by Euc. VI. 3, $\dfrac{BC}{AC} = \dfrac{BL}{AL}$;

$$\therefore \frac{\sqrt{x^2 - 28x + 2500}}{50} = \frac{x - 30}{30}.$$

By squaring, $9(x^2 - 28x + 2500) = 25(x^2 - 60x + 900)$.

On reduction, we obtain $16x^2 - 1248x = 0$; whence $x = 78$.

Thus the tower is 78 ft. high.

EXAMPLES. XVII. b.

1. At one side of a road is a flagstaff 25 ft. high fixed on the top of a wall 15 ft. high. On the other side of the road, at a point on the ground directly opposite, the flagstaff and wall subtend equal angles: find the width of the road.

2. A statue a feet high stands on a column $3a$ feet high. To an observer on a level with the top of the statue, the column and statue subtend equal angles: find the distance of the observer from the top of the statue.

3. A flagstaff a feet high placed on the top of a tower b feet high subtends the same angle as the tower to an observer h feet high standing on the horizontal plane at a distance d feet from the foot of the tower: shew that

$$(a-b)\,d^2=(a+b)\,b^2-2b^2h-(a-b)\,h^2.$$

Example II. A flagstaff is fixed on the top of a wall standing upon a horizontal plane. An observer finds that the angles subtended at a point on this plane by the wall and the flagstaff are α and β. He then walks a distance c directly towards the wall and finds that the flagstaff again subtends an angle β. Find the heights of the wall and flagstaff.

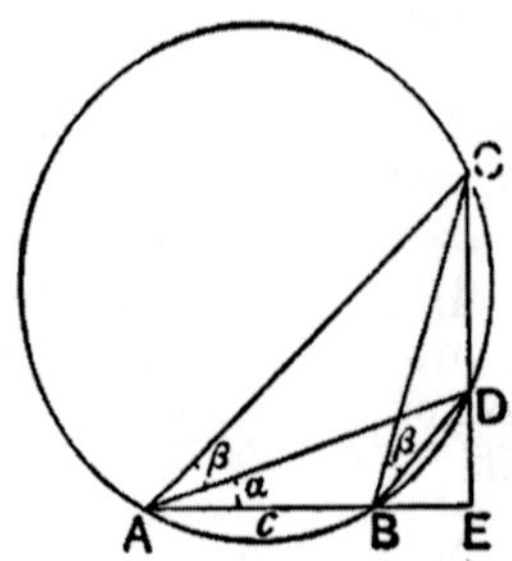

Let ED be the wall, DC the flagstaff, A and B the points of observation.

Then $\angle CAD=\beta=\angle CBD$, so that the four points C, A, B, D are concyclic.

$$\therefore\ \angle ABD=\text{supp}^{t}.\text{ of }\angle ACE$$
$$=90^\circ+(\alpha+\beta),\text{ from }\triangle CAE.$$

Hence in $\triangle ADB$,

$$\angle ADB=180^\circ-\alpha-\{90^\circ+(\alpha+\beta)\}$$
$$=90^\circ-(2\alpha+\beta).$$

$$\therefore\ AD=\frac{AB\sin ABD}{\sin ADB}=\frac{c\cos(\alpha+\beta)}{\cos(2\alpha+\beta)}.$$

Hence in $\triangle ADE$,

$$DE=AD\sin\alpha=\frac{c\sin\alpha\cos(\alpha+\beta)}{\cos(2\alpha+\beta)}.$$

And in $\triangle CAD$,

$$CD=\frac{AD\sin CAD}{\sin ACD}=\frac{AD\sin\beta}{\cos(\alpha+\beta)}=\frac{c\sin\beta}{\cos(2\alpha+\beta)}.$$

Example III. A man walking towards a tower AB on which a flagstaff BC is fixed observes that when he is at a point E, distant c ft. from the tower, the flagstaff subtends its greatest angle. If $\angle BEC = \alpha$, prove that the heights of the tower and flagstaff are $c \tan\left(\frac{\pi}{4} - \frac{\alpha}{2}\right)$ and $2c \tan \alpha$ ft. respectively.

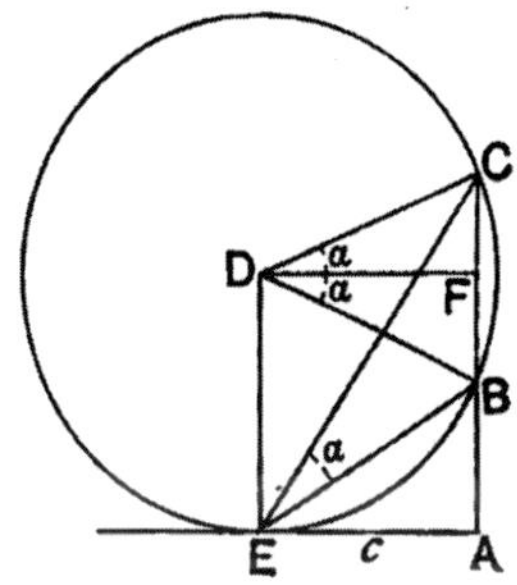

Since E is the point in the horizontal line AE at which BC subtends a maximum angle, it can easily be proved that AE touches the circle passing round the triangle CBE.

[See Hall and Stevens' *Geometry*, p. 315.]

The centre D of this circle lies in the vertical line through E. Draw DF perpendicular to BC, then DF bisects BC and also $\angle CDB$.

By Euc. III. 20,

$$\angle CDB = 2 \angle CEB = 2\alpha;$$

$$\therefore \angle CDF = \angle BDF = \alpha.$$

$$\therefore CB = 2CF = 2DF \tan \alpha = 2c \tan \alpha.$$

Again, $\angle AEB = \angle ECB$ in alternate segment

$$= \frac{1}{2} \angle EDB \text{ at centre}$$

$$= \frac{1}{2}\left(\frac{\pi}{2} - \alpha\right).$$

$$\therefore AB = c \tan AEB = c \tan\left(\frac{\pi}{4} - \frac{\alpha}{2}\right).$$

4. A tower standing on a cliff subtends an angle β at each of two stations in the same horizontal line passing through the base of the cliff and at distances of a feet and b feet from the cliff. Prove that the height of the tower is $(a+b)\tan\beta$ feet.

5. A column placed on a pedestal 20 feet high subtends an angle of 45° at a point on the ground, and it also subtends an angle of 45° at a point which is 20 feet nearer the pedestal: find the height of the column.

6. A flagstaff on a tower subtends the same angle at each of two places A and B on the ground. The elevations of the top of the flagstaff as seen from A and B are α and β respectively. If $AB = a$, shew that the length of the flagstaff is

$$a \sin(\alpha + \beta - 90°) \operatorname{cosec}(\alpha - \beta).$$

7. A pillar stands on a pedestal. At a distance of 60 feet from the base of the pedestal the pillar subtends its greatest angle 30°: shew that the length of the pillar is $40\sqrt{3}$ feet, and that the pedestal also subtends 30° at the point of observation.

8. A person walking along a canal observes that two objects are in the same line which is inclined at an angle α to the canal. He walks a distance c further and observes that the objects subtend their greatest angle β: shew that their distance apart is

$$2c \sin\alpha \sin\beta \,/\, (\cos\alpha + \cos\beta).$$

9. A tower with a flagstaff stands on a horizontal plane. Shew that the distances from the base at which the flagstaff subtends the same angle and that at which it subtends the greatest possible angle are in geometrical progression.

10. The line joining two stations A and B subtends equal angles at two other stations C and D: prove that

$$AB \sin CBD = CD \sin ADB.$$

11. Two straight lines ABC, DEC meet at C. If

$$\angle DAE = \angle DBE = \alpha, \text{ and } \angle EAB = \beta, \ \angle EBC = \gamma,$$

shew that
$$BC = \frac{AB \sin\beta \sin(\alpha+\beta)}{\sin(\gamma-\beta)\sin(\alpha+\beta+\gamma)}.$$

12. Two objects P and Q subtend an angle of 30° at A. Lengths of 20 feet and 10 feet are measured from A at right angles to AP and AQ respectively to points R and S at each of which PQ subtends angles of 30°: find the length of PQ.

13. A ship sailing N.E. is in a line with two beacons which are 5 miles apart, and of which one is due N. of the other. In 3 minutes and also in ·21 minutes the beacons are found to subtend a right angle at the ship. Prove that the ship is sailing at the rate of 10 miles an hour, and that the beacons subtend their greatest angle at the ship at the end of $3\sqrt{7}$ minutes.

14. A flagstaff stands on the top of a tower. A man walking along a straight road towards the tower observes that the angle of elevation of the top of the flagstaff is β; after walking a distance a further along the road he notices that the flagstaff subtends its maximum angle α; shew that the height of the flagstaff is

$$\frac{2a \sin\alpha \sin\beta}{\cos\beta + \sin(\alpha-\beta)}.$$

Measurements in more than one plane.

201. In Art. 199 the base line AB was measured *directly towards* the object. If this is not possible we may proceed as follows.

From A measure a base line AB in any convenient direction in the horizontal plane. At A observe the two angles PAB and PAC; and at B observe the angle PBA.

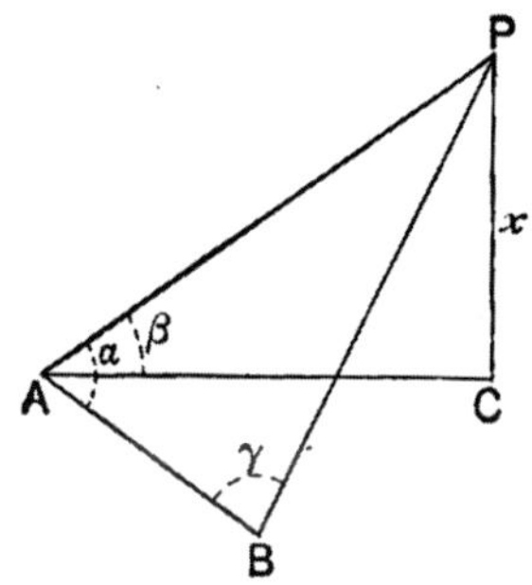

Let $\angle PAB = \alpha$, $\angle PAC = \beta$,

$$\angle PBA = \gamma,$$

$$AB = a, \quad PC = x.$$

From ΔPAC,

$$x = PA \sin \beta.$$

From ΔPAB,

$$PA = \frac{AB \sin PBA}{\sin APB} = \frac{a \sin \gamma}{\sin (\alpha + \gamma)};$$

$$\therefore \quad x = a \sin \beta \sin \gamma \operatorname{cosec} (\alpha + \gamma).$$

202. *To shew how to find the distance between two inaccessible objects.*

Let P and Q be the objects.

Take any two convenient stations A and B in the same horizontal plane, and measure the distance between them.

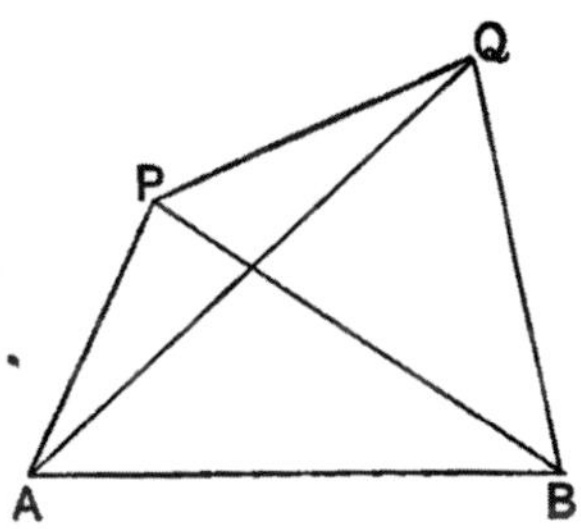

At A observe the angles PAQ and QAB. Also if AP, AQ, AB are not in the same plane, measure the angle PAB.

At B observe the angles ABP and ABQ.

In ΔPAB, we know $\angle PAB$, $\angle PBA$, and AB;
so that AP may be found.

In ΔQAB, we know $\angle QAB$, $\angle QBA$, and AB;
so that AQ may be found.

In ΔPAQ, we know AP, AQ, and $\angle PAQ$;
so that PQ may be found.

Example 1. The angular elevation of a tower CD at a place A due South of it is 30°, and at a place B due West of A the elevation is 18°. If $AB=a$, shew that the height of the tower is $\dfrac{a}{\sqrt{2+2\sqrt{5}}}$.

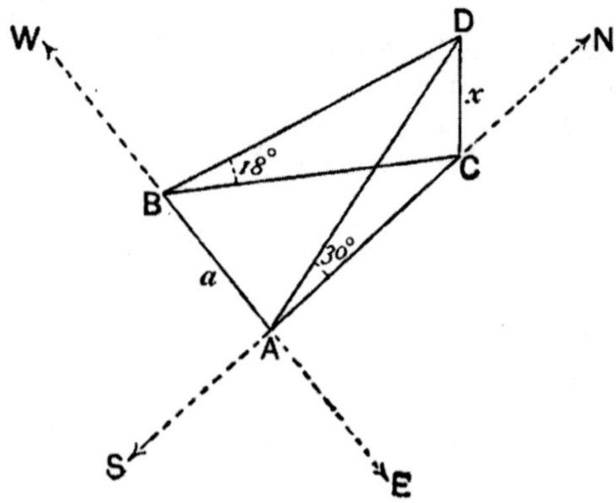

Let $CD=x$.

From the right-angled triangle DCA, $AC=x\cot 30°$.

From the right-angled triangle DCB, $BC=x\cot 18°$.

But $\angle BAC$ is a right angle,

$$\therefore\ BC^2-AC^2=a^2;$$

$$\therefore\ x^2(\cot^2 18°-\cot^2 30°)=a^2;$$

$$\therefore\ x^2(\operatorname{cosec}^2 18°-\operatorname{cosec}^2 30°)=a^2;$$

$$\therefore\ x^2\left\{\left(\frac{4}{\sqrt{5}-1}\right)^2-4\right\}=a^2;$$

$$\therefore\ x^2\{(\sqrt{5}+1)^2-4\}=a^2;$$

$$\therefore\ x^2(2+2\sqrt{5})=a^2,$$

which gives the height required.

Example 2. A hill of inclination 1 in 5 faces South. Shew that a road on it which takes a N.E. direction has an inclination 1 in 7.

Let AD running East and West be the ridge of the hill, and let $ABFD$ be a vertical plane through AD. Let C be a point at the foot of the hill, and ABC a section made by a vertical plane running North and South. Draw CG in a N.E. direction in the horizontal plane and let it meet BF in G; draw GH parallel to BA; then if CH is joined it will represent the direction of the road.

Since the inclination of CA is 1 in 5, we may take $AB=a$, and $AC=5a$, so that $BC^2=24a^2$.

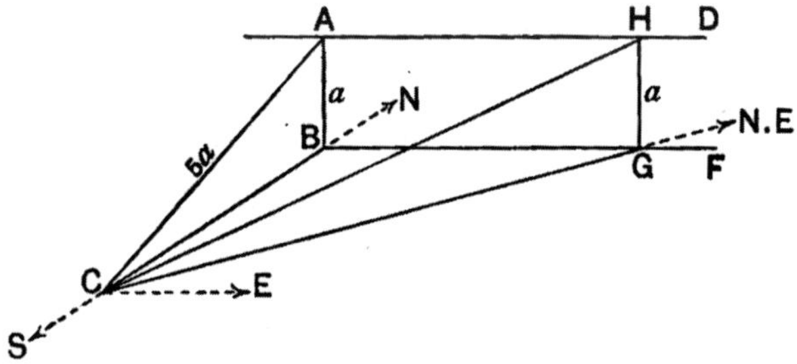

Since CBG is a right-angled isosceles triangle,

$$CG^2=2CB^2=48a^2.$$

Hence in the right-angled triangle CGH,

$$CH^2=48a^2+a^2=49a^2;$$

$$\therefore\ CH=7a=7GH.$$

Thus the slope of the road is 1 in 7.

EXAMPLES. XVII. c.

1. The elevation of a hill at a place P due East of it is 45°, and at a place Q due South of P the elevation is 30°. If the distance from P to Q is 500 yards, find the height of the hill in feet.

2. The elevation of a spire at a point A due West of it is 60°, and at point B due South of A the elevation is 30°. If the spire is 250 feet high, find the distance between A and B.

3. A river flows due North, and a tower stands on its left bank. From a point A up-stream and on the same bank as the tower the elevation of the tower is 60°, and from a point B just opposite on the other bank the elevation is 45°. If the tower is 360 feet high, find the breadth of the river.

4. The elevation of a steeple at a place A due S. of it is 45°, and at a place B due W. of A the elevation is 15°. If $AB=2a$, shew that the height of the steeple is $a(3^{\frac{1}{4}}-3^{-\frac{1}{4}})$.

5. A person due S. of a lighthouse observes that his shadow cast by the light at the top is 24 feet long. On walking 100 yards due E. he finds his shadow to be 30 feet long. Supposing him to be 6 feet high, find the height of the light from the ground.

6. The angles of elevation of a balloon from two stations a mile apart and from a point halfway between them are observed to be 60°, 30°, and 45° respectively. Prove that the height of the balloon is $440\sqrt{6}$ yards.

[*If AD is a median of the triangle ABC,*
then $2AD^2+2BD^2=AB^2+AC^2$.]

7. At each end of a base of length $2a$, the angular elevation of a mountain is θ, and at the middle point of the base the elevation is ϕ. Prove that the height of the mountain is

$$a\sin\theta\sin\phi\sqrt{\operatorname{cosec}(\phi+\theta)\operatorname{cosec}(\phi-\theta)}.$$

8. Two vertical poles, whose heights are a and b, subtend the same angle α at a point in the line joining their feet. If they subtend angles β and γ at any point in the horizontal plane at which the line joining their feet subtends a right angle, prove that

$$(a+b)^2\cot^2\alpha=a^2\cot^2\beta+b^2\cot^2\gamma.$$

9. From the top of a hill a person finds that the angles of depression of three consecutive milestones on a straight level road are α, β, γ. Shew that the height of the hill is

$$5280\sqrt{2}\,\Big/\sqrt{\cot^2\alpha-2\cot^2\beta+\cot^2\gamma}\text{ feet.}$$

10. Two chimneys AB and CD are of equal height. A person standing between them in the line AC joining their bases observes the elevation of the one nearer to him to be 60°. After walking 80 feet in a direction at right angles to AC he observes their elevations to be 45° and 30°: find their height and distance apart.

11. Two persons who are 500 yards apart observe the bearing and angular elevation of a balloon at the same instant. One finds the elevation 60° and the bearing S.W., the other finds the elevation 45° and the bearing W. Find the height of the balloon.

12. The side of a hill faces due S. and is inclined to the horizon at an angle α. A straight railway upon it is inclined at an angle β to the horizon: if the bearing of the railway be x degrees E. of N., shew that $\cos x=\cot\alpha\tan\beta$.

EXAMPLES. XVII. d.

[*In the following examples the logarithms are to be taken from Four-Figure Tables.*]

1. A man in a balloon observes that two churches which he knows to be one mile apart subtend an angle of 11° 24′ when he is exactly over the middle point between them: find the height of the balloon in miles.

2. There are three points A, B, C in a straight line on a level piece of ground. A vertical pole erected at C has an elevation of 5° 30′ from A and 10° 45′ from B. If AB is 100 yards, find the height of the pole and the distance BC.

3. The angular altitude of a lighthouse seen from a point on the shore is 12° 32′, and from a point 500 feet nearer the altitude is 26° 34′: find its height above the sea-level.

4. From a boat the angles of elevation of the highest and lowest points of a flagstaff 30 ft. high on the edge of a cliff are 46° 14′ and 44° 8′: find the height and distance of the cliff.

5. From the top of a hill the angles of depression of two successive milestones on level ground, and in the same vertical plane as the observer, are 5° and 10°. Find the height of the hill in feet and the distance of the nearer milestone in miles.

6. An observer whose eye is 15 feet above the roadway finds that the angle of elevation of the top of a telegraph post is 17° 19′, and that the angle of depression of the foot of the post is 8° 36′: find the height of the post and its distance from the observer.

7. Two straight railroads are inclined at an angle of 20° 16′. At the same instant two engines start from the point of intersection, one along each line; one travels at the rate of 20 miles an hour: at what rate must the other travel so that after 3 hours the distance between them shall be 30 miles?

8. An observer finds that from the doorstep of his house the elevation of the top of a spire is $5a$, and that from the roof above the doorstep it is $4a$. If h be the height of the roof above the doorstep, prove that the height of the spire above the doorstep and the horizontal distance of the spire from the house are respectively

$$h \operatorname{cosec} a \cos 4a \sin 5a \quad \text{and} \quad h \operatorname{cosec} a \cos 4a \cos 5a.$$

If $h=39$ feet, and $a=7° 19′$, calculate the height and the distance.

CHAPTER XVIII.

PROPERTIES OF TRIANGLES AND POLYGONS.

203. *To find the area of a triangle.*

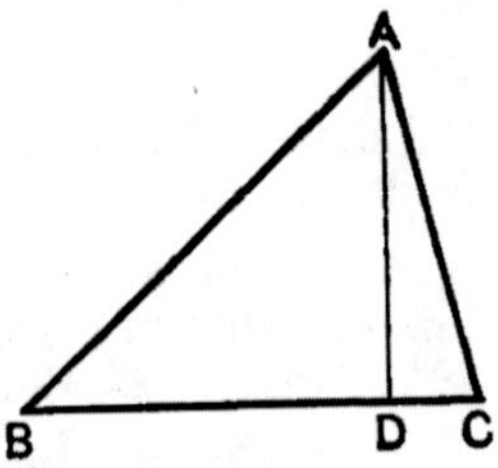

Let Δ denote the area of the triangle ABC. Draw AD perpendicular to BC.

By Euc. I. 41, the area of a triangle is half the area of a rectangle on the same base and of the same altitude.

$$\therefore \quad \Delta = \frac{1}{2} \text{(base} \times \text{altitude)}$$

$$= \frac{1}{2} BC \,.\, AD = \frac{1}{2} BC \,.\, AB \sin B$$

$$= \frac{1}{2} ca \sin B.$$

Similarly, it may be proved that

$$\Delta = \frac{1}{2} ab \sin C, \text{ and } \Delta = \frac{1}{2} bc \sin A.$$

These three expressions for the area are comprised in the single statement

$$\Delta = \frac{1}{2} \textit{(product of two sides)} \times \textit{(sine of included angle).}$$

Again, $\quad \Delta = \dfrac{1}{2} bc \sin A = bc \sin \dfrac{A}{2} \cos \dfrac{A}{2}$

$$= bc \sqrt{\frac{(s-b)(s-c)}{bc}} \sqrt{\frac{s(s-a)}{bc}}$$

$$= \sqrt{s(s-a)(s-b)(s-c)},$$

which gives the area in terms of the sides.

Again, $\Delta=\frac{1}{2}bc\sin A=\frac{1}{2}\sin A\,.\,\frac{a\sin B}{\sin A}\,.\,\frac{a\sin C}{\sin A}$

$$=\frac{a^2\sin B\sin C}{2\sin A}$$

$$=\frac{a^2\sin B\sin C}{2\sin(B+C)},$$

which gives the area in terms of one side and the functions of the adjacent angles.

NOTE. Many writers use the symbol S for the area of a triangle, but to avoid confusion between S and s in manuscript work the symbol Δ is preferable.

Example 1. The sides of a triangle are 17, 25, 28: find the lengths of the perpendiculars from the angles upon the opposite sides.

From the formula $\Delta=\frac{1}{2}$ (base × altitude),

it is evident that the three perpendiculars are found by dividing 2Δ by the three sides in turn.

Now $\Delta=\sqrt{s(s-a)(s-b)(s-c)}=\sqrt{35\times 18\times 10\times 7}$

$$=5\times 7\times 6=210.$$

Thus the perpendiculars are $\frac{420}{17}$, $\frac{420}{25}$, $\frac{420}{28}$, or $\frac{420}{17}$, $\frac{84}{5}$, 15.

Example 2. Two angles of a triangular field are 22·5° and 45°, and the length of the side opposite to the latter is one furlong: find the area.

Let $A=22\frac{1}{2}^\circ$, $B=45^\circ$, then $b=220$ yds., and $C=112\frac{1}{2}^\circ$.

From the formula $\Delta=\frac{b^2\sin A\sin C}{2\sin B}$,

the area in sq. yds. $=\frac{220\times 220\times\sin 22\frac{1}{2}^\circ\times\sin 112\frac{1}{2}^\circ}{2\sin 45^\circ}$

$$=\frac{220\times 220\times\sin 22\frac{1}{2}^\circ\times\cos 22\frac{1}{2}^\circ}{2\times 2\sin 22\frac{1}{2}^\circ\cos 22\frac{1}{2}^\circ}$$

$$=110\times 110.$$

Expressed in acres, the area $=\frac{110\times 110}{4840}=2\frac{1}{2}$.

204. *To find the radius of the circle circumscribing a triangle.*

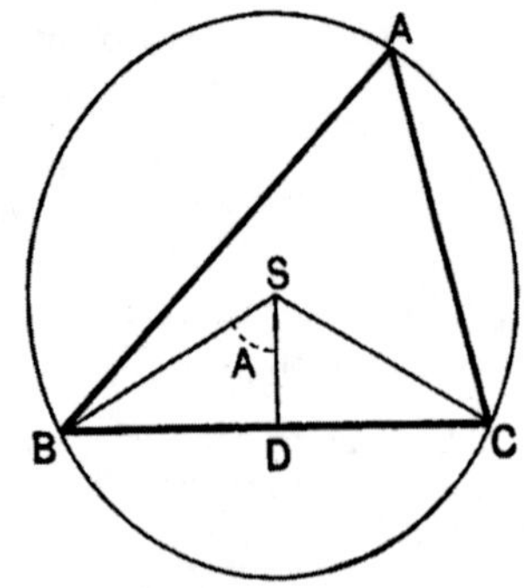

Let S be the centre of the circle circumscribing the triangle ABC, and R its radius.

Bisect $\angle BSC$ by SD, which will also bisect BC at right angles.

Now by Euc. III. 20,

$$\angle BSC \text{ at centre} = \text{twice } \angle BAC = 2A;$$

and $$\frac{a}{2} = BD = BS \sin BSD = R \sin A;$$

$$\therefore \quad R = \frac{a}{2 \sin A}.$$

Thus $$\frac{a}{\sin A} = \frac{b}{\sin B} = \frac{c}{\sin C} = 2R,$$

or $$a = 2R \sin A, \quad b = 2R \sin B, \quad c = 2R \sin C.$$

Example. Shew that $2R^2 \sin A \sin B \sin C = \Delta$.

The first side $= \frac{1}{2} \cdot 2R \sin A \cdot 2R \sin B \cdot \sin C$

$$= \frac{1}{2} ab \sin C$$

$$= \Delta.$$

205. From the result of the last article we deduce the following important theorem:

If a chord of length l *subtend an angle* θ *at the circumference of a circle whose radius is* R, *then* $l = 2R \sin \theta$.

206. For shortness, the circle circumscribing a triangle may be called the *Circum-circle*, its centre the *Circum-centre*, and its radius the *Circum-radius*.

The circum-radius may be expressed in a form not involving the angles, for

$$R = \frac{a}{2 \sin A} = \frac{abc}{2bc \sin A} = \frac{abc}{4\Delta}.$$

207. *To find the radius of the circle inscribed in a triangle.*

Let I be the centre of the circle inscribed in the triangle ABC, and D, E, F the points of contact; then ID, IE, IF are perpendicular to the sides.

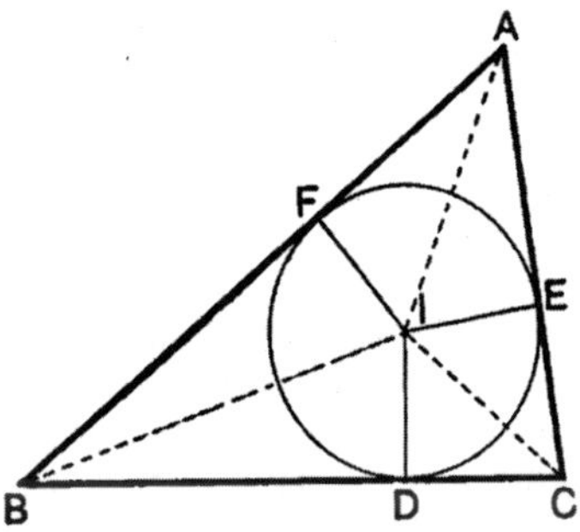

Now Δ=sum of the areas of the triangles BIC, CIA, AIB

$$=\frac{1}{2}ar+\frac{1}{2}br+\frac{1}{2}cr=\frac{1}{2}(a+b+c)r$$

$$=sr;$$

whence $$r=\frac{\Delta}{s}.$$

208. *To express the radius of the inscribed circle in terms of one side and the functions of the half-angles.*

In the figure of the previous article, we know from Euc. IV. 4 that I is the point of intersection of the lines bisecting the angles, so that

$$\angle IBD=\frac{B}{2},\quad \angle ICD=\frac{C}{2}.$$

$$\therefore\ BD=r\cot\frac{B}{2},\quad CD=r\cot\frac{C}{2}.$$

$$\therefore\ r\left(\cot\frac{B}{2}+\cot\frac{C}{2}\right)=a;$$

$$\therefore\ r\sin\frac{B+C}{2}=a\sin\frac{B}{2}\sin\frac{C}{2};$$

$$\therefore\ r=\frac{a\sin\frac{B}{2}\sin\frac{C}{2}}{\cos\frac{A}{2}}.$$

209. Definition. A circle which touches one side of a triangle and the other two sides produced is said to be an **escribed circle** of the triangle.

Thus the triangle ABC has *three* escribed circles, one touching BC, and AB, AC produced; a second touching CA, and BC, BA produced; a third touching AB, and CA, CB produced.

We shall assume that the student is familiar with the construction of the escribed circles.

[See Hall and Stevens' *Geometry*, p. 195.]

For shortness, we shall call the circle inscribed in a triangle the *In-circle*, its centre the *In-centre*, and its radius the *In-radius*; and similarly the escribed circles may be called the *Ex-circles*, their centres the *Ex-centres*, and their radii the *Ex-radii.*

210. *To find the radius of an escribed circle of a triangle.*

Let I_1 be the centre of the circle touching the side BC and the two sides AB and AC produced. Let D_1, E_1, F_1 be the points of contact; then the lines joining I_1 to these points are perpendicular to the sides.

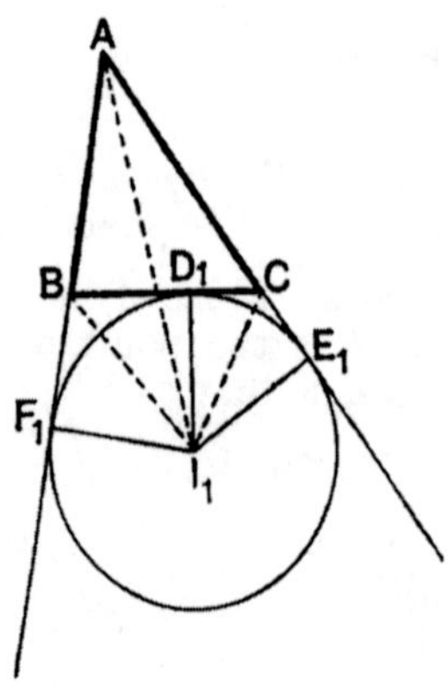

Let r_1 be the radius; then

$$\Delta = \text{area } ABC$$
$$= \text{area } ABI_1C - \text{area } BI_1C$$
$$= \text{area } BI_1A + \text{area } CI_1A - \text{area } BI_1C$$
$$= \frac{1}{2}cr_1 + \frac{1}{2}br_1 - \frac{1}{2}ar_1$$
$$= \frac{1}{2}(c+b-a)r_1$$
$$= (s-a)r_1;$$

$$\therefore \quad r_1 = \frac{\Delta}{s-a}.$$

Similarly, if r_2, r_3 be the radii of the escribed circles opposite to the angles B and C respectively,

$$r_2 = \frac{\Delta}{s-b}, \quad r_3 = \frac{\Delta}{s-c}.$$

211. *To find the radii of the escribed circles in terms of one side and the functions of the half-angles.*

In the figure of the last article, I_1 is the point of intersection of the lines bisecting the angles B and C externally; so that

$$\angle I_1BD_1 = 90^\circ - \frac{B}{2}, \quad \angle I_1CD_1 = 90^\circ - \frac{C}{2}.$$

$$\therefore \quad BD_1 = r_1 \cot\left(90^\circ - \frac{B}{2}\right) = r_1 \tan\frac{B}{2},$$

$$CD_1 = r_1 \cot\left(90^\circ - \frac{C}{2}\right) = r_1 \tan\frac{C}{2};$$

$$\therefore \quad r_1\left(\tan\frac{B}{2} + \tan\frac{C}{2}\right) = a;$$

$$\therefore \quad r_1 \sin\frac{B+C}{2} = a\cos\frac{B}{2}\cos\frac{C}{2};$$

$$\therefore \quad r_1 = \frac{a\cos\frac{B}{2}\cos\frac{C}{2}}{\cos\frac{A}{2}}.$$

Similarly,

$$r_2 = \frac{b\cos\frac{C}{2}\cos\frac{A}{2}}{\cos\frac{B}{2}}, \qquad r_3 = \frac{c\cos\frac{A}{2}\cos\frac{B}{2}}{\cos\frac{C}{2}}.$$

212. By substituting

$$a = 2R\sin A, \quad b = 2R\sin B, \quad c = 2R\sin C,$$

in the formulæ of Art. 208 and Art. 211, we have

$$r = 4R\sin\frac{A}{2}\sin\frac{B}{2}\sin\frac{C}{2},$$

$$r_1 = 4R\sin\frac{A}{2}\cos\frac{B}{2}\cos\frac{C}{2},$$

$$r_2 = 4R\cos\frac{A}{2}\sin\frac{B}{2}\cos\frac{C}{2},$$

$$r_3 = 4R\cos\frac{A}{2}\cos\frac{B}{2}\sin\frac{C}{2}.$$

Example 1. Shew that $\dfrac{r_1-r}{a}+\dfrac{r_2-r}{b}=\dfrac{c}{r_3}$.

$$\text{The first side}=\frac{1}{a}\left(\frac{\Delta}{s-a}-\frac{\Delta}{s}\right)+\frac{1}{b}\left(\frac{\Delta}{s-b}-\frac{\Delta}{s}\right)$$

$$=\frac{\Delta}{s(s-a)}+\frac{\Delta}{s(s-b)}=\frac{\Delta(2s-a-b)}{s(s-a)(s-b)}$$

$$=\frac{\Delta c}{s(s-a)(s-b)}=\frac{\Delta c(s-c)}{s(s-a)(s-b)(s-c)}$$

$$=\frac{\Delta c(s-c)}{\Delta^2}=\frac{c(s-c)}{\Delta}$$

$$=\frac{c}{r_3}.$$

Example 2. If $r_1=r_2+r_3+r$, prove that the triangle is right-angled.

By transposition, $\qquad r_1-r=r_2+r_3;$

$$\therefore\ 4R\sin\frac{A}{2}\cos\frac{B}{2}\cos\frac{C}{2}-4R\sin\frac{A}{2}\sin\frac{B}{2}\sin\frac{C}{2}$$

$$=4R\cos\frac{A}{2}\sin\frac{B}{2}\cos\frac{C}{2}+4R\cos\frac{A}{2}\cos\frac{B}{2}\sin\frac{C}{2};$$

$$\therefore\ \sin\frac{A}{2}\left(\cos\frac{B}{2}\cos\frac{C}{2}-\sin\frac{B}{2}\sin\frac{C}{2}\right)$$

$$=\cos\frac{A}{2}\left(\sin\frac{B}{2}\cos\frac{C}{2}+\cos\frac{B}{2}\sin\frac{C}{2}\right);$$

$$\therefore\ \sin\frac{A}{2}\cos\frac{B+C}{2}=\cos\frac{A}{2}\sin\frac{B+C}{2};$$

$$\therefore\ \sin^2\frac{A}{2}=\cos^2\frac{A}{2};$$

whence $\dfrac{A}{2}=45°$, and $A=90°$.

213. Many important relations connecting a triangle and its circles may be established by elementary geometry.

With the notation of previous articles, since tangents to a circle from the same point are equal, we have $$AF=AE,\quad BD=BF,\quad CD=CE;$$

$$\therefore\ AF+(BD+CD)=\text{half the sum of the sides};$$

$$\therefore\ AF+a=s.$$

$$\therefore\ AF=s-a=AE.$$

Similarly, $BD=BF=s-b$, $CD=CE=s-c$.

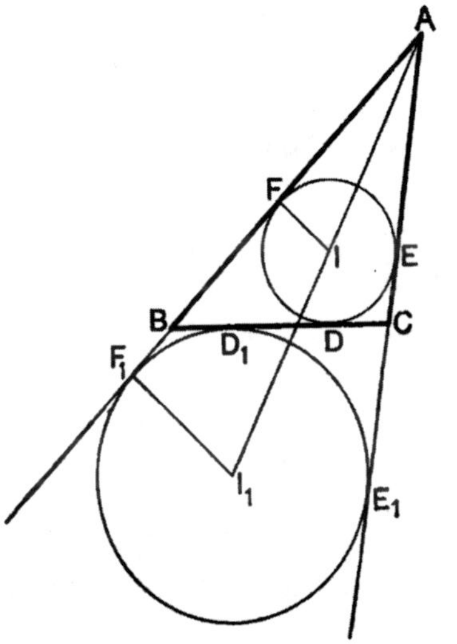

Also $$r=AF\tan\frac{A}{2}=(s-a)\tan\frac{A}{2}.$$

Similarly, $r=(s-b)\tan\dfrac{B}{2}$, $r=(s-c)\tan\dfrac{C}{2}$.

Again, $AF_1=AE_1$, $BF_1=BD_1$, $CE_1=CD_1$;

$$\therefore\ 2AF_1=AF_1+AE_1=(AB+BD_1)+(AC+CD_1)$$
$$=\text{sum of the sides};$$

$$\therefore\ AF_1=s=AE_1.$$

$$\therefore\ BD_1=BF_1=s-c,\quad CD_1=CE_1=s-b.$$

Also $$r_1=AF_1\tan\frac{A}{2}=s\tan\frac{A}{2}.$$

Similarly, $r_2=s\tan\dfrac{B}{2}$, $r_3=s\tan\dfrac{C}{2}$.

EXAMPLES. XVIII. a.

1. Two sides of a triangle are 300 ft. and 120 ft., and the included angle is 150°; find the area.

2. Find the area of the triangle whose sides are 171, 204, 195.

3. Find the sine of the greatest angle of a triangle whose sides are 70, 147, and 119.

4. If the sides of a triangle are 39, 40, 25, find the lengths of the three perpendiculars from the angular points on the opposite sides.

5. One side of a triangle is 30 ft. and the adjacent angles are $22\frac{1}{2}°$ and $112\frac{1}{2}°$, find the area.

6. Find the area of a parallelogram two of whose adjacent sides are 42 and 32 ft., and include an angle of 30°.

7. The area of a rhombus is 648 sq. yds. and one of the angles is 150°: find the length of each side.

8. In a triangle if $a=13$, $b=14$, $c=15$, find r and R.

9. Find r_1, r_2, r_3 in the case of a triangle whose sides are 17, 10, 21.

10. If the area of a triangle is 96, and the radii of the escribed circles are 8, 12, 24, find the sides.

Prove the following formulæ:

11. $\sqrt{rr_1r_2r_3}=\Delta$.

12. $s(s-a)\tan\frac{A}{2}=\Delta$.

13. $rr_1\cot\frac{A}{2}=\Delta$.

14. $4Rrs=abc$.

15. $r_1r_2r_3=rs^2$.

16. $r\cot\frac{B}{2}\cot\frac{C}{2}=r_1$.

17. $Rr(\sin A+\sin B+\sin C)=\Delta$.

18. $r_1r_2+rr_3=ab$.

19. $\cos\frac{A}{2}\sqrt{bc(s-b)(s-c)}=\Delta$.

20. $r_1+r_2=c\cot\frac{C}{2}$.

21. $(r_1-r)(r_2+r_3)=a^2$.

22. $r_1 \cot\frac{A}{2} = r_2 \cot\frac{B}{2} = r_3 \cot\frac{C}{2} = r \cot\frac{A}{2}\cot\frac{B}{2}\cot\frac{C}{2}$.

23. $\frac{1}{r_1}+\frac{1}{r_2}+\frac{1}{r_3}=\frac{1}{r}$.

24. $r_2r_3+r_3r_1+r_1r_2=s^2$.

25. $r_1+r_2+r_3-r=4R$.

26. $r+r_1+r_2-r_3=4R\cos C$.

27. $b^2\sin 2C+c^2\sin 2B=4\Delta$.

28. $4R\cos\frac{C}{2}=(a+b)\sec\frac{A-B}{2}$.

29. $a^2-b^2=2Rc\sin(A-B)$.

30. $\frac{a^2-b^2}{2}\cdot\frac{\sin A\sin B}{\sin(A-B)}=\Delta$.

31. If the perpendiculars from A, B, C to the opposite sides are p_1, p_2, p_3 respectively, prove that

(1) $\frac{1}{p_1}+\frac{1}{p_2}+\frac{1}{p_3}=\frac{1}{r}$; (2) $\frac{1}{p_1}+\frac{1}{p_2}-\frac{1}{p_3}=\frac{1}{r_3}$

Prove the following identities:

32. $(r_1-r)(r_2-r)(r_3-r)=4Rr^2$.

33. $\left(\frac{1}{r}-\frac{1}{r_1}\right)\left(\frac{1}{r}-\frac{1}{r_2}\right)\left(\frac{1}{r}-\frac{1}{r_3}\right)=\frac{4R}{r^2s^2}$.

34. $4\Delta(\cot A+\cot B+\cot C)=a^2+b^2+c^2$.

35. $\frac{b-c}{r_1}+\frac{c-a}{r_2}+\frac{a-b}{r_3}=0$.

36. $a^2b^2c^2(\sin 2A+\sin 2B+\sin 2C)=32\Delta^3$.

37. $a\cos A+b\cos B+c\cos C=4R\sin A\sin B\sin C$.

38. $a\cot A+b\cot B+c\cot C=2(R+r)$.

39. $(b+c)\tan\frac{A}{2}+(c+a)\tan\frac{B}{2}+(a+b)\tan\frac{C}{2}$
$=4R(\cos A+\cos B+\cos C)$.

40. $r(\sin A+\sin B+\sin C)=2R\sin A\sin B\sin C$.

41. $\cos^2\frac{A}{2}+\cos^2\frac{B}{2}+\cos^2\frac{C}{2}=2+\frac{r}{2R}$.

Inscribed and circumscribed Polygons.

214. *To find the perimeter and area of a regular polygon of n sides inscribed in a circle.*

Let r be the radius of the circle, and AB a side of the polygon.

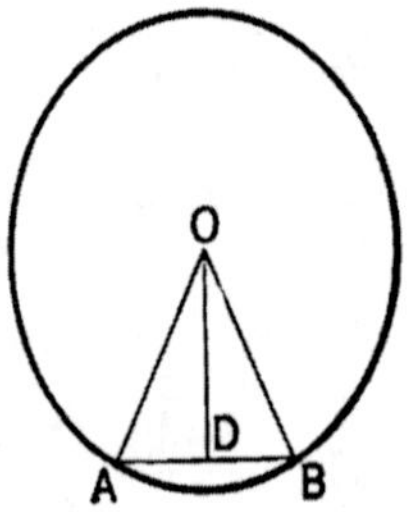

Join OA, OB, and draw OD bisecting $\angle AOB$; then AB is bisected at right angles in D.

And $\angle AOB = \frac{1}{n}$ (four right angles)

$$= \frac{2\pi}{n}.$$

$$\text{Perimeter of polygon} = nAB = 2nAD = 2nOA \sin AOD$$

$$= 2nr \sin \frac{\pi}{n}.$$

$$\text{Area of polygon} = n\,(\text{area of triangle } AOB)$$

$$= \frac{1}{2} nr^2 \sin \frac{2\pi}{n}.$$

215. *To find the perimeter and area of a regular polygon of n sides circumscribed about a given circle.*

Let r be the radius of the circle, and AB a side of the polygon. Let AB touch the circle at D. Join OA, OB, OD; then OD bisects AB at right angles, and also bisects $\angle AOB$.

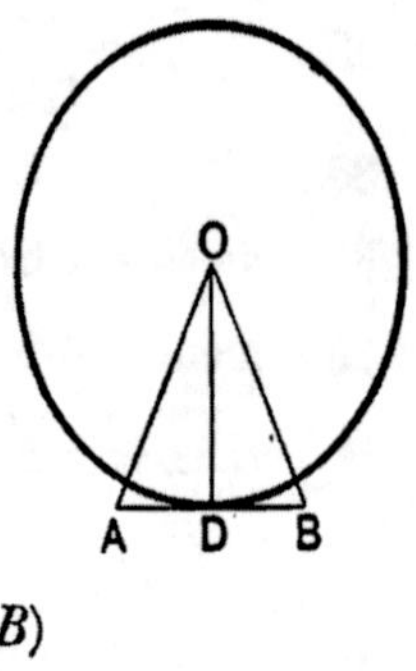

Perimeter of polygon

$$= nAB = 2nAD = 2nOD \tan AOD$$

$$= 2nr \tan \frac{\pi}{n}.$$

$$\text{Area of polygon} = n\,(\text{area of triangle } AOB)$$

$$= nOD \, . \, AD$$

$$= nr^2 \tan \frac{\pi}{n}.$$

216. There is no need to burden the memory with the formulæ of the last two articles, as in any particular instance they are very readily obtained.

Example 1. The side of a regular dodecagon is 2 ft., find the radius of the circumscribed circle.

Let r be the required radius. In the adjoining figure we have

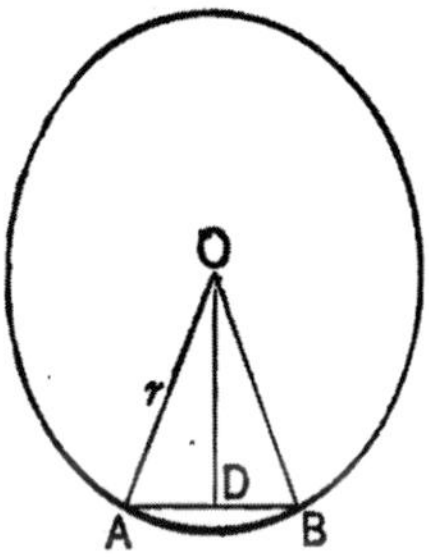

$$AB=2,\quad \angle AOB=\frac{2\pi}{12}.$$

$$AB=2AD=2r\sin\frac{\pi}{12};$$

$$\therefore\ 2r\sin 15^\circ=2;$$

$$\therefore\ r=\frac{1}{\sin 15^\circ}=\frac{2\sqrt{2}}{\sqrt{3}-1}=\sqrt{2}\,(\sqrt{3}+1).$$

Thus the radius is $\sqrt{6}+\sqrt{2}$ feet.

Example 2. A regular pentagon and a regular decagon have the same perimeter, prove that their areas are as 2 to $\sqrt{5}$.

Let AB be one of the n sides of a regular polygon, O the centre of the circumscribed circle, OD perpendicular to AB.

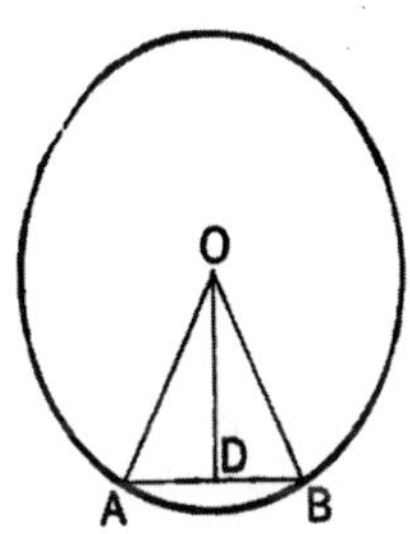

Then if $AB=a$,

$$\text{area of polygon}=nAD\,.\,OD$$

$$=nAD\,.\,AD\cot\frac{\pi}{n}$$

$$=\frac{na^2}{4}\cot\frac{\pi}{n}.$$

Denote the perimeter of the pentagon and decagon by $10c$. Then each side of the pentagon is $2c$, and its area is $5c^2\cot\frac{\pi}{5}$.

Each side of the decagon is c, and its area is $\frac{5}{2}c^2\cot\frac{\pi}{10}$.

$$\therefore\ \frac{\text{Area of pentagon}}{\text{Area of decagon}}=\frac{2\cot 36^\circ}{\cot 18^\circ}=\frac{2\cos 36^\circ\sin 18^\circ}{\sin 36^\circ\cos 18^\circ}=\frac{2\cos 36^\circ}{2\cos^2 18^\circ}$$

$$=\frac{2\cos 36^\circ}{1+\cos 36^\circ}=\frac{2(\sqrt{5}+1)}{4}\div\left(1+\frac{\sqrt{5}+1}{4}\right)$$

$$=\frac{2(\sqrt{5}+1)}{5+\sqrt{5}}=\frac{2}{\sqrt{5}}.$$

217. *To find the area of a circle.*

Let r be the radius of the circle, and let a regular polygon of n sides be described about it. Then from the adjoining figure, we have

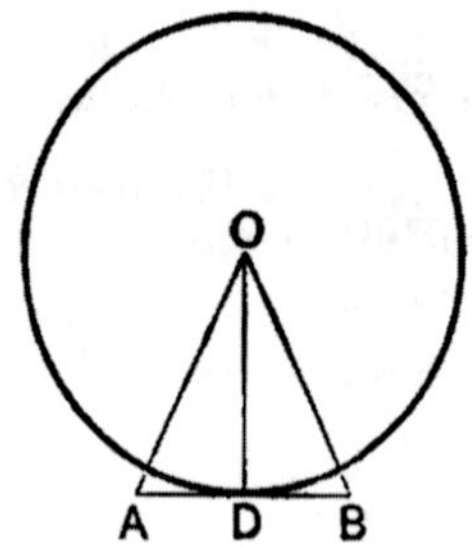

area of polygon $= n$ (area of triangle AOB)

$$= n\left(\frac{1}{2}AB \cdot OD\right)$$

$$= \frac{1}{2}OD \cdot nAB$$

$$= \frac{r}{2} \times \text{perimeter of polygon.}$$

By increasing the number of sides without limit, the area and the perimeter of the polygon may be made to differ as little as we please from the area and the circumference of the circle. Hence

$$\text{area of a circle} = \frac{r}{2} \times \text{circumference}$$

$$= \frac{r}{2} \times 2\pi r \quad \text{[Art. 59.]}$$

$$= \pi r^2.$$

218. *To find the area of the sector of a circle.*

Let θ be the circular measure of the angle of the sector; then by Geometry,

$$\frac{\text{area of sector}}{\text{area of circle}} = \frac{\theta}{2\pi};$$

$$\therefore \text{ area of sector} = \frac{\theta}{2\pi} \times \pi r^2 = \frac{1}{2}r^2\theta.$$

EXAMPLES. XVIII. b.

[*In this Exercise take* $\pi = \frac{22}{7}$.]

1. Find the area of a regular decagon inscribed in a circle whose radius is 3 feet; given $\sin 36° = \cdot 588$.

2. Find the perimeter and area of a regular quindecagon described about a circle whose diameter is 3 yards; given

$$\tan 12° = \cdot 213.$$

3. Shew that the areas of the inscribed and circumscribed circles of a regular hexagon are in the ratio of 3 to 4.

4. Find the area of a circle inscribed in a regular pentagon whose area is 250 sq. ft.; given $\cot 36° = 1\cdot376$.

5. Find the perimeter of a regular octagon inscribed in a circle whose area is 1386 sq. inches; given $\sin 22° \, 30' = \cdot382$.

6. Find the perimeter of a regular pentagon described about a circle whose area is 616 sq. ft.; given $\tan 36° = \cdot727$.

7. Find the diameter of the circle circumscribing a regular quindecagon, whose inscribed circle has an area of 2464 sq. ft.; given $\sec 12° = 1\cdot022$.

8. Find the area of a regular dodecagon in a circle about a regular pentagon 50 sq. ft. in area; given $\operatorname{cosec} 72° = 1\cdot0515$.

9. A regular pentagon and a regular decagon have the same area, prove that the ratio of their perimeters is $\sqrt[4]{5} : \sqrt{2}$.

10. Two regular polygons of n sides and $2n$ sides have the same perimeter; shew that the ratio of their areas is

$$2\cos\frac{\pi}{n} : 1 + \cos\frac{\pi}{n}.$$

11. If $2a$ be the side of a regular polygon of n sides, R and r the radii of the circumscribed and inscribed circles, prove that

$$R + r = a\cot\frac{\pi}{2n}.$$

12. Prove that the square of the side of a regular pentagon inscribed in a circle is equal to the sum of the squares of the sides of a regular hexagon and decagon inscribed in the same circle.

13. With reference to a given circle, A_1 and B_1 are the areas of the inscribed and circumscribed regular polygons of n sides, A_2 and B_2 are corresponding quantities for regular polygons of $2n$ sides: prove that

(1) A_2 is a geometric mean between A_1 and B_1;

(2) B_2 is a harmonic mean between A_2 and B_1.

The Ex-central Triangle.

***219.** Let ABC be a triangle, I_1, I_2, I_3 its ex-centres; then $I_1I_2I_3$ is called the **Ex-central triangle** of ABC.

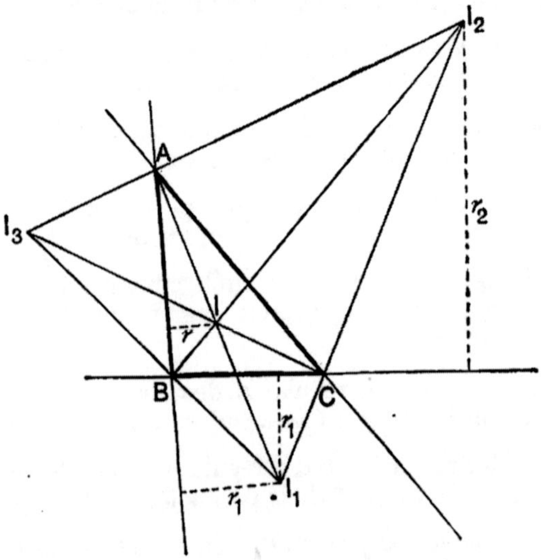

Let I be the in-centre; then from the construction for finding the positions of the in-centre and ex-centres, it follows that:

(i) The points I, I_1 lie on the line bisecting the angle BAC; the points I, I_2 lie on the line bisecting the angle ABC; the points I, I_3 lie on the line bisecting the angle ACB.

(ii) The points I_2, I_3 lie on the line bisecting the angle BAC externally; the points I_3, I_1 lie on the line bisecting the angle ABC externally; the points I_1, I_2 lie on the line bisecting the angle ACB externally.

(iii) The line AI_1 is perpendicular to I_2I_3; the line BI_2 is perpendicular to I_3I_1; the line CI_3 is perpendicular to I_1I_2. Thus the triangle ABC is the *Pedal triangle* of its ex-central triangle $I_1I_2I_3$. [See Art. 223.]

(iv) The angles IBI_1 and ICI_1 are right angles; hence the points B, I, C, I_1 are concyclic. Similarly, the points C, I, A, I_2, and the points A, I, B, I_3 are concyclic.

(v) The lines AI_1, BI_2, CI_3 meet at the in-centre I, which is therefore the *Orthocentre* of the ex-central triangle $I_1I_2I_3$.

(vi) Each of the four points I, I_1, I_2, I_3 is the orthocentre of the triangle formed by joining the other three points.

***220.** *To find the sides and angles of the ex-central triangle.*

With the figure of the last article,

$$\angle BI_1C = \angle BI_1I + \angle CI_1I$$
$$= \angle BCI + \angle CBI$$
$$= \frac{C}{2} + \frac{B}{2} = 90^\circ - \frac{A}{2}.$$

Thus the angles are

$$90^\circ - \frac{A}{2}, \quad 90^\circ - \frac{B}{2}, \quad 90^\circ - \frac{C}{2}.$$

Again, the points B, I_3, I_2, C are concyclic;

$$\therefore \angle I_1I_2I_3 = \text{supplement of } \angle I_3BC = \angle I_1BC;$$

$\therefore$ the triangles $I_1I_2I_3$, I_1BC are similar;

$$\therefore \frac{I_2I_3}{BC} = \frac{I_3I_1}{I_1C} = \sec\left(90^\circ - \frac{A}{2}\right) = \operatorname{cosec}\frac{A}{2};$$

$$\therefore I_2I_3 = a \operatorname{cosec}\frac{A}{2} = 4R\cos\frac{A}{2}.$$

Thus the sides are

$$4R\cos\frac{A}{2}, \quad 4R\cos\frac{B}{2}, \quad 4R\cos\frac{C}{2}.$$

***221.** *To find the area and circum-radius of the ex-central triangle.*

The area $= \frac{1}{2}$ (product of two sides) $\times$ (sine of included angle)

$$= \frac{1}{2} \times 4R\cos\frac{B}{2} \times 4R\cos\frac{C}{2} \times \sin\left(90^\circ - \frac{A}{2}\right)$$
$$= 8R^2 \cos\frac{A}{2}\cos\frac{B}{2}\cos\frac{C}{2}.$$

The circum-radius $= \dfrac{I_2I_3}{2\sin I_2I_1I_3} = \dfrac{4R\cos\frac{A}{2}}{2\sin\left(90^\circ - \frac{A}{2}\right)} = 2R.$

***222.** *To find the distances between the in-centre and ex-centres.*

With the figure of Art. 219,

$$\text{the } \angle\text{s } IBI_1,\ ICI_1 \text{ are right angles;}$$

$\therefore$ II_1 is the diameter of the circum-circle of the triangle BCI_1;

$$\therefore\ II_1 = \frac{BC}{\sin BI_1C} = \frac{a}{\cos\frac{A}{2}}$$

$$= 4R\sin\frac{A}{2}.$$

Thus the distances are

$$4R\sin\frac{A}{2},\quad 4R\sin\frac{B}{2},\quad 4R\sin\frac{C}{2}.$$

The Pedal Triangle.

***223.** Let G, H, K be the feet of the perpendiculars from the angular points on the opposite sides of the triangle ABC; then GHK is called the **Pedal triangle** of ABC.

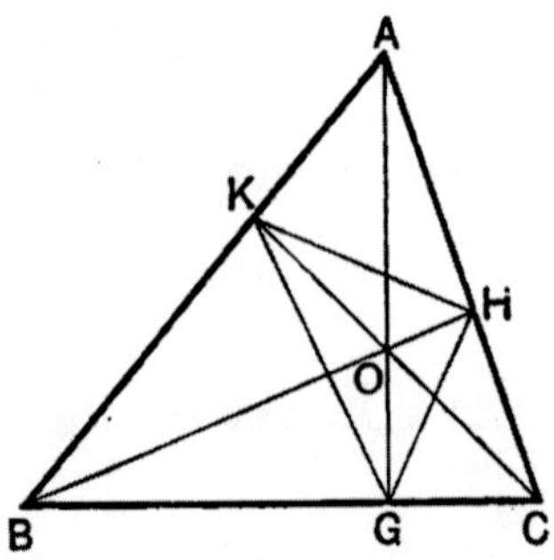

The three perpendiculars AG, BH, CK meet in a point O which is called the **Orthocentre** of the triangle ABC.

***224.** *To find the sides and angles of the pedal triangle.*

In the figure of the last article, the points K, O, G, B are concyclic;

$$\therefore\ \angle OGK = \angle OBK = 90^\circ - A.$$

Also the points H, O, G, C are concyclic;

$$\therefore\ \angle OGH = \angle OCH = 90^\circ - A;$$

$$\therefore\ \angle KGH = 180^\circ - 2A.$$

Thus the angles of the pedal triangle are

$$180^\circ - 2A,\quad 180^\circ - 2B,\quad 180^\circ - 2C.$$

Again, the triangles AKH, ABC are similar;

$$\therefore \frac{HK}{BC} = \frac{AK}{AC} = \cos A;$$

$$\therefore HK = a\cos A.$$

Thus the sides of the pedal triangle are

$$a\cos A, \quad b\cos B, \quad c\cos C.$$

In terms of R, the equivalent forms become

$$R\sin 2A, \quad R\sin 2B, \quad R\sin 2C.$$

If the angle ACB of the given triangle is obtuse, the expressions $180° - 2C$ and $c\cos C$ are both negative, and the values we have obtained require some modification. We leave the student to shew that in this case the angles are $2A$, $2B$, $2C - 180°$, and the sides $a\cos A$, $b\cos B$, $-c\cos C$.

***225.** *To find the area and circum-radius of the pedal triangle.*

The area $= \frac{1}{2}$ (product of two sides) × (sine of included angle)

$$= \frac{1}{2} R\sin 2B \,.\, R\sin 2C \,.\, \sin(180 - 2A)$$

$$= \frac{1}{2} R^2 \sin 2A \sin 2B \sin 2C.$$

The circum-radius $= \dfrac{HK}{2\sin HGK} = \dfrac{R\sin 2A}{2\sin(180° - 2A)} = \dfrac{R}{2}.$

NOTE. The circum-circle of the pedal triangle is the nine points circle of the triangle ABC. Thus the radius of the nine points circle of the triangle ABC is $\frac{R}{2}$. [See Hall and Stevens' *Geometry*, p. 217.]

***226.** In Art. 224, we have proved that OG, OH, OK bisect the angles HGK, KHG, GKH respectively, so that O is the in-centre of the triangle GHK. Thus the orthocentre of a triangle is the in-centre of the pedal triangle.

Again, the line CGB which is at right angles to OG bisects $\angle HGK$ externally. Similarly the lines AHC and BKA bisect $\angle KHG$ and $\angle GKH$ externally, so that ABC is the ex-central triangle of its pedal triangle GHK.

***227.** In Art. 219, we have seen that ABC is the pedal triangle of its ex-central triangle $I_1I_2I_3$. Certain theorems depending on this connection are more evident from the adjoining figure, in which the fact that ABC is the pedal triangle of $I_1I_2I_3$ is brought more prominently into view. For instance, the circum-circle of the triangle ABC is the nine points circle of the triangle $I_1I_2I_3$, and passes through the middle points of II_1, II_2, II_3 and of I_2I_3, I_3I_1, I_1I_2.

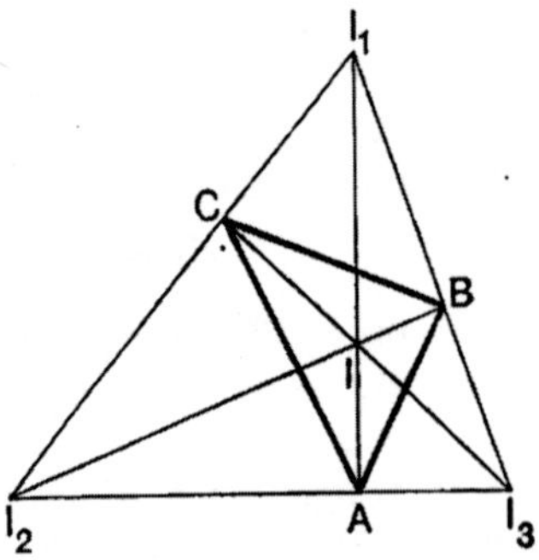

***228.** *To find the distance between the in-centre and circum-centre.*

Let S be the circum-centre and I the in-centre. Produce AI to meet the circum-circle in H; join CH and CI.

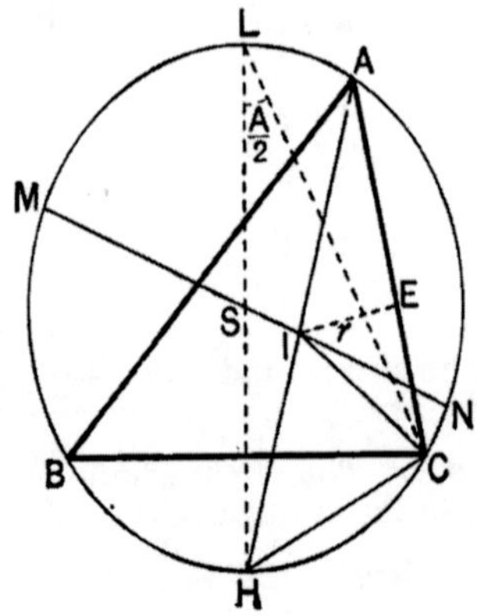

Draw IE perpendicular to AC. Produce HS to meet the circumference in L, and join CL. Then

$$\angle HIC = \angle IAC + \angle ICA$$
$$= \frac{A}{2} + \frac{C}{2};$$
$$\angle HCI = \angle ICB + \angle BCH$$
$$= \frac{C}{2} + \angle BAH$$
$$= \frac{C}{2} + \frac{A}{2};$$

$$\therefore \angle HCI = \angle HIC;$$

$$\therefore HI = HC = 2R \sin\frac{A}{2}.$$

Also $$AI = IE \operatorname{cosec}\frac{A}{2} = r \operatorname{cosec}\frac{A}{2};$$

$$\therefore AI \,.\, IH = 2Rr.$$

Produce SI to meet the circumference in M and N.

Since MIN, AIH are chords of the circle,

$$AI \cdot IH = MI \cdot IN = (R+SI)(R-SI);$$

$$\therefore \; 2Rr = R^2 - SI^2;$$

that is,
$$SI^2 = R^2 - 2Rr.$$

***229.** *To find the distance of an ex-centre from the circum-centre.*

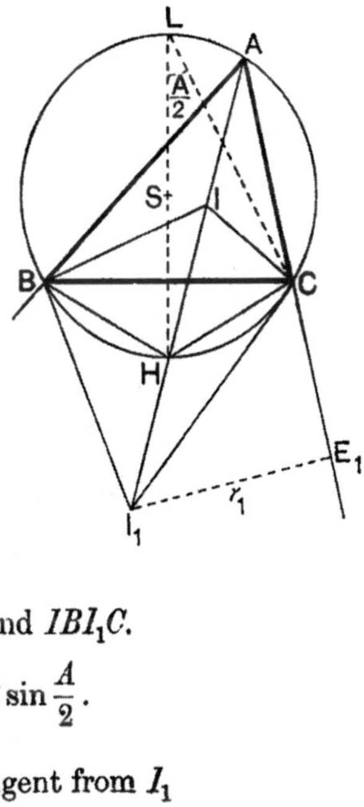

Let S be the circum-centre, and I the in-centre; then AI produced passes through the ex-centre I_1.

Let AI_1 meet the circum-circle in H; join CI, BI, CH, BH, CI_1, BI_1. Draw I_1E_1 perpendicular to AC.

Produce HS to meet the circumference in L, and join CL.

The angles IBI_1 and ICI_1 are right angles; hence the circle on II_1 as diameter passes through B and C.

The chords BH and CH of the circum-circle subtend equal angles at A, and are therefore equal.

But from the last article, $HC = HI$;

$$\therefore \; HB = HC = HI;$$

hence H is the centre of the circle round IBI_1C.

$$\therefore \; HI_1 = HC = 2R\sin\frac{A}{2}.$$

Now
$$SI_1^2 - R^2 = \text{square of tangent from } I_1$$

$$= I_1H \cdot I_1A$$

$$= 2R\sin\frac{A}{2} \cdot r_1 \operatorname{cosec}\frac{A}{2}$$

$$= 2Rr_1.$$

$$\therefore \; SI_1^2 = R^2 + 2Rr_1.$$

***230.** *To find the distance of the orthocentre from the circumcentre.*

With the usual notation, we have

$SO^2 = SA^2 + AO^2 - 2SA \cdot AO \cos SAO.$

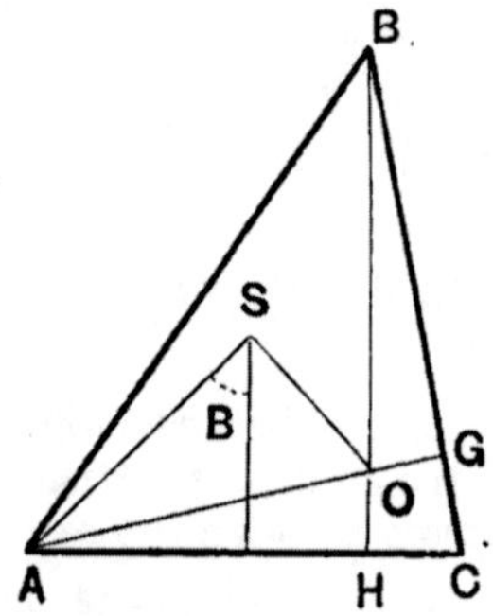

Now $AS = R$;

$$\begin{aligned} AO &= AH \operatorname{cosec} C \\ &= c \cos A \operatorname{cosec} C \\ &= 2R \sin C \cos A \operatorname{cosec} C \\ &= 2R \cos A; \end{aligned}$$

$$\begin{aligned} \angle SAO &= \angle SAC - \angle OAC \\ &= (90^\circ - B) - (90^\circ - C) \\ &= C - B. \end{aligned}$$

$$\begin{aligned} \therefore \; SO^2 &= R^2 + 4R^2 \cos^2 A - 4R^2 \cos A \cos (C - B) \\ &= R^2 - 4R^2 \cos A \{\cos (B + C) + \cos (C - B)\} \\ &= R^2 - 8R^2 \cos A \cos B \cos C. \end{aligned}$$

The student may apply a similar method to establish the results of the last two articles.

*EXAMPLES. XVIII. c.

1. Shew that the distance of the in-centre from A is

$$4R \sin \frac{B}{2} \sin \frac{C}{2}.$$

2. Shew that the distances of the ex-centre I_1 from the angular points A, B, C are

$$4R \cos \frac{B}{2} \cos \frac{C}{2}, \quad 4R \sin \frac{A}{2} \cos \frac{C}{2}, \quad 4R \sin \frac{A}{2} \cos \frac{B}{2}.$$

3. Prove that the area of the ex-central triangle is equal to

(1) $2Rs$; (2) $\frac{1}{2} \Delta \operatorname{cosec} \frac{A}{2} \operatorname{cosec} \frac{B}{2} \operatorname{cosec} \frac{C}{2}$.

4. Shew that

$$r \cdot II_1 \cdot II_2 \cdot II_3 = 4R \cdot IA \cdot IB \cdot IC.$$

5. Shew that the perimeter and in-radius of the pedal triangle are respectively

$$4R \sin A \sin B \sin C \quad \text{and} \quad 2R \cos A \cos B \cos C.$$

6. If g, h, k denote the sides of the pedal triangle, prove that

(1) $\dfrac{g}{a^2}+\dfrac{h}{b^2}+\dfrac{k}{c^2}=\dfrac{a^2+b^2+c^2}{2abc}$;

(2) $\dfrac{(b^2-c^2)\,g}{a^2}+\dfrac{(c^2-a^2)\,h}{b^2}+\dfrac{(a^2-b^2)\,k}{c^2}=0.$

7. Prove that the ex-radii of the pedal triangle are
$2R\cos A\sin B\sin C$, $2R\sin A\cos B\sin C$, $2R\sin A\sin B\cos C$.

8. Prove that any formula which connects the sides and angles of a triangle holds if we replace

(1) a, b, c by $a\cos A$, $b\cos B$, $c\cos C$,
and A, B, C by $180°-2A$, $180°-2B$, $180°-2C$;

(2) a, b, c by $a\operatorname{cosec}\dfrac{A}{2}$, $b\operatorname{cosec}\dfrac{B}{2}$, $c\operatorname{cosec}\dfrac{C}{2}$,

and A, B, C by $90°-\dfrac{A}{2}$, $90°-\dfrac{B}{2}$, $90°-\dfrac{C}{2}$.

9. Prove that the radius of the circum-circle is never less than the diameter of the in-circle.

10. If $R=2r$, shew that the triangle is equilateral.

11. Prove that

$$SI^2+SI_1^2+SI_2^2+SI_3^2=12R^2.$$

12. Prove that

(1) $a\,.\,AI^2+b\,.\,BI^2+c\,.\,CI^2=abc$;

(2) $a\,.\,AI_1^2-b\,.\,BI_1^2-c\,.\,CI_1^2=abc.$

13. If GHK be the pedal triangle, and O the orthocentre, prove that

(1) $\dfrac{OG}{AG}+\dfrac{OH}{BH}+\dfrac{OK}{CK}=1$;

(2) $\dfrac{OG}{OG+a\cot A}+\dfrac{OH}{OH+b\cot B}+\dfrac{OK}{OK+c\cot C}=1.$

14. If GHK be the pedal triangle, shew that the sum of the circum-radii of the triangles AHK, BKG, CGH is equal to $R+r$.

15. If $A_1B_1C_1$ is the ex-central triangle of ABC, and $A_2B_2C_2$ the ex-central triangle of $A_1B_1C_1$, and $A_3B_3C_3$ the ex-central triangle of $A_2B_2C_2$, and so on: find the angles of the triangle $A_nB_nC_n$, and prove that when n is indefinitely increased the triangle becomes equilateral.

16. Prove that

(1) $OS^2 = 9R^2 - a^2 - b^2 - c^2$;

(2) $OI^2 = 2r^2 - 4R^2 \cos A \cos B \cos C$;

(3) $OI_1^2 = 2r_1^2 - 4R^2 \cos A \cos B \cos C$.

17. If f, g, h denote the distances of the circum-centre of the pedal triangle from the angular points of the original triangle, shew that

$$4(f^2+g^2+h^2) = 11R^2 + 8R^2 \cos A \cos B \cos C.$$

Quadrilaterals.

***231.** *To prove that the area of a quadrilateral is equal to*

$$\frac{1}{2}\,(\textit{product of the diagonals}) \times (\textit{sine of included angle}).$$

Let the diagonals AC, BD intersect at P, and let $\angle DPA = \alpha$, and let S denote the area of the quadrilateral.

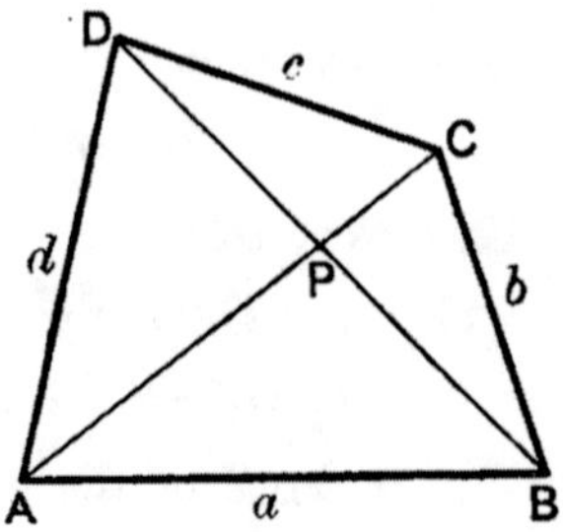

$$\Delta DAC = \Delta APD + \Delta CPD$$

$$= \frac{1}{2} DP \,.\, AP \sin \alpha + \frac{1}{2} DP \,.\, PC \sin(\pi - \alpha)$$

$$= \frac{1}{2} DP(AP + PC) \sin \alpha$$

$$= \frac{1}{2} DP \,.\, AC \sin \alpha.$$

Similarly $\quad \Delta ABC = \frac{1}{2} BP \,.\, AC \sin \alpha.$

$$\therefore\ S = \frac{1}{2}(DP + BP)\, AC \sin \alpha$$

$$= \frac{1}{2} DB \,.\, AC \sin \alpha.$$

***232.** *To find the area of a quadrilateral in terms of the sides and the sum of two opposite angles.*

Let $ABCD$ be the quadrilateral, and let a, b, c, d be the lengths of its sides, S its area.

By equating the two values of BD^2 found from the triangles BAD, BCD, we have

$$a^2+d^2-2ad\cos A=b^2+c^2-2bc\cos C;$$

$$\therefore\ a^2+d^2-b^2-c^2=2ad\cos A-2bc\cos C \quad \text{.........(1)}.$$

Also $\quad S=$ sum of areas of triangles BAD, BCD

$$=\frac{1}{2}ad\sin A+\frac{1}{2}bc\sin C;$$

$$\therefore\ 4S=2ad\sin A+2bc\sin C \quad \text{..............(2)}.$$

Square (2) and add to the square of (1);

$$\therefore\ 16S^2+(a^2+d^2-b^2-c^2)^2=4a^2d^2+4b^2c^2-8abcd\cos(A+C).$$

Let $A+C=2\alpha$; then

$$\cos(A+C)=\cos 2\alpha=2\cos^2\alpha-1;$$

$$\therefore\ 16S^2=4(ad+bc)^2-(a^2+d^2-b^2-c^2)^2-16abcd\cos^2\alpha.$$

But the first two terms on the right

$$=(2ad+2bc+a^2+d^2-b^2-c^2)(2ad+2bc-a^2-d^2+b^2+c^2)$$

$$=\{(a+d)^2-(b-c)^2\}\{(b+c)^2-(a-d)^2\}$$

$$=(a+d+b-c)(a+d-b+c)(b+c+a-d)(b+c-a+d)$$

$$=(2\sigma-2c)(2\sigma-2b)(2\sigma-2d)(2\sigma-2a),$$

where $a+b+c+d=2\sigma$,

$$=16(\sigma-a)(\sigma-b)(\sigma-c)(\sigma-d).$$

Thus $\quad S^2=(\sigma-a)(\sigma-b)(\sigma-c)(\sigma-d)-abcd\cos^2\alpha$,
where 2σ denotes the sum of the sides, 2α the sum of either pair of opposite angles.

***233.** In the case of a *cyclic quadrilateral*, $A+C=180°$, so that $\alpha=90°$; hence

$$S=\sqrt{(\sigma-a)(\sigma-b)(\sigma-c)(\sigma-d)}.$$

This formula may be obtained directly as in the last article by making use of the condition $A+C=180^\circ$ during the course of the work. In this case $\cos C=-\cos A$, and $\sin C=\sin A$, so that the expressions (1) and (2) become

$$a^2+d^2-b^2-c^2=2(ad+bc)\cos A,$$

and
$$4S=2(ad+bc)\sin A;$$

whence by eliminating A we obtain

$$16S^2+(a^2+d^2-b^2-c^2)^2=4(ad+bc)^2.$$

***234.** *To find the diagonals and the circum-radius of a cyclic quadrilateral.*

If $ABCD$ is a cyclic quadrilateral, we have just proved that

$$2(ad+bc)\cos A=a^2+d^2-b^2-c^2.$$

Now
$$BD^2=a^2+d^2-2ad\cos A$$

$$=a^2+d^2-\frac{ad(a^2+d^2-b^2-c^2)}{ad+bc}$$

$$=\frac{bc(a^2+d^2)+ad(b^2+c^2)}{ad+bc}$$

$$=\frac{(ab+cd)(ac+bd)}{ad+bc}.$$

Similarly, we may prove that

$$AC^2=\frac{(ad+bc)(ac+bd)}{ab+cd}.$$

Thus
$$AC\,.\,BD=ac+bd,$$

and
$$\frac{AC}{BD}=\frac{ad+bc}{ab+cd}.$$

The circle passing round the quadrilateral circumscribes the triangle ABD; hence

$$\text{the circum-radius}=\frac{BD}{2\sin A}$$

$$=\frac{(ad+bc)BD}{2(ad+bc)\sin A}=\frac{(ad+bc)BD}{4S}$$

$$=\frac{1}{4S}\sqrt{(ab+cd)(ac+bd)(ad+bc)}.$$

Example. A quadrilateral $ABCD$ is such that one circle can be inscribed in it and another circle circumscribed about it; shew that $\tan^2 \frac{A}{2} = \frac{bc}{ad}$.

If a circle can be inscribed in a quadrilateral, the sum of one pair of the opposite sides is equal to that of the other pair;

$$\therefore a + c = b + d.$$

Since the quadrilateral is cyclic,

$$\cos A = \frac{a^2 + d^2 - b^2 - c^2}{2(ad + bc)}. \qquad \text{[Art. 233.]}$$

But $a - d = b - c$, so that $a^2 - 2ad + d^2 = b^2 - 2bc + c^2$;

$$\therefore a^2 + d^2 - b^2 - c^2 = 2(ad - bc);$$

$$\therefore \cos A = \frac{ad - bc}{ad + bc};$$

$$\therefore \tan^2 \frac{A}{2} = \frac{1 - \cos A}{1 + \cos A} = \frac{bc}{ad}.$$

*EXAMPLES. XVIII. d.

1. If a circle can be inscribed in a quadrilateral, shew that its radius is S/σ where S is the area and 2σ the sum of the sides of the quadrilateral.

2. If the sides of a cyclic quadrilateral be 3, 3, 4, 4, shew that a circle can be inscribed in it, and find the radii of the inscribed and circumscribed circles.

3. If the sides of a cyclic quadrilateral be 1, 2, 4, 3, shew that the cosine of the angle between the two greatest sides is $\frac{5}{7}$, and that the radius of the inscribed circle is ·98 nearly.

4. The sides of a cyclic quadrilateral are 60, 25, 52, 39: shew that two of the angles are right angles, and find the diagonals and the area.

5. The sides of a quadrilateral are 4, 5, 8, 9, and one diagonal is 9: find the area.

6. If a circle can be inscribed in a cyclic quadrilateral, shew that the area of the quadrilateral is $\sqrt{abcd}$, and that the radius of the circle is

$$2\sqrt{abcd}/(a + b + c + d).$$

7. If the sides of a quadrilateral are given, shew that the area is a maximum when the quadrilateral can be inscribed in a circle.

8. If the sides of a quadrilateral are 23, 29, 37, 41 inches, prove that the maximum area is 7 sq. ft.

9. If $ABCD$ is a cyclic quadrilateral, prove that

$$\tan^2\frac{B}{2}=\frac{(\sigma-a)(\sigma-b)}{(\sigma-c)(\sigma-d)}.$$

10. If f, g denote the diagonals of a quadrilateral and β the angle between them, prove that

$$2fg\cos\beta=(a^2+c^2)\sim(b^2+d^2).$$

11. If β is the angle between the diagonals of any quadrilateral, prove that the area is

$$\frac{1}{4}\{(a^2+c^2)\sim(b^2+d^2)\}\tan\beta.$$

12. Prove that the area of a quadrilateral in which a circle can be inscribed is

$$\sqrt{abcd}\sin\frac{A+C}{2}.$$

13. If a circle can be inscribed in a quadrilateral whose diagonals are f and g, prove that

$$4S^2=f^2g^2-(ac-bd)^2.$$

14. If β is the angle between the diagonals of a cyclic quadrilateral, prove that

(1) $(ac+bd)\sin\beta=(ad+bc)\sin A$;

(2) $\cos\beta=\dfrac{(a^2+c^2)\sim(b^2+d^2)}{2(ac+bd)}$;

(3) $\tan^2\dfrac{\beta}{2}=\dfrac{(\sigma-b)(\sigma-d)}{(\sigma-a)(\sigma-c)}$ or $\dfrac{(\sigma-a)(\sigma-c)}{(\sigma-b)(\sigma-d)}$.

15. If f, g are the diagonals of a quadrilateral, shew that

$$S=\frac{1}{4}\sqrt{4f^2g^2-(a^2+c^2-b^2-d^2)^2}.$$

16. In a cyclic quadrilateral, prove that the product of the segments of a diagonal is

$$abcd\,(ac+bd)/(ab+cd)(ad+bc).$$

235. The following exercise consists of miscellaneous questions involving the properties of triangles.

EXAMPLES. XVIII. e.

1. If the sides of a triangle are 242, 1212, 1450 yards, shew that the area is 6 acres.

2. One of the sides of a triangle is 200 yards and the adjacent angles are 22·5° and 67·5° : find the area.

3. If $r_1 = 2r_2 = 2r_3$, shew that $3a = 4b$.

4. If a, b, c are in A. P., shew that r_1, r_2, r_3 are in H. P.

5. Find the area of a triangle whose sides are

$$\frac{y}{z}+\frac{z}{x}, \quad \frac{z}{x}+\frac{x}{y}, \quad \frac{x}{y}+\frac{y}{z}.$$

6. If $\sin A : \sin C = \sin(A-B) : \sin(B-C)$, shew that a^2, b^2, c^2 are in A. P.

Prove that

7. $$\frac{a\sin A+b\sin B+c\sin C}{4\cos\frac{A}{2}\cos\frac{B}{2}\cos\frac{C}{2}}=\frac{a^2+b^2+c^2}{2s}.$$

8. $$\left(\frac{a^2}{\sin A}+\frac{b^2}{\sin B}+\frac{c^2}{\sin C}\right)\sin\frac{A}{2}\sin\frac{B}{2}\sin\frac{C}{2}=\Delta.$$

9. $$(r_2+r_3)(r_3+r_1)(r_1+r_2)=4R(r_2r_3+r_3r_1+r_1r_2).$$

10. $$\tan\frac{A}{2}+\tan\frac{B}{2}+\tan\frac{C}{2}=\frac{r_1+r_2+r_3}{(r_2r_3+r_3r_1+r_1r_2)^{\frac{1}{2}}}.$$

11. $$bc\cot\frac{A}{2}+ca\cot\frac{B}{2}+ab\cot\frac{C}{2}=4Rs^2\left(\frac{1}{a}+\frac{1}{b}+\frac{1}{c}-\frac{3}{s}\right).$$

12. $$\left(\frac{1}{r}+\frac{1}{r_1}+\frac{1}{r_2}+\frac{1}{r_3}\right)^2=\frac{4}{r}\left(\frac{1}{r_1}+\frac{1}{r_2}+\frac{1}{r_3}\right).$$

13. The perimeter of a right-angled triangle is 70, and the in-radius is 6 : find the sides.

14. If f, g, h are the perpendiculars from the circum-centre on the sides, prove that

$$\frac{a}{f}+\frac{b}{g}+\frac{c}{h}=\frac{abc}{4fgh}.$$

15. An equilateral triangle and a regular hexagon have the same perimeter: shew that the areas of their inscribed circles are as 4 to 9.

***16.** Shew that the perimeter of the pedal triangle is equal to

$$abc/2R^2.$$

***17.** Shew that the area of the ex-central triangle is equal to

$$abc\,(a+b+c)/4\Delta.$$

18. In the ambiguous case, if A, a, b are the given parts, and c_1, c_2 the two values of the third side, shew that the distance between the circum-centres of the two triangles is $\dfrac{c_1 \sim c_2}{2 \sin A}$.

***19.** If β be the angle between the diagonals of a cyclic quadrilateral, shew that

$$\sin\beta=\frac{2S}{ac+bd}.$$

***20.** Shew that

$$r^3 . II_1 . II_2 . II_3 = IA^2 . IB^2 . IC^2.$$

***21.** Shew that the sum of the squares of the sides of the ex-central triangle is equal to $8R\,(4R+r)$.

***22.** If circles can be inscribed in and circumscribed about a quadrilateral, and if β be the angle between the diagonals, shew that

$$\cos\beta=(ac \sim bd)/(ac+bd).$$

23. If l, m, n are the lengths of the medians of a triangle, prove that

(1) $4\,(l^2+m^2+n^2)=3\,(a^2+b^2+c^2)$;

(2) $(b^2-c^2)\,l^2+(c^2-a^2)\,m^2+(a^2-b^2)\,n^2=0$;

(3) $16\,(l^4+m^4+n^4)=9\,(a^4+b^4+c^4)$.

24. Shew that the radii of the escribed circles are the roots of the equation

$$x^3-(4R+r)\,x^2+s^2x-s^2r=0.$$

25. If Δ_1, Δ_2, Δ_3 be the areas of the triangles cut off by tangents to the in-circle parallel to the sides of a triangle, prove that

$$\frac{\Delta_1}{(s-a)^2}=\frac{\Delta_2}{(s-b)^2}=\frac{\Delta_3}{(s-c)^2}=\frac{\Delta}{s^2}.$$

*26. The triangle LMN is formed by joining the points of contact of the in-circle; shew that it is similar to the ex-central triangle, and that their areas are as r^2 to $4R^2$.

27. In the triangle PQR formed by drawing tangents at A, B, C to the circum-circle, prove that the angles and sides are

$$180^\circ - 2A, \quad 180^\circ - 2B, \quad 180^\circ - 2C;$$

and $$\frac{a}{2\cos B\cos C}, \quad \frac{b}{2\cos C\cos A}, \quad \frac{c}{2\cos A\cos B}.$$

28. If p, q, r be the lengths of the bisectors of the angles of a triangle, prove that

(1) $$\frac{1}{p}\cos\frac{A}{2}+\frac{1}{q}\cos\frac{B}{2}+\frac{1}{r}\cos\frac{C}{2}=\frac{1}{a}+\frac{1}{b}+\frac{1}{c};$$

(2) $$\frac{pqr}{4\Delta}=\frac{abc\,(a+b+c)}{(b+c)(c+a)(a+b)}.$$

29. If the perpendiculars AG, BH, CK are produced to meet the circum-circle in L, M, N, prove that

(1) area of triangle $LMN = 8\Delta\cos A\cos B\cos C$;

(2) $AL\sin A + BM\sin B + CN\sin C = 8R\sin A\sin B\sin C$.

30. If r_a, r_b, r_c be the radii of the circles inscribed between the in-circle and the sides containing the angles A, B, C respectively, shew that

(1) $r_a = r\tan^2\dfrac{\pi - A}{4}$; (2) $\sqrt{r_b r_c}+\sqrt{r_c r_a}+\sqrt{r_a r_b}=r$.

*31. Lines drawn through the angular points of a triangle ABC parallel to the sides of the pedal triangle form a triangle XYZ: shew that the perimeter and area of XYZ are respectively

$$2R\tan A\tan B\tan C \quad\text{and}\quad R^2\tan A\tan B\tan C.$$

*32. A straight line cuts three concentric circles in A, B, C and passes at a distance p from their centre: shew that the area of the triangle formed by the tangents at A, B, C is

$$\frac{BC\,.\,CA\,.\,AB}{2p}.$$

MISCELLANEOUS EXAMPLES. F.

1. If $\alpha+\beta+\gamma+\delta=180°$, shew that

$$\cos\alpha\cos\beta+\cos\gamma\cos\delta=\sin\alpha\sin\beta+\sin\gamma\sin\delta.$$

2. Prove that

$$\cos(15°-A)\sec 15°-\sin(15°-A)\operatorname{cosec} 15°=4\sin A.$$

3. Shew that in a triangle

$$\cot A+\sin A\operatorname{cosec} B\operatorname{cosec} C$$

retains the same value if any two of the angles A, B, C are interchanged.

4. If $a=2$, $b=\sqrt{8}$, $A=30°$, solve the triangle.

5. Shew that

(1) $\cot 18°=\sqrt{5}\cot 36°$;

(2) $16\sin 36°\sin 72°\sin 108°\sin 144°=5$.

6. Find the number of ciphers before the first significant digit in $(\cdot 0396)^{90}$, given

$$\log 2=\cdot 30103,\ \log 3=\cdot 47712,\ \log 11=1\cdot 04139.$$

7. An observer finds that the angle subtended by the line joining two points A and B on the horizontal plane is 30°. On walking 50 yards directly towards A the angle increases to 75°: find his distance from B at each observation.

8. Prove that $\cos^2\alpha+\cos^2\beta+\cos^2\gamma+\cos^2(\alpha+\beta+\gamma)$

$$=2+2\cos(\beta+\gamma)\cos(\gamma+\alpha)\cos(\alpha+\beta).$$

9. Shew that

(1) $\tan 40°+\cot 40°=2\sec 10°$;

(2) $\tan 70°+\tan 20°=2\operatorname{cosec} 40°$.

10. Prove that

(1) $2\sin 4\alpha-\sin 10\alpha+\sin 2\alpha=16\sin\alpha\cos\alpha\cos 2\alpha\sin^2 3\alpha$;

(2) $\sin\frac{2\pi}{7}+\sin\frac{4\pi}{7}-\sin\frac{6\pi}{7}=4\sin\frac{\pi}{7}\sin\frac{3\pi}{7}\sin\frac{5\pi}{7}$.

11. If $B=30°$, $b=3\sqrt{2}-\sqrt{6}$, $c=6-2\sqrt{3}$, solve the triangle.

12. From a ship which is sailing N.E., the bearing of a rock is N.N.W. After the ship has sailed 10 miles the rock bears due W.: find the distance of the ship from the rock at each observation.

13. Shew that in any triangle

$$\frac{b^2-c^2}{\cos B+\cos C}+\frac{c^2-a^2}{\cos C+\cos A}+\frac{a^2-b^2}{\cos A+\cos B}=0.$$

14. If $\cos(\theta-a)$, $\cos\theta$, $\cos(\theta+a)$ are in harmonical progression, shew that

$$\cos\theta=\sqrt{2}\cos\frac{a}{2}.$$

15. If $\sin\beta$ be the geometric mean between $\sin a$ and $\cos a$, prove that $$\cos 2\beta=2\cos^2\left(\frac{\pi}{4}+a\right).$$

16. Shew that the distances of the orthocentre from the sides are $2R\cos B\cos C$, $2R\cos C\cos A$, $2R\cos A\cos B$.

17. If $$\cos\theta=\frac{\cos u-e}{1-e\cos u},$$

prove that $$\tan\frac{\theta}{2}=\sqrt{\frac{1+e}{1-e}}\tan\frac{u}{2}.$$

18. If the sides of a right-angled triangle are

$$2(1+\sin\theta)+\cos\theta \quad\text{and}\quad 2(1+\cos\theta)+\sin\theta,$$

prove that the hypotenuse is

$$3+2(\cos\theta+\sin\theta).$$

***19.** Prove that the distances of the in-centre of the ex-central triangle $I_1I_2I_3$ from its ex-centres are

$$8R\sin\frac{B+C}{4},\quad 8R\sin\frac{C+A}{4},\quad 8R\sin\frac{A+B}{4}.$$

***20.** Prove that the distances between the ex-centres of the ex-central triangle $I_1I_2I_3$ are

$$8R\cos\frac{B+C}{4},\quad 8R\cos\frac{C+A}{4},\quad 8R\cos\frac{A+B}{4}.$$

21. If

$(1+\cos\alpha)(1+\cos\beta)(1+\cos\gamma)=(1-\cos\alpha)(1-\cos\beta)(1-\cos\gamma)$,
shew that each expression is equal to $\pm\sin\alpha\sin\beta\sin\gamma$.

22. If the sum of four angles is 180°, shew that the sum of the products of their sines taken two together is equal to the sum of the products of their cosines taken two together.

***23.** In a triangle, shew that

(1) $II_1 . II_2 . II_3 = 16R^2r$; (2) $II_1^2 + I_2I_3^2 = 16R^2$.

24. Find the angles of a triangle whose sides are proportional to

(1) $\cos\frac{A}{2}$, $\cos\frac{B}{2}$, $\cos\frac{C}{2}$;

(2) $\sin 2A$, $\sin 2B$, $\sin 2C$.

25. Prove that the expression

$$\sin^2(\theta+\alpha)+\sin^2(\theta+\beta)-2\cos(\alpha-\beta)\sin(\theta+\alpha)\sin(\theta+\beta)$$

is independent of θ.

***26.** If a, b, c, d are the sides of a quadrilateral described about a circle, prove that

$$ad\sin^2\frac{A}{2}=bc\sin^2\frac{C}{2}.$$

27. Tangents parallel to the three sides are drawn to the in-circle. If p, q, r be the lengths of the parts of the tangents within the triangle, prove that $\frac{p}{a}+\frac{q}{b}+\frac{r}{c}=1$.

[*The Tables will be required for Examples* 28 *and* 29.]

28. From the top of a cliff 1566 ft. in height a train, which is travelling at a uniform speed in a straight line to a tunnel immediately below the observer, is seen to pass two consecutive stations at an interval of 3 minutes. The angles of depression of the two stations are 13° 14′ 12″ and 56° 24′ 36″ respectively; how fast is the train travelling?

29. A harbour lies in a direction 46° 8′ 8·6″ South of West from a fort, and at a distance of 27·23 miles from it. A ship sets out from the harbour at noon and sails due East at 10 miles an hour; when will the ship be 20 miles from the fort?

CHAPTER XIX.

GENERAL VALUES AND INVERSE FUNCTIONS.

236. THE equation $\sin\theta=\frac{1}{2}$ is satisfied by $\theta=\frac{\pi}{6}$, and by $\theta=\pi-\frac{\pi}{6}$, and all angles coterminal with these will have the same sine. This example shews that there are an infinite number of angles whose sine is equal to a given quantity. Similar remarks apply to the other functions.

We proceed to shew how to express by a single formula all angles which have a given sine, cosine, or tangent.

237. From the results proved in Chap. IX., it is easily seen that in going once through the four quadrants, there are two and only two positions of the boundary line which give angles with the same sine, cosine, or tangent.

Thus if $\sin\alpha$ has a given value, the positions of the radius vector are OP and OP' bounding the angles α and $\pi-\alpha$. [Art. 92.]

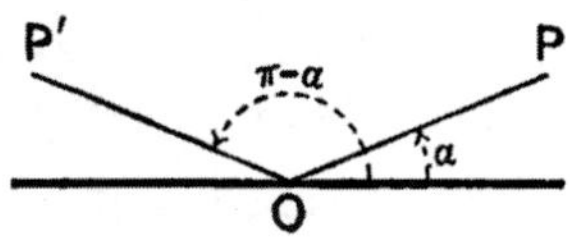

If $\cos\alpha$ has a given value, the positions of the radius vector are OP and OP' bounding the angles α and $2\pi-\alpha$. [Art. 105.]

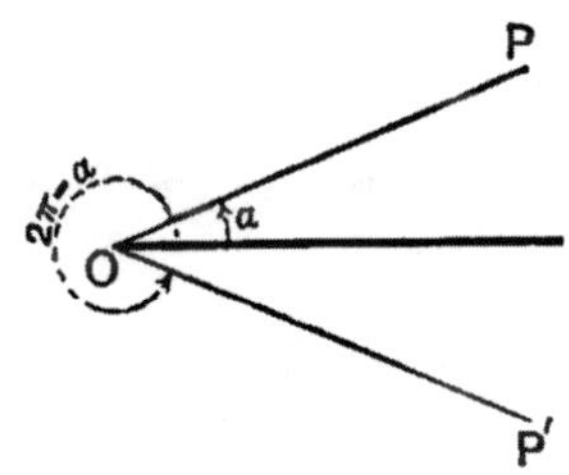

If $\tan\alpha$ has a given value, the positions of the radius vector are OP and OP' bounding the angles α and $\pi+\alpha$. [Art. 97.]

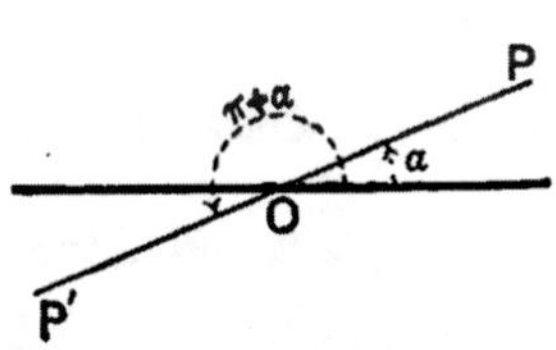

238. *To find a formula for all the angles which have a given sine.*

Let a be the smallest positive angle which has a given sine. Draw OP and OP' bounding the angles a and $\pi - a$; then the required angles are those coterminal with OP and OP'.

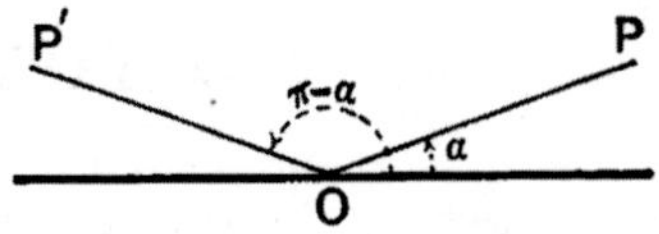

The positive angles are

$$2p\pi + a \quad \text{and} \quad 2p\pi + (\pi - a),$$

where p is zero, or any positive integer.

The negative angles are

$$-(\pi + a) \quad \text{and} \quad -(2\pi - a),$$

and those which may be obtained from them by the addition of any negative multiple of 2π; that is, angles denoted by

$$2q\pi - (\pi + a) \quad \text{and} \quad 2q\pi - (2\pi - a),$$

where q is zero, or any negative integer.

These angles may be grouped as follows:

$$\left.\begin{array}{r} 2p\pi + a, \\ (2q-2)\pi + a, \end{array}\right\} \text{ and } \left\{\begin{array}{l} (2p+1)\pi - a, \\ (2q-1)\pi - a, \end{array}\right.$$

and it will be noticed that even multiples of π are followed by $+a$, and odd multiples of π by $-a$.

Thus all angles equi-sinal with a are included in the formula

$$n\pi + (-1)^n a,$$

where n is zero, or any integer positive or negative.

This is also the formula for all angles which have the same cosecant as a.

Example 1. Write down the general solution of $\sin\theta = \frac{\sqrt{3}}{2}$.

The least value of θ which satisfies the equation is $\frac{\pi}{3}$; therefore the general solution is $n\pi + (-1)^n \frac{\pi}{3}$.

Example 2. Find the general solution of $\sin^2\theta = \sin^2 a$.

This equation gives either $\sin\theta = +\sin a$..........................(1),

or $\sin\theta = -\sin a = \sin(-a)$ ....(2).

From (1), $\theta = n\pi + (-1)^n \alpha$;
and from (2), $\theta = n\pi + (-1)^n (-\alpha)$.

Both values are included in the formula $\theta = n\pi \pm \alpha$.

239. *To find a formula for all the angles which have a given cosine.*

Let α be the smallest positive angle which has a given cosine. Draw OP and OP' bounding the angles α and $2\pi - \alpha$; then the required angles are those coterminal with OP and OP'.

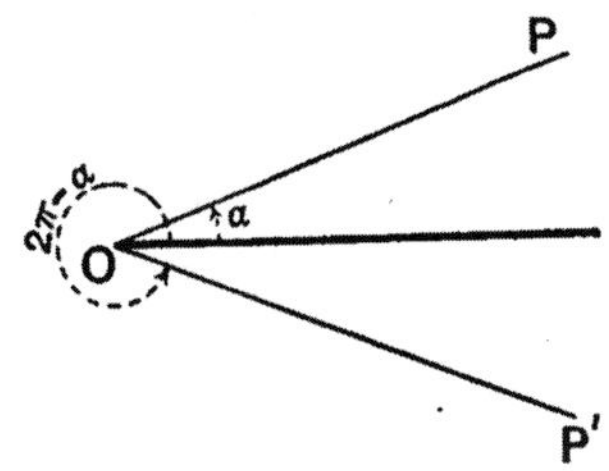

The positive angles are

$2p\pi + \alpha$ and $2p\pi + (2\pi - \alpha)$,

where p is zero, or any positive integer.

The negative angles are

$$-\alpha \text{ and } -(2\pi - \alpha),$$

and those which may be obtained from them by the addition of any negative multiple of 2π; that is, angles denoted by

$$2q\pi - \alpha \text{ and } 2q\pi - (2\pi - \alpha),$$

where q is zero, or any negative integer.

The angles may be grouped as follows:

$$\left.\begin{array}{l} 2p\pi + \alpha, \\ 2q\pi - \alpha, \end{array}\right\} \text{ and } \left\{\begin{array}{l} (2p+2)\pi - \alpha, \\ (2q-2)\pi + \alpha, \end{array}\right.$$

and it will be noticed that the multiples of π are always even, but may be followed by $+\alpha$ or by $-\alpha$.

Thus all angles equi-cosinal with α are included in the formula

$$2n\pi \pm \alpha,$$

where n is zero, or any integer positive or negative.

This is also the formula for all angles which have the same secant as α.

Example 1. Find the general solution of $\cos\theta = -\frac{1}{2}$.

The least value of θ is $\pi - \frac{\pi}{3}$, or $\frac{2\pi}{3}$; hence the general solution is $2n\pi \pm \frac{2\pi}{3}$.

240. *To find a formula for all the angles which have a given tangent.*

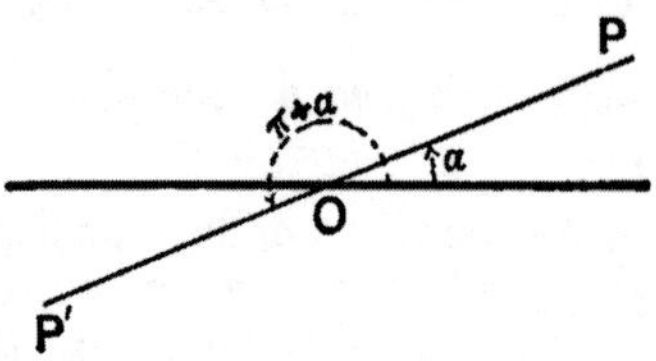

Let α be the smallest positive angle which has a given tangent. Draw OP and OP' bounding the angles α and $\pi+\alpha$; then the required angles are those coterminal with OP and OP'.

The positive angles are

$$2p\pi+\alpha \text{ and } 2p\pi+(\pi+\alpha),$$

where p is zero, or any positive integer.

The negative angles are

$$-(\pi-\alpha) \text{ and } -(2\pi-\alpha),$$

and those which may be obtained from them by the addition of any negative multiple of 2π; that is, angles denoted by

$$2q\pi-(\pi-\alpha) \text{ and } 2q\pi-(2\pi-\alpha),$$

where q is zero, or any negative integer.

The angles may be grouped as follows:

$$\left.\begin{matrix} 2p\pi+\alpha, \\ (2q-2)\pi+\alpha, \end{matrix}\right\} \text{ and } \left\{\begin{matrix} (2p+1)\pi+\alpha, \\ (2q-1)\pi+\alpha, \end{matrix}\right.$$

and it will be noticed that whether the multiple of π is even or odd, it is always followed by $+\alpha$. Thus all angles equi-tangential with α are included in the formula

$$\theta=n\pi+\alpha.$$

This is also the formula for all the angles which have the same cotangent as α.

Example. Solve the equation $\cot 4\theta=\cot\theta$.

The general solution is $4\theta=n\pi+\theta$;

whence $3\theta=n\pi$, or $\theta=\frac{n\pi}{3}$.

241. *All angles which are both equi-sinal and equi-cosinal with α are included in the formula $2n\pi+\alpha$.*

All angles equi-cosinal with α are included in the formula $2n\pi\pm\alpha$; so that the multiple of π is even. But in the formula $n\pi+(-1)^n\alpha$, which includes all angles equi-sinal with α, when the multiple of π is even, α must be preceded by the $+$ sign. Thus the formula is $2n\pi+\alpha$.

242. In the solution of equations, the general value of the angle should always be given.

Example. Solve the equation $\cos 9\theta = \cos 5\theta - \cos\theta$.

By transposition, $(\cos 9\theta + \cos\theta) - \cos 5\theta = 0$;

$$\therefore\ 2\cos 5\theta \cos 4\theta - \cos 5\theta = 0;$$

$$\therefore\ \cos 5\theta\,(2\cos 4\theta - 1) = 0;$$

$\therefore$ either $\cos 5\theta = 0$, or $2\cos 4\theta - 1 = 0$.

From the first equation, $5\theta = 2n\pi \pm \frac{\pi}{2}$, or $\theta = \frac{(4n \pm 1)\pi}{10}$;

and from the second, $4\theta = 2n\pi \pm \frac{\pi}{3}$, or $\theta = \frac{(6n \pm 1)\pi}{12}$.

EXAMPLES. XIX. a.

Find the general solution of the equations:

1. $\sin\theta = \frac{1}{2}$.
2. $\sin\theta = \frac{1}{\sqrt{2}}$.
3. $\cos\theta = \frac{1}{2}$.
4. $\tan\theta = \sqrt{3}$.
5. $\cot\theta = -\sqrt{3}$.
6. $\sec\theta = -\sqrt{2}$.
7. $\cos^2\theta = \frac{1}{2}$.
8. $\tan^2\theta = \frac{1}{3}$.
9. $\operatorname{cosec}^2\theta = \frac{4}{3}$.
10. $\cos\theta = \cos\alpha$.
11. $\tan^2\theta = \tan^2\alpha$.
12. $\sec^2\theta = \sec^2\alpha$.
13. $\tan 2\theta = \tan\theta$.
14. $\operatorname{cosec} 3\theta = \operatorname{cosec} 3\alpha$.
15. $\cos 3\theta = \cos 2\theta$.
16. $\sin 5\theta + \sin\theta = \sin 3\theta$.
17. $\cos\theta - \cos 7\theta = \sin 4\theta$.
18. $\sin 4\theta - \sin 3\theta + \sin 2\theta - \sin\theta = 0$.
19. $\cos\theta + \cos 3\theta + \cos 5\theta + \cos 7\theta = 0$.
20. $\sin 5\theta \cos\theta = \sin 6\theta \cos 2\theta$.
21. $\sin 11\theta \sin 4\theta + \sin 5\theta \sin 2\theta = 0$.
22. $\sqrt{2}\cos 3\theta - \cos\theta = \cos 5\theta$.
23. $\sin 7\theta - \sqrt{3}\cos 4\theta = \sin\theta$.
24. $1 + \cos\theta = 2\sin^2\theta$.
25. $\tan^2\theta + \sec\theta = 1$.
26. $\cot^2\theta - 1 = \operatorname{cosec}\theta$.
27. $\cot\theta - \tan\theta = 2$.
28. If $2\cos\theta = -1$ and $2\sin\theta = \sqrt{3}$, find θ.
29. If $\sec\theta = \sqrt{2}$ and $\tan\theta = -1$, find θ.

243. In the following examples, the solution is simplified by the use of some particular artifice.

Example 1. Solve the equation $\cos m\theta = \sin n\theta$.

Here $$\cos m\theta = \cos\left(\frac{\pi}{2} - n\theta\right):$$

$$\therefore\ m\theta = 2k\pi \pm \left(\frac{\pi}{2} - n\theta\right),$$

where k is zero, or any integer.

By transposition, we obtain

$$(m+n)\,\theta = \left(2k + \frac{1}{2}\right)\pi, \quad \text{or} \quad (m-n)\,\theta = \left(2k - \frac{1}{2}\right)\pi.$$

This equation may also be solved through the medium of the sine. For we have

$$\sin\left(\frac{\pi}{2} - m\theta\right) = \sin n\theta;$$

$$\therefore\ \frac{\pi}{2} - m\theta = p\pi + (-1)^p n\theta,$$

where p is zero or any integer;

$$\therefore\ \{m + (-1)^p n\}\,\theta = \left(\frac{1}{2} - p\right)\pi.$$

NOTE. The general solution can frequently be obtained in several ways. The various forms which the result takes are merely different modes of expressing the same series of angles.

Example 2. Solve $\sqrt{3}\cos\theta + \sin\theta = 1$.

Multiply every term by $\frac{1}{2}$, then

$$\frac{\sqrt{3}}{2}\cos\theta + \frac{1}{2}\sin\theta = \frac{1}{2},$$

$$\therefore\ \cos\frac{\pi}{6}\cos\theta + \sin\frac{\pi}{6}\sin\theta = \frac{1}{2};$$

$$\because\ \cos\left(\theta - \frac{\pi}{6}\right) = \frac{1}{2};$$

$$\therefore\ \theta - \frac{\pi}{6} = 2n\pi \pm \frac{\pi}{3};$$

$$\therefore\ \theta = 2n\pi + \frac{\pi}{2} \quad \text{or} \quad 2n\pi - \frac{\pi}{6}.$$

NOTE. In examples of this type, it is a common mistake to square the equation; but this process is objectionable, because it introduces solutions which do not belong to the given equation. Thus in the present instance,

$$\sqrt{3}\cos\theta = 1 - \sin\theta;$$

by squaring, $$3\cos^2\theta = (1 - \sin\theta)^2.$$

But the solutions of this equation include the solutions of

$$-\sqrt{3}\cos\theta = 1 - \sin\theta,$$

as well as those of the given equation.

Example 3. Solve $\cos 2\theta = \cos\theta + \sin\theta$.

From this equation, $\cos^2\theta - \sin^2\theta = \cos\theta + \sin\theta$;

$$\therefore (\cos\theta + \sin\theta)(\cos\theta - \sin\theta) = \cos\theta + \sin\theta;$$

$\therefore$ either $$\cos\theta + \sin\theta = 0 \quad\text{.........................(1)},$$

or $$\cos\theta - \sin\theta = 1 \quad\text{.........................(2)}.$$

From (1), $$\tan\theta = -1,$$

$$\therefore \theta = n\pi - \frac{\pi}{4}.$$

From (2), $$\frac{1}{\sqrt{2}}\cos\theta - \frac{1}{\sqrt{2}}\sin\theta = \frac{1}{\sqrt{2}};$$

$$\therefore \cos\theta\cos\frac{\pi}{4} - \sin\theta\sin\frac{\pi}{4} = \frac{1}{\sqrt{2}}.$$

$$\therefore \cos\left(\theta + \frac{\pi}{4}\right) = \frac{1}{\sqrt{2}};$$

$$\therefore \theta + \frac{\pi}{4} = 2n\pi \pm \frac{\pi}{4};$$

$$\therefore \theta = 2n\pi \text{ or } 2n\pi - \frac{\pi}{2}.$$

EXAMPLES. XIX. b.

Find the general solution of the equations:

1. $\tan p\theta = \cot q\theta$.
2. $\sin m\theta + \cos n\theta = 0$.
3. $\cos\theta - \sqrt{3}\sin\theta = 1$.
4. $\sin\theta - \sqrt{3}\cos\theta = 1$.
5. $\cos\theta = \sqrt{3}(1 - \sin\theta)$.
6. $\sin\theta + \sqrt{3}\cos\theta = \sqrt{2}$.

Find the general solution of the equations:

7. $\cos\theta - \sin\theta = \frac{1}{\sqrt{2}}$. **8.** $\cos\theta + \sin\theta + \sqrt{2} = 0$.

9. $\operatorname{cosec}\theta + \cot\theta = \sqrt{3}$. **10.** $\cot\theta - \cot 2\theta = 2$.

11. $2\sin\theta\sin 3\theta = 1$. **12.** $\sin 3\theta = 8\sin^3\theta$.

13. $\tan\theta + \tan 3\theta = 2\tan 2\theta$. **14.** $\cos\theta - \sin\theta = \cos 2\theta$.

15. $\operatorname{cosec}\theta + \sec\theta = 2\sqrt{2}$. **16.** $\sec\theta - \operatorname{cosec}\theta = 2\sqrt{2}$.

17. $\sec 4\theta - \sec 2\theta = 2$. **18.** $\cos 3\theta + 8\cos^3\theta = 0$.

19. $1 + \sqrt{3}\tan^2\theta = (1 + \sqrt{3})\tan\theta$.

20. $\tan^3\theta + \cot^3\theta = 8\operatorname{cosec}^3 2\theta + 12$.

21. $\sin\theta = \sqrt{2}\sin\phi$, $\sqrt{3}\cos\theta = \sqrt{2}\cos\phi$.

22. $\operatorname{cosec}\theta = \sqrt{3}\operatorname{cosec}\phi$, $\cot\theta = 3\cot\phi$.

23. $\sec\phi = \sqrt{2}\sec\theta$, $\cot\theta = \sqrt{3}\cot\phi$.

24. Explain why the same two series of angles are given by the equations

$$\theta + \frac{\pi}{4} = n\pi + (-1)^n\frac{\pi}{6} \quad \text{and} \quad \theta - \frac{\pi}{4} = 2n\pi \pm \frac{\pi}{3}.$$

25. Shew that the formulæ

$$\left(2n + \frac{1}{4}\right)\pi \pm a \quad \text{and} \quad \left(n - \frac{1}{4}\right)\pi + (-1)^n\left(\frac{\pi}{2} - a\right)$$

comprise the same angles, and illustrate by a figure.

Inverse Circular Functions.

244. If $\sin\theta = s$, we know that θ may be *any* angle whose sine is s. It is often convenient to express this statement *inversely* by writing $\theta = \sin^{-1}s$.

In this *inverse notation* θ stands alone on one side of the equation, and may be regarded as an angle whose value is only known through the medium of its sine. Similarly, $\tan^{-1}\sqrt{3}$ indicates in a concise form any one of the angles whose tangent is $\sqrt{3}$. But all these angles are given by the formula $n\pi + \frac{\pi}{3}$. Thus

$$\theta = \tan^{-1}\sqrt{3} \quad \text{and} \quad \theta = n\pi + \frac{\pi}{3}$$

are equivalent statements expressed in different forms.

245. Expressions of the form $\cos^{-1}x$, $\sin^{-1}a$, $\tan^{-1}b$ are called **Inverse Circular Functions.**

It must be remembered that these expressions denote angles, and that -1 *is not an algebraical index*; that is,

$$\sin^{-1}x \textit{ is not the same as } (\sin x)^{-1} \textit{ or } \frac{1}{\sin x}.$$

246. From Art. 244, we see that an inverse function has an infinite number of values.

If f denote any one of the circular functions, and $f^{-1}(x)=A$, the **principal value** of $f^{-1}(x)$ is the smallest numerical value of A. Thus the principal values of

$$\cos^{-1}\frac{1}{2},\quad \sin^{-1}\left(-\frac{1}{2}\right),\quad \cos^{-1}\left(-\frac{1}{\sqrt{2}}\right),\quad \tan^{-1}(-1)$$

are $60°$, $-30°$, $135°$, $-45°$.

Hence if x be positive, the principal values of $\sin^{-1}x$, $\cos^{-1}x$, $\tan^{-1}x$ all lie between 0 and 90°.

If x be negative, the principal values of $\sin^{-1}x$ and $\tan^{-1}x$ lie between 0 and $-90°$, and the principal value of $\cos^{-1}x$ lies between 90° and 180°.

In numerical instances we shall usually suppose that the *principal value* is selected.

247. If $\sin\theta=x$, we have $\cos\theta=\sqrt{1-x^2}$.

Expressed in the inverse notation, these equations become

$$\theta=\sin^{-1}x,\quad \theta=\cos^{-1}\sqrt{1-x^2}.$$

In each of these two statements, θ has an infinite number of values; but, as the formulæ for the general values of the sine and cosine are not identical, we cannot assert that the equation

$$\sin^{-1}x=\cos^{-1}\sqrt{1-x^2}$$

is identically true. This will be seen more clearly from a numerical instance. If $x=\frac{1}{2}$, then $\sqrt{1-x^2}=\frac{\sqrt{3}}{2}$.

Here $\sin^{-1}x$ may be any one of the angles

$$30°,\ 150°,\ 390°,\ 510°,\ \ldots;$$

and $\cos^{-1}\sqrt{1-x^2}$ may be any one of the angles

$$30°,\ 330°,\ 390°,\ 690°,\ \ldots.$$

248. From the relations established in the previous chapters, we may deduce corresponding relations connecting the inverse functions. Thus in the identity

$$\cos 2\theta = \frac{1-\tan^2\theta}{1+\tan^2\theta},$$

let $\tan\theta = a$, so that $\theta = \tan^{-1} a$; then

$$\cos(2\tan^{-1} a) = \frac{1-a^2}{1+a^2};$$

$$\therefore \quad 2\tan^{-1} a = \cos^{-1}\frac{1-a^2}{1+a^2}.$$

Similarly, the formula

$$\cos 3\theta = 4\cos^3\theta - 3\cos\theta$$

when expressed in the inverse notation becomes

$$3\cos^{-1} a = \cos^{-1}(4a^3 - 3a).$$

249. *To prove that*

$$\tan^{-1} x + \tan^{-1} y = \tan^{-1}\frac{x+y}{1-xy}.$$

Let $\tan^{-1} x = \alpha$, so that $\tan\alpha = x$;

and $\tan^{-1} y = \beta$, so that $\tan\beta = y$.

We require $\alpha + \beta$ in the form of an inverse tangent.

Now
$$\tan(\alpha+\beta) = \frac{\tan\alpha + \tan\beta}{1-\tan\alpha\tan\beta}$$

$$= \frac{x+y}{1-xy};$$

$$\therefore \quad \alpha+\beta = \tan^{-1}\frac{x+y}{1-xy};$$

that is,
$$\tan^{-1} x + \tan^{-1} y = \tan^{-1}\frac{x+y}{1-xy}.$$

By putting $y = x$, we obtain

$$2\tan^{-1} x = \tan^{-1}\frac{2x}{1-x^2}.$$

NOTE. It is useful to remember that

$$\tan(\tan^{-1} x + \tan^{-1} y) = \frac{x+y}{1-xy}.$$

Example 1. Prove that

$$\tan^{-1}5-\tan^{-1}3+\tan^{-1}\frac{7}{9}=n\pi+\frac{\pi}{4}.$$

The first side $=\tan^{-1}\dfrac{5-3}{1+15}+\tan^{-1}\dfrac{7}{9}$

$$=\tan^{-1}\frac{1}{8}+\tan^{-1}\frac{7}{9}$$

$$=\tan^{-1}\frac{\frac{1}{8}+\frac{7}{9}}{1-\frac{7}{72}}=\tan^{-1}1$$

$$=n\pi+\frac{\pi}{4}.$$

NOTE. The value of n cannot be assigned until we have selected some particular values for the angles $\tan^{-1}5$, $\tan^{-1}3$, $\tan^{-1}\frac{7}{9}$. If we choose the *principal values*, then $n=0$.

Example 2. Prove that

$$\sin^{-1}\frac{4}{5}+\cos^{-1}\frac{12}{13}+\sin^{-1}\frac{16}{65}=\frac{\pi}{2}.$$

We may write this identity in the form

$$\sin^{-1}\frac{4}{5}+\cos^{-1}\frac{12}{13}=\frac{\pi}{2}-\sin^{-1}\frac{16}{65}=\cos^{-1}\frac{16}{65}.$$

Let $\alpha=\sin^{-1}\frac{4}{5}$, so that $\sin\alpha=\frac{4}{5}$;

and $\beta=\cos^{-1}\frac{12}{13}$, so that $\cos\beta=\frac{12}{13}$.

We have to express $\alpha+\beta$ as an inverse cosine.

Now $\cos(\alpha+\beta)=\cos\alpha\cos\beta-\sin\alpha\sin\beta$; whence by reading off the values of the functions from the figures in the margin, we have

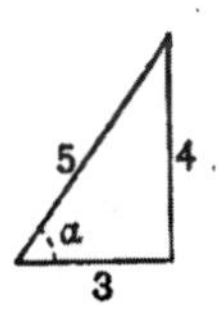

$$\cos(\alpha+\beta)=\frac{3}{5}\cdot\frac{12}{13}-\frac{4}{5}\cdot\frac{5}{13}$$

$$=\frac{16}{65};$$

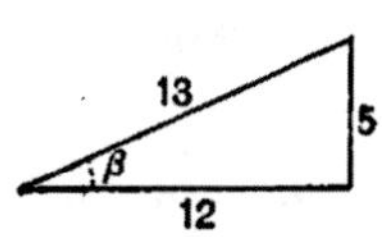

$$\therefore\ \alpha+\beta=\cos^{-1}\frac{16}{65}.$$

It is sometimes convenient to work entirely in terms of the tangent or cotangent.

Example 3. Prove that

$$2\cot^{-1}7+\cos^{-1}\frac{3}{5}=\operatorname{cosec}^{-1}\frac{125}{117}.$$

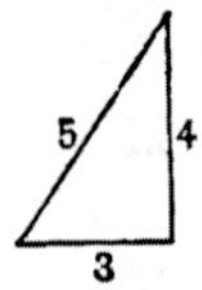

The first side $=\cot^{-1}\dfrac{7^2-1}{2\times 7}+\cot^{-1}\dfrac{3}{4}$

$$=\cot^{-1}\frac{24}{7}+\cot^{-1}\frac{3}{4}$$

$$=\cot^{-1}\frac{\frac{24}{7}\times\frac{3}{4}-1}{\frac{24}{7}+\frac{3}{4}}$$

$$=\cot^{-1}\frac{44}{117}=\operatorname{cosec}^{-1}\frac{125}{117}.$$

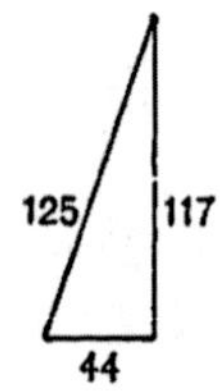

EXAMPLES. XIX. c.

Prove the following statements :

1. $\sin^{-1}\dfrac{12}{13}=\cot^{-1}\dfrac{5}{12}.$

2. $\operatorname{cosec}^{-1}\dfrac{17}{8}=\tan^{-1}\dfrac{8}{15}.$

3. $\sec(\tan^{-1}x)=\sqrt{1+x^2}.$

4. $2\tan^{-1}\dfrac{1}{3}=\tan^{-1}\dfrac{3}{4}.$

5. $\tan^{-1}\dfrac{4}{3}-\tan^{-1}1=\tan^{-1}\dfrac{1}{7}.$

6. $\tan^{-1}\dfrac{2}{11}+\cot^{-1}\dfrac{24}{7}=\tan^{-1}\dfrac{1}{2}.$

7. $\cot^{-1}\dfrac{4}{3}-\cot^{-1}\dfrac{15}{8}=\cot^{-1}\dfrac{84}{13}.$

8. $2\tan^{-1}\dfrac{1}{5}+\tan^{-1}\dfrac{1}{4}=\tan^{-1}\dfrac{32}{43}.$

9. $\tan^{-1}\dfrac{1}{2}+\tan^{-1}\dfrac{1}{3}=\tan^{-1}\dfrac{5}{6}+\tan^{-1}\dfrac{1}{11}.$

10. $\tan^{-1}\dfrac{1}{7}+\tan^{-1}\dfrac{1}{8}+\tan^{-1}\dfrac{1}{18}=\cot^{-1}3.$

11. $\tan^{-1}\frac{3}{5}+\sin^{-1}\frac{3}{5}=\tan^{-1}\frac{27}{11}$.

12. $2\cot^{-1}\frac{5}{4}=\tan^{-1}\frac{40}{9}$. **13.** $2\tan^{-1}\frac{8}{15}=\sin^{-1}\frac{240}{289}$.

14. $\sin(2\sin^{-1}x)=2x\sqrt{1-x^2}$.

15. $\cos^{-1}x=2\sin^{-1}\sqrt{\frac{1-x}{2}}$.

16. $2\tan^{-1}\sqrt{\frac{x}{a}}=\cos^{-1}\frac{a-x}{a+x}$.

17. $2\tan^{-1}\frac{1}{8}+\tan^{-1}\frac{1}{7}+2\tan^{-1}\frac{1}{5}=\frac{\pi}{4}$.

18. $\sin^{-1}a-\cos^{-1}b=\cos^{-1}\{b\sqrt{1-a^2}+a\sqrt{1-b^2}\}$.

19. $\sin^{-1}\frac{4}{5}+\cos^{-1}\frac{2}{\sqrt{5}}=\cot^{-1}\frac{2}{11}$.

20. $\cos^{-1}\frac{63}{65}+2\tan^{-1}\frac{1}{5}=\sin^{-1}\frac{3}{5}$.

21. $\tan^{-1}m+\tan^{-1}n=\cos^{-1}\frac{1-mn}{\sqrt{(1+m^2)(1+n^2)}}$.

22. $\cos^{-1}\frac{20}{29}-\tan^{-1}\frac{16}{63}=\cos^{-1}\frac{1596}{1885}$.

23. $\cos^{-1}\sqrt{\frac{2}{3}}-\cos^{-1}\frac{\sqrt{6}+1}{2\sqrt{3}}=\frac{\pi}{6}$.

24. $\tan(2\tan^{-1}x)=2\tan(\tan^{-1}x+\tan^{-1}x^3)$.

25. $\tan^{-1}a=\tan^{-1}\frac{a-b}{1+ab}+\tan^{-1}\frac{b-c}{1+bc}+\tan^{-1}c$.

26. If $\tan^{-1}x+\tan^{-1}y+\tan^{-1}z=\pi$, prove that

$$x+y+z=xyz.$$

27. If $u=\cot^{-1}\sqrt{\cos a}-\tan^{-1}\sqrt{\cos a}$, prove that

$$\sin u=\tan^2\frac{a}{2}.$$

250. We shall now shew how to solve equations expressed in the inverse notation.

Example 1. Solve $\tan^{-1}2x+\tan^{-1}3x=n\pi+\frac{3\pi}{4}$.

We have $$\tan^{-1}\frac{2x+3x}{1-6x^2}=n\pi+\frac{3\pi}{4};$$

$$\therefore \frac{2x+3x}{1-6x^2}=\tan\left(n\pi+\frac{3\pi}{4}\right)=-1;$$

$$\therefore 6x^2-5x-1=0, \text{ or } (6x+1)(x-1)=0;$$

$$\therefore x=1, \text{ or } -\frac{1}{6}.$$

Example 2. Solve $\sin^{-1}x+\sin^{-1}(1-x)=\cos^{-1}x$.

By transposition, $\sin^{-1}(1-x)=\cos^{-1}x-\sin^{-1}x$.

Let $\cos^{-1}x=\alpha$, and $\sin^{-1}x=\beta$; then

$$\sin^{-1}(1-x)=\alpha-\beta;$$

$$\therefore 1-x=\sin(\alpha-\beta)=\sin\alpha\cos\beta-\cos\alpha\sin\beta.$$

But $\cos\alpha=x$, and therefore $\sin\alpha=\sqrt{1-x^2}$;

also $\sin\beta=x$, and therefore $\cos\beta=\sqrt{1-x^2}$;

$$\therefore 1-x=(1-x^2)-x^2=1-2x^2;$$

$$\therefore 2x^2-x=0;$$

whence $$x=0, \text{ or } \frac{1}{2}.$$

EXAMPLES. XIX. d.

Solve the equations:

1. $\sin^{-1}x=\cos^{-1}x$. 2. $\tan^{-1}x=\cot^{-1}x$.

3. $\tan^{-1}(x+1)-\tan^{-1}(x-1)=\cot^{-1}2$.

4. $\cot^{-1}x+\cot^{-1}2x=\frac{3\pi}{4}$.

5. $\sin^{-1}x-\cos^{-1}x=\sin^{-1}(3x-2)$.

6. $\cos^{-1}x-\sin^{-1}x=\cos^{-1}x\sqrt{3}$.

7. $\tan^{-1}\dfrac{x-1}{x-2}+\tan^{-1}\dfrac{x+1}{x+2}=\dfrac{\pi}{4}$.

8. $2\cot^{-1}2+\cos^{-1}\dfrac{3}{5}=\operatorname{cosec}^{-1}x$.

9. $\tan^{-1}x+\tan^{-1}(1-x)=2\tan^{-1}\sqrt{x-x^2}$.

10. $\cos^{-1}\dfrac{1-a^2}{1+a^2}-\cos^{-1}\dfrac{1-b^2}{1+b^2}=2\tan^{-1}x$.

11. $\sin^{-1}\dfrac{2a}{1+a^2}+\tan^{-1}\dfrac{2x}{1-x^2}=\cos^{-1}\dfrac{1-b^2}{1+b^2}$.

12. $\cot^{-1}\dfrac{x^2-1}{2x}+\tan^{-1}\dfrac{2x}{x^2-1}+\dfrac{4\pi}{3}=0$.

13. Shew that we can express

$$\sin^{-1}\frac{2ab}{a^2+b^2}+\sin^{-1}\frac{2cd}{c^2+d^2}\text{ in the form }\sin^{-1}\frac{2xy}{x^2+y^2}$$

where x and y are rational functions of a, b, c, d.

14. If $\sin[2\cos^{-1}\{\cot(2\tan^{-1}x)\}]=0$, find x.

15. If $2\tan^{-1}(\cos\theta)=\tan^{-1}(2\operatorname{cosec}\theta)$, find θ.

16. If $\sin(\pi\cos\theta)=\cos(\pi\sin\theta)$, shew that

$$2\theta=\pm\sin^{-1}\frac{3}{4}.$$

17. If $\sin(\pi\cot\theta)=\cos(\pi\tan\theta)$, and n is any integer, shew that either $\cot 2\theta$ or $\operatorname{cosec} 2\theta$ is of the form $\dfrac{4n+1}{4}$.

18. If $\tan(\pi\cot\theta)=\cot(\pi\tan\theta)$, and n is any integer, shew that

$$\tan\theta=\frac{2n+1}{4}\pm\frac{\sqrt{4n^2+4n-15}}{4}.$$

19. Find all the positive integral solutions of

$$\tan^{-1}x+\cot^{-1}y=\tan^{-1}3.$$

MISCELLANEOUS EXAMPLES. G.

1. If the sines of the angles of a triangle are in the ratio of 4 : 5 : 6, shew that the cosines are in the ratio of 12 : 9 : 2.

2. Solve the equations:

(1) $2\cos^3\theta+\sin^2\theta-1=0$; (2) $\sec^3\theta-2\tan^2\theta=2$.

3. If $\tan\beta=2\sin\alpha\sin\gamma\operatorname{cosec}(\alpha+\gamma)$, prove that $\cot\alpha$, $\cot\beta$, $\cot\gamma$ are in arithmetical progression.

4. In a triangle shew that

$$4r(r_1+r_2+r_3)=2(bc+ca+ab)-(a^2+b^2+c^2).$$

5. Prove that

(1) $\tan^{-1}\frac{1}{3}-\tan^{-1}\frac{1}{5}+\tan^{-1}\frac{1}{7}=\tan^{-1}\frac{3}{11}$;

(2) $\sin^{-1}\frac{3}{5}+\sin^{-1}\frac{8}{17}+\sin^{-1}\frac{36}{85}=\frac{\pi}{2}$.

6. Find the greatest angle of the triangle whose sides are 185, 222, 259; given $\log 6=\cdot 7781513$,

$$L\cos 39^\circ\, 14'=9\cdot 8890644,\ \text{diff. for } 1'=1032.$$

7. If $\tan(\alpha+\theta)=n\tan(\alpha-\theta)$, prove that $\dfrac{\sin 2\theta}{\sin 2\alpha}=\dfrac{n-1}{n+1}$.

8. If in a triangle $8R^2=a^2+b^2+c^2$, prove that one of the angles is a right angle.

9. The area of a regular polygon of n sides inscribed in a circle is three-fourths of the area of the circumscribed regular polygon with the same number of sides: find n.

10. $ABCD$ is a straight sea-wall. From B the straight lines drawn to two boats are each inclined at 45° to the direction of the wall, and from C the angles of inclination are 15° and 75°. If $BC=400$ yards, find the distance between the boats, and the distance of each from the sea-wall.

CHAPTER XX.

FUNCTIONS OF SUBMULTIPLE ANGLES.

251. Trigonometrical ratios of $22\frac{1}{2}^\circ$ or $\frac{\pi}{8}$.

From the identity

$$2\sin^2 22\tfrac{1}{2}^\circ = 1 - \cos 45^\circ,$$

we have $$4\sin^2 22\tfrac{1}{2}^\circ = 2 - 2\cos 45^\circ = 2 - \sqrt{2};$$

$$\therefore\ 2\sin 22\tfrac{1}{2}^\circ = \sqrt{2-\sqrt{2}} \quad \text{..................(1)}.$$

In like manner from

$$2\cos^2 22\tfrac{1}{2}^\circ = 1 + \cos 45^\circ,$$

we obtain $$2\cos 22\tfrac{1}{2}^\circ = \sqrt{2+\sqrt{2}} \quad \text{..................(2)}.$$

In each of these cases the positive sign must be taken before the radical, since $22\frac{1}{2}^\circ$ is an acute angle.

Again, $$\tan 22\tfrac{1}{2}^\circ = \frac{1-\cos 45^\circ}{\sin 45^\circ} = \operatorname{cosec} 45^\circ - \cot 45^\circ;$$

$$\therefore\ \tan 22\tfrac{1}{2}^\circ = \sqrt{2} - 1.$$

252. We have seen that $2\cos\frac{\pi}{8} = \sqrt{2+\sqrt{2}}$;

but $$4\cos^2\frac{\pi}{16} = 2 + 2\cos\frac{\pi}{8};$$

$$\therefore\ 4\cos^2\frac{\pi}{16} = 2 + \sqrt{2+\sqrt{2}};$$

$$\therefore\ 2\cos\frac{\pi}{16} = \sqrt{2+\sqrt{2+\sqrt{2}}}.$$

Similarly, $$2\cos\frac{\pi}{32} = \sqrt{2+\sqrt{2+\sqrt{2+\sqrt{2}}}};$$

and so on.

253. Suppose that $\cos A = \frac{1}{2}$ and that it is required to find $\sin\frac{A}{2}$.

$$\sin\frac{A}{2} = \sqrt{\frac{1-\cos A}{2}} = \sqrt{\frac{1}{2}\left(1-\frac{1}{2}\right)} = \sqrt{\frac{1}{4}} = \pm\frac{1}{2}.$$

This case differs from those of the two previous articles in that the datum is less precise. All we know of the angle A is contained in the statement that its cosine is equal to $\frac{1}{2}$, and without some further knowledge respecting A we cannot remove the ambiguity of sign in the value found for $\sin\frac{A}{2}$.

We now proceed to a more general discussion.

254. *Given* $\cos A$ *to find* $\sin\frac{A}{2}$ *and* $\cos\frac{A}{2}$ *and to explain the presence of the two values in each case.*

From the identities

$$2\sin^2\frac{A}{2} = 1-\cos A, \quad \text{and} \quad 2\cos^2\frac{A}{2} = 1+\cos A,$$

we have

$$\sin\frac{A}{2} = \pm\sqrt{\frac{1-\cos A}{2}}, \quad \text{and} \quad \cos\frac{A}{2} = \pm\sqrt{\frac{1+\cos A}{2}}.$$

Thus corresponding to *one* value of $\cos A$, there are *two* values for $\sin\frac{A}{2}$, and *two* values for $\cos\frac{A}{2}$.

The presence of these two values may be explained as follows. If $\cos A$ is given and nothing further is stated about the angle A, all we know is that A belongs to a certain group of *equi-cosinal angles.* Let α be the smallest positive angle belonging to this group, then $A = 2n\pi \pm \alpha$. Thus in finding $\sin\frac{A}{2}$ and $\cos\frac{A}{2}$ we are really finding the values of

$$\sin\frac{1}{2}(2n\pi \pm \alpha) \quad \text{and} \quad \cos\frac{1}{2}(2n\pi \pm \alpha).$$

Now $\sin \frac{1}{2}(2n\pi \pm a) = \sin\left(n\pi \pm \frac{a}{2}\right)$

$$= \sin n\pi \cos \frac{a}{2} \pm \cos n\pi \sin \frac{a}{2}$$

$$= \pm \sin \frac{a}{2},$$

for $\sin n\pi = 0$ and $\cos n\pi = \pm 1$.

Again, $\cos \frac{1}{2}(2n\pi \pm a) = \cos n\pi \cos \frac{a}{2} \mp \sin n\pi \sin \frac{a}{2}$

$$= \pm \cos \frac{a}{2}.$$

Thus there are two values for $\sin \frac{A}{2}$ and two values for $\cos \frac{A}{2}$ when $\cos A$ is given and nothing further is known respecting A.

255. Geometrical Illustration. Let a be the smallest positive angle which has the same cosine as A; then

$$A = 2n\pi \pm a,$$

and we have to find the sine and cosine of $\frac{A}{2}$, that is of

$$n\pi \pm \frac{a}{2}.$$

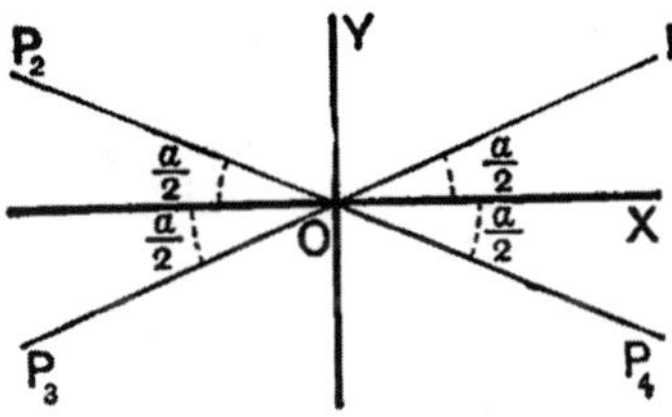

Each of the angles denoted by this formula is bounded by one of the lines OP_1, OP_2, OP_3, OP_4. Now

$$\sin XOP_2 = \sin \frac{a}{2}, \quad \sin XOP_3 = -\sin \frac{a}{2}, \quad \sin XOP_4 = -\sin \frac{a}{2},$$

$$\cos XOP_2 = -\cos \frac{a}{2}, \quad \cos XOP_3 = -\cos \frac{a}{2}, \quad \cos XOP_4 = \cos \frac{a}{2}.$$

Thus the values of $\sin \frac{A}{2}$ are $\pm \sin \frac{a}{2}$, and the values of $\cos \frac{A}{2}$ are $\pm \cos \frac{a}{2}$.

256. If $\cos A$ is given, and A lies between certain known limits, the ambiguities of sign in the formulæ of Art. 254 may be removed.

Example. If $\cos A = -\frac{7}{25}$, and A lies between 450° and 540°, find $\sin\frac{A}{2}$ and $\cos\frac{A}{2}$.

$$\sin\frac{A}{2} = \sqrt{\frac{1-\cos A}{2}} = \sqrt{\frac{1}{2}\left(1+\frac{7}{25}\right)} = \sqrt{\frac{16}{25}} = \pm\frac{4}{5};$$

$$\cos\frac{A}{2} = \sqrt{\frac{1+\cos A}{2}} = \sqrt{\frac{1}{2}\left(1-\frac{7}{25}\right)} = \sqrt{\frac{9}{25}} = \pm\frac{3}{5}.$$

Now $\frac{A}{2}$ lies between 225° and 270°, so that $\sin\frac{A}{2}$ and $\cos\frac{A}{2}$ are both negative;

$$\therefore \sin\frac{A}{2} = -\frac{4}{5}, \text{ and } \cos\frac{A}{2} = -\frac{3}{5}.$$

257. *To find* $\sin\frac{A}{2}$ *and* $\cos\frac{A}{2}$ *in terms of* $\sin A$ *and to explain the presence of four values in each case.*

We have $$\sin^2\frac{A}{2} + \cos^2\frac{A}{2} = 1,$$

and $$2\sin\frac{A}{2}\cos\frac{A}{2} = \sin A.$$

By addition, $$\left(\sin\frac{A}{2} + \cos\frac{A}{2}\right)^2 = 1 + \sin A;$$

by subtraction, $$\left(\sin\frac{A}{2} - \cos\frac{A}{2}\right)^2 = 1 - \sin A.$$

$$\therefore \sin\frac{A}{2} + \cos\frac{A}{2} = \pm\sqrt{1+\sin A} \quad \ldots\ldots\ldots\ldots(1),$$

and $$\sin\frac{A}{2} - \cos\frac{A}{2} = \pm\sqrt{1-\sin A} \quad \ldots\ldots\ldots\ldots(2).$$

By addition and subtraction, we obtain $\sin\frac{A}{2}$ and $\cos\frac{A}{2}$; and since there is a double sign before each radical, there are *four* values for $\sin\frac{A}{2}$, and *four* values for $\cos\frac{A}{2}$ corresponding to *one* value of $\sin A$.

The presence of these four values may be explained as follows.

If $\sin A$ is given and nothing else is stated about the angle A all we know is that A belongs to a certain group of *equi-sinal angles.* Let α be the smallest positive angle belonging to this group, then $A = n\pi + (-1)^n \alpha$. Thus in finding $\sin\frac{A}{2}$ and $\cos\frac{A}{2}$ we are really finding

$$\sin\frac{1}{2}\{n\pi + (-1)^n \alpha\}, \text{ and } \cos\frac{1}{2}\{n\pi + (-1)^n \alpha\}.$$

First suppose n even and equal to $2m$; then

$$\sin\frac{1}{2}\{n\pi + (-1)^n \alpha\} = \sin\left(m\pi + \frac{\alpha}{2}\right)$$

$$= \sin m\pi \cos\frac{\alpha}{2} + \cos m\pi \sin\frac{\alpha}{2}$$

$$= \pm\sin\frac{\alpha}{2},$$

since $\sin m\pi = 0$, and $\cos m\pi = \pm 1$.

Next suppose n odd and equal to $2m+1$; then

$$\sin\frac{1}{2}\{n\pi + (-1)^n \alpha\} = \sin\left(m\pi + \frac{\pi}{2} - \frac{\alpha}{2}\right)$$

$$= \sin m\pi \cos\left(\frac{\pi}{2} - \frac{\alpha}{2}\right) + \cos m\pi \sin\left(\frac{\pi}{2} - \frac{\alpha}{2}\right)$$

$$= \pm\sin\left(\frac{\pi}{2} - \frac{\alpha}{2}\right).$$

Thus we have *four* values for $\sin\frac{A}{2}$ when $\sin A$ is given and nothing further is known respecting A.

In like manner it may be shewn that

$$\cos\frac{A}{2} \text{ has the } \textit{four} \text{ values } \pm\cos\frac{\alpha}{2}, \quad \pm\cos\left(\frac{\pi}{2} - \frac{\alpha}{2}\right).$$

258. Geometrical Illustration. Let a be the smallest positive angle which has the same sine as A; then

$$A = n\pi + (-1)^n a,$$

and we have to find the sine and cosine of $\frac{A}{2}$, that is of

$$\frac{1}{2}\{n\pi + (-1)^n a\}.$$

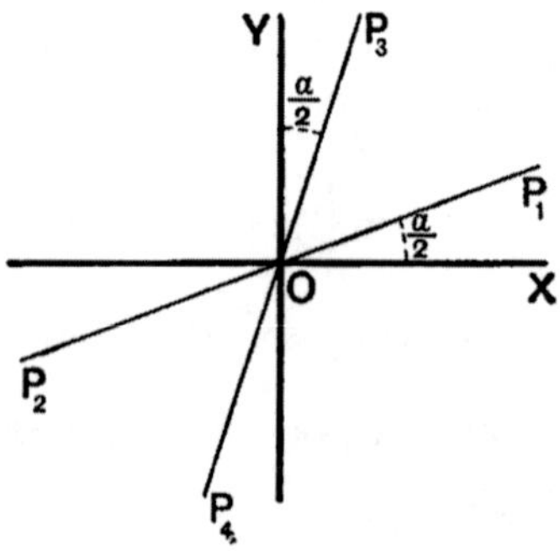

If n is even and equal to $2m$, this expression becomes $m\pi + \frac{a}{2}$.

If n is odd and equal to $2m+1$, the expression becomes

$$m\pi + \left(\frac{\pi}{2} - \frac{a}{2}\right).$$

The angles denoted by the formula $m\pi + \frac{a}{2}$ are bounded by one of the lines OP_1 or OP_2; and those denoted by the formula $m\pi + \left(\frac{\pi}{2} - \frac{a}{2}\right)$ are bounded by one of the lines OP_3 or OP_4.

Now $\qquad \sin XOP_1 = \sin \frac{a}{2}$;

$$\sin XOP_2 = -\sin XOP_1 = -\sin \frac{a}{2};$$

$$\sin XOP_3 = \sin\left(\frac{\pi}{2} - \frac{a}{2}\right);$$

$$\sin XOP_4 = -\sin XOP_3 = -\sin\left(\frac{\pi}{2} - \frac{a}{2}\right).$$

Thus the values of $\sin\frac{A}{2}$ are

$$\pm \sin\frac{a}{2} \text{ and } \pm \sin\left(\frac{\pi}{2} - \frac{a}{2}\right).$$

Similarly the values of $\cos\frac{A}{2}$ are $\pm\cos\frac{a}{2}$ and $\pm\cos\left(\frac{\pi}{2} - \frac{a}{2}\right)$.

259. If in addition to the value of $\sin A$ we know that A lies between certain limits, the ambiguities of sign in equations (1) and (2) of Art. 257 may be removed.

Example 1. Find $\sin\frac{A}{2}$ and $\cos\frac{A}{2}$ in terms of $\sin A$ when A lies between 450° and 630°.

In this case $\frac{A}{2}$ lies between 225° and 315°. From the adjoining figure it is evident that between these limits $\sin\frac{A}{2}$ is greater than $\cos\frac{A}{2}$ and is negative.

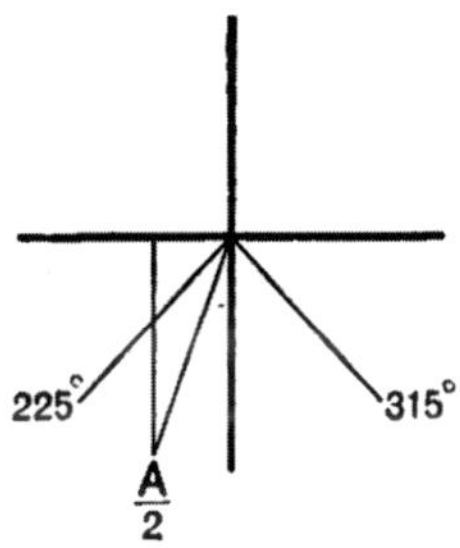

$$\therefore \ \sin\frac{A}{2}+\cos\frac{A}{2} = -\sqrt{1+\sin A},$$

and $$\sin\frac{A}{2}-\cos\frac{A}{2} = -\sqrt{1-\sin A}.$$

$$\therefore \ 2\sin\frac{A}{2} = -\sqrt{1+\sin A}-\sqrt{1-\sin A},$$

and $$2\cos\frac{A}{2} = -\sqrt{1+\sin A}+\sqrt{1-\sin A}.$$

Example 2. Determine the limits between which A must lie in order that

$$2\cos A = -\sqrt{1+\sin 2A}-\sqrt{1-\sin 2A}.$$

The given relation is obtained by combining

$$\sin A+\cos A = -\sqrt{1+\sin 2A} \quad \text{.................(1)},$$

and $$\sin A-\cos A = +\sqrt{1-\sin 2A} \quad \text{.................(2)}.$$

From (1), we see that of $\sin A$ and $\cos A$ *the numerically greater is negative.*

From (2), we see that the cosine is the greater.

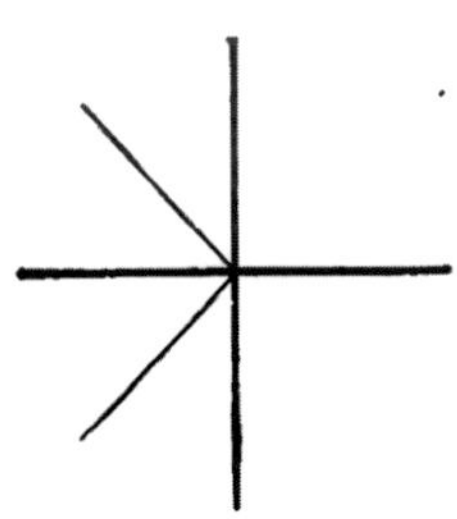

Hence we have to choose limits between which $\cos A$ is numerically greater than $\sin A$ and is negative. From the figure we see that A lies between $2n\pi+\frac{3\pi}{4}$ and $2n\pi+\frac{5\pi}{4}$.

Example 3. Trace the changes of $\cos\theta - \sin\theta$ in sign and magnitude as θ increases from 0 to 2π.

$$\cos\theta - \sin\theta = \sqrt{2}\left(\frac{1}{\sqrt{2}}\cos\theta - \frac{1}{\sqrt{2}}\sin\theta\right)$$

$$= \sqrt{2}\left(\cos\theta\cos\frac{\pi}{4} - \sin\theta\sin\frac{\pi}{4}\right)$$

$$= \sqrt{2}\cos\left(\theta + \frac{\pi}{4}\right).$$

As θ increases from 0 to $\frac{\pi}{4}$, the expression is positive and decreases from 1 to 0.

As θ increases from $\frac{\pi}{4}$ to $\frac{3\pi}{4}$, the expression is negative and increases numerically from 0 to $-\sqrt{2}$.

As θ increases from $\frac{3\pi}{4}$ to $\frac{5\pi}{4}$, the expression is negative and decreases numerically from $-\sqrt{2}$ to 0.

As θ increases from $\frac{5\pi}{4}$ to $\frac{7\pi}{4}$, the expression is positive and increases from 0 to $\sqrt{2}$.

As θ increases from $\frac{7\pi}{4}$ to 2π, the expression is positive and decreases from $\sqrt{2}$ to 1.

260. *To find the sine and cosine of* 9°.

Since $\cos 9° > \sin 9°$ and is positive, we have

$$\sin 9° + \cos 9° = +\sqrt{1 + \sin 18°},$$

and $$\sin 9° - \cos 9° = -\sqrt{1 - \sin 18°}.$$

$$\therefore\ \sin 9° + \cos 9° = +\sqrt{1 + \frac{\sqrt{5}-1}{4}} = +\frac{1}{2}\sqrt{3+\sqrt{5}},$$

and $$\sin 9° - \cos 9° = -\sqrt{1 - \frac{\sqrt{5}-1}{4}} = -\frac{1}{2}\sqrt{5-\sqrt{5}}.$$

$$\therefore\ \sin 9° = \frac{1}{4}\left\{\sqrt{3+\sqrt{5}} - \sqrt{5-\sqrt{5}}\right\},$$

and $$\cos 9° = \frac{1}{4}\left\{\sqrt{3+\sqrt{5}} + \sqrt{5-\sqrt{5}}\right\}.$$

EXAMPLES. XX. a.

1. When A lies between $-270°$ and $-360°$, prove that

$$\sin\frac{A}{2} = -\sqrt{\frac{1-\cos A}{2}}.$$

2. If $\cos A = \frac{119}{169}$, find $\sin\frac{A}{2}$ and $\cos\frac{A}{2}$ when A lies between $270°$ and $360°$.

3. If $\cos A = -\frac{161}{289}$, find $\sin\frac{A}{2}$ and $\cos\frac{A}{2}$ when A lies between $540°$ and $630°$.

4. Find $\sin\frac{A}{2}$ and $\cos\frac{A}{2}$ in terms of $\sin A$ when A lies between $270°$ and $450°$.

5. Find $\sin\frac{A}{2}$ and $\cos\frac{A}{2}$ in terms of $\sin A$ when $\frac{A}{2}$ lies between $225°$ and $315°$.

6. Find $\sin\frac{A}{2}$ and $\cos\frac{A}{2}$ in terms of $\sin A$ when A lies between $-450°$ and $-630°$.

7. If $\sin A = \frac{24}{25}$, find $\sin\frac{A}{2}$ and $\cos\frac{A}{2}$ when A lies between $90°$ and $180°$.

8. If $\sin A = -\frac{240}{289}$, find $\sin\frac{A}{2}$ and $\cos\frac{A}{2}$ when A lies between $270°$ and $360°$.

9. Determine the limits between which A must lie in order that

(1) $2\sin A = \sqrt{1+\sin 2A} - \sqrt{1-\sin 2A}$;

(2) $2\cos A = -\sqrt{1+\sin 2A} + \sqrt{1-\sin 2A}$;

(3) $2\sin A = -\sqrt{1+\sin 2A} + \sqrt{1-\sin 2A}$.

10. If $A=240°$, is the following statement correct?

$$2\sin\frac{A}{2}=\sqrt{1+\sin A}-\sqrt{1-\sin A}.$$

If not, how must it be modified?

11. Prove that

(1) $\tan 7\frac{1}{2}°=\sqrt{6}-\sqrt{3}+\sqrt{2}-2$;

(2) $\cot 142\frac{1}{2}°=\sqrt{2}+\sqrt{3}-2-\sqrt{6}$.

12. Shew that $\sin 9°$ lies between ·156 and ·157.

13. Prove that

(1) $2\sin 11°\ 15'=\sqrt{2-\sqrt{2+\sqrt{2}}}$;

(2) $\tan 11°\ 15'=\sqrt{4+2\sqrt{2}}-(\sqrt{2}+1)$.

14. When θ varies from 0 to 2π trace the changes in sign and magnitude of

(1) $\cos\theta+\sin\theta$; (2) $\sin\theta-\sqrt{3}\cos\theta$.

15. When θ varies from 0 to π, trace the changes in sign and magnitude of

(1) $\dfrac{\tan\theta+\cot\theta}{\tan\theta-\cot\theta}$; (2) $\dfrac{2\sin\theta-\sin 2\theta}{2\sin\theta+\sin 2\theta}$.

261. *To find* $\tan\frac{A}{2}$ *when* $\tan A$ *is given and to explain the presence of the two values.*

Denote $\tan A$ by t; then

$$t=\tan A=\frac{2\tan\frac{A}{2}}{1-\tan^2\frac{A}{2}};$$

$$\therefore\ t\tan^2\frac{A}{2}+2\tan\frac{A}{2}-t=0;$$

$$\therefore\ \tan\frac{A}{2}=\frac{-2\pm\sqrt{4+4t^2}}{2t}=\frac{-1\pm\sqrt{1+t^2}}{t}.$$

The presence of these two values may be explained as follows.

If a be the smallest positive angle which has the given tangent, then $A = n\pi + a$, and we are really finding the value of

$$\tan \frac{1}{2}(n\pi + a).$$

(1) Let n be even and equal to $2m$; then

$$\tan \frac{1}{2}(n\pi + a) = \tan\left(m\pi + \frac{a}{2}\right) = \tan \frac{a}{2}.$$

(2) Let n be odd and equal to $2m+1$; then

$$\tan \frac{1}{2}(n\pi + a) = \tan\left(m\pi + \frac{\pi}{2} + \frac{a}{2}\right) = \tan\left(\frac{\pi}{2} + \frac{a}{2}\right).$$

Thus $\tan \frac{A}{2}$ has the two values $\tan \frac{a}{2}$ and $\tan\left(\frac{\pi}{2} + \frac{a}{2}\right)$.

Example 1. If $A = 170°$, prove that $\tan \frac{A}{2} = \frac{-1 - \sqrt{1 + \tan^2 A}}{\tan A}$.

Here $\frac{A}{2}$ is an acute angle, so that $\tan \frac{A}{2}$ must be positive. Hence in the formula $\frac{-1 \pm \sqrt{1 + \tan^2 A}}{\tan A}$ the numerator must have the same sign as the denominator. But when $A = 170°$, $\tan A$ is negative, and therefore we must choose the sign which will make the numerator negative; thus $\tan \frac{A}{2} = \frac{-1 - \sqrt{1 + \tan^2 A}}{\tan A}$.

Example 2. Given $\cos A = \cdot 6$, find $\tan \frac{A}{2}$, and explain the double answer.

$$\tan^2 \frac{A}{2} = \frac{1 - \cos A}{1 + \cos A} = \frac{\cdot 4}{1 \cdot 6} = \frac{1}{4};$$

$$\therefore \tan \frac{A}{2} = \pm \frac{1}{2}.$$

Here all we know of the angle A is that it must be one of a group of equi-cosinal angles. Let a be the smallest positive angle of this group; then $A = 2n\pi \pm a$.

$$\therefore \tan \frac{A}{2} = \tan\left(n\pi \pm \frac{a}{2}\right) = \tan\left(\pm \frac{a}{2}\right) = \pm \tan \frac{a}{2}.$$

Thus we have two values differing only in sign.

262. When any one of the functions of an acute angle A is given, we may in some cases conveniently obtain the functions of $\frac{A}{2}$, as in the following example.

Example. Given $\cos A = \frac{b}{a}$, to find the functions of $\frac{A}{2}$.

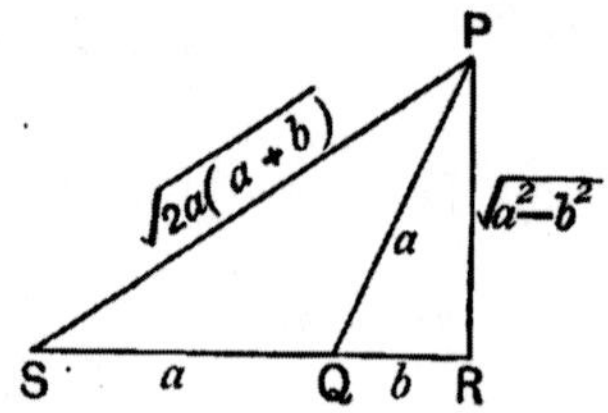

Make a right-angled triangle PQR in which the hypotenuse $PQ=a$, and base $QR=b$; then

$$\cos PQR = \frac{QR}{PQ} = \frac{b}{a} = \cos A;$$

$$\therefore \angle PQR = A.$$

Produce RQ to S making $QS=QP$;

$$\therefore \angle PSQ = \angle SPQ = \frac{1}{2}\angle PQR = \frac{A}{2}.$$

Now $$SR = a+b, \text{ and } PR = \sqrt{a^2-b^2},$$

$$\therefore PS^2 = (a+b)^2 + (a^2-b^2) = 2a^2 + 2ab;$$

$$\therefore PS = \sqrt{2a(a+b)}.$$

The functions of $\frac{A}{2}$ may now be written down in terms of the sides of the triangle PRS.

263. From Art. 125, we have

$$\cos A = 4\cos^3\frac{A}{3} - 3\cos\frac{A}{3}.$$

Thus it appears that if $\cos A$ be given we have a *cubic* equation to find $\cos\frac{A}{3}$; so that $\cos\frac{A}{3}$ has *three* values.

Similarly, from the equation

$$\sin A = 3\sin\frac{A}{3} - 4\sin^3\frac{A}{3}$$

it appears that corresponding to *one* value of $\sin A$ there are *three* values of $\sin\frac{A}{3}$.

It will be a useful exercise to prove these two statements analytically as in Arts. 254 and 257. In the next article we shall give a geometrical explanation for the case of the cosine.

264. *Given* $\cos A$ *to find* $\cos\frac{A}{3}$, *and to explain the presence of the three values.*

Let a be the smallest positive angle with the given cosine; then $A=2n\pi\pm a$, and we have to find all the values of

$$\cos\frac{1}{3}(2n\pi\pm a).$$

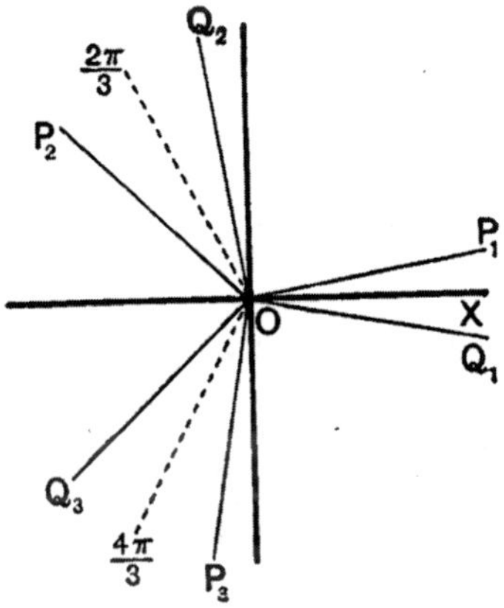

Consider the angles denoted by the formula

$$\frac{1}{3}(2n\pi\pm a),$$

and ascribe to n in succession the values 0, 1, 2, 3, ...

When $n=0$, the angles are $\pm\frac{a}{3}$, bounded by OP_1 and OQ_1;

when $n=1$, the angles are $\frac{2\pi}{3}\pm\frac{a}{3}$, bounded by OP_2 and OQ_2

when $n=2$, the angles are $\frac{4\pi}{3}\pm\frac{a}{3}$, bounded by OP_3 and OQ_3.

By giving to n the values 3, 4, 5, ... we obtain a series of angles coterminal with those indicated in the figure.

Thus OP_1, OQ_1, OP_2, OQ_2, OP_3, OQ_3 bound all the angles included in the formula $\frac{1}{3}(2n\pi\pm a)$.

Now $\qquad \cos XOQ_1=\cos XOP_1=\cos\frac{a}{3}$;

$$\cos XOP_3=\cos XOQ_2=\cos\left(\frac{2\pi}{3}-\frac{a}{3}\right);$$

$$\cos XOQ_3=\cos XOP_2=\cos\left(\frac{2\pi}{3}+\frac{a}{3}\right).$$

Thus the values of $\cos\frac{A}{3}$ are $\cos\frac{a}{3}$, $\cos\frac{2\pi+a}{3}$, $\cos\frac{2\pi-a}{3}$.

EXAMPLES. XX. b.

1. If $A=320°$, prove that

$$\tan\frac{A}{2}=\frac{-1+\sqrt{1+\tan^2 A}}{\tan A}.$$

2. Shew that

$$\tan A=-\frac{1+\sqrt{1+\tan^2 2A}}{\tan 2A} \text{ when } A=110°.$$

3. Find $\tan A$ when $\cos 2A=\frac{12}{13}$ and A lies between 180° and 225°.

4. Find $\cot\frac{A}{2}$ when $\cos A=-\frac{4}{5}$ and A lies between 180° and 270°.

5. If $\cot 2\theta=\cot 2\alpha$, shew that $\cot\theta$ has the two values $\cot\alpha$ and $-\tan\alpha$.

6. Given that $\sin\theta=\sin\alpha$, shew that the values of $\sin\frac{\theta}{3}$ are

$$\sin\frac{\alpha}{3},\quad \sin\frac{\pi-\alpha}{3},\quad -\sin\frac{\pi+\alpha}{3}.$$

7. If $\tan\theta=\tan\alpha$, shew that the values of $\tan\frac{\theta}{3}$ are

$$\tan\frac{\alpha}{3},\quad \tan\frac{\pi+\alpha}{3},\quad -\tan\frac{\pi-\alpha}{3}.$$

8. Given that $\cos 3\theta=\cos 3\alpha$, shew that the values of $\sin\theta$ are

$$\pm\sin\alpha,\quad -\sin\left(\frac{\pi}{3}\pm\alpha\right),\quad \sin\left(\frac{2\pi}{3}\pm\alpha\right).$$

9. Given that $\sin 3\theta=\sin 3\alpha$, shew that the values of $\cos\theta$ are

$$\pm\cos\alpha,\quad \cos\left(\frac{\pi}{3}\pm\alpha\right),\quad \cos\left(\frac{2\pi}{3}\pm\alpha\right).$$

CHAPTER XXI.

LIMITS AND APPROXIMATIONS.

265. *If θ be the radian measure of an angle less than a right angle, to shew that* $\sin\theta$, θ, $\tan\theta$ *are in ascending order of magnitude.*

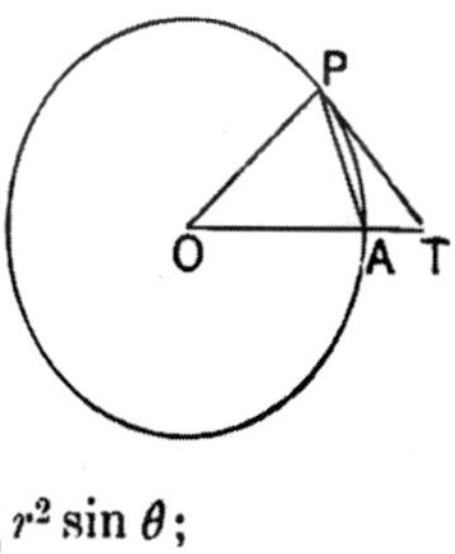

Let the angle θ be represented by AOP.

With centre O and radius OA describe a circle. Draw PT at right angles to OP to meet OA produced in T, and join PA.

Let r be the radius of the circle.

Area of $\Delta AOP = \frac{1}{2} AO \,.\, OP \sin AOP = \frac{1}{2} r^2 \sin\theta$;

area of sector $AOP = \frac{1}{2} r^2\theta$;

area of $\Delta OPT = \frac{1}{2} OP \,.\, PT = \frac{1}{2} r \,.\, r\tan\theta = \frac{1}{2} r^2 \tan\theta$.

But the areas of the triangle AOP, the sector AOP, and the triangle OTP are in ascending order of magnitude; that is,

$$\frac{1}{2} r^2 \sin\theta, \quad \frac{1}{2} r^2\theta, \quad \frac{1}{2} r^2 \tan\theta$$

are in ascending order of magnitude;

$\therefore$ $\sin\theta$, θ, $\tan\theta$ are in ascending order of magnitude.

266. *When θ is indefinitely diminished, to prove that $\frac{\sin\theta}{\theta}$ and $\frac{\tan\theta}{\theta}$ each have unity for their limit.*

In the last article, we have proved that $\sin\theta$, θ, $\tan\theta$ are in ascending order of magnitude. Divide each of these quantities by $\sin\theta$; then

$$1, \frac{\theta}{\sin\theta}, \frac{1}{\cos\theta} \text{ are in ascending order of magnitude;}$$

that is, $\frac{\theta}{\sin\theta}$ lies between 1 and $\sec\theta$.

But when θ is indefinitely diminished, the limit of $\sec\theta$ is 1; hence the limit of $\frac{\theta}{\sin\theta}$ is 1; that is, the limit of $\frac{\sin\theta}{\theta}$ is unity.

Again, by dividing each of the quantities $\sin\theta$, θ, $\tan\theta$ by $\tan\theta$, we find that $\cos\theta$, $\frac{\theta}{\tan\theta}$, 1 are in ascending order of magnitude. Hence the limit of $\frac{\tan\theta}{\theta}$ is unity.

These results are often written concisely in the forms

$$\underset{\theta=0}{Lt.}\left(\frac{\sin\theta}{\theta}\right)=1, \qquad \underset{\theta=0}{Lt.}\left(\frac{\tan\theta}{\theta}\right)=1.$$

Example. Find the limit of $n\sin\frac{\theta}{n}$ when $n=\infty$.

$$n\sin\frac{\theta}{n}=\theta\,.\,\frac{n}{\theta}\,.\,\sin\frac{\theta}{n}=\theta\left(\sin\frac{\theta}{n}\div\frac{\theta}{n}\right);$$

but since $\frac{\theta}{n}$ is indefinitely small, the limit of $\sin\frac{\theta}{n}\div\frac{\theta}{n}$ is unity;

$$\therefore \underset{n=\infty}{Lt.}\left(n\sin\frac{\theta}{n}\right)=\theta.$$

Similarly $$\underset{n=\infty}{Lt.}\left(n\tan\frac{\theta}{n}\right)=\theta.$$

267. It is important to remember that the conclusions of the foregoing articles only hold when the angle is expressed in radian measure. If any other system of measurement is used, the results will require modification.

Example. Find the value of $\underset{n=0}{Lt.}\left(\frac{\sin n^\circ}{n}\right)$.

Let θ be the number of radians in n°; then

$$\frac{n}{180}=\frac{\theta}{\pi}, \text{ and } n=\frac{180\theta}{\pi}; \text{ also } \sin n^\circ=\sin\theta;$$

$$\therefore \frac{\sin n^\circ}{n}=\frac{\pi\sin\theta}{180\theta}=\frac{\pi}{180}\cdot\frac{\sin\theta}{\theta}.$$

When n is indefinitely small, θ is indefinitely small;

$$\therefore \underset{n=0}{Lt.}\left(\frac{\sin n^\circ}{n}\right)=\frac{\pi}{180}\cdot\underset{\theta=0}{Lt.}\left(\frac{\sin\theta}{\theta}\right);$$

$$\therefore \underset{n=0}{Lt.}\left(\frac{\sin n^\circ}{n}\right)=\frac{\pi}{180}.$$

268. When θ is the radian measure of a very small angle, we have shewn that

$$\frac{\sin\theta}{\theta}=1,\quad \cos\theta=1,\quad \frac{\tan\theta}{\theta}=1;$$

that is, $\sin\theta=\theta,\quad \cos\theta=1,\quad \tan\theta=\theta.$

Hence $r\tan\theta=r\theta$, and therefore in the figure of Art. 265, the tangent PT is equal to the arc PA, when $\angle AOP$ is very small.

In Art. 270, it will be shewn that these results hold so long as θ is so small that its square may be neglected. When this is the case, we have

$$\sin(a+\theta)=\sin a\cos\theta+\cos a\sin\theta$$
$$=\sin a+\theta\cos a;$$
$$\cos(a+\theta)=\cos a\cos\theta-\sin a\sin\theta$$
$$=\cos a-\theta\sin a.$$

Example 1. The inclination of a railway to the horizontal plane is 52′ 30″, find how many feet it rises in a mile.

Let OA be the horizontal plane, and OP a mile of the railway. Draw PN perpendicular to OA.

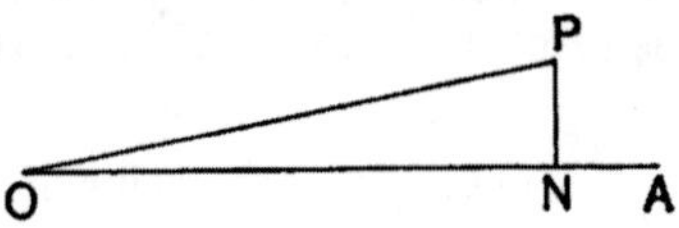

Let $PN = x$ feet, $\angle PON = \theta$;

then $$\frac{PN}{OP} = \sin\theta = \theta \text{ approximately.}$$

But θ = radian measure of $52'\,30'' = \dfrac{52\frac{1}{2}}{60} \times \dfrac{\pi}{180} = \dfrac{7}{8} \times \dfrac{\pi}{180}$;

$$\therefore \frac{x}{1760 \times 3} = \frac{7}{8} \times \frac{22}{7} \times \frac{1}{180};$$

$$\therefore x = \frac{1760 \times 3 \times 22}{8 \times 180} = \frac{242}{3} = 80\tfrac{2}{3}.$$

Thus the rise is $80\frac{2}{3}$ feet.

Example 2. A pole 6 ft. long stands on the top of a tower 54 ft. high: find the angle subtended by the pole at a point on the ground which is at a distance of 180 yds. from the foot of the tower.

Let A be the point on the ground, BC the tower, CD the pole.

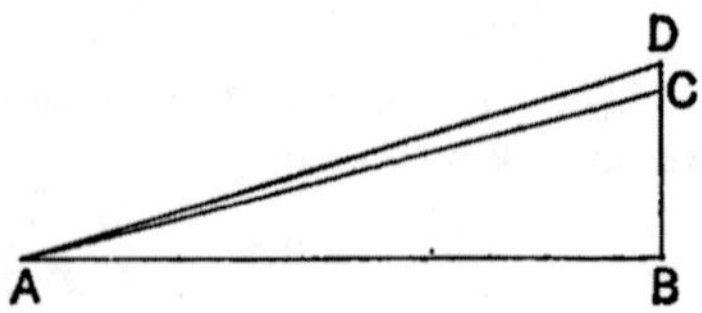

Let $\angle BAC = \alpha$, $\angle CAD = \theta$;

then $\tan\alpha = \dfrac{BC}{AB} = \dfrac{54}{540} = \dfrac{1}{10}$;

$$\tan(\alpha + \theta) = \frac{BD}{AB} = \frac{60}{540} = \frac{1}{9}.$$

But $$\tan(\alpha+\theta) = \frac{\tan\alpha + \tan\theta}{1 - \tan\alpha\tan\theta} = \frac{\tan\alpha + \theta}{1 - \theta\tan\alpha} \text{ approximately;}$$

$$\therefore \frac{1}{9} = \frac{\frac{1}{10} + \theta}{1 - \frac{\theta}{10}} = \frac{1 + 10\theta}{10 - \theta};$$

whence $\theta = \dfrac{1}{91}$; that is, the angle is $\dfrac{1}{91}$ of a radian, and therefore contains $\dfrac{1}{91} \times \dfrac{180}{\pi}$ degrees.

On reduction, we find that the angle is 37′ 46″ nearly.

269. *If θ be the number of radians in an acute angle, to prove that*

$$\cos\theta > 1 - \frac{\theta^2}{2}, \text{ and } \sin\theta > \theta - \frac{\theta^3}{4}.$$

Since $\cos\theta = 1 - 2\sin^2\frac{\theta}{2}$, and $\sin\frac{\theta}{2} < \frac{\theta}{2}$;

$$\therefore \cos\theta > 1 - 2\left(\frac{\theta}{2}\right)^2;$$

that is,

$$\cos\theta > 1 - \frac{\theta^2}{2}.$$

Again, $\sin\theta = 2\sin\frac{\theta}{2}\cos\frac{\theta}{2} = 2\tan\frac{\theta}{2}\cos^2\frac{\theta}{2}$;

but

$$\tan\frac{\theta}{2} > \frac{\theta}{2};$$

$$\therefore \sin\theta > 2\,\frac{\theta}{2}\cos^2\frac{\theta}{2};$$

$$\therefore \sin\theta > \theta\left(1 - \sin^2\frac{\theta}{2}\right).$$

But $\sin\frac{\theta}{2} < \frac{\theta}{2}$, and therefore

$$1 - \sin^2\frac{\theta}{2} > 1 - \left(\frac{\theta}{2}\right)^2;$$

$$\therefore \sin\theta > \theta\left\{1 - \left(\frac{\theta}{2}\right)^2\right\};$$

$$\therefore \sin\theta > \theta - \frac{\theta^3}{4}.$$

270. From the propositions established in this chapter, it follows that if θ is an acute angle,

$$\cos\theta \text{ lies between } 1 \text{ and } 1 - \frac{\theta^2}{2},$$

and

$$\sin\theta \text{ lies between } \theta \text{ and } \theta - \frac{\theta^3}{4}.$$

Thus $\cos\theta = 1 - k\theta^2$ and $\sin\theta = \theta - k'\theta^3$, where k and k' are proper fractions less than $\frac{1}{2}$ and $\frac{1}{4}$ respectively.

Hence if θ be so small that its square can be neglected,

$$\cos\theta = 1, \quad \sin\theta = \theta.$$

Example. Find the approximate value of $\sin 10''$.

The circular measure of $10''$ is $\dfrac{10\pi}{180\times 60\times 60}$ or $\dfrac{\pi}{64800}$;

$$\therefore \sin 10'' < \frac{\pi}{64800} \text{ and } > \frac{\pi}{64800} - \frac{1}{4}\left(\frac{\pi}{64800}\right)^3.$$

But $$\frac{\pi}{64800} = \frac{3\cdot 1415926535\ldots}{64800} = \cdot 000048481368\ldots\ldots;$$

$$\therefore \frac{\pi}{64800} < \cdot 00005 \text{ and } \left(\frac{\pi}{64800}\right)^3 < \cdot 000000000000125;$$

$$\therefore \sin 10'' < \frac{\pi}{64800} \text{ and } > \frac{\pi}{64800} - \frac{1}{4}(\cdot 000000000000125).$$

Hence to 12 places of decimals,

$$\sin 10'' = \frac{\pi}{64800} = \cdot 000048481368\ldots.$$

271. *To shew that when* n *is an indefinitely large integer, the limit of*

$$\cos\frac{\theta}{2}\cos\frac{\theta}{4}\cos\frac{\theta}{8}\ldots\cos\frac{\theta}{2^n} = \frac{\sin\theta}{\theta}.$$

We have
$$\begin{aligned}\sin\theta &= 2\sin\frac{\theta}{2}\cos\frac{\theta}{2}\\ &= 2^2\sin\frac{\theta}{4}\cos\frac{\theta}{4}\cos\frac{\theta}{2}\\ &= 2^3\sin\frac{\theta}{8}\cos\frac{\theta}{8}\cos\frac{\theta}{4}\cos\frac{\theta}{2}\\ &\ldots\ldots\ldots\ldots\ldots\ldots\\ &= 2^n\sin\frac{\theta}{2^n}\cos\frac{\theta}{2^n}\ldots\cos\frac{\theta}{8}\cos\frac{\theta}{4}\cos\frac{\theta}{2}.\end{aligned}$$

$$\therefore \cos\frac{\theta}{2}\cos\frac{\theta}{4}\cos\frac{\theta}{8}\ldots\cos\frac{\theta}{2^n} = \frac{\sin\theta}{2^n\sin\dfrac{\theta}{2^n}}.$$

But the limit of $2^n\sin\dfrac{\theta}{2^n}$ is θ, and thus the proposition is established. [See Art. 266.]

272. *To shew that* $\frac{\sin\theta}{\theta}$ *continually decreases from* 1 *to* $\frac{2}{\pi}$ *as* θ *continually increases from* 0 *to* $\frac{\pi}{2}$.

We shall first shew that the fraction

$$\frac{\sin\theta}{\theta} - \frac{\sin(\theta+h)}{\theta+h} \text{ is positive,}$$

h denoting the radian measure of a small positive angle.

$$\text{This fraction} = \frac{(\theta+h)\sin\theta - \theta(\sin\theta\cos h + \cos\theta\sin h)}{\theta(\theta+h)}$$

$$= \frac{\theta\sin\theta(1-\cos h) + (h\sin\theta - \theta\cos\theta\sin h)}{\theta(\theta+h)}.$$

Now $\tan\theta > \theta$, that is $\sin\theta > \theta\cos\theta$, and $h > \sin h$;

$$\therefore\ h\sin\theta > \theta\cos\theta\sin h.$$

Also $1-\cos h$ is positive; hence the numerator is positive, and therefore the fraction is positive;

$$\therefore\ \frac{\sin(\theta+h)}{\theta+h} < \frac{\sin\theta}{\theta};$$

$\therefore\ \frac{\sin\theta}{\theta}$ continually decreases as θ continually increases.

When $\theta=0$, $\frac{\sin\theta}{\theta}=1$; and when $\theta=\frac{\pi}{2}$, $\frac{\sin\theta}{\theta}=\frac{2}{\pi}$.

Thus the proposition is established.

EXAMPLES. XXI. a.

[*In this Exercise take* $\pi=\frac{22}{7}$.]

1. A tower 44 feet high subtends an angle of 35′ at a point A on the ground: find the distance of A from the tower.

2. From the top of a wall 7 ft. 4 in. high the angle of depression of an object on the ground is 24′ 30″: find its distance from the wall.

3. Find the height of an object whose angle of elevation at a distance of 840 yards is $1° 30'$.

4. Find the angle subtended by a pole 10 ft. 1 in. high at a distance of a mile.

5. Find the angle subtended by a circular target 4 feet in diameter at a distance of 1000 yards.

6. Taking the diameter of a penny as 1·25 inches, find at what distance it must be held from the eye so as just to hide the moon, supposing the diameter of the moon to be half a degree.

7. Find the distance at which a globe 11 inches in diameter subtends an angle of $5'$.

8. Two places on the same meridian are 11 miles apart: find the difference in their latitudes, taking the radius of the earth as 3960 miles.

9. A man 6 ft. high stands on a tower whose height is 120 ft.: shew that at a point 24 ft. from the tower the man subtends an angle of $31·5'$ nearly.

10. A flagstaff standing on the top of a cliff 490 feet high subtends an angle of ·04 radians at a point 980 feet from the base of the cliff: find the height of the flagstaff.

11. When $n=0$, find the limit of

$$(1)\quad \frac{\sin n'}{n}; \qquad (2)\quad \frac{\sin n''}{n}.$$

12. When $n=\infty$, find the limit of $\frac{1}{2}nr^2 \sin\frac{2\pi}{n}$.

When $\theta=0$, find the limit of

13. $\dfrac{1-\cos\theta}{\theta\sin\theta}$. 14. $\dfrac{m\sin m\theta - n\sin n\theta}{\tan m\theta+\tan n\theta}$.

15. If $\theta=·01$ of a radian, calculate $\cos\left(\frac{\pi}{3}+\theta\right)$.

16. Find the value of $\sin 30°\ 10'\ 30''$.

17. Given $\cos\left(\frac{\pi}{3}+\theta\right)=·49$, find the sexagesimal value of θ.

Distance and Dip of the Visible Horizon.

273. Let A be a point above the earth's surface, BCD a section of the earth by a plane passing through its centre E and A.

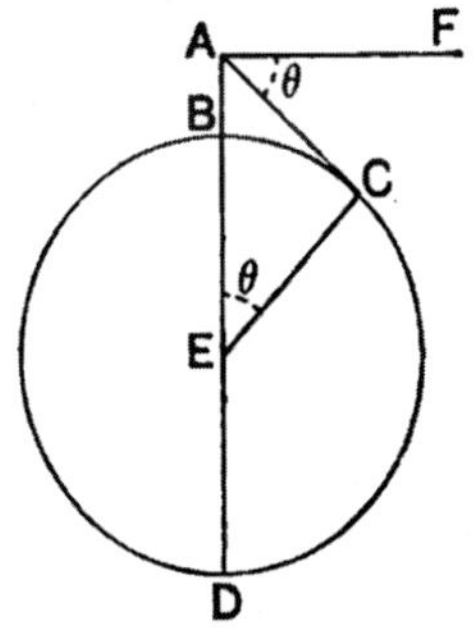

Let AE cut the circumference in B and D.

From A draw AC to touch the circle BCD in C, and join EC.

Draw AF at right angles to AD; then $\angle FAC$ is called the **dip of the horizon** as seen from A.

Thus the *dip of the horizon* is the angle of depression of any point on the horizon visible from A.

274. *To find the distance of the horizon.*

In the figure of the last article, let

$$AB = h, \quad EB = ED = r, \quad AC = x;$$

then by Euc. III. 36, $\quad AC^2 = AB \, . \, AD$;

that is, $\quad x^2 = h(2r + h) = 2hr + h^2.$

For ordinary altitudes h^2 is very small in comparison with $2hr$; hence approximately

$$x^2 = 2hr \quad \text{and} \quad x = \sqrt{2hr}.$$

In this formula, suppose the measurements are made in *miles*, and let a be the number of *feet* in AB; then

$$a = 1760 \times 3 \times h.$$

By taking $r = 3960$, we have

$$x^2 = \frac{2 \times 3960 \times a}{1760 \times 3} = \frac{3a}{2}.$$

Thus we have the following rule:

Twice the square of the distance of the horizon measured in miles *is equal to three times the height of the place of observation measured in* feet.

Hence a man whose eye is 6 feet from the ground can see to a distance of 3 miles on a horizontal plane.

Example. The top of a ship's mast is $66\frac{2}{3}$ ft. above the sea-level, and from it the lamp of a lighthouse can just be seen. After the ship has sailed directly towards the lighthouse for half-an-hour the lamp can be seen from the deck, which is 24 ft. above the sea. Find the rate at which the ship is sailing.

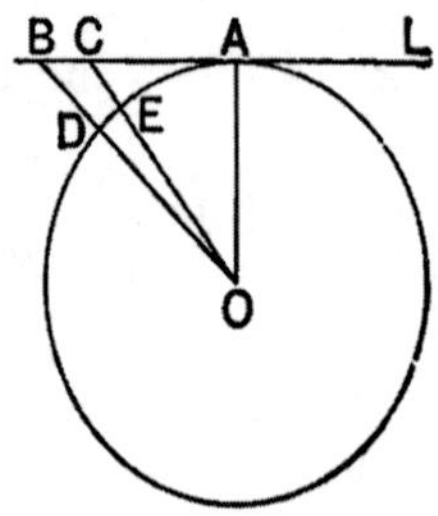

Let L denote the lamp, D and E the two positions of the ship, B the top of the mast, C the point on the deck from which the lamp is seen; then LCB is a tangent to the earth's surface at A.

[In problems like this some of the lines must necessarily be greatly out of proportion.]

Let AB and AC be expressed in miles; then since $DB = 66\frac{2}{3}$ feet and $EC = 24$ feet, we have by the rule

$$AB^2 = \frac{3}{2} \times 66\tfrac{2}{3} = 100;$$

$$\therefore\ AB = 10 \text{ miles.}$$

$$AC^2 = \frac{3}{2} \times 24 = 36;$$

$$\therefore\ AC = 6 \text{ miles.}$$

But the angles subtended by AB and AC at O the centre of the earth are very small;

$$\therefore\ \text{arc } AD = AB, \text{ and arc } AC = AE. \qquad \text{[Art. 268.]}$$

$$\therefore\ \text{arc } DE = AD - AE = AB - AC = 4 \text{ miles.}$$

Thus the ship sails 4 miles in half-an-hour, or 8 miles per hour.

275. Let θ be the number of radians in the dip of the horizon; then with the figure of Art. 273, we have

$$\cos\theta = \frac{EC}{EA} = \frac{r}{h+r} = \left(1 + \frac{h}{r}\right)^{-1};$$

$$\therefore\ 1 - 2\sin^2\frac{\theta}{2} = 1 - \frac{h}{r} + \frac{h^2}{r^2} - \dots;$$

$$\therefore\ 2\sin^2\frac{\theta}{2} = \frac{h}{r} - \frac{h^2}{r^2} + \dots.$$

Since θ and $\frac{h}{r}$ are small, we may replace $\sin\frac{\theta}{2}$ by $\frac{\theta}{2}$ and neglect the terms on the right after the first.

Thus $$\frac{\theta^2}{2}=\frac{h}{r}, \text{ or } \theta=\sqrt{\frac{2h}{r}}.$$

Let N be the number of degrees in θ radians; then

$$N=\frac{180\theta}{\pi}=\frac{180}{\pi}\sqrt{\frac{2h}{r}}.$$

Now $\sqrt{r}=63$ nearly; hence we have approximately

$$N=\frac{180\times 7\times\sqrt{2h}}{22\times 63},$$

or $$N=\frac{10}{11}\sqrt{2h},$$

a formula connecting the dip of the horizon in degrees and the height of the place of observation in miles.

EXAMPLES. XXI. b.

[Here $\pi=\frac{22}{7}$, *and radius of earth*$=3960$ *miles.*]

1. Find the greatest distance at which the lamp of a lighthouse can be seen, the light being 96 feet above the sea-level.

2. If the lamp of a lighthouse begins to be seen at a distance of 15 miles, find its height above the sea-level.

3. The tops of the masts of two ships are 32 ft. 8 in. and 42 ft. 8 in. above the sea-level: find the greatest distance at which one mast can be seen from the other.

4. Find the height of a ship's mast which is just visible at a distance of 20 miles from a point on the mast of another ship which is 54 ft. above the sea-level.

5. From the mast of a ship 73 ft. 6 in. high the lamp of a lighthouse is just visible at a distance of 28 miles: find the height of the lamp.

6. Find in minutes and seconds the dip of the horizon from a hill 2640 feet high.

7. Along a straight coast there are lighthouses at intervals of 24 miles: find at what height the lamp must be placed so that the light of one at least may be visible at a distance of $3\frac{1}{2}$ miles from any point of the coast.

8. From the top of a mountain the dip of the horizon is $1\frac{9}{11}°$: find its height in feet.

9. The distance of the horizon as seen from the top of a hill is 30·25 miles: find the height of the hill and the dip of the horizon.

10. If x miles be the distance of the visible horizon and N degrees the dip, shew that

$$N = \frac{x}{66}\sqrt{\frac{10}{11}}.$$

When $\theta = 0$, find the limit of

11. $\dfrac{\sin 4\theta \cot \theta}{\operatorname{vers} 2\theta \cot^2 2\theta}$. 12. $\dfrac{1 - \cos\theta + \sin\theta}{1 - \cos\theta - \sin\theta}$.

13. When $\theta = a$, find the limit of

(1) $\dfrac{\sin\theta - \sin a}{\theta - a}$; (2) $\dfrac{\cos\theta - \cos a}{\theta - a}$.

14. Two sides of a triangle are 31 and 32, and they include a right angle: find the other angles.

15. A person walks directly towards a distant object P, and observes that at the three points A, B, C, the elevations of P are a, $2a$, $3a$ respectively: shew that $AB = 3BC$ nearly.

16. Shew that $\dfrac{\tan\theta}{\theta}$ continually increases from 1 to ∞ as θ continually increases from 0 to $\dfrac{\pi}{2}$.

CHAPTER XXII.

GEOMETRICAL PROOFS.

276. *To find the expansion of* $\tan(A+B)$ *geometrically.*

Let $\angle LOM = A$, and $\angle MON = B$; then $\angle LON = A + B$.

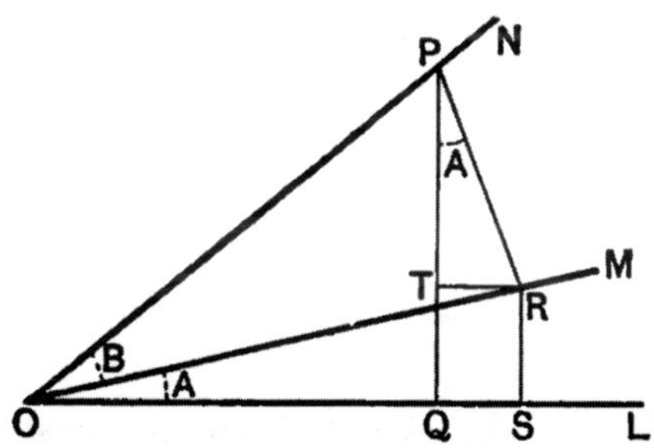

In ON take any point P, and draw PQ and PR perpendicular to OL and OM respectively. Also draw RS and RT perpendicular to OL and PQ respectively.

$$\tan(A+B) = \frac{PQ}{OQ} = \frac{RS+PT}{OS-TR}$$

$$= \frac{\dfrac{RS}{OS} + \dfrac{PT}{OS}}{1 - \dfrac{TR}{OS}} = \frac{\dfrac{RS}{OS} + \dfrac{PT}{OS}}{1 - \dfrac{TR}{TP} \cdot \dfrac{TP}{OS}}.$$

Now $\qquad \dfrac{RS}{OS} = \tan A, \text{ and } \dfrac{TR}{TP} = \tan A;$

also the triangles ROS and TPR are similar, and therefore

$$\frac{TP}{OS} = \frac{PR}{OR} = \tan B.$$

$$\therefore \tan(A+B) = \frac{\tan A + \tan B}{1 - \tan A \tan B}.$$

In like manner, with the help of the figure on page 95, we may obtain the expansion of $\tan(A-B)$ geometrically.

277. *To prove geometrically the formulæ for transformation of sums into products.*

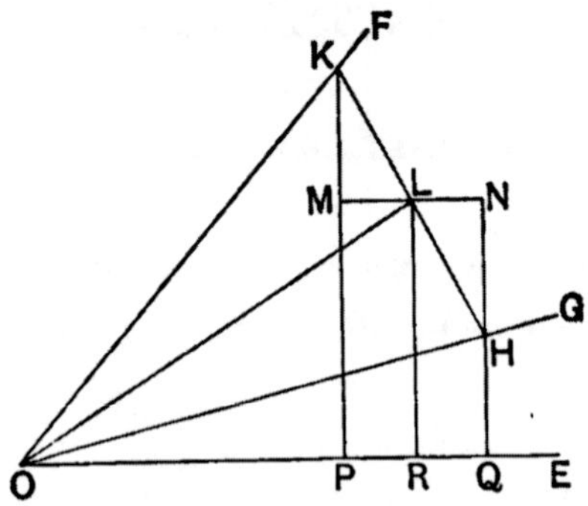

Let $\angle EOF$ be denoted by A, and $\angle EOG$ by B.

With centre O and any radius describe an arc of a circle meeting OG in H and OF in K.

Bisect $\angle KOH$ by OL; then OL bisects HK at right angles.

Draw KP, HQ, LR perpendicular to OE, and through L draw MLN parallel to OE meeting KP in M and QH in N.

It is easy to prove that the triangles MKL and NHL are equal in all respects, so that $KM=NH$, $ML=LN$, $PR=RQ$.

Also $\angle GOF=A-B$, and therefore

$$\angle HOL=\angle KOL=\frac{A-B}{2};$$

$$\therefore \quad \angle EOL=B+\frac{A-B}{2}=\frac{A+B}{2}.$$

$$\sin A+\sin B=\frac{KP}{OK}+\frac{HQ}{OH}=\frac{KP+HQ}{OK}$$

$$=\frac{(KM+LR)+(LR-NH)}{OK}=2\,\frac{LR}{OK};$$

$$\therefore \sin A + \sin B = 2\frac{LR}{OL}\cdot\frac{OL}{OK} = 2\sin ROL \cos KOL$$

$$= 2\sin\frac{A+B}{2}\cos\frac{A-B}{2}.$$

$$\cos A + \cos B = \frac{OP}{OK} + \frac{OQ}{OH} = \frac{OP+OQ}{OK}$$

$$= \frac{(OR-PR)+(OR+RQ)}{OK} = 2\frac{OR}{OK}$$

$$= 2\frac{OR}{OL}\cdot\frac{OL}{OK} = 2\cos ROL \cos KOL$$

$$= 2\cos\frac{A+B}{2}\cos\frac{A-B}{2}.$$

$$\sin A - \sin B = \frac{KP}{OK} - \frac{HQ}{OH} = \frac{KP-HQ}{OK}$$

$$= \frac{(KM+LR)-(LR-NH)}{OK} = 2\frac{KM}{OK}$$

$$= 2\frac{KM}{KL}\cdot\frac{KL}{OK} = 2\cos LKM \sin KOL$$

$$= 2\cos\frac{A+B}{2}\sin\frac{A-B}{2},$$

since $\angle LKM$ = comp^t of $\angle KLM = \angle MLO = \angle LOE = \dfrac{A+B}{2}$.

$$\cos B - \cos A = \frac{OQ}{OH} - \frac{OP}{OK} = \frac{OQ-OP}{OK}$$

$$= \frac{(OR+RQ)-(OR-PR)}{OK} = 2\frac{PR}{OK} = 2\frac{ML}{OK}$$

$$= 2\frac{ML}{KL}\cdot\frac{KL}{OK} = 2\sin LKM \sin KOL$$

$$= 2\sin\frac{A+B}{2}\sin\frac{A-B}{2}.$$

278. *Geometrical proof of the 2A formulæ.*

Let BPD be a semicircle, BD the diameter, C the centre.

On the circumference, take any point P, and join PB, PC, PD.

Draw PN perpendicular to BD.

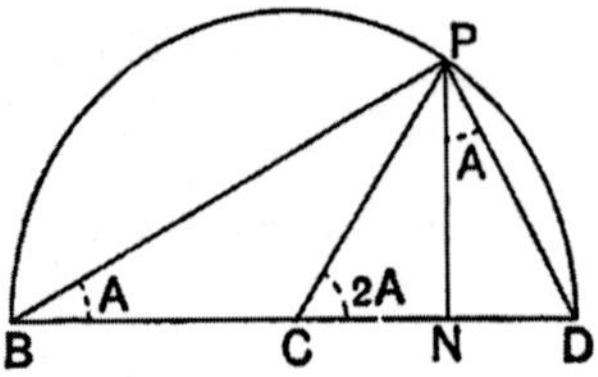

Let $\angle PBD = A$, then

$$\angle PCD = 2A.$$

And $\angle NPD = \text{comp}^{t}$ of $\angle PDN = \angle PBD = A$.

$$\sin 2A = \frac{PN}{CP} = \frac{2PN}{2CP} = \frac{2PN}{BD} = 2\frac{PN}{BP}\cdot\frac{BP}{BD}$$

$$= 2 \sin PBN \cos PBD$$

$$= 2 \sin A \cos A.$$

$$\cos 2A = \frac{CN}{CP} = \frac{2CN}{BD} = \frac{CN+CN}{BD}$$

$$= \frac{(BN-BC)+(CD-ND)}{BD} = \frac{BN-ND}{BD}$$

$$= \frac{BN}{BP}\cdot\frac{BP}{BD} - \frac{ND}{PD}\cdot\frac{PD}{BD}$$

$$= \cos A \,.\, \cos A - \sin A \,.\, \sin A$$

$$= \cos^2 A - \sin^2 A.$$

$$\cos 2A = \frac{CN}{CP} = \frac{CD-DN}{CP} = 1 - \frac{DN}{CP} = 1 - \frac{2DN}{BD}$$

$$= 1 - 2\frac{DN}{DP}\cdot\frac{DP}{BD} = 1 - 2\sin A \,.\, \sin A$$

$$= 1 - 2\sin^2 A.$$

$$\cos 2A = \frac{CN}{CP} = \frac{BN-BC}{CP} = \frac{BN}{CP} - 1 = \frac{2BN}{BD} - 1$$

$$= 2\frac{BN}{BP}\cdot\frac{BP}{BD} - 1 = 2\cos A \,.\, \cos A - 1$$

$$= 2\cos^2 A - 1.$$

$$\tan 2A = \frac{PN}{CN} = \frac{2PN}{2CN} = \frac{2PN}{BN - ND}$$

$$= \frac{2\frac{PN}{BN}}{1 - \frac{ND}{BN}} = \frac{2\frac{PN}{BN}}{1 - \frac{ND}{PN} \cdot \frac{PN}{BN}}$$

$$= \frac{2 \tan A}{1 - \tan A \,.\, \tan A}$$

$$= \frac{2 \tan A}{1 - \tan^2 A}.$$

279. *To find the value of* sin 18° *geometrically.*

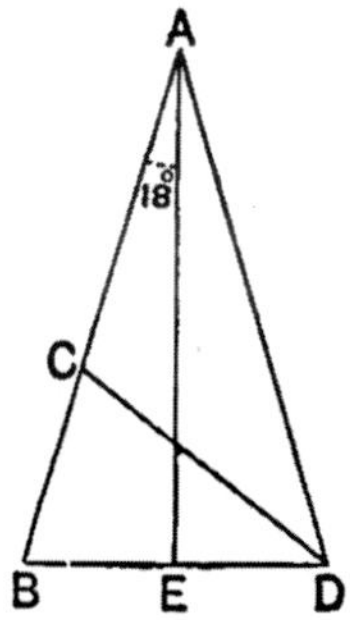

Let ABD be an isosceles triangle in which each angle at the base BD is double the vertical angle A; then

$$A + 2A + 2A = 180°,$$

and therefore $A = 36°$.

Bisect $\angle BAD$ by AE; then AE bisects BD at right angles;

$$\therefore \angle BAE = 18°.$$

Thus $$\sin 18° = \frac{BE}{AB} = \frac{x}{a},$$

where $AB = a$, and $BE = x$.

From the construction given in Euc. IV. 10,

$$AC = BD = 2BE = 2x,$$

and $$AB \,.\, BC = AC^2;$$

$$\therefore a(a - 2x) = (2x)^2;$$

$$\therefore 4x^2 + 2ax - a^2 = 0;$$

$$\therefore x = \frac{-2a \pm \sqrt{20a^2}}{8} = \frac{-1 \pm \sqrt{5}}{4} a.$$

The upper sign must be taken, since x is positive. Thus

$$\sin 18° = \frac{\sqrt{5} - 1}{4}.$$

Proofs by Projection.

280. Definition. If from any two points A and B, lines AC and BD are drawn perpendicular to OX, then the intercept CD is called the **projection** of AB upon OX.

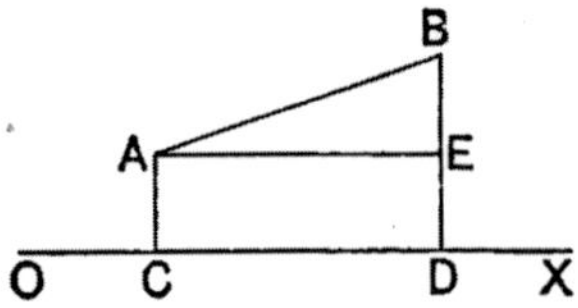

Through A draw AE parallel to OX; then

$$CD = AE = AB \cos BAE;$$

that is, $$CD = AB \cos \alpha,$$

where α is the angle of inclination of the lines AB and OX.

281. *To shew that the projection of a straight line is equal to the projection of an equal and parallel straight line drawn from a fixed point.*

Let AB be any straight line, O a fixed point, which we shall call the origin, OP a straight line equal and parallel to AB.

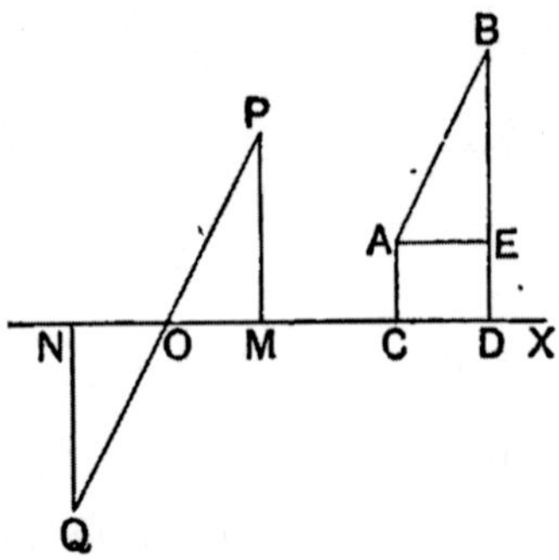

Let CD and OM be the projections of AB and OP upon any straight line OX drawn through the origin.

Draw AE parallel to OX.

The two triangles AEB and OMP are identically equal;

$$\therefore \; OM = AE = CD;$$

that is, projection of OP = projection of AB.

282. In the figure of the last article, *two* straight lines OP and OQ can be drawn from O equal and parallel to AB; it is therefore necessary to have some means of fixing the *direction* in which the line from O is to be drawn. Accordingly it is agreed to consider that

the direction of a line is fixed by the order of the letters.

Thus AB denotes a line drawn from A to B, and BA denotes a line drawn from B to A.

Hence OP denotes a line drawn from the origin parallel to AB, and OQ denotes a line drawn from the origin parallel to BA.

Similarly the direction of a projected line is fixed by the order of the letters.

Thus CD is drawn to the right from C to D and is positive, while DC is drawn to the left from D to C and is negative.

Hence in sign as well as in magnitude

$$OM = CD, \text{ and } ON = DC;$$

that is, projection of OP = projection of AB,

and projection of OQ = projection of BA.

Thus *the projection of a straight line can be represented both in sign and magnitude by the projection of an equal and parallel straight line drawn from the origin.*

283. Whatever be the direction of AB, the line OP will fall within one of the four quadrants.

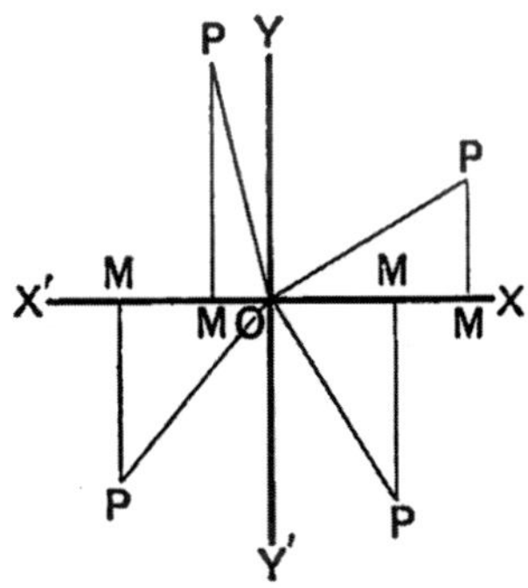

Also from the definitions given in Art. 75, we have

$$\frac{OM}{OP} = \cos XOP,$$

that is,

$$OM = OP \cos XOP,$$

whatever be the magnitude of the angle XOP. We shall always suppose, unless the contrary is stated, that the angles are measured in the positive direction.

284. Let O be the origin, P and Q any two points.

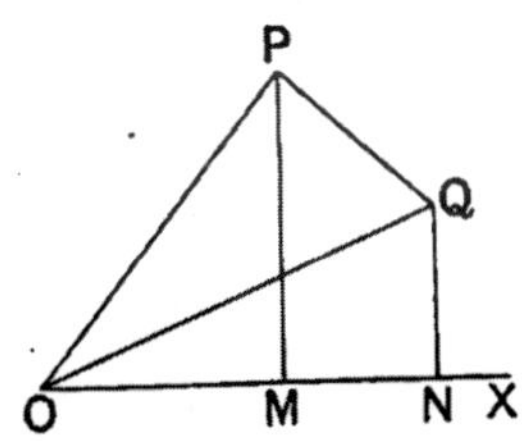

Join OP, OQ, PQ, and draw PM and QN perpendicular to OX.

We have

$$OM = ON + NM,$$

since the line NM is to be regarded as negative; that is,

the projection of OP = projection of OQ + projection of QP.

Hence, *the projection of one side of a triangle is equal to the sum of the projections of the other two sides taken in order.* Thus

projection of OQ = projection of OP + projection of PQ;

projection of QP = projection of QO + projection of OP.

General Proof of the Addition Formulæ.

285. In Fig. 1, let a line starting from OX revolve until it has traced the angle A, taking up the position OM, and then let it further revolve until it has traced the angle B, taking up the final position ON. Thus XON is the angle $A+B$.

Fig. 1.

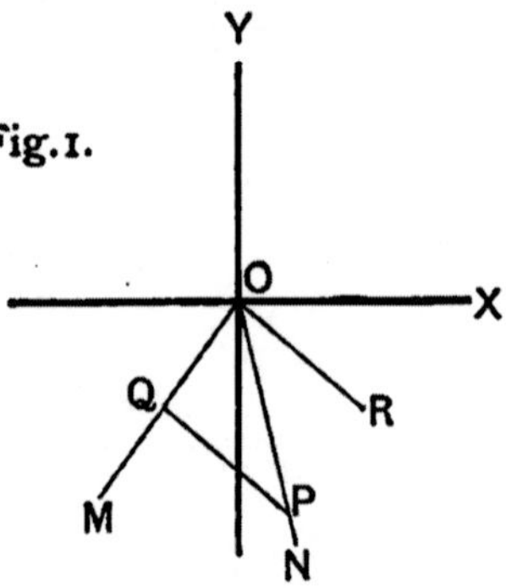

Fig. 2.

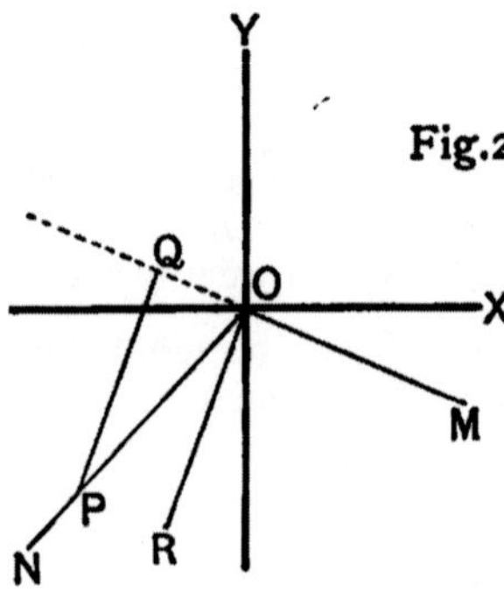

In ON take any point P, and draw PQ perpendicular to OM; also draw OR equal and parallel to QP.

Projecting upon OX, we have

projection of OP = projection of OQ + projection of QP

= projection of OQ + projection of OR.

$$\therefore \quad OP \cos XOP = OQ \cos XOQ + OR \cos XOR \quad \text{..........(1)}$$

$$= OP \cos B \cos XOQ + OP \sin B \cos XOR;$$

$$\therefore \quad \cos XOP = \cos B \cos XOQ + \sin B \cos XOR;$$

that is, $\quad \cos(A+B) = \cos B \cos A + \sin B \cos(90^\circ + A)$

$$= \cos A \cos B - \sin A \sin B.$$

Projecting upon OY, we have only to write Y for X in (1);

thus $OP\cos YOP = OQ\cos YOQ + OR\cos YOR$

$$= OP\cos B\cos YOQ + OP\sin B\cos YOR;$$

$$\therefore\ \cos YOP = \cos B\cos YOQ + \sin B\cos YOR;$$

that is,

$$\cos(A+B-90^\circ) = \cos B\cos(A-90^\circ) + \sin B\cos A;$$

$$\therefore\ \sin(A+B) = \sin A\cos B + \cos A\sin B.$$

In Fig. 2, let a line starting from OX revolve until it has traced the angle A, taking up the position OM, and then let it revolve *back again* until it has traced the angle B, taking up the final position ON. Thus XON is the angle $A-B$.

In ON take any point P, and draw PQ perpendicular to MO produced; also draw OR equal and parallel to QP.

Projecting upon OX, we have as in the previous case

$$OP\cos XOP = OQ\cos XOQ + OR\cos XOR$$

$$= OP\cos(180^\circ - B)\cos XOQ$$

$$+ OP\sin(180^\circ - B)\cos XOR;$$

$$\therefore\ \cos XOP = -\cos B\cos XOQ + \sin B\cos XOR;$$

that is,

$$\cos(A-B) = -\cos B\cos(A-180^\circ) + \sin B\cos(A-90^\circ)$$

$$= -\cos B(-\cos A) + \sin B\sin A$$

$$= \cos A\cos B + \sin A\sin B.$$

Projecting upon OY, we have

$$OP\cos YOP = OQ\cos YOQ + OR\cos YOR;$$

$$= OP\cos(180^\circ - B)\cos YOQ$$

$$+ OP\sin(180^\circ - B)\cos YOR;$$

$$\therefore\ \cos YOP = -\cos B\cos YOQ + \sin B\cos YOR;$$

that is,

$$\cos(A-B-90^\circ) = -\cos B\cos(A-270^\circ) + \sin B\cos(A-180^\circ);$$

$$\therefore\ \sin(A-B) = -\cos B(-\sin A) + \sin B(-\cos A)$$

$$= \sin A\cos B - \cos A\sin B.$$

286. The above method of proof is applicable to every case, and therefore the Addition Formulæ are universally established.

The universal truth of the Addition Formulæ may also be deduced from the special geometrical investigations of Arts. 110 and 111 by analysis, as in the next article.

287. When each of the angles A, B, $A+B$ is less than $90°$, we have shewn that

$$\cos(A+B)=\cos A\cos B-\sin A\sin B\text{............}(1).$$

But $\cos(A+B)=\sin(\overline{A+B}+90°)=\sin(\overline{A+90°}+B)$;

also $$\cos A=\sin(A+90°),$$

and $$-\sin A=\cos(A+90°).\quad\text{[Art. 98.]}$$

Hence by substitution in (1), we have

$$\sin(\overline{A+90°}+B)=\sin(A+90°)\cos B+\cos(A+90°)\sin B.$$

In like manner, it may be proved that

$$\cos(\overline{A+90°}+B)=\cos(A+90°)\cos B-\sin(A+90°)\sin B.$$

Thus the formulæ for the sine and cosine of $A+B$ hold when A is increased by $90°$. Similarly we may shew that they hold when B is increased by $90°$.

By repeated applications of the same process it may be proved that the formulæ are true when either or both of the angles A and B is increased by any multiple of $90°$.

Again, $$\cos(A+B)=\cos A\cos B-\sin A\sin B\text{............}(1).$$

But $\cos(A+B)=-\sin(\overline{A+B}-90°)=-\sin(\overline{A-90°}+B)$;

also $$\cos A=-\sin(A-90°),$$

and $$\sin A=\cos(A-90°).\quad\text{[Arts. 99 and 102.]}$$

Hence by substitution in (1), we have

$$\sin(\overline{A-90°}+B)=\sin(A-90°)\cos B+\cos(A-90°)\sin B.$$

Similarly we may shew that

$$\cos(\overline{A-90°}+B)=\cos(A-90°)\cos B-\sin(A-90°)\sin B.$$

Thus the formulæ for the sine and cosine of $A+B$ hold when A is diminished by $90°$. In like manner we may prove that they are true when B is diminished by $90°$.

By repeated applications of the same process it may be shewn that the formulæ hold when either or both of the angles A and B is diminished by any multiple of $90°$. Further, it will be seen that the formulæ are true if either of the angles A or B is increased by a multiple of $90°$ and the other is diminished by a multiple of $90°$.

Thus $\sin(P+Q)=\sin P\cos Q+\cos P\sin Q$,

and $\cos(P+Q)=\cos P\cos Q-\sin P\sin Q$,

where $P=A\pm m.90°$, and $Q=B\pm n.90°$,

m and n being any positive integers, and A and B any acute angles.

Thus the Addition Formulæ are true for the algebraical sum of any two angles.

MISCELLANEOUS EXAMPLES. H.

1. If the sides of a right-angled triangle are

$$\cos 2\alpha+\cos 2\beta+2\cos(\alpha+\beta) \text{ and } \sin 2\alpha+\sin 2\beta+2\sin(\alpha+\beta),$$

shew that the hypotenuse is $4\cos^2\dfrac{\alpha-\beta}{2}$.

2. If the in-centre and circum-centre be at equal distances from BC, prove that

$$\cos B+\cos C=1.$$

3. The shadow of a tower is observed to be half the known height of the tower, and some time afterwards to be equal to the height: how much will the sun have gone down in the interval? Given log 2,

$$L\tan 63°\,26'=10{\cdot}3009994, \text{ diff. for } 1'=3159.$$

4. If $(1+\sin\alpha)(1+\sin\beta)(1+\sin\gamma)$

$$=(1-\sin\alpha)(1-\sin\beta)(1-\sin\gamma),$$

shew that each expression is equal to $\pm\cos\alpha\cos\beta\cos\gamma$.

5. Two parallel chords of a circle lying on the same side of the centre subtend $72°$ and $144°$ at the centre: prove that the distance between them is one-half of the radius.

Also shew that the sum of the squares of the chords is equal to five times the square of the radius.

6. Two straight railways are inclined at an angle of 60°. From their point of intersection two trains P and Q start at the same time, one along each line. P travels at the rate of 48 miles per hour, at what rate must Q travel so that after one hour they shall be 43 miles apart?

7. If
$$\alpha = \cos^{-1}\frac{x}{a} + \cos^{-1}\frac{y}{b},$$
shew that
$$\sin^2\alpha = \frac{x^2}{a^2} - \frac{2xy}{ab}\cos\alpha + \frac{y^2}{b^2}.$$

8. If p, q, r denote the sides of the ex-central triangle, prove that
$$\frac{a^2}{p^2} + \frac{b^2}{q^2} + \frac{c^2}{r^2} + \frac{2abc}{pqr} = 1.$$

9. A tower is situated within the angle formed by two straight roads OA and OB, and subtends angles α and β at the points A and B where the roads are nearest to it. If $OA = a$, and $OB = b$, shew that the height of the tower is
$$\sqrt{a^2 - b^2}\sin\alpha\sin\beta \,/\, \sqrt{\sin(\alpha+\beta)\sin(\alpha-\beta)}.$$

10. In a triangle, shew that
$$r^2 + r_1^2 + r_2^2 + r_3^2 = 16R^2 - a^2 - b^2 - c^2.$$

11. If AD be a median of the triangle ABC, shew that

(1) $\cot BAD = 2\cot A + \cot B$;

(2) $2\cot ADC = \cot B - \cot C$.

12. If p, q, r are the distances of the orthocentre from the sides, prove that
$$4\left(\frac{a}{p} + \frac{b}{q} + \frac{c}{r}\right) = \left(\frac{a}{p} + \frac{b}{q} - \frac{c}{r}\right)\left(\frac{b}{q} + \frac{c}{r} - \frac{a}{p}\right)\left(\frac{c}{r} + \frac{a}{p} - \frac{b}{q}\right).$$

Graphical Representation of the Circular Functions.

288. Definition. Let $f(x)$ be a function of x which has a single value for all values of x, and let the values of x be represented by lines measured from O along OX or OX', and the values of $f(x)$ by lines drawn perpendicular to XX'. Then with the figure of the next article, if OM represent any value of x, and MP the corresponding value of $f(x)$, the curve traced out by the point P is called the **Graph** of $f(x)$.

Graphs of sin θ and cos θ.

289. Suppose that the unit of length is chosen to represent a radian; then any angle of θ radians will be represented by a line OM which contains θ units of length.

Graph of $\sin\theta$.

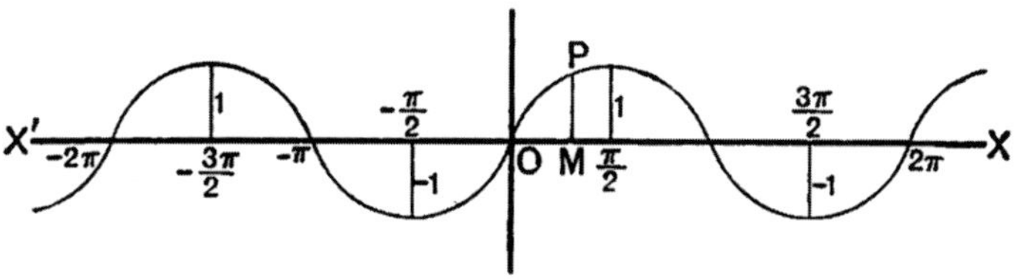

Let MP, drawn perpendicular to OX, represent the value of $\sin\theta$ corresponding to the value OM of θ; then the curve traced out by the point P represents the graph of $\sin\theta$.

As OM or θ increases from 0 to $\frac{\pi}{2}$, MP or $\sin\theta$ increases from 0 to 1, which is its greatest value.

As OM increases from $\frac{\pi}{2}$ to π, MP decreases from 0 to 1.

As OM increases from π to $\frac{3\pi}{2}$, MP increases numerically from 0 to -1.

As OM increases from $\frac{3\pi}{2}$ to 2π, MP decreases numerically from -1 to 0.

As OM increases from 2π to 4π, from 4π to 6π, from 6π to 8π, ..., MP passes through the same series of values as when OM increases from 0 to 2π.

Since $\sin(-\theta) = -\sin\theta$, the values of MP lying to the left of O are equal in magnitude but are of opposite sign to values of MP lying at an equal distance to the right of O.

Thus the graph of $\sin\theta$ is a *continuous* waving line extending to an infinite distance on each side of O.

The graph of $\cos\theta$ is the same as that of $\sin\theta$, the origin being at the point marked $\frac{\pi}{2}$ in the figure.

Graphs of tan θ and cot θ.

290. As before, suppose that the unit of length is chosen to represent a radian; then any angle of θ radians will be represented by a line OM which contains θ units of length.

Let MP, drawn perpendicular to OX, represent the value of $\tan\theta$ corresponding to the value OM of θ; then the curve traced out by the point P represents the graph of $\tan\theta$.

By tracing the changes in the value of $\tan\theta$ as θ varies from 0 to 2π, from 2π to 4π,......, it will be seen that the graph of $\tan\theta$ consists of an infinite number of *discontinuous* equal branches as represented in the figure below. The part of each branch beneath XX' is convex towards XX', and the part of each branch above XX' is also convex towards XX'; hence at the point where any branch cuts XX' there is what is called a *point of inflexion*, where the direction of curvature changes. The proof of these statements is however beyond the range of the present work.

The various branches touch the dotted lines passing through the points marked

$$\pm\frac{\pi}{2}, \quad \pm\frac{3\pi}{2}, \quad \pm\frac{5}{2}, \ldots\ldots,$$

at an infinite distance from XX'.

Graph of tan θ.

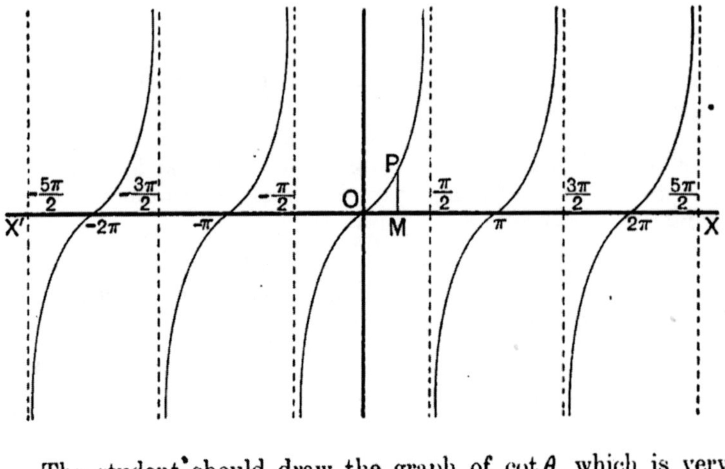

The student should draw the graph of $\cot\theta$, which is very similar to that of $\tan\theta$.

Graphs of sec θ and cosec θ.

291. The graph of $\sec\theta$ is represented in the figure below. It consists of an infinite number of equal festoons lying alternately above and below XX', the vertex of each being at the unit of distance from XX'. The various festoons touch the dotted lines passing through the points marked

$$\pm\frac{\pi}{2},\quad \pm\frac{3\pi}{2},\quad \pm\frac{5\pi}{2},\ldots\ldots,$$

at an infinite distance from XX'.

Graph of sec θ.

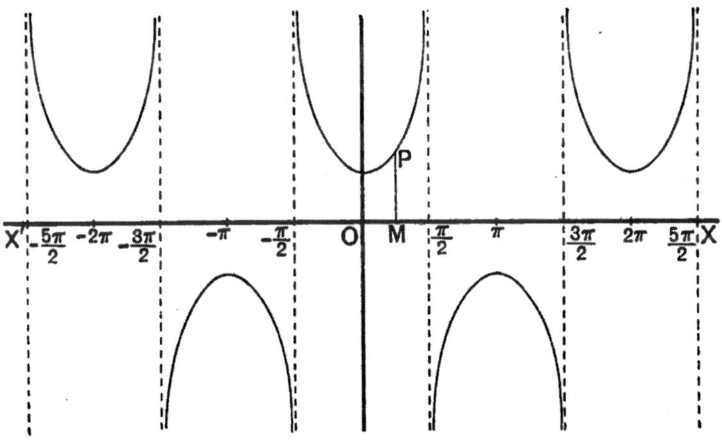

The graph of $\operatorname{cosec}\theta$ is the same as that of $\sec\theta$, the origin being at the point marked $-\frac{\pi}{2}$ in the figure.

CHAPTER XXIII.

SUMMATION OF FINITE SERIES.

292. An expression in which the successive terms are formed by some regular law is called a **series.** If the series ends at some assigned term it is called a **finite series** ; if the number of terms is unlimited it is called an **infinite series.**

A series may be denoted by an expression of the form

$$u_1+u_2+u_3+\ldots+u_{n-1}+u_n+u_{n+1}+\ldots,$$

where u_{n+1}, the $(n+1)^{\text{th}}$ term, is obtained from u_n, the nth term, by replacing n by $n+1$.

Thus if $u_n=\cos(a+n\beta)$, then $u_{n+1}=\cos\{a+(n+1)\beta\}$;

and if $\qquad u_n=\cot 2^{n-1}a$, then $u_{n+1}=\cot 2^n a$.

293. If the rth term of a series can be expressed as the difference of two quantities one of which is the same function of r that the other is of $r+1$, the sum of the series may be readily found.

For let the series be denoted by

$$u_1+u_2+u_3+\ldots\ldots+u_n,$$

and its sum by S, and suppose that any term

$$u_r=v_{r+1}-v_r;$$

then $S=(v_2-v_1)+(v_3-v_2)+(v_4-v_3)+\ldots+(v_n-v_{n-1})+(v_{n+1}-v_n)$
$=v_{n+1}-v_1$.

Example. Find the sum of the series

$$\operatorname{cosec} a+\operatorname{cosec} 2a+\operatorname{cosec} 4a+\ldots\ldots+\operatorname{cosec} 2^{n-1}a.$$

$$\operatorname{cosec} a=\frac{1}{\sin a}=\frac{\sin\frac{a}{2}}{\sin\frac{a}{2}\sin a}=\frac{\sin\left(a-\frac{a}{2}\right)}{\sin\frac{a}{2}\sin a}.$$

Hence $$\operatorname{cosec} a = \cot\frac{a}{2} - \cot a.$$

If we replace a by $2a$, we obtain
$$\operatorname{cosec} 2a = \cot a - \cot 2a.$$

Similarly, $$\operatorname{cosec} 4a = \cot 2a - \cot 4a,$$
....
$$\operatorname{cosec} 2^{n-1} a = \cot 2^{n-2} a - \cot 2^{n-1} a.$$

By addition, $$S = \cot\frac{a}{2} - \cot 2^{n-1} a.$$

294. *To find the sum of the sines of a series of* n *angles which are in arithmetical progression.*

Let the sine-series be denoted by
$$\sin a + \sin(a+\beta) + \sin(a+2\beta) + \ldots\ldots + \sin\{a+(n-1)\beta\}.$$

We have the identities
$$2\sin a \sin\frac{\beta}{2} = \cos\left(a-\frac{\beta}{2}\right) - \cos\left(a+\frac{\beta}{2}\right),$$
$$2\sin(a+\beta)\sin\frac{\beta}{2} = \cos\left(a+\frac{\beta}{2}\right) - \cos\left(a+\frac{3\beta}{2}\right),$$
$$2\sin(a+2\beta)\sin\frac{\beta}{2} = \cos\left(a+\frac{3\beta}{2}\right) - \cos\left(a+\frac{5\beta}{2}\right),$$
....
$$2\sin\{a+(n-1)\beta\}\sin\frac{\beta}{2} = \cos\left(a+\frac{2n-3}{2}\beta\right) - \cos\left(a+\frac{2n-1}{2}\beta\right).$$

By addition,
$$2S\sin\frac{\beta}{2} = \cos\left(a-\frac{\beta}{2}\right) - \cos\left(a+\frac{2n-1}{2}\beta\right)$$
$$= 2\sin\left(a+\frac{n-1}{2}\beta\right)\sin\frac{n\beta}{2};$$
$$\therefore\ S = \frac{\sin\frac{n\beta}{2}}{\sin\frac{\beta}{2}}\sin\left(a+\frac{n-1}{2}\beta\right).$$

295. In like manner we may shew that the sum of the cosine-series

$$\cos \alpha + \cos(\alpha+\beta) + \cos(\alpha+2\beta) + \ldots + \cos\{\alpha+(n-1)\beta\}$$

$$= \frac{\sin \frac{n\beta}{2}}{\sin \frac{\beta}{2}} \cos\left(\alpha + \frac{n-1}{2}\beta\right).$$

296. The formulæ of the two last articles may be expressed verbally as follows.

The sum of the sines of a series of n *angles in* A.P.

$$= \frac{\sin \frac{\text{n } diff.}{2}}{\sin \frac{diff.}{2}} \sin \frac{first\ angle + last\ angle}{2}.$$

The sum of the cosines of a series of n *angles in* A.P.

$$= \frac{\sin \frac{\text{n } diff.}{2}}{\sin \frac{diff.}{2}} \cos \frac{first\ angle + last\ angle}{2}.$$

Example. Find the sum of the series

$$\cos a + \cos 3a + \cos 5a + \ldots\ldots + \cos(2n-1)a.$$

Here the common difference of the angles is $2a$;

$$\therefore S = \frac{\sin na}{\sin a} \cos \frac{a + (2n-1)a}{2}$$

$$= \frac{\sin na \cos na}{\sin a} = \frac{\sin 2na}{2 \sin a}.$$

297. If $\sin \frac{n\beta}{2} = 0$, each of the expressions found in Arts. 294 and 295 for the sum vanishes. In this case

$$\frac{n\beta}{2} = k\pi, \quad \text{or} \quad \beta = \frac{2k\pi}{n}, \text{ where } k \text{ is any integer.}$$

Hence *the sum of the sines and the sum of the cosines of* n *angles in arithmetical progression are each equal to zero, when the common difference of the angles is an even multiple of* $\frac{\pi}{n}$.

298. Some series may be brought under the rule of Art. 296 by a simple transformation.

Example 1. Find the sum of n terms of the series

$$\cos\alpha - \cos(\alpha+\beta) + \cos(\alpha+2\beta) - \cos(\alpha+3\beta) + \ldots\ldots .$$

This series is equal to

$$\cos\alpha + \cos(\alpha+\beta+\pi) + \cos(\alpha+2\beta+2\pi) + \cos(\alpha+3\beta+3\pi) + \ldots\ldots,$$

a series in which the common difference of the angles is $\beta+\pi$, and the last angle is $\alpha+(n-1)(\beta+\pi)$.

$$\therefore S = \frac{\sin\dfrac{n(\beta+\pi)}{2}}{\sin\dfrac{\beta+\pi}{2}} \cos\left\{\alpha + \frac{(n-1)(\beta+\pi)}{2}\right\}.$$

Example 2. Find the sum of n terms of the series

$$\sin\alpha + \cos(\alpha+\beta) - \sin(\alpha+2\beta) - \cos(\alpha+3\beta) + \sin(\alpha+4\beta) + \ldots\ldots .$$

This series is equal to

$$\sin\alpha + \sin\left(\alpha+\beta+\frac{\pi}{2}\right) + \sin(\alpha+2\beta+\pi) + \sin\left(\alpha+3\beta+\frac{3\pi}{2}\right) + \ldots\ldots,$$

a series in which the common difference of the angles is $\beta + \dfrac{\pi}{2}$.

$$\therefore S = \frac{\sin\dfrac{n(2\beta+\pi)}{4}}{\sin\dfrac{2\beta+\pi}{4}} \sin\left\{\alpha + \frac{(n-1)(2\beta+\pi)}{4}\right\}.$$

EXAMPLES. XXIII. a.

Sum each of the following series to n terms:

1. $\sin\alpha + \sin 3\alpha + \sin 5\alpha + \ldots\ldots .$

2. $\cos\alpha + \cos(\alpha-\beta) + \cos(\alpha-2\beta) + \ldots\ldots .$

3. $\sin\alpha + \sin\left(\alpha - \dfrac{\pi}{n}\right) + \sin\left(\alpha - \dfrac{2\pi}{n}\right) + \ldots\ldots .$

4. $\cos\dfrac{\pi}{k} + \cos\dfrac{2\pi}{k} + \cos\dfrac{3\pi}{k} + \ldots\ldots .$

Find the sum of each of the following series :

5. $\cos\frac{\pi}{19}+\cos\frac{3\pi}{19}+\cos\frac{5\pi}{19}+\ldots\ldots+\cos\frac{17\pi}{19}$.

6. $\cos\frac{2\pi}{21}+\cos\frac{4\pi}{21}+\cos\frac{6\pi}{21}+\ldots\ldots+\cos\frac{20\pi}{21}$.

7. $\sin\frac{\pi}{n}+\sin\frac{2\pi}{n}+\sin\frac{3\pi}{n}+\ldots\ldots$ to $n-1$ terms.

8. $\cos\frac{\pi}{n}+\cos\frac{3\pi}{n}+\cos\frac{5\pi}{n}+\ldots\ldots$ to $2n-1$ terms.

9. $\sin n\alpha+\sin(n-1)\alpha+\sin(n-2)\alpha+\ldots\ldots$ to $2n$ terms.

Sum each of the following series to n terms :

10. $\sin\theta-\sin 2\theta+\sin 3\theta-\sin 4\theta+\ldots\ldots$.

11. $\cos\alpha-\cos(\alpha-\beta)+\cos(\alpha-2\beta)-\cos(\alpha-3\beta)+\ldots\ldots$.

12. $\cos\alpha-\sin(\alpha-\beta)-\cos(\alpha-2\beta)+\sin(\alpha-3\beta)+\ldots\ldots$.

13. $\sin 2\theta\sin\theta+\sin 3\theta\sin 2\theta+\sin 4\theta\sin 3\theta+\ldots\ldots$.

14. $\sin\alpha\cos 3\alpha+\sin 3\alpha\cos 5\alpha+\sin 5\alpha\cos 7\alpha+\ldots\ldots$.

15. $\sec\alpha\sec 2\alpha+\sec 2\alpha\sec 3\alpha+\sec 3\alpha\sec 4\alpha+\ldots\ldots$.

16. $\operatorname{cosec}\theta\operatorname{cosec}3\theta+\operatorname{cosec}3\theta\operatorname{cosec}5\theta$
$+\operatorname{cosec}5\theta\operatorname{cosec}7\theta+\ldots\ldots$.

17. $\tan\frac{\alpha}{2}\sec\alpha+\tan\frac{\alpha}{2^2}\sec\frac{\alpha}{2}+\tan\frac{\alpha}{2^3}\sec\frac{\alpha}{2^2}+\ldots\ldots$.

18. $\cos 2\alpha\operatorname{cosec}3\alpha+\cos 6\alpha\operatorname{cosec}9\alpha+\cos 18\alpha\operatorname{cosec}27\alpha+\ldots\ldots$.

19. $\sin\alpha\sec 3\alpha+\sin 3\alpha\sec 9\alpha+\sin 9\alpha\sec 27\alpha+\ldots\ldots$.

20. The circumference of a semicircle of radius a is divided into n equal arcs. Shew that the sum of the distances of the several points of section from either extremity of the diameter is

$$a\left(\cot\frac{\pi}{4n}-1\right).$$

21. From the angular points of a regular polygon, perpendiculars are drawn to XX' and YY' the horizontal and vertical diameter of the circumscribing circle: shew that the algebraical sums of each of the two sets of perpendiculars are equal to zero.

299. By means of the identities

$$2\sin^2 a = 1 - \cos 2a, \qquad 2\cos^2 a = 1 + \cos 2a,$$

$$4\sin^3 a = 3\sin a - \sin 3a, \qquad 4\cos^3 a = 3\cos a + \cos 3a,$$

we can find the sum of the squares and cubes of the sines and cosines of a series of angles in arithmetical progression.

Example 1. Find the sum of n terms of the series

$$\sin^2 a + \sin^2(a+\beta) + \sin^2(a+2\beta) + \ldots\ldots.$$

$$2S = \{1 - \cos 2a\} + \{1 - \cos(2a+2\beta)\} + \{1 - \cos(2a+4\beta)\} + \ldots\ldots$$

$$= n - \{\cos 2a + \cos(2a+2\beta) + \cos(2a+4\beta) + \ldots\ldots\};$$

$$= n - \frac{\sin n\beta}{\sin\beta}\cos\frac{2a + \{2a + (n-1)\,2\beta\}}{2};$$

$$\therefore\ S = \frac{n}{2} - \frac{\sin n\beta}{2\sin\beta}\cos\{2a + (n-1)\beta\}.$$

Example 2. Find the sum of the series

$$\cos^3 a + \cos^3 3a + \cos^3 5a + \ldots\ldots + \cos^3(2n-1)\,a.$$

$$4S = (3\cos a + \cos 3a) + (3\cos 3a + \cos 9a) + (3\cos 5a + \cos 15a) + \ldots\ldots$$

$$= 3(\cos a + \cos 3a + \cos 5a + \ldots\ldots) + (\cos 3a + \cos 9a + \cos 15a + \ldots\ldots)$$

$$= \frac{3\sin na}{\sin a}\cos\left\{\frac{a + (2n-1)\,a}{2}\right\} + \frac{\sin 3na}{\sin 3a}\cos\left\{\frac{3a + (2n-1)\,3a}{2}\right\};$$

$$\therefore\ S = \frac{3\sin na\cos na}{4\sin a} + \frac{\sin 3na\cos 3na}{4\sin 3a}.$$

300. The following further examples illustrate the principle of Art. 293.

Example 1. Find the sum of the series

$$\tan^{-1}\frac{x}{1+1\,.\,2\,.\,x^2} + \tan^{-1}\frac{x}{1+2\,.\,3\,.\,x^2} + \ldots\ldots + \tan^{-1}\frac{x}{1+n(n+1)\,x^2}.$$

As in Art. 249, we have

$$\tan^{-1}\frac{x}{1+r(r+1)\,x^2} = \tan^{-1}(r+1)\,x - \tan^{-1}rx;$$

$$\therefore\ S = \tan^{-1}(n+1)\,x - \tan^{-1}x.$$

Example 2. Find the sum of n terms of the series

$$\tan\alpha+\frac{1}{2}\tan\frac{\alpha}{2}+\frac{1}{2^2}\tan\frac{\alpha}{2^2}+\frac{1}{2^3}\tan\frac{\alpha}{2^3}+\dots\dots.$$

We have $$\tan\alpha=\cot\alpha-2\cot2\alpha.$$

Replacing α by $\frac{\alpha}{2}$ and dividing by 2, we obtain

$$\frac{1}{2}\tan\frac{\alpha}{2}=\frac{1}{2}\cot\frac{\alpha}{2}-\cot\alpha.$$

Similarly, $$\frac{1}{2^2}\tan\frac{\alpha}{2^2}=\frac{1}{2^2}\cot\frac{\alpha}{2^2}-\frac{1}{2}\cot\frac{\alpha}{2};$$

....

$$\frac{1}{2^{n-1}}\tan\frac{\alpha}{2^{n-1}}=\frac{1}{2^{n-1}}\cot\frac{\alpha}{2^{n-1}}-\frac{1}{2^{n-2}}\cot\frac{\alpha}{2^{n-2}}.$$

By addition, $$S=\frac{1}{2^{n-1}}\cot\frac{\alpha}{2^{n-1}}-2\cot2\alpha.$$

EXAMPLES. XXIII. b.

Sum each of the following series to n terms:

1. $\cos^2\theta+\cos^2 3\theta+\cos^2 5\theta+\dots\dots.$

2. $\sin^2\alpha+\sin^2\left(\alpha+\frac{\pi}{n}\right)+\sin^2\left(\alpha+\frac{2\pi}{n}\right)+\dots\dots.$

3. $\cos^2\alpha+\cos^2\left(\alpha-\frac{\pi}{n}\right)+\cos^2\left(\alpha-\frac{2\pi}{n}\right)+\dots\dots.$

4. $\sin^3\theta+\sin^3 2\theta+\sin^3 3\theta+\dots\dots.$

5. $\sin^3\alpha+\sin^3\left(\alpha+\frac{2\pi}{n}\right)+\sin^3\left(\alpha+\frac{4\pi}{n}\right)+\dots\dots.$

6. $\cos^3\alpha+\cos^3\left(\alpha-\frac{2\pi}{n}\right)+\cos^3\left(\alpha-\frac{4\pi}{n}\right)+\dots\dots.$

7. $\tan\theta+2\tan2\theta+2^2\tan2^2\theta+\dots\dots$

8. $\frac{1}{\cos\alpha+\cos3\alpha}+\frac{1}{\cos\alpha+\cos5\alpha}+\frac{1}{\cos\alpha+\cos7\alpha}+\dots\dots.$

9. $\sin^2\theta \sin 2\theta + \frac{1}{2}\sin^2 2\theta \sin 4\theta + \frac{1}{4}\sin^2 4\theta \sin 8\theta + \ldots\ldots$

10. $2\cos\theta \sin^2\frac{\theta}{2} + 2^2\cos\frac{\theta}{2}\sin^2\frac{\theta}{2^2} + 2^3\cos\frac{\theta}{2^2}\sin^2\frac{\theta}{2^3} + \ldots\ldots$

11. $\tan^{-1}\frac{x}{1.2+x^2} + \tan^{-1}\frac{x}{2.3+x^2} + \tan^{-1}\frac{x}{3.4+x^2} + \ldots\ldots$

12. $\tan^{-1}\frac{1}{1+1+1^2} + \tan^{-1}\frac{1}{1+2+2^2} + \tan^{-1}\frac{1}{1+3+3^2} + \ldots\ldots$

13. $\tan^{-1}\frac{2}{2+1^2+1^4} + \tan^{-1}\frac{4}{2+2^2+2^4} + \tan^{-1}\frac{6}{2+3^2+3^4} + \ldots$

14. $\tan^{-1}\frac{2}{1-1^2+1^4} + \tan^{-1}\frac{4}{1-2^2+2^4} + \tan^{-1}\frac{6}{1-3^2+3^4} + \ldots$

15. From any point on the circumference of a circle of radius r, chords are drawn to the angular points of the regular inscribed polygon of n sides: shew that the sum of the squares of the chords is $2nr^2$.

16. From a point P within a regular polygon of $2n$ sides, perpendiculars $PA_1, PA_2, PA_3, \ldots PA_{2n}$ are drawn to the sides: shew that

$$PA_1 + PA_3 + \ldots + PA_{2n-1} = PA_2 + PA_4 + \ldots + PA_{2n} = nr,$$

where r is the radius of the inscribed circle.

17. If $A_1A_2A_3\ldots A_{2n+1}$ is a regular polygon and P a point on the circumscribed circle lying on the arc A_1A_{2n+1}, shew that

$$PA_1 + PA_3 + \ldots + PA_{2n+1} = PA_2 + PA_4 + \ldots + PA_{2n}.$$

18. From any point on the circumference of a circle, perpendiculars are drawn to the sides of the regular circumscribing polygon of n sides: shew that

(1) the sum of the squares of the perpendiculars is $\frac{3nr^2}{2}$;

(2) the sum of the cubes of the perpendiculars is $\frac{5nr^3}{2}$.

CHAPTER XXIV.

MISCELLANEOUS TRANSFORMATIONS AND IDENTITIES.

Symmetrical Expressions.

301. An expression is said to be *symmetrical* with respect to certain of the letters it contains, if the value of the expression remains unaltered when any pair of these letters are interchanged. Thus

$$\cos\alpha+\cos\beta+\cos\gamma, \qquad \sin\alpha\sin\beta\sin\gamma,$$
$$\tan(\alpha-\theta)+\tan(\beta-\theta)+\tan(\gamma-\theta),$$

are expressions which are symmetrical with respect to the letters α, β, γ.

302. A symmetrical expression involving the *sum* of a number of quantities may be concisely denoted by writing down one of the *terms* and prefixing the symbol Σ. Thus $\Sigma\cos\alpha$ stands for the sum of all the terms of which $\cos\alpha$ is the type, $\Sigma\sin\alpha\sin\beta$ stands for the sum of all the terms of which $\sin\alpha\sin\beta$ is the type; and so on.

For instance, if the expression is symmetrical with respect to the three letters α, β, γ,

$$\Sigma\cos\beta\cos\gamma=\cos\beta\cos\gamma+\cos\gamma\cos\alpha+\cos\alpha\cos\beta;$$
$$\Sigma\sin(\alpha-\theta)=\sin(\alpha-\theta)+\sin(\beta-\theta)+\sin(\gamma-\theta).$$

303. A symmetrical expression involving the *product* of a number of quantities may be denoted by writing down one of the *factors* and prefixing the symbol Π. Thus $\Pi\sin\alpha$ stands for the product of all the factors of which $\sin\alpha$ is the type.

For instance, if the expression is symmetrical with respect to the three letters α, β, γ,

$$\Pi\tan(\alpha+\theta)=\tan(\alpha+\theta)\tan(\beta+\theta)\tan(\gamma+\theta);$$
$$\Pi(\cos\beta+\cos\gamma)=(\cos\beta+\cos\gamma)(\cos\gamma+\cos\alpha)(\cos\alpha+\cos\beta).$$

304. With the notation just explained, certain theorems in Chap. XII. involving the three angles A, B, C, which are connected by the relation $A+B+C=180°$, may be written more concisely. For instance

$$\Sigma \sin 2A = 4\Pi \sin A;$$

$$\Sigma \sin A = 4\Pi \cos \frac{A}{2};$$

$$\Sigma \tan A = \Pi \tan A;$$

$$\Sigma \tan \frac{B}{2} \tan \frac{C}{2} = 1.$$

Example 1. Find the ratios of $a : b : c$ from the equations

$$a \cos \theta + b \sin \theta = c \quad \text{and} \quad a \cos \phi + b \sin \phi = c.$$

From the given equations, we have

$$a \cos \theta + b \sin \theta - c = 0,$$

and $$a \cos \phi + b \sin \phi - c = 0;$$

whence by *cross multiplication*

$$\frac{a}{\sin \phi - \sin \theta} = \frac{b}{\cos \theta - \cos \phi} = \frac{c}{\sin \phi \cos \theta - \cos \phi \sin \theta};$$

$$\therefore \frac{a}{2 \cos \frac{\phi+\theta}{2} \sin \frac{\phi-\theta}{2}} = \frac{b}{2 \sin \frac{\phi+\theta}{2} \sin \frac{\phi-\theta}{2}} = \frac{c}{\sin (\phi - \theta)}.$$

Dividing each denominator by $2 \sin \frac{\phi - \theta}{2}$, we have

$$\frac{a}{\cos \frac{\theta+\phi}{2}} = \frac{b}{\sin \frac{\theta+\phi}{2}} = \frac{c}{\cos \frac{\theta-\phi}{2}}.$$

NOTE. This result is important in Analytical Geometry.

It should be remarked that $\cos (\theta - \phi)$ is a symmetrical function of θ and ϕ, for $\cos (\theta - \phi) = \cos (\phi - \theta)$; hence the values obtained for $a : b : c$ involve θ and ϕ symmetrically.

Example 2. If α and β are two different values of θ which satisfy the equation $a \cos \theta + b \sin \theta = c$, find the values of

$$4 \cos^2 \frac{\alpha}{2} \cos^2 \frac{\beta}{2}, \qquad \sin \alpha + \sin \beta, \qquad \sin \alpha \ \sin \beta.$$

From the given equation, by transposing and squaring,

$$(a\cos\theta - c)^2 = b^2\sin^2\theta = b^2(1-\cos^2\theta);$$

$$\therefore\ (a^2+b^2)\cos^2\theta - 2ac\cos\theta + c^2 - b^2 = 0.$$

The roots of this quadratic in $\cos\theta$ are $\cos\alpha$ and $\cos\beta$;

$$\therefore\ \cos\alpha + \cos\beta = \frac{2ac}{a^2+b^2} \quad\text{.......................(1)},$$

and

$$\cos\alpha\cos\beta = \frac{c^2-b^2}{a^2+b^2} \quad\text{.......................(2)}.$$

And $$4\cos^2\frac{\alpha}{2}\cos^2\frac{\beta}{2} = (1+\cos\alpha)(1+\cos\beta)$$

$$= 1 + \frac{2ac}{a^2+b^2} + \frac{c^2-b^2}{a^2+b^2}$$

$$= \frac{(a+c)^2}{a^2+b^2}.$$

From the data, we see that $\frac{\pi}{2}-\alpha$ and $\frac{\pi}{2}-\beta$ are values of θ which satisfy the equation $a\sin\theta + b\cos\theta = c$.

By writing a for b and b for a, equation (1) becomes

$$\cos\left(\frac{\pi}{2}-\alpha\right) + \cos\left(\frac{\pi}{2}-\beta\right) = \frac{2bc}{b^2+a^2},$$

or

$$\sin\alpha + \sin\beta = \frac{2bc}{a^2+b^2}.$$

Similarly, from equation (2) we have

$$\sin\alpha\sin\beta = \frac{c^2-a^2}{a^2+b^2}.$$

These last two results may also be derived from the equation

$$(b\sin\theta - c)^2 = a^2\cos^2\theta = a^2(1-\sin^2\theta).$$

Example 3. If α and β are two different values of θ which satisfy the equation $a\cos\theta + b\sin\theta = c$, prove that $\tan\frac{\alpha+\beta}{2} = \frac{b}{a}$. Also if the values of α and β are equal, shew that $a^2+b^2=c^2$.

By substituting $\cos\theta = \dfrac{1-\tan^2\frac{\theta}{2}}{1+\tan^2\frac{\theta}{2}}$ and $\sin\theta = \dfrac{2\tan\frac{\theta}{2}}{1+\tan^2\frac{\theta}{2}}$

in the given equation $a\cos\theta+b\sin\theta=c$, we have

$$a\left(1-\tan^2\frac{\theta}{2}\right)+2b\tan\frac{\theta}{2}=c\left(1+\tan^2\frac{\theta}{2}\right);$$

that is, $$(c+a)\tan^2\frac{\theta}{2}-2b\tan\frac{\theta}{2}+(c-a)=0 \quad\ldots\ldots\ldots\ldots(1).$$

The roots of this equation are $\tan\frac{\alpha}{2}$ and $\tan\frac{\beta}{2}$;

$$\therefore\ \tan\frac{\alpha}{2}+\tan\frac{\beta}{2}=\frac{2b}{c+a},\quad\text{and}\quad \tan\frac{\alpha}{2}\tan\frac{\beta}{2}=\frac{c-a}{c+a};$$

$$\therefore\ \tan\frac{\alpha+\beta}{2}=\frac{2b}{c+a}\Big/\left(1-\frac{c-a}{c+a}\right)=\frac{b}{a}.$$

If the roots of equation (1) are equal, we have

$$b^2=(c+a)(c-a);$$

whence $$a^2+b^2=c^2.$$

NOTE. The substitution here employed is frequently used in Analytical Geometry.

Example 4. If $\cos\theta+\cos\phi=a$ and $\sin\theta+\sin\phi=b$, find the values of $\cos(\theta+\phi)$ and $\sin 2\theta+\sin 2\phi$.

From the given equations, we have

$$\frac{\sin\theta+\sin\phi}{\cos\theta+\cos\phi}=\frac{b}{a};$$

$$\therefore\ \tan\frac{\theta+\phi}{2}=\frac{b}{a}.$$

For shortness write t instead of $\tan\frac{\theta+\phi}{2}$; then

$$\cos(\theta+\phi)=\frac{1-t^2}{1+t^2}=\frac{a^2-b^2}{a^2+b^2},$$

and $$\sin(\theta+\phi)=\frac{2t}{1+t^2}=\frac{2ab}{a^2+b^2}.$$

Multiplying the two given equations together, we have

$$\sin 2\theta+\sin 2\phi+2\sin(\theta+\phi)=2ab;$$

$$\therefore\ \sin 2\theta+\sin 2\phi=2ab\left(1-\frac{2}{a^2+b^2}\right).$$

Example 5. Resolve into factors the expression

$$\cos^2\alpha+\cos^2\beta+\cos^2\gamma+2\cos\alpha\cos\beta\cos\gamma-1,$$

and shew that it vanishes if any one of the four angles $\alpha\pm\beta\pm\gamma$ is an odd multiple of two right angles.

The expression $=\cos^2\alpha+(\cos^2\beta+\cos^2\gamma-1)+2\cos\alpha\cos\beta\cos\gamma$

$$=\cos^2\alpha+(\cos^2\beta-\sin^2\gamma)+2\cos\alpha\cos\beta\cos\gamma$$

$$=\cos^2\alpha+\cos(\beta+\gamma)\cos(\beta-\gamma)+\cos\alpha\{\cos(\beta+\gamma)+\cos(\beta-\gamma)\}$$

$$=\{\cos\alpha+\cos(\beta+\gamma)\}\{\cos\alpha+\cos(\beta-\gamma)\}$$

$$=4\cos\frac{\alpha+\beta+\gamma}{2}\cos\frac{\alpha-\beta-\gamma}{2}\cos\frac{\alpha+\beta-\gamma}{2}\cos\frac{\alpha-\beta+\gamma}{2}.$$

The expression vanishes if one of the quantities $\cos\dfrac{\alpha\pm\beta\pm\gamma}{2}=0$;

that is, if one of the four angles $\dfrac{\alpha\pm\beta\pm\gamma}{2}=(2n+1)\dfrac{\pi}{2}$;

that is, if $\alpha\pm\beta\pm\gamma=(2n+1)\pi$, where n is any integer.

Example 6. If $\qquad \tan\theta=\dfrac{\sin\alpha\sin\beta}{\cos\alpha+\cos\beta}$,

prove that one value of $\tan\dfrac{\theta}{2}$ is $\tan\dfrac{\alpha}{2}\tan\dfrac{\beta}{2}$.

From the given equation, we have

$$\sec^2\theta=1+\frac{\sin^2\alpha\sin^2\beta}{(\cos\alpha+\cos\beta)^2}=\frac{(\cos\alpha+\cos\beta)^2+(1-\cos^2\alpha)(1-\cos^2\beta)}{(\cos\alpha+\cos\beta)^2}$$

$$=\frac{1+2\cos\alpha\cos\beta+\cos^2\alpha\cos^2\beta}{(\cos\alpha+\cos\beta)^2}.$$

Taking the positive root, $\sec\theta=\dfrac{1+\cos\alpha\cos\beta}{\cos\alpha+\cos\beta}$;

$$\therefore\ \cos\theta=\frac{\cos\alpha+\cos\beta}{1+\cos\alpha\cos\beta}.$$

$$\therefore\ \frac{1-\cos\theta}{1+\cos\theta}=\frac{1-\cos\alpha-\cos\beta+\cos\alpha\cos\beta}{1+\cos\alpha+\cos\beta+\cos\alpha\cos\beta}=\frac{(1-\cos\alpha)(1-\cos\beta)}{(1+\cos\alpha)(1+\cos\beta)};$$

$$\therefore\ \tan^2\frac{\theta}{2}=\tan^2\frac{\alpha}{2}\tan^2\frac{\beta}{2};$$

and therefore one value of $\tan\dfrac{\theta}{2}$ is $\tan\dfrac{\alpha}{2}\tan\dfrac{\beta}{2}$.

Example 7. In any triangle, shew that

$$\Sigma a^3 \cos A = abc\,(1 + 4\Pi \cos A).$$

Let $$k = \frac{a}{\sin A} = \frac{b}{\sin B} = \frac{c}{\sin C};$$

so that $$a = k \sin A, \quad b = k \sin B, \quad c = k \sin C.$$

By substituting these values in the given identity, and dividing by k^3, we have to prove that

$$\Sigma \sin^3 A \cos A = \sin A \sin B \sin C\,(1 + 4\Pi \cos A).$$

Now $$8\Sigma \sin^3 A \cos A = 4\Sigma \sin^2 A \sin 2A$$
$$= 2\Sigma\,(1 - \cos 2A) \sin 2A$$
$$= 2\Sigma \sin 2A - \Sigma \sin 4A;$$

and it has been shewn in Example 1, Art. 135, that

$$\Sigma \sin 2A = 4\Pi \sin A;$$

and it is easy to prove that

$$\Sigma \sin 4A = -4\Pi \sin 2A = -32\Pi \sin A\,.\,\Pi \cos A;$$

$$\therefore\ 8\,\Sigma \sin^3 A \cos A = 8\Pi \sin A + 32\Pi \sin A\,.\,\Pi \cos A;$$

$$\therefore\ \Sigma \sin^3 A \cos A = \Pi \sin A\,(1 + 4\Pi \cos A).$$

EXAMPLES. XXIV. a.

1. If $\theta = \alpha$, and $\theta = \beta$ satisfy the equation

$$\frac{1}{a} \cos \theta + \frac{1}{b} \sin \theta = \frac{1}{c},$$

prove that $$a \cos \frac{\alpha + \beta}{2} = b \sin \frac{\alpha + \beta}{2} = c \cos \frac{\alpha - \beta}{2}.$$

Solve the simultaneous equations:

2. $\frac{x}{a} \cos \alpha + \frac{y}{b} \sin \alpha = 1, \qquad \frac{x}{a} \cos \beta + \frac{y}{b} \sin \beta = 1.$

3. $\frac{x}{a} \cos \alpha + \frac{y}{b} \sin \alpha = 1, \qquad \frac{x}{a} \sin \alpha - \frac{y}{b} \cos \alpha = 1.$

If α and β are two different solutions of $a\cos\theta + b\sin\theta = c$, prove that

4. $\cos(\alpha+\beta) = \dfrac{a^2-b^2}{a^2+b^2}$. 5. $\cos^2\dfrac{\alpha-\beta}{2} = \dfrac{c^2}{a^2+b^2}$.

6. $\sin 2\alpha + \sin 2\beta = \dfrac{4ab(2c^2-a^2-b^2)}{(a^2+b^2)^2}$.

7. $\sin^2\alpha + \sin^2\beta = \dfrac{2a^2(a^2+b^2) - 2c^2(a^2-b^2)}{(a^2+b^2)^2}$.

8. If $a\cos\alpha + b\sin\alpha = a\cos\beta + b\sin\beta = c$, prove that

$$\sin(\alpha+\beta) = \frac{2ab}{a^2+b^2}, \quad \text{and} \quad \cot\alpha + \cot\beta = \frac{2ab}{c^2-a^2}.$$

If $\cos\theta + \cos\phi = a$ and $\sin\theta + \sin\phi = b$, prove that

9. $\cos\theta\cos\phi = \dfrac{(a^2+b^2)^2 - 4b^2}{4(a^2+b^2)}$.

10. $\cos 2\theta + \cos 2\phi = \dfrac{(a^2-b^2)(a^2+b^2-2)}{a^2+b^2}$.

11. $\tan\theta + \tan\phi = \dfrac{8ab}{(a^2+b^2)^2 - 4b^2}$.

12. $\tan\dfrac{\theta}{2} + \tan\dfrac{\phi}{2} = \dfrac{4b}{a^2+b^2+2a}$.

13. Express

$$1 - \cos^2\alpha - \cos^2\beta - \cos^2\gamma + 2\cos\alpha\cos\beta\cos\gamma$$

as the product of four sines, and shew that it vanishes if any one of the four angles $\alpha \pm \beta \pm \gamma$ is zero or an even multiple of π.

14. Express

$$\sin^2\alpha + \sin^2\beta - \sin^2\gamma + 2\sin\alpha\sin\beta\cos\gamma$$

as the product of two sines and two cosines.

15. Express

$$\sin^2\alpha + \sin^2\beta + \sin^2\gamma - 2\sin\alpha\sin\beta\sin\gamma - 1$$

as the product of four cosines.

16. If $$\cos\theta=\frac{\cos a-\cos\beta}{1-\cos a\cos\beta},$$
prove that one value of $\tan\frac{\theta}{2}$ is $\tan\frac{a}{2}\cot\frac{\beta}{2}$.

17. If $$\tan^2\theta\cos^2\frac{a+\beta}{2}=\sin a\sin\beta,$$
prove that one value of $\tan^2\frac{\theta}{2}$ is $\tan\frac{a}{2}\tan\frac{\beta}{2}$.

18. If $$\tan\theta(\cos a+\sin\beta)=\sin a\cos\beta,$$
prove that one value of $\tan\frac{\theta}{2}$ is $\tan\frac{a}{2}\tan\left(\frac{\pi}{4}-\frac{\beta}{2}\right)$.

In any triangle, shew that

19. $\Sigma a^3\sin B\sin C=2abc(1+\cos A\cos B\cos C)$.

20. $\Sigma a\cos^3 A=\frac{abc}{4R^2}(1-4\cos A\cos B\cos C)$.

21. $\Sigma a^3\cos(B-C)=3abc$.

22. If a and β are roots of the equation $a\cos\theta+b\sin\theta=c$, form the equations whose roots are

(1) $\sin a$ and $\sin\beta$; (2) $\cos 2a$ and $\cos 2\beta$.

Alternating Expressions.

305. An expression is said to be *alternating* with respect to certain of the letters it contains, if the sign of the expression but not its numerical value is altered when any pair of these letters are interchanged.

Thus $\cos a-\cos\beta$, $\sin(a-\beta)$, $\tan(a-\beta)$,
$$\cos^2 a\sin(\beta-\gamma)+\cos^2\beta\sin(\gamma-a)+\cos^2\gamma\sin(a-\beta)$$
are alternating expressions.

306. Alternating expressions may be abridged by means of the symbols Σ and Π. Thus

$$\Sigma\sin^2 a\sin(\beta-\gamma)=\sin^2 a\sin(\beta-\gamma)+\sin^2\beta\sin(\gamma-a)+\sin^2\gamma\sin(a-\beta);$$

$$\Pi\tan(\beta-\gamma)=\tan(\beta-\gamma)\tan(\gamma-a)\tan(a-\beta).$$

We shall confine our attention chiefly to alternating expressions involving the three letters α, β, γ, and we shall adopt the *cyclical arrangement* $\beta-\gamma$, $\gamma-\alpha$, $\alpha-\beta$ in which β follows α, γ follows β, and α follows γ.

Example 1. Prove that $\Sigma \cos(\alpha+\theta)\sin(\beta-\gamma)=0$.

$$\begin{aligned}\Sigma \cos(\alpha+\theta)\sin(\beta-\gamma) &= \Sigma(\cos\alpha\cos\theta-\sin\alpha\sin\theta)\sin(\beta-\gamma)\\ &= \cos\theta\,\Sigma\cos\alpha\sin(\beta-\gamma)-\sin\theta\,\Sigma\sin\alpha\sin(\beta-\gamma)\\ &=0,\end{aligned}$$

since $\qquad \Sigma\cos\alpha\sin(\beta-\gamma)=0$ and $\Sigma\sin\alpha\sin(\beta-\gamma)=0$.

Example 2. Shew that $\Sigma\sin 2(\beta-\gamma)=-4\Pi\sin(\beta-\gamma)$.

$$\begin{aligned}&\sin 2(\beta-\gamma)+\sin 2(\gamma-\alpha)+\sin 2(\alpha-\beta)\\ &\quad=2\sin(\beta-\alpha)\cos(\alpha+\beta-2\gamma)+2\sin(\alpha-\beta)\cos(\alpha-\beta)\\ &\quad=2\sin(\alpha-\beta)\{\cos(\alpha-\beta)-\cos(\alpha+\beta-2\gamma)\}\\ &\quad=4\sin(\alpha-\beta)\sin(\alpha-\gamma)\sin(\beta-\gamma)\\ &\quad=-4\Pi\sin(\beta-\gamma).\end{aligned}$$

Example 3. Prove that

(1) $\Sigma\tan(\beta-\gamma)=\Pi\tan(\beta-\gamma)$;

(2) $\Sigma\tan\beta\tan\gamma\tan(\beta-\gamma)=-\Pi\tan(\beta-\gamma)$.

(1) From Art. 118, if $A+B+C=0$, we see that

$$\tan A+\tan B+\tan C=\tan A\tan B\tan C.$$

Hence by writing $A=\beta-\gamma$, $B=\gamma-\alpha$, $C=\alpha-\beta$, we have

$$\Sigma\tan(\beta-\gamma)=\Pi\tan(\beta-\gamma).$$

(2) From the formulæ for $\tan(\beta-\gamma)$, $\tan(\gamma-\alpha)$, $\tan(\alpha-\beta)$, we have

$$\Sigma(1+\tan\beta\tan\gamma)\tan(\beta-\gamma)=\Sigma(\tan\beta-\tan\gamma)=0;$$

whence by transposition

$$\begin{aligned}\Sigma\tan\beta\tan\gamma\tan(\beta-\gamma)&=-\Sigma\tan(\beta-\gamma)\\ &=-\Pi\tan(\beta-\gamma).\end{aligned}$$

Example 4. Shew that

$$\Sigma \cos 3\alpha \sin(\beta-\gamma) = 4\cos(\alpha+\beta+\gamma)\,\Pi \sin(\beta-\gamma).$$

Since $2\cos 3\alpha \sin(\beta-\gamma) = \sin(3\alpha+\beta-\gamma) - \sin(3\alpha-\beta+\gamma)$,

we have

$$2\Sigma \cos 3\alpha \sin(\beta-\gamma) = \sin(3\alpha+\beta-\gamma) - \sin(3\alpha-\beta+\gamma) + \sin(3\beta+\gamma-\alpha)$$
$$- \sin(3\beta-\gamma+\alpha) + \sin(3\gamma+\alpha-\beta) - \sin(3\gamma-\alpha+\beta).$$

Combining the second and third terms, the fourth and fifth terms, the sixth and first terms, and dividing by 2, we have

$$\Sigma \cos 3\alpha \sin(\beta-\gamma)$$
$$= \cos(\alpha+\beta+\gamma)\,\{\sin 2(\beta-\alpha) + \sin 2(\gamma-\beta) + \sin 2(\alpha-\gamma)\}$$
$$= 4\cos(\alpha+\beta+\gamma)\,\Pi \sin(\beta-\gamma).$$ [See Example 2.]

307. The following example is given as a specimen of a concise solution.

Example. If $(y+z)\tan\alpha + (z+x)\tan\beta + (x+y)\tan\gamma = 0$,

and $x\tan\beta\tan\gamma + y\tan\gamma\tan\alpha + z\tan\alpha\tan\beta = x+y+z$,

prove that $x\sin 2\alpha + y\sin 2\beta + z\sin 2\gamma = 0$.

From the given equations, we have

$$x(1-\tan\beta\tan\gamma) + y(1-\tan\gamma\tan\alpha) + z(1-\tan\alpha\tan\beta) = 0,$$

and $$x(\tan\beta+\tan\gamma) + y(\tan\gamma+\tan\alpha) + z(\tan\alpha+\tan\beta) = 0.$$

If we find the values of $x : y : z$ by cross multiplication, the denominator of x

$$= (1-\tan\gamma\tan\alpha)(\tan\alpha+\tan\beta) - (1-\tan\alpha\tan\beta)(\tan\gamma+\tan\alpha)$$
$$= (\tan\beta-\tan\gamma) + \tan^2\alpha\,(\tan\beta-\tan\gamma)$$
$$= (1+\tan^2\alpha)(\tan\beta-\tan\gamma)$$
$$= \sec^2\alpha\,(\tan\beta-\tan\gamma)$$
$$= \frac{\sec\alpha\sin(\beta-\gamma)}{\cos\alpha\cos\beta\cos\gamma}.$$

Hence $$\frac{x}{\sec\alpha\sin(\beta-\gamma)} = \frac{y}{\sec\beta\sin(\gamma-\alpha)} = \frac{z}{\sec\gamma\sin(\alpha-\beta)} = k \text{ say}.$$

$$\therefore\ x\sin 2\alpha + y\sin 2\beta + z\sin 2\gamma = k\Sigma \sin 2\alpha \sec\alpha \sin(\beta-\gamma)$$
$$= 2k\Sigma \sin\alpha \sin(\beta-\gamma)$$
$$= 0.$$

Allied formulæ in Algebra and Trigonometry.

308. From well-known algebraical identities we can deduce some interesting trigonometrical identities.

Example 1. In the identity

$$(x-a)(b-c)+(x-b)(c-a)+(x-c)(a-b)=0,$$

put $\qquad x=\cos 2\theta, \quad a=\cos 2\alpha, \quad b=\cos 2\beta, \quad c=\cos 2\gamma;$

then $\qquad x-a=\cos 2\theta-\cos 2\alpha=2\sin(\alpha+\theta)\sin(\alpha-\theta),$

and $\qquad b-c=\cos 2\beta-\cos 2\gamma=-2\sin(\beta+\gamma)\sin(\beta-\gamma);$

$$\therefore\ \Sigma\sin(\alpha+\theta)\sin(\alpha-\theta)\sin(\beta+\gamma)\sin(\beta-\gamma)=0.$$

Example 2. In the identity

$$\Sigma a^2(b-c)=-\Pi(b-c),$$

put $\qquad a=\sin^2\alpha, \quad b=\sin^2\beta, \quad c=\sin^2\gamma;$

then $\qquad b-c=\sin^2\beta-\sin^2\gamma=\sin(\beta+\gamma)\sin(\beta-\gamma);$

$$\therefore\ \Sigma\sin^4\alpha\sin(\beta+\gamma)\sin(\beta-\gamma)=-\Pi\sin(\beta+\gamma)\,.\,\Pi\sin(\beta-\gamma).$$

Example 3. In the identity

$$\Sigma a^3(b-c)=-(a+b+c)\,\Pi(b-c),$$

put $\qquad a=\cos\alpha, \quad b=\cos\beta, \quad c=\cos\gamma;$

$$\therefore\ \Sigma\cos^3\alpha(\cos\beta-\cos\gamma)=-(\cos\alpha+\cos\beta+\cos\gamma)\,\Pi(\cos\beta-\cos\gamma).$$

But $\qquad \Sigma\cos\alpha(\cos\beta-\cos\gamma)=0;$

$$\therefore\ \Sigma(4\cos^3\alpha-3\cos\alpha)(\cos\beta-\cos\gamma)$$
$$=-4(\cos\alpha+\cos\beta+\cos\gamma)\,\Pi(\cos\beta-\cos\gamma);$$

that is,

$$\Sigma\cos 3\alpha(\cos\beta-\cos\gamma)=-4(\cos\alpha+\cos\beta+\cos\gamma)\,\Pi(\cos\beta-\cos\gamma).$$

Example 4. If $a+b+c=0$, then $a^3+b^3+c^3=3abc$.

Here a, b, c may be any three quantities whose sum is zero; this condition is satisfied if we put $a=\cos(\alpha+\theta)\sin(\beta-\gamma)$, and b and c equal to corresponding quantities.

Thus $\quad \Sigma\cos^3(\alpha+\theta)\sin^3(\beta-\gamma)=3\Pi\cos(\alpha+\theta)\sin(\beta-\gamma).$

309. An algebraical identity may sometimes be established by the aid of Trigonometry.

Example. If $x+y+z=xyz$, prove that

$$x(1-y^2)(1-z^2)+y(1-z^2)(1-x^2)+z(1-x^2)(1-y^2)=4xyz.$$

By putting $x=\tan\alpha$, $y=\tan\beta$, $z=\tan\gamma$, we have

$$\tan\alpha+\tan\beta+\tan\gamma=\tan\alpha\tan\beta\tan\gamma;$$

whence $$\tan\alpha=-\frac{\tan\beta+\tan\gamma}{1-\tan\beta\tan\gamma}=-\tan(\beta+\gamma);$$

$$\therefore\ \alpha=n\pi-(\beta+\gamma),\text{ where } n \text{ is an integer;}$$

$$\therefore\ \alpha+\beta+\gamma=n\pi;$$

$$\therefore\ 2\alpha+2\beta+2\gamma=2n\pi.$$

From this relation it is easy to shew that

$$\tan 2\alpha+\tan 2\beta+\tan 2\gamma=\tan 2\alpha\tan 2\beta\tan 2\gamma;$$

$$\therefore\ \frac{2x}{1-x^2}+\frac{2y}{1-y^2}+\frac{2z}{1-z^2}=\frac{8xyz}{(1-x^2)(1-y^2)(1-z^2)};$$

$$\therefore\ x(1-y^2)(1-z^2)+y(1-z^2)(1-x^2)+z(1-x^2)(1-y^2)=4xyz.$$

EXAMPLES. XXIV. b.

Prove the following identities:

1. $\Sigma\sin(\alpha-\theta)\sin(\beta-\gamma)=0.$
2. $\Sigma\cos\beta\cos\gamma\sin(\beta-\gamma)=\Sigma\sin\beta\sin\gamma\sin(\beta-\gamma).$
3. $\Sigma\sin(\beta-\gamma)\cos(\beta+\gamma+\theta)=0.$
4. $\Sigma\cos 2(\beta-\gamma)=4\Pi\cos(\beta-\gamma)-1.$
5. $\Sigma\sin\beta\sin\gamma\sin(\beta-\gamma)=-\Pi\sin(\beta-\gamma).$
6. $\Sigma\cot(\alpha-\beta)\cot(\alpha-\gamma)+1=0.$
7. $\Sigma\sin 3\alpha\sin(\beta-\gamma)=4\sin(\alpha+\beta+\gamma)\,\Pi\sin(\beta-\gamma).$
8. $\Sigma\cos^3\alpha\sin(\beta-\gamma)=\cos(\alpha+\beta+\gamma)\,\Pi\sin(\beta-\gamma).$
9. $\Sigma\cos(\theta+\alpha)\cos(\beta+\gamma)\sin(\theta-\alpha)\sin(\beta-\gamma)=0.$
10. $\Sigma\sin^2\beta\sin^2\gamma\sin(\beta+\gamma)\sin(\beta-\gamma)$
$$=-\Pi\sin(\beta+\gamma)\,.\,\Pi\sin(\beta-\gamma).$$

Prove the following identities:

11. $\Sigma \cos 2\beta \cos 2\gamma \sin(\beta+\gamma)\sin(\beta-\gamma)$

$$= -4\Pi \sin(\beta+\gamma) \,.\, \Pi \sin(\beta-\gamma).$$

12. $\Sigma \cos 4\alpha \sin(\beta+\gamma)\sin(\beta-\gamma)$

$$= -8\Pi \sin(\beta+\gamma) \,.\, \Pi \sin(\beta-\gamma).$$

13. $\Sigma \sin 3\alpha (\sin\beta - \sin\gamma)$

$$= 4(\sin\alpha + \sin\beta + \sin\gamma)\, \Pi(\sin\beta - \sin\gamma).$$

14. $\Sigma \sin^3(\beta+\gamma)\sin^3(\beta-\gamma) = 3\Pi \sin(\beta+\gamma) \,.\, \Pi \sin(\beta-\gamma).$

15. $\Sigma \cos^3(\beta+\gamma+\theta)\sin^3(\beta-\gamma)$

$$= 3\Pi \cos(\beta+\gamma+\theta) \,.\, \Pi \sin(\beta-\gamma).$$

16. If $x+y+z=xyz$, prove that

$$\Sigma \frac{3x-x^3}{1-3x^2} = \Pi \frac{3x-x^3}{1-3x^2}.$$

17. If $yz+zx+xy=1$, prove that

$$\Sigma x(1-y^2)(1-z^2) = 4xyz.$$

310. From a trigonometrical identity many others may be derived by various substitutions.

For instance, if A, B, C are *any* angles, positive or negative, connected by the relation $A+B+C=\pi$, we know that

$$\sin A + \sin B + \sin C = 4\cos\frac{A}{2}\cos\frac{B}{2}\cos\frac{C}{2}.$$

Let $\qquad A=\pi-2\alpha, \quad B=\pi-2\beta, \quad C=\pi-2\gamma;$

then $\qquad \sin A = \sin 2\alpha$, and $\cos\frac{A}{2} = \sin\alpha.$

Also $\qquad 2(\alpha+\beta+\gamma) = 3\pi - (A+B+C) = 2\pi;$

$$\therefore\ \alpha+\beta+\gamma=\pi,$$

and $\qquad \sin 2\alpha + \sin 2\beta + \sin 2\gamma = 4\sin\alpha\sin\beta\sin\gamma.$

Again, let $\qquad A=\frac{\pi}{2}-\frac{\alpha}{2}, \quad B=\frac{\pi}{2}-\frac{\beta}{2}, \quad C=\frac{\pi}{2}-\frac{\gamma}{2};$

then $\qquad \sin A = \cos\frac{\alpha}{2}$, and $\cos\frac{A}{2} = \cos\frac{\pi-\alpha}{4}.$

Also $\alpha+\beta+\gamma=3\pi-2(A+B+C)=3\pi-2\pi$;

$$\therefore\ \alpha+\beta+\gamma=\pi,$$

and $$\cos\frac{\alpha}{2}+\cos\frac{\beta}{2}+\cos\frac{\gamma}{2}=4\cos\frac{\pi-\alpha}{4}\cos\frac{\pi-\beta}{4}\cos\frac{\pi-\gamma}{4}$$

Example. If $A+B+C=\pi$, shew that

$$\cos\frac{A}{2}+\cos\frac{B}{2}-\cos\frac{C}{2}=4\cos\frac{\pi+A}{4}\cos\frac{\pi+B}{4}\cos\frac{\pi-C}{4}.$$

Put $$\frac{\pi+A}{4}=\frac{\alpha}{2},\quad \frac{\pi+B}{4}=\frac{\beta}{2},\quad \frac{C-\pi}{4}=\frac{\gamma}{2};$$

then $\cos\frac{A}{2}=\cos\left(\alpha-\frac{\pi}{2}\right)=\sin\alpha$, and $\cos\frac{C}{2}=\cos\left(\gamma+\frac{\pi}{2}\right)=-\sin\gamma$,

so that the above identity becomes

$$\sin\alpha+\sin\beta+\sin\gamma=4\cos\frac{\alpha}{2}\cos\frac{\beta}{2}\cos\frac{\gamma}{2},$$

which is clearly true since

$$\alpha+\beta+\gamma=\frac{\pi}{2}+\frac{A+B+C}{2}=\frac{\pi}{2}+\frac{\pi}{2}=\pi.$$

311. When $A+B+C=n\pi$,

$$\tan(A+B)=\tan(n\pi-C)=-\tan C;$$

whence we obtain $\Sigma\tan A=\Pi\tan A$.

When $n=0$, the given condition is satisfied in the case of any three angles whose sum is 0; as for instance if

$$A=\beta+\gamma-2\alpha,\quad B=\gamma+\alpha-2\beta,\quad C=\alpha+\beta-2\gamma.$$

Hence $\Sigma\tan(\beta+\gamma-2\alpha)=\Pi\tan(\beta+\gamma-2\alpha)$.

Example. If $\alpha+\beta+\gamma=0$, shew that

$$\Sigma\cot(\gamma+\alpha-\beta)\cot(\alpha+\beta-\gamma)=1.$$

Put $\beta+\gamma-\alpha=A,\quad \gamma+\alpha-\beta=B,\quad \alpha+\beta-\gamma=C$;

then, by addition,

$$A+B+C=\alpha+\beta+\gamma=0;$$

$$\therefore\ \cot(A+B)=-\cot C;$$

whence $$\Sigma\cot A\cot B=1,$$

that is, $$\Sigma\cot(\gamma+\alpha-\beta)\cot(\alpha+\beta-\gamma)=1.$$

312. The following example is a further illustration of the manner in which an identity may be established by appropriate substitutions in some simpler identity.

Example. Prove that

$$2\Pi \cos(\beta+\gamma) + \Pi \cos 2\alpha = \Sigma \cos 2\alpha \cos^2(\beta+\gamma).$$

In Example 5, Art. 133, we have proved that

$$4\cos\alpha \cos\beta \cos\gamma = \Sigma \cos(\beta+\gamma-\alpha) + \cos(\alpha+\beta+\gamma).$$

In this identity first replace α, β, γ by $\beta+\gamma$, $\gamma+\alpha$, $\alpha+\beta$ respectively, and secondly replace α, β, γ by 2α, 2β, 2γ respectively.

Thus $\quad 8\Pi \cos(\beta+\gamma) = 2\Sigma \cos 2\alpha + 2\cos 2(\alpha+\beta+\gamma)$,

and $\quad 4\Pi \cos 2\alpha = \Sigma \cos 2(\beta+\gamma-\alpha) + \cos 2(\alpha+\beta+\gamma)$;

whence by addition

$$\begin{aligned} 8\Pi \cos(\beta+\gamma) + 4\Pi \cos 2\alpha \\ &= 2\Sigma \cos 2\alpha + \Sigma \cos 2(\beta+\gamma-\alpha) + 3\cos 2(\alpha+\beta+\gamma) \\ &= 2\Sigma \cos 2\alpha + \Sigma \{\cos 2(\beta+\gamma-\alpha) + \cos 2(\alpha+\beta+\gamma)\} \\ &= 2\Sigma \cos 2\alpha + 2\Sigma \cos 2(\beta+\gamma) \cos 2\alpha \\ &= 2\Sigma \cos 2\alpha \{1 + \cos 2(\beta+\gamma)\} \\ &= 4\Sigma \cos 2\alpha \cos^2(\beta+\gamma); \end{aligned}$$

$$\therefore\ 2\Pi \cos(\beta+\gamma) + \Pi \cos 2\alpha = \Sigma \cos 2\alpha \cos^2(\beta+\gamma).$$

313. Suppose that $A'B'C'$ is the pedal triangle of ABC, and let the sides and angles of the pedal triangle be denoted by a', b', c', and A', B', C', and its circum-radius by R'. Then from Arts. 224 and 225, we have

$$a' = a\cos A, \quad b' = b\cos B, \quad c' = c\cos C, \quad R' = \frac{R}{2},$$

$$A' = 180^\circ - 2A, \quad B' = 180^\circ - 2B, \quad C' = 180^\circ - 2C.$$

By means of these relations, we may from any identity proved for the triangle ABC derive another, as in the following case.

In the triangle ABC, we know that

$$\Sigma a \cos A = 4R \sin A \sin B \sin C;$$

hence in the pedal triangle $A'B'C'$,

$$\Sigma a' \cos A' = 4R' \sin A' \sin B' \sin C';$$

$$\therefore\ \Sigma a \cos A \cos(180^\circ - 2A) = 2R\Pi \sin(180^\circ - 2A);$$

that is, $\quad -\Sigma a \cos A \cos 2A = 2R \sin 2A \sin 2B \sin 2C.$

Example. In any triangle ABC, shew that

$$\frac{a^2\cos^2 A - b^2\cos^2 B - c^2\cos^2 C}{2bc\cos B\cos C} = \cos 2A.$$

In the pedal triangle $A'B'C'$, we have

$$\frac{b'^2 + c'^2 - a'^2}{2b'c'} = \cos A';$$

hence, by substituting the equivalents of a', b', c', A', we have

$$\frac{b^2\cos^2 B + c^2\cos^2 C - a^2\cos^2 A}{2bc\cos B\cos C} = \cos(180° - 2A) = -\cos 2A;$$

whence the required identity follows at once.

314. If $A_1B_1C_1$ be the ex-central triangle of ABC, we may, as in the preceding article, from any identity proved for the triangle ABC derive another by means of the relations

$$a_1 = a\operatorname{cosec}\frac{A}{2},\quad b_1 = b\operatorname{cosec}\frac{B}{2},\quad c_1 = c\operatorname{cosec}\frac{C}{2},\quad R_1 = 2R,$$

$$A_1 = 90° - \frac{A}{2},\quad B_1 = 90° - \frac{B}{2},\quad C_1 = 90° - \frac{C}{2}.$$

315. The following Exercise consists of miscellaneous questions on the subject of this Chapter.

EXAMPLES. XXIV. c.

1. Shew that

$$\Sigma\cot(2\alpha + \beta - 3\gamma)\cot(2\beta + \gamma - 3\alpha) = 1.$$

2. Shew that

(1) $2\Pi\sin(\beta + \gamma) + \Pi\sin 2\alpha = \Sigma\sin 2\alpha\sin^2(\beta + \gamma)$;

(2) $\Pi\sin(\beta + \gamma - \alpha) + 2\Pi\sin\alpha = \Sigma\sin^2\alpha\sin(\beta + \gamma - \alpha)$.

3. In any triangle, prove that

(1) $a^2\cos^2 A - b^2\cos^2 B = Rc\cos C\sin 2(B - A)$;

(2) $a^2\operatorname{cosec}^2\frac{A}{2} - b^2\operatorname{cosec}^2\frac{B}{2} = 4Rc\operatorname{cosec}\frac{C}{2}\sin\frac{B - A}{2}$;

(3) $\Sigma(b\cos B + c\cos C)\cot A = -2R\Sigma\cos 2A$.

4. If $$\sin 2\theta = 2 \sin \alpha \sin \gamma,$$
and $$\cos 2\theta = \cos 2\alpha \cos 2\beta = \cos 2\gamma \cos 2\delta,$$
prove that one value of $\tan\theta$ is $\tan\beta\tan\delta$.

5. If $$\tan\frac{\theta}{2}\tan\frac{\phi}{2} = \tan\frac{\gamma}{2},$$
and $$\sec\alpha\cos\theta = \sec\beta\cos\phi = \cos\gamma,$$
prove that $$\sin^2\gamma = (\sec\alpha - 1)(\sec\beta - 1).$$

6. If $$\frac{\cos\theta - \cos\alpha}{\cos\theta - \cos\beta} = \frac{\sin^2\alpha\cos\beta}{\sin^2\beta\cos\alpha},$$
prove that one value of $\tan\frac{\theta}{2}$ is $\tan\frac{\alpha}{2}\tan\frac{\beta}{2}$.

7. If $\sin\theta = \cot\alpha\tan\gamma$ and $\tan\theta = \cos\alpha\tan\beta$, prove that one value of $\cos\theta$ is $\cos\beta\sec\gamma$.

8. If α and β are two different values of θ which satisfy
$$bc\cos\theta\cos\phi + ac\sin\theta\sin\phi = ab,$$
prove that
$$(b^2+c^2-a^2)\cos\alpha\cos\beta + (c^2+a^2-b^2)\sin\alpha\sin\beta = a^2+b^2-c^2.$$

9. If β and γ are two different values of θ which satisfy
$$\sin\alpha\cos\theta + \cos\alpha\sin\theta = \cos\alpha\sin\alpha,$$
prove that $$\frac{\cos\beta\cos\gamma}{\cos^2\alpha} + \frac{\sin\beta\sin\gamma}{\sin^2\alpha} = 1.$$

10. If β and γ are two different values of θ which satisfy
$$k^2\cos\alpha\cos\theta + k(\sin\alpha + \sin\theta) + 1 = 0,$$
prove that $$k^2\cos\beta\cos\gamma + k(\sin\beta + \sin\gamma) + 1 = 0.$$

11. If β and γ are two different values of θ which satisfy
$$\frac{\cos\theta\cos\phi}{\cos^2\alpha} + \frac{\sin\theta\sin\phi}{\sin^2\alpha} + 1 = 0,$$
prove that $$\frac{\cos\beta\cos\gamma}{\cos^2\alpha} + \frac{\sin\beta\sin\gamma}{\sin^2\alpha} + 1 = 0.$$

CHAPTER XXV.

MISCELLANEOUS THEOREMS AND EXAMPLES.

Inequalities. Maxima and Minima.

316. THE methods of proving trigonometrical inequalities are in many cases identical with those by which algebraical inequalities are established.

Example 1. Shew that $a^2 \tan^2\theta + b^2 \cot^2\theta > 2ab$.

We have $a^2 \tan^2\theta + b^2 \cot^2\theta = (a \tan\theta - b \cot\theta)^2 + 2ab$;

$$\therefore\ a^2 \tan^2\theta + b^2 \cot^2\theta > 2ab,$$

unless $$a \tan\theta - b \cot\theta = 0, \text{ or } a \tan^2\theta = b.$$

In this case the inequality becomes an equality.

This proposition may be otherwise expressed by saying that the *minimum value* of $a^2 \tan^2\theta + b^2 \cot^2\theta$ is $2ab$.

Example 2. Shew that

$$1 + \sin^2\alpha + \sin^2\beta > \sin\alpha + \sin\beta + \sin\alpha \sin\beta.$$

Since $(1 - \sin\alpha)^2$ is positive,

$$1 + \sin^2\alpha > 2 \sin\alpha;$$

similarly $$1 + \sin^2\beta > 2 \sin\beta,$$

and $$\sin^2\alpha + \sin^2\beta > 2 \sin\alpha \sin\beta.$$

Adding and dividing by 2, we have

$$1 + \sin^2\alpha + \sin^2\beta > \sin\alpha + \sin\beta + \sin\alpha \sin\beta.$$

Example 3. When is $12 \sin\theta - 9 \sin^2\theta$ a maximum?

The expression $= 4 - (2 - 3 \sin\theta)^2$, and is therefore a maximum when $2 - 3 \sin\theta = 0$, so that its maximum value is 4.

317. *To find the numerically greatest values of*

$$a \cos \theta + b \sin \theta.$$

Let $$a = r \cos \alpha \text{ and } b = r \sin \alpha,$$

so that $$r^2 = a^2 + b^2 \text{ and } \tan \alpha = \frac{b}{a};$$

then $$a \cos \theta + b \sin \theta = r(\cos \theta \cos \alpha + \sin \theta \sin \alpha)$$
$$= r \cos(\theta - \alpha).$$

Thus the expression is numerically greatest when

$$\cos(\theta - \alpha) = \pm 1;$$

that is, the greatest positive value $= r = \sqrt{a^2 + b^2}$,
and the numerically greatest negative value $= -r = -\sqrt{a^2 + b^2}$.

Hence, if $$c^2 > a^2 + b^2,$$

the maximum value of $a \cos \theta + b \sin \theta + c$ is $c + \sqrt{a^2 + b^2}$,
and the minimum value is $c - \sqrt{a^2 + b^2}$.

318. The expression $a \cos(\alpha + \theta) + b \cos(\beta + \theta)$

$$= (a \cos \alpha + b \cos \beta) \cos \theta - (a \sin \alpha + b \sin \beta) \sin \theta;$$

and therefore its numerically greatest values are equal to the positive and negative square roots of

$$(a \cos \alpha + b \cos \beta)^2 + (a \sin \alpha + b \sin \beta)^2;$$

that is, are equal to

$$\pm \sqrt{a^2 + b^2 + 2ab \cos(\alpha - \beta)}.$$

In like manner, we may find the maximum and minimum values of the sum of any number of expressions of the form $a \cos(\alpha + \theta)$ or $a \sin(\alpha + \theta)$.

319. *If α and β are two angles, each lying between* 0 *and* $\frac{\pi}{2}$, *whose sum is given, to find the maximum value of* $\cos \alpha \cos \beta$ *and of* $\cos \alpha + \cos \beta$.

Suppose that $$\alpha + \beta = \sigma;$$

then $$2 \cos \alpha \cos \beta = \cos(\alpha + \beta) + \cos(\alpha - \beta)$$
$$= \cos \sigma + \cos(\alpha - \beta),$$

and is therefore a maximum when $\alpha-\beta=0$, or $\alpha=\beta=\frac{\sigma}{2}$.

Thus the maximum value of $\cos\alpha\cos\beta$ is $\cos^2\frac{\sigma}{2}$.

Again,
$$\cos\alpha+\cos\beta=2\cos\frac{\alpha+\beta}{2}\cos\frac{\alpha-\beta}{2}$$
$$=2\cos\frac{\sigma}{2}\cos\frac{\alpha-\beta}{2},$$

and is therefore a maximum when $\alpha=\beta=\frac{\sigma}{2}$.

Thus the maximum value of $\cos\alpha+\cos\beta$ is $2\cos\frac{\sigma}{2}$.

Similar theorems hold in case of the sine.

Example 1. If A, B, C are the angles of a triangle, find the maximum value of

$$\sin A+\sin B+\sin C \text{ and of } \sin A\sin B\sin C.$$

Let us suppose that C remains constant, while A and B vary.

$$\sin A+\sin B+\sin C=2\sin\frac{A+B}{2}\cos\frac{A-B}{2}+\sin C$$
$$=2\cos\frac{C}{2}\cos\frac{A-B}{2}+\sin C.$$

This expression is a maximum when $A=B$.

Hence, so long as any two of the angles A, B, C are unequal, the expression $\sin A+\sin B+\sin C$ is not a maximum; that is, the expression is a maximum when $A=B=C=60°$.

Thus the maximum value $=3\sin 60°=\frac{3\sqrt{3}}{2}$.

Again,
$$2\sin A\sin B\sin C=\{\cos(A-B)-\cos(A+B)\}\sin C$$
$$=\{\cos(A-B)+\cos C\}\sin C.$$

This expression is a maximum when $A=B$.

Hence, by reasoning as before, $\sin A\sin B\sin C$ has its maximum value when $A=B=C=60°$.

Thus the maximum value $=\sin^3 60°=\frac{3\sqrt{3}}{8}$.

Example 2. If α and β are two angles, each lying between 0 and $\frac{\pi}{2}$, whose sum is constant, find the minimum value of $\sec\alpha+\sec\beta$.

We have $\sec\alpha+\sec\beta=\dfrac{1}{\cos\alpha}+\dfrac{1}{\cos\beta}=\dfrac{\cos\alpha+\cos\beta}{\cos\alpha\cos\beta}$

$$=\frac{4\cos\frac{\alpha+\beta}{2}\cos\frac{\alpha-\beta}{2}}{\cos(\alpha+\beta)+\cos(\alpha-\beta)}=\frac{2\cos\frac{\alpha+\beta}{2}\cos\frac{\alpha-\beta}{2}}{\cos^2\frac{\alpha-\beta}{2}-\sin^2\frac{\alpha+\beta}{2}}$$

$$=\cos\frac{\alpha+\beta}{2}\left(\frac{1}{\cos\frac{\alpha-\beta}{2}+\sin\frac{\alpha+\beta}{2}}+\frac{1}{\cos\frac{\alpha-\beta}{2}-\sin\frac{\alpha+\beta}{2}}\right).$$

Since $\alpha+\beta$ is constant, this expression is least when the denominators are greatest; that is, when $\alpha=\beta=\dfrac{\alpha+\beta}{2}$.

Thus the minimum value is $2\sec\dfrac{\alpha+\beta}{2}$.

320. *If $\alpha, \beta, \gamma, \delta, \ldots\ldots$ are n angles, each lying between* 0 *and $\frac{\pi}{2}$, whose sum is constant, to find the maximum value of*

$$\cos\alpha\cos\beta\cos\gamma\cos\delta\ldots\ldots$$

Let $$\alpha+\beta+\gamma+\delta+\ldots\ldots=\sigma.$$

Suppose that any two of the angles, say α and β, are unequal; then if in the given product we replace the two unequal factors $\cos\alpha$ and $\cos\beta$ by the two equal factors $\cos\dfrac{\alpha+\beta}{2}$ and $\cos\dfrac{\alpha+\beta}{2}$, the value of the product is increased while the sum of the angles remains unaltered. Hence so long as any two of the angles $\alpha, \beta, \gamma, \delta, \ldots$ are unequal the product is not a maximum; that is, the product is a maximum when all the angles are equal. In this case each angle$=\dfrac{\sigma}{n}$.

Thus the maximum value is $\cos^n\dfrac{\sigma}{n}$.

In like manner we may shew that

the maximum value of $\cos\alpha+\cos\beta+\cos\gamma+\ldots\ldots=n\cos\dfrac{\sigma}{n}$.

321. The methods of solution used in the following examples are worthy of notice.

Example 1. Shew that $\tan 3\alpha \cot \alpha$ cannot lie between 3 and $\frac{1}{3}$.

We have $$\tan 3\alpha \cot \alpha = \frac{\tan 3\alpha}{\tan \alpha} = \frac{3 - \tan^2 \alpha}{1 - 3\tan^2 \alpha} = n \text{ say};$$

$$\therefore \tan^2 \alpha = \frac{n-3}{3n-1} = \frac{3-n}{1-3n}.$$

These two fractional values of $\tan^2 \alpha$ must be positive, and therefore n must be greater than 3 or less than $\frac{1}{3}$.

Example 2. If a and b are positive quantities, of which a is the greater, find the minimum value of $a \sec \theta - b \tan \theta$.

Denote the expression by x, and put $\tan \theta = t$;

then $$x = a\sqrt{1+t^2} - bt;$$

$$\therefore b^2t^2 + 2bxt + x^2 = a^2(1+t^2);$$

$$\therefore t^2(b^2 - a^2) + 2bxt + x^2 - a^2 = 0.$$

In order that the values of t found from this equation may be real,

$$b^2x^2 > (b^2 - a^2)(x^2 - a^2);$$

$$\therefore 0 > a^2(a^2 - b^2 - x^2);$$

$$\therefore x^2 > a^2 - b^2.$$

Thus the minimum value is $\sqrt{a^2 - b^2}$.

Example 3. If a, b, c, k are constant quantities and α, β, γ variable quantities subject to the relation $a \tan \alpha + b \tan \beta + c \tan \gamma = k$, find the minimum value of $\tan^2 \alpha + \tan^2 \beta + \tan^2 \gamma$.

By multiplying out and re-arranging the terms, we have

$$(a^2+b^2+c^2)(\tan^2\alpha + \tan^2\beta + \tan^2\gamma) - (a\tan\alpha + b\tan\beta + c\tan\gamma)^2$$
$$= (b\tan\gamma - c\tan\beta)^2 + (c\tan\alpha - a\tan\gamma)^2 + (a\tan\beta - b\tan\alpha)^2.$$

But the minimum value of the right side of this equation is zero; hence the minimum value of

$$(a^2+b^2+c^2)(\tan^2\alpha + \tan^2\beta + \tan^2\gamma) - k^2 = 0;$$

that is, the minimum value of

$$\tan^2\alpha + \tan^2\beta + \tan^2\gamma = \frac{k^2}{a^2+b^2+c^2}.$$

EXAMPLES. XXV. a.

When θ is variable find the minimum value of the following expressions:

1. $p \cot \theta + q \tan \theta$.
2. $4 \sin^2 \theta + \operatorname{cosec}^2 \theta$.
3. $8 \sec^2 \theta + 18 \cos^2 \theta$.
4. $3 - 2 \cos \theta + \cos^2 \theta$.

Prove the following inequalities:

5. $\tan^2 a + \tan^2 \beta + \tan^2 \gamma > \tan \beta \tan \gamma + \tan \gamma \tan a + \tan a \tan \beta$.

6. $\sin^2 a + \sin^2 \beta > 2 (\sin a + \sin \beta - 1)$.

When θ is variable, find the maximum value of

7. $\sin \theta + \cos \theta$.
8. $\cos \theta + \sqrt{3} \sin \theta$.
9. $a \cos (a + \theta) + b \sin \theta$.
10. $p \cos \theta + q \sin (a + \theta)$.

If $\sigma = a + \beta$, where a and β are two angles each lying between 0 and $\frac{\pi}{2}$, and σ is constant, find the maximum or minimum value of

11. $\sin a + \sin \beta$.
12. $\sin a \sin \beta$.
13. $\tan a + \tan \beta$.
14. $\operatorname{cosec} a + \operatorname{cosec} \beta$.

If A, B, C are the angles of a triangle, find the maximum or minimum value of

15. $\cos A \cos B \cos C$.
16. $\cot A + \cot B + \cot C$.
17. $\sin^2 \frac{A}{2} + \sin^2 \frac{B}{2} + \sin^2 \frac{C}{2}$.
18. $\sec A + \sec B + \sec C$.
19. $\tan^2 \frac{A}{2} + \tan^2 \frac{B}{2} + \tan^2 \frac{C}{2}$. $\left[Use\ \Sigma \tan \frac{B}{2} \tan \frac{C}{2} = 1.\right]$
20. $\cot^2 A + \cot^2 B + \cot^2 C$. [*Use* $\Sigma \cot B \cot C = 1$.]
21. If $b^2 < 4ac$, find the maximum and minimum values of
$$a \sin^2 \theta + b \sin \theta \cos \theta + c \cos^2 \theta.$$

22. If α, β, γ lie between 0 and $\frac{\pi}{2}$, shew that

$$\sin\alpha+\sin\beta+\sin\gamma>\sin(\alpha+\beta+\gamma).$$

23. If a and b are two positive quantities of which a is the greater, shew that $a\operatorname{cosec}\theta>b\cot\theta+\sqrt{a^2-b^2}$.

24. Shew that $\dfrac{\sec^2\theta-\tan\theta}{\sec^2\theta+\tan\theta}$ lies between 3 and $\dfrac{1}{3}$.

25. Find the maximum value of $\dfrac{\tan^2\theta-\cot^2\theta+1}{\tan^2\theta+\cot^2\theta-1}$.

26. If a, b, c, k are constant positive quantities, and α, β, γ variable quantities subject to the relation

$$a\cos\alpha+b\cos\beta+c\cos\gamma=k,$$

find the minimum value of

$$\cos^2\alpha+\cos^2\beta+\cos^2\gamma \text{ and of } a\cos^2\alpha+b\cos^2\beta+c\cos^2\gamma.$$

Elimination.

322. No general rules can be given for the elimination of some assigned quantity or quantities from two or more trigonometrical equations. The form of the equations will often suggest special methods, and in addition to the usual algebraical artifices we shall always have at our disposal the identical relations subsisting between the trigonometrical functions. Thus suppose it is required to eliminate θ from the equations

$$x\cos\theta=a, \quad y\cot\theta=b.$$

Here $$\sec\theta=\frac{x}{a}, \text{ and } \tan\theta=\frac{y}{b};$$

but for all values of θ, we have

$$\sec^2\theta-\tan^2\theta=1.$$

$\therefore$ by substitution,

$$\frac{x^2}{a^2}-\frac{y^2}{b^2}=1.$$

From this example we see that since θ satisfies *two* equations (either of which is sufficient to determine θ) there is a relation, independent of θ, which subsists between the coefficients and constants of the equations. To determine this relation we *eliminate* θ, and the result is called the *eliminant* of the given equations.

323. The following examples will illustrate some useful methods of elimination.

Example 1. Eliminate θ between the equations

$$l\cos\theta+m\sin\theta+n=0 \text{ and } p\cos\theta+q\sin\theta+r=0.$$

From the given equations, we have by cross multiplication

$$\frac{\cos\theta}{mr-nq}=\frac{\sin\theta}{np-lr}=\frac{1}{lq-mp};$$

$$\therefore \cos\theta=\frac{mr-nq}{lq-mp}, \text{ and } \sin\theta=\frac{np-lr}{lq-mp};$$

whence by squaring, adding, and clearing of fractions, we obtain

$$(mr-nq)^2+(np-lr)^2=(lq-mp)^2.$$

The particular instance in which $q=l$ and $p=-m$ is of frequent occurrence in Analytical Geometry. In this case the eliminant may be written down at once; for we have

$$l\cos\theta+m\sin\theta=-n,$$

and

$$l\sin\theta-m\cos\theta=-r;$$

whence by squaring and adding, we obtain

$$l^2+m^2=n^2+r^2.$$

Example 2. Eliminate θ between the equations

$$\frac{ax}{\cos\theta}-\frac{by}{\sin\theta}=c^2 \text{ and } l\tan\theta=m.$$

From the second equation, we have

$$\frac{\sin\theta}{m}=\frac{\cos\theta}{l}=\frac{\sqrt{\sin^2\theta+\cos^2\theta}}{\sqrt{m^2+l^2}}=\frac{1}{\sqrt{m^2+l^2}};$$

$$\therefore \sin\theta=\frac{m}{\sqrt{m^2+l^2}}, \text{ and } \cos\theta=\frac{l}{\sqrt{m^2+l^2}}.$$

By substituting in the first equation, we obtain

$$\frac{ax}{l}-\frac{by}{m}=\frac{c^2}{\sqrt{m^2+l^2}}.$$

Example 3. **Eliminate θ between the equations**

$$x=\cot\theta+\tan\theta \text{ and } y=\sec\theta-\cos\theta.$$

From the given equations, we have

$$x=\frac{1}{\tan\theta}+\tan\theta=\frac{1+\tan^2\theta}{\tan\theta}$$

$$=\frac{\sec^2\theta}{\tan\theta},$$

and

$$y=\sec\theta-\frac{1}{\sec\theta}=\frac{\sec^2\theta-1}{\sec\theta}$$

$$=\frac{\tan^2\theta}{\sec\theta}.$$

From these values of x and y we obtain

$$x^2y=\sec^3\theta \text{ and } xy^2=\tan^3\theta.$$

But

$$\sec^2\theta-\tan^2\theta=1;$$

$$\therefore (x^2y)^{\frac{2}{3}}-(xy^2)^{\frac{2}{3}}=1;$$

that is,

$$x^{\frac{4}{3}}y^{\frac{2}{3}}-x^{\frac{2}{3}}y^{\frac{4}{3}}=1.$$

Example 4. Eliminate θ from the equations

$$\frac{x}{a}=\cos\theta+\cos 2\theta \text{ and } \frac{y}{b}=\sin\theta+\sin 2\theta.$$

From the given equations, we have

$$\frac{x}{a}=2\cos\frac{3\theta}{2}\cos\frac{\theta}{2},$$

and

$$\frac{y}{b}=2\sin\frac{3\theta}{2}\cos\frac{\theta}{2};$$

whence by squaring and adding, we obtain

$$\frac{x^2}{a^2}+\frac{y^2}{b^2}=4\cos^2\frac{\theta}{2}.$$

But

$$\frac{x}{a}=2\cos\frac{\theta}{2}\left(4\cos^3\frac{\theta}{2}-3\cos\frac{\theta}{2}\right)$$

$$=2\cos^2\frac{\theta}{2}\left(4\cos^2\frac{\theta}{2}-3\right);$$

$$\therefore \frac{2x}{a}=\left(\frac{x^2}{a^2}+\frac{y^2}{b^2}\right)\left(\frac{x^2}{a^2}+\frac{y^2}{b^2}-3\right).$$

324. The following examples are instances of the elimination of two quantities.

Example 1. Eliminate θ and ϕ from the equations

$$a\sin^2\theta+b\cos^2\theta=m,\quad b\sin^2\phi+a\cos^2\phi=n,\quad a\tan\theta=b\tan\phi.$$

From the first equation, we have

$$a\sin^2\theta+b\cos^2\theta=m(\sin^2\theta+\cos^2\theta);$$

$$\therefore\ (a-m)\sin^2\theta=(m-b)\cos^2\theta;$$

$$\therefore\ \tan^2\theta=\frac{m-b}{a-m}.$$

From the second equation, we have

$$b\sin^2\phi+a\cos^2\phi=n(\sin^2\phi+\cos^2\phi);$$

$$\therefore\ \tan^2\phi=\frac{n-a}{b-n}.$$

From the third equation,

$$a^2\tan^2\theta=b^2\tan^2\phi;$$

$$\therefore\ \frac{a^2(m-b)}{a-m}=\frac{b^2(n-a)}{b-n};$$

$$\therefore\ a^2(bm-b^2-mn+bn)=b^2(an-a^2-mn+am);$$

$$\therefore\ mab(a-b)+nab(a-b)=mn(a^2-b^2);$$

$$\therefore\ mab+nab=mn(a+b);$$

$$\therefore\ \frac{1}{n}+\frac{1}{m}=\frac{1}{a}+\frac{1}{b}.$$

Example 2. Eliminate θ and ϕ from the equations

$$x\cos\theta+y\sin\theta=x\cos\phi+y\sin\phi=2a,\quad 2\sin\frac{\theta}{2}\sin\frac{\phi}{2}=1.$$

From the data, we see that θ and ϕ are the roots of the equation

$$x\cos\alpha+y\sin\alpha=2a;$$

$$\therefore\ (x\cos\alpha-2a)^2=y^2\sin^2\alpha=y^2(1-\cos^2\alpha);$$

$$\therefore\ (x^2+y^2)\cos^2\alpha-4ax\cos\alpha+4a^2-y^2=0,$$

which is a quadratic in $\cos\alpha$ with roots $\cos\theta$ and $\cos\phi$.

But $$1=4\sin^2\frac{\theta}{2}\sin^2\frac{\phi}{2}=(1-\cos\theta)(1-\cos\phi);$$

whence $\cos\theta + \cos\phi = \cos\theta\cos\phi$;

$$\therefore \frac{4ax}{x^2+y^2} = \frac{4a^2-y^2}{x^2+y^2};$$

$$\therefore y^2 = 4a(a-x).$$

325. The method exhibited in the following example is one frequently used in Analytical Geometry.

Example. If a, b, c are unequal, find the relations that hold between the coefficients, when

$$a\cos\theta + b\sin\theta = c,$$

and $$a\cos^2\theta + 2a\cos\theta\sin\theta + b\sin^2\theta = c.$$

The required relation will be obtained by eliminating θ from the given equations. This is most conveniently done by making each equation homogeneous in $\sin\theta$ and $\cos\theta$.

From the first equation, we have

$$a\cos\theta + b\sin\theta = c\sqrt{\cos^2\theta + \sin^2\theta};$$

whence, by squaring and transposing,

$$(a^2-c^2)\cos^2\theta + 2ab\cos\theta\sin\theta + (b^2-c^2)\sin^2\theta = 0 \quad \text{......(1)}.$$

From the second equation, we have

$$a\cos^2\theta + 2a\cos\theta\sin\theta + b\sin^2\theta = c(\cos^2\theta + \sin^2\theta);$$

$$\therefore (a-c)\cos^2\theta + 2a\cos\theta\sin\theta + (b-c)\sin^2\theta = 0 \quad \text{......(2)}.$$

From (1) and (2) we have by cross-multiplication,

$$\frac{\cos^2\theta}{2ab(b-c) - 2a(b^2-c^2)} = \frac{\cos\theta\sin\theta}{(b^2-c^2)(a-c) - (a^2-c^2)(b-c)}$$

$$= \frac{\sin^2\theta}{2a(a^2-c^2) - 2ab(a-c)};$$

or $$\frac{\cos^2\theta}{-2ac(b-c)} = \frac{\cos\theta\sin\theta}{(b-c)(a-c)(b-a)} = \frac{\sin^2\theta}{2a(a-c)(a+c-b)};$$

$$\therefore -4a^2c(b-c)(a-c)(a+c-b) = (b-c)^2(a-c)^2(b-a)^2.$$

By supposition, the quantities a, b, c are unequal; hence dividing by $(b-c)(a-c)$, we obtain

$$4a^2c(a+c-b) + (b-c)(a-c)(a-b)^2 = 0.$$

EXAMPLES. XXV. b.

Eliminate θ between the equations:

1. $\frac{x}{a}\cos\theta+\frac{y}{b}\sin\theta=1, \quad \frac{x}{a}\sin\theta-\frac{y}{b}\cos\theta=1.$

2. $a\sec\theta-x\tan\theta=y, \quad b\sec\theta+y\tan\theta=x.$

3. $\cos\theta+\sin\theta=a, \quad \cos 2\theta=b,$

4. $x=\sin\theta+\cos\theta, \quad y=\tan\theta+\cot\theta.$

5. $a=\cot\theta+\cos\theta, \quad b=\cot\theta-\cos\theta.$

Find the eliminant in each of the following cases:

6. $x=\cot\theta+\tan\theta, \quad y=\operatorname{cosec}\theta-\sin\theta.$

7. $\operatorname{cosec}\theta-\sin\theta=a^3, \quad \sec\theta-\cos\theta=b^3.$

8. $4x=3a\cos\theta+a\cos 3\theta, \quad 4y=3a\sin\theta-a\sin 3\theta.$

9. $x=\tan^2\theta\,(a\tan\theta-x), \quad y=\sec^2\theta\,(y-a\sec\theta).$

10. $x=a\cos\theta\,(2\cos 2\theta-1), \quad y=b\sin\theta\,(4\cos^2\theta-1).$

11. If $\cos(\theta-\alpha)=a$, and $\sin(\theta-\beta)=b$,
shew that $a^2-2ab\sin(\alpha-\beta)+b^2=\cos^2(\alpha-\beta)$.

Find the relation that must hold between x and y if

12. $x+y=3-\cos 4\theta, \quad x-y=4\sin 2\theta.$

13. $x=\sin\theta+\cos\theta\sin 2\theta, \quad y=\cos\theta+\sin\theta\sin 2\theta.$

14. If $\sin\theta+\cos\theta=a$, and $\sin 2\theta+\cos 2\theta=b$,
shew that $(a^2-b-1)^2=a^2(2-a^2)$.

15. If $\cos\theta-\sin\theta=b$, and $\cos 3\theta+\sin 3\theta=a$,
shew that $a=3b-2b^3$.

16. Eliminate θ from the equations:
$$a\cos\theta-b\sin\theta=c, \quad 2ab\cos 2\theta+(a^2-b^2)\sin 2\theta=2c^2.$$

17. If $x=a\cos\theta+b\cos 2\theta$, and $y=a\sin\theta+b\sin 2\theta$,
shew that $a^2\{(x+b)^2+y^2\}=(x^2+y^2-b^2)^2$.

18. If $\dfrac{\tan(\theta+a)}{\tan(\theta-a)}=\dfrac{a+b}{a-b}$, and $a\cos 2a+b\cos 2\theta=c$,

shew that $$a^2+c^2-2ac\cos 2a=b^2.$$

19. If $x=a(\sin 3\theta-\sin\theta)$, and $y=a(\cos\theta-\cos 3\theta)$,

shew that $$(x^2+y^2)(2a^2-x^2-y^2)^2=4a^4x^2.$$

Eliminate θ from the equations:

20. $x\cos\theta-y\sin\theta=a\cos 2\theta,\quad x\sin\theta+y\cos\theta=2a\sin 2\theta.$

21. $x\sin\theta-y\cos\theta=\sqrt{x^2+y^2},\quad \dfrac{\cos^2\theta}{a^2}+\dfrac{\sin^2\theta}{b^2}=\dfrac{1}{x^2+y^2}.$

22. $\dfrac{x\cos\theta}{a}+\dfrac{y\sin\theta}{b}=1,\quad x\sin\theta-y\cos\theta=\sqrt{a^2\sin^2\theta+b^2\cos^2\theta}.$

23. If $\cos(a-3\theta)=m\cos^3\theta$, and $\sin(a-3\theta)=m\sin^3\theta$,

shew that $$m^2+m\cos a=2.$$

Eliminate θ and ϕ from the equations:

24. $\tan\theta+\tan\phi=x,\quad \cot\theta+\cot\phi=y,\quad \theta+\phi=a.$

25. $\sin\theta+\sin\phi=a,\quad \cos\theta+\cos\phi=b,\quad \theta-\phi=a.$

26. $a\sin^2\theta+b\cos^2\theta=a\cos^2\phi+b\sin^2\phi=1,\quad a\tan\theta=b\tan\phi.$

27. If $\dfrac{x}{a}\cos\theta+\dfrac{y}{b}\sin\theta=\dfrac{x}{a}\cos\phi+\dfrac{y}{b}\sin\phi=1$, and $\theta-\phi=a$,

shew that $$\frac{x^2}{a^2}+\frac{y^2}{b^2}=\sec^2\frac{a}{2}.$$

28. If $\tan\theta+\tan\phi=a,\quad \cot\theta+\cot\phi=b,\quad \theta-\phi=a,$

shew that $$ab(ab-4)=(a+b)^2\tan^2 a.$$

Eliminate θ and ϕ between the equations:

29. $a\cos^2\theta+b\sin^2\theta=m\cos^2\phi,\quad a\sin^2\theta+b\cos^2\theta=n\sin^2\phi,$
$$m\tan^2\theta-n\tan^2\phi=0.$$

30. $x\cos\theta+y\sin\theta=2a\sqrt{3},\quad x\cos(\theta+\phi)+y\sin(\theta+\phi)=4a,$
$$x\cos(\theta-\phi)+y\sin(\theta-\phi)=2a.$$

31. $c\sin\theta=a\sin(\theta+\phi),\quad a\sin\phi=b\sin\theta,\quad \cos\theta-\cos\phi=2m.$

Application of Trigonometry to the Theory of Equations.

326. In the Theory of Equations it is shewn that the solution of *any* cubic equation may be made to depend on the solution of a cubic equation of the form $x^3+ax+b=0$. In certain cases the solution is very conveniently obtained by Trigonometry.

327. Consider the equation

$$x^3-qx-r=0 \quad \text{.........................(1)},$$

in which each of the letters q and r represents a positive quantity.

From the identity $\cos 3\theta=4\cos^3\theta-3\cos\theta$,

we have
$$\cos^3\theta-\frac{3}{4}\cos\theta-\frac{\cos 3\theta}{4}=0 \quad \text{.................(2)}.$$

Let $x=y\cos\theta$, where y is a positive quantity; then from (1),

$$\cos^3\theta-\frac{q}{y^2}\cos\theta-\frac{r}{y^3}=0 \text{.......................(3)}.$$

If the equations (2) and (3) are identical, we have $\frac{q}{y^2}=\frac{3}{4}$, so that

$y=+\sqrt{\frac{4q}{3}}$, since y is positive; and

$$\frac{\cos 3\theta}{4}=\frac{r}{y^3}=\sqrt{\frac{27r^2}{64q^3}};$$

whence
$$\cos 3\theta=\sqrt{\frac{27r^2}{4q^3}}.$$

Hence the values of θ are real if $27r^2<4q^3$;

that is, if
$$\left(\frac{r}{2}\right)^2<\left(\frac{q}{3}\right)^3.$$

Let α be the smallest angle whose cosine is equal to $\sqrt{\frac{27r^2}{4q^3}}$; then $\cos 3\theta=\cos\alpha$; whence $3\theta=2n\pi\pm\alpha$.

Thus the values of $\cos\theta$ are

$$\cos\frac{\alpha}{3},\quad \cos\frac{2\pi+\alpha}{3},\quad \cos\frac{2\pi-\alpha}{3}.$$ [See Art. 264.]

But $$x = y\cos\theta = \sqrt{\frac{4q}{3}}\cos\theta,$$

and therefore the roots of $x^3 - qx - r = 0$ are

$$\sqrt{\frac{4q}{3}}\cos\alpha,\quad \sqrt{\frac{4q}{3}}\cos\frac{2\pi+\alpha}{3},\quad \sqrt{\frac{4q}{3}}\cos\frac{2\pi-\alpha}{3}.$$

328. Following the method explained in the preceding article, we may use the identity

$$\sin^3\theta - \frac{3}{4}\sin\theta + \frac{\sin 3\theta}{4} = 0$$

to obtain the solution of the equation

$$x^3 - qx + r = 0,$$

each of the quantities represented by q and r being positive.

Example. Solve the equation $x^3 - 12x + 8 = 0$.

We have $$\sin^3\theta - \frac{3}{4}\sin\theta + \frac{\sin 3\theta}{4} = 0.$$

In the given equation put $x = y\sin\theta$, where y is positive; then

$$\sin^3\theta - \frac{12}{y^2}\sin\theta + \frac{8}{y^3} = 0.$$

$$\therefore\ \frac{3}{4} = \frac{12}{y^2};\ \text{whence } y = 4;$$

and $$\frac{\sin 3\theta}{4} = \frac{8}{y^3} = \frac{1}{8};\ \text{whence } \sin 3\theta = \frac{1}{2}.$$

Suppose that θ is estimated in sexagesimal measure; then

$$3\theta = n.180^\circ + (-1)^n 30^\circ.$$

By ascribing to n the values 0, 1, 2, 3, 4 we obtain

$$\theta = 10^\circ,\quad \theta = 50^\circ,\quad \theta = 130^\circ,\quad \theta = 170^\circ,\quad \theta = 250^\circ;$$

and by further ascribing to n the values 5, 6, 7,... it will easily be seen that the values of $\sin\theta$ are equal to some one of the three quantities

$$\sin 10^\circ,\quad \sin 50^\circ,\quad -\sin 70^\circ.$$

But $x = y\sin\theta = 4\sin\theta$, and therefore the roots are

$$4\sin 10^\circ,\quad 4\sin 50^\circ,\quad -4\sin 70^\circ.$$

Application of the Theory of Equations to Trigonometry.

329. In the Theory of Equations it is shewn that the equation whose roots are a_1, a_2, a_3, ..., a_n is

$$(x-a_1)(x-a_2)(x-a_3)\ldots\ldots\ldots\ldots(x-a_n)=0,$$

or $$x^n - S_1x^{n-1} + S_2x^{n-2} - S_3x^{n-3} + \ldots\ldots + (-1)^n S_n = 0,$$

where S_1 = sum of the roots;

S_2 = sum of the products of the roots taken two at a time;

S_3 = sum of the products of the roots taken three at a time;

....

S_n = product of the roots.

[See Hall and Knight's *Higher Algebra*, Art. 538 and Art. 539.]

Example 1. If α, β, γ are the values of θ which satisfy the equation

$$a\tan^3\theta + (2a-x)\tan\theta + y = 0 \quad \ldots\ldots\ldots\ldots\ldots\ldots(1),$$

shew that (i) if $\tan\alpha + \tan\beta = h$, then $ah^3 + (2a-x)h = y$;

(ii) if $\tan\alpha\tan\beta = k$, then $y^2 + (2a-x)ak^2 = a^2k^3$.

(i) From the theory of equations, we have from (1),

$$\tan\alpha + \tan\beta + \tan\gamma = 0;$$

$$\therefore\ h + \tan\gamma = 0, \text{ or } \tan\gamma = -h.$$

But $$a\tan^3\gamma + (2a-x)\tan\gamma + y = 0;$$

$$\therefore\ ah^3 + (2a-x)h - y = 0.$$

(ii) From the theory of equations, we have from (1),

$$\tan\alpha\tan\beta\tan\gamma = -\frac{y}{a};$$

$$\therefore\ k\tan\gamma = -\frac{y}{a}, \text{ or } \tan\gamma = -\frac{y}{ak}.$$

Substituting in $a\tan^3\gamma + (2a-x)\tan\gamma + y = 0$, we have

$$-\frac{ay^3}{a^3k^3} - (2a-x)\frac{y}{ak} + y = 0;$$

$$\therefore\ y^2 + (2a-x)ak^2 - a^2k^3 = 0.$$

Example 2. Shew that

$$\cos^2 \alpha + \cos^2\left(\frac{2\pi}{3}+\alpha\right) + \cos^2\left(\frac{2\pi}{3}-\alpha\right) = \frac{3}{2}.$$

Suppose that $\cos 3\theta = k$;

then $4\cos^3\theta - 3\cos\theta = \cos 3\theta = k$;

$$\therefore\ \cos^3\theta - \frac{3}{4}\cos\theta - \frac{k}{4} = 0.$$

The roots of this cubic in $\cos\theta$ are

$$\cos\alpha,\ \cos\left(\frac{2\pi}{3}+\alpha\right),\ \text{and}\ \cos\left(\frac{2\pi}{3}-\alpha\right),$$

where α is any angle which satisfies the equation $\cos 3\alpha = k$. For shortness, denote the roots by a, b, c; then

$$a^2+b^2+c^2 = (a+b+c)^2 - 2(bc+ca+ab)$$

$$= 0 - 2\left(-\frac{3}{4}\right);$$

$$\therefore\ \cos^2 \alpha + \cos^2\left(\frac{2\pi}{3}+\alpha\right) + \cos^2\left(\frac{2\pi}{3}-\alpha\right) = \frac{3}{2}.$$

330. If $5\theta = 2n\pi$, where n is any integer, we have

$$3\theta = 2n\pi - 2\theta;$$

$$\therefore\ \sin 3\theta = -\sin 2\theta.$$

The values of $\sin\theta$ found from this equation are

$$0,\ \sin\frac{2\pi}{5},\ \sin\frac{4\pi}{5},\ \sin\frac{6\pi}{5},\ \sin\frac{8\pi}{5},$$

being obtained by giving to n the values 0, 1, 2, 3, 4. It will easily be seen that no new values of $\sin\theta$ are obtained by ascribing to n the values 5, 6, 7, ...

But $$\sin\frac{6\pi}{5} = -\sin\frac{4\pi}{5} = -\sin\frac{\pi}{5},$$

and $$\sin\frac{8\pi}{5} = -\sin\frac{2\pi}{5};$$

hence rejecting the zero solution, the values of $\sin\theta$ found from the equation $\sin 3\theta = -\sin 2\theta$ are

$$\pm\sin\frac{\pi}{5},\ \text{and}\ \pm\sin\frac{2\pi}{5}.$$

If we put $\sin\theta=x$, the equation $\sin 3\theta=-\sin 2\theta$ becomes

$$3x-4x^3=-2x\sqrt{1-x^2}.$$

Dividing by x, and thus removing the solution $x=0$, we have

$$(3-4x^2)^2=4(1-x^2),$$

or
$$16x^4-20x^2+5=0.$$

This is a quadratic in x^2, and as we have just seen the values of x^2 are

$$\sin^2\frac{\pi}{5} \text{ and } \sin^2\frac{2\pi}{5}.$$

From the theory of quadratic equations, we have

$$\sin^2\frac{\pi}{5}+\sin^2\frac{2\pi}{5}=\frac{20}{16}=\frac{5}{4};$$

$$\sin^2\frac{\pi}{5}\sin^2\frac{2\pi}{5}=\frac{5}{16}.$$

Example. Shew that

$$\sin\frac{2\pi}{7}+\sin\frac{4\pi}{7}+\sin\frac{8\pi}{7}=\frac{1}{2}\sqrt{7}.$$

If $7\theta=2n\pi$, where n is any integer, we have

$$\sin 4\theta=-\sin 3\theta.$$

The values of $\sin\theta$ found from this equation are

$$0,\quad \pm\sin\frac{2\pi}{7},\quad \pm\sin\frac{4\pi}{7},\quad \pm\sin\frac{8\pi}{7},$$

since
$$\sin\frac{6\pi}{7}=-\sin\frac{8\pi}{7}.$$

If $\sin\theta=x$, the equation $\sin 4\theta=-\sin 3\theta$ becomes

$$4x(1-2x^2)\sqrt{1-x^2}=4x^3-3x;$$

whence rejecting the solution $x=0$, we obtain

$$16(1-4x^2+4x^4)(1-x^2)=16x^4-24x^2+9,$$

or
$$64x^6-112x^4+56x^2-7=0 \quad\text{.....................(1).}$$

The values of x^2 found from this equation are

$$\sin^2\frac{2\pi}{7},\quad \sin^2\frac{4\pi}{7},\quad \sin^2\frac{8\pi}{7};$$

hence $$\sin^2\frac{2\pi}{7}+\sin^2\frac{4\pi}{7}+\sin^2\frac{8\pi}{7}=\frac{112}{64}=\frac{7}{4}.$$

But $$\sin\frac{2\pi}{7}\sin\frac{4\pi}{7}+\sin\frac{2\pi}{7}\sin\frac{8\pi}{7}+\sin\frac{4\pi}{7}\sin\frac{8\pi}{7}$$

$$=\frac{1}{2}\left\{\left(\cos\frac{2\pi}{7}-\cos\frac{6\pi}{7}\right)+\left(\cos\frac{6\pi}{7}-\cos\frac{10\pi}{7}\right)+\left(\cos\frac{4\pi}{7}-\cos\frac{12\pi}{7}\right)\right\}$$

$$=0.$$

$$\therefore\ \left(\sin\frac{2\pi}{7}+\sin\frac{4\pi}{7}+\sin\frac{8\pi}{7}\right)^2=\sin^2\frac{2\pi}{7}+\sin^2\frac{4\pi}{7}+\sin^2\frac{8\pi}{7}=\frac{7}{4};$$

$$\therefore\ \sin\frac{2\pi}{7}+\sin\frac{4\pi}{7}+\sin\frac{8\pi}{7}=\frac{1}{2}\sqrt{7}.$$

331. If $7\theta=2n\pi$, where n is any integer, we have

$$4\theta=2n\pi-3\theta;$$

$$\therefore\ \cos 4\theta=\cos 3\theta.$$

By giving to n the values 0, 1, 2, 3, the values of $\cos\theta$ obtained from this equation are

$$1,\ \cos\frac{2\pi}{7},\ \cos\frac{4\pi}{7},\ \cos\frac{6\pi}{7}.$$

It will easily be seen that no new values of $\cos\theta$ are found by ascribing to n the values 4, 5, 6, 7, ...; for

$$\cos\frac{8\pi}{7}=\cos\frac{6\pi}{7},\quad \cos\frac{10\pi}{7}=\cos\frac{4\pi}{7},\quad \ldots\ldots\ldots\ldots$$

Now $$\cos 4\theta=8\cos^4\theta-8\cos^2\theta+1,$$

and therefore if $x=\cos\theta$, the equation $\cos 4\theta=\cos 3\theta$ becomes

$$8x^4-8x^2+1=4x^3-3x,$$

or $$8x^4-4x^3-8x^2+3x+1=0.$$

Removing the factor $x-1$, which corresponds to the root $\cos\theta=1$, we obtain

$$8x^3+4x^2-4x-1=0,$$

the roots of which equation are

$$\cos\frac{2\pi}{7},\ \cos\frac{4\pi}{7},\ \cos\frac{6\pi}{7}.$$

Example 1. Find the values of

$$\tan^2\frac{\pi}{7}+\tan^2\frac{2\pi}{7}+\tan^2\frac{3\pi}{7} \text{ and } \tan\frac{\pi}{7}\tan\frac{2\pi}{7}\tan\frac{3\pi}{7}.$$

If $7\theta=n\pi$, where n is any integer, we have

$$\tan 4\theta=-\tan 3\theta.$$

By writing $\tan\theta=t$, this equation becomes

$$\frac{4t-4t^3}{1-6t^2+t^4}=-\frac{3t-t^3}{1-3t^2},$$

or

$$t^6-21t^4+35t^2-7=0.$$

The roots of this cubic in t^2 are

$$\tan^2\frac{\pi}{7},\quad \tan^2\frac{2\pi}{7},\quad \tan^2\frac{3\pi}{7}.$$

$$\therefore\ \tan^2\frac{\pi}{7}+\tan^2\frac{2\pi}{7}+\tan^2\frac{3\pi}{7}=21,$$

and

$$\tan\frac{\pi}{7}\tan\frac{2\pi}{7}\tan\frac{3\pi}{7}=\sqrt{7},$$

the positive value of the square root being taken, since each of the factors on the left is positive.

Example 2. Shew that

$$\cos^4\frac{\pi}{7}+\cos^4\frac{2\pi}{7}+\cos^4\frac{3\pi}{7}=\frac{13}{16};$$

and

$$\sec^4\frac{\pi}{7}+\sec^4\frac{2\pi}{7}+\sec^4\frac{3\pi}{7}=416.$$

Let y denote any one of the quantities

$$\cos^2\frac{\pi}{7},\quad \cos^2\frac{2\pi}{7},\quad \cos^2\frac{3\pi}{7};$$

then $2y=1+x$, where x denotes one of the quantities

$$\cos\frac{2\pi}{7},\quad \cos\frac{4\pi}{7},\quad \cos\frac{6\pi}{7}.$$

From Art. 331, the equation whose roots are

$$\cos\frac{2\pi}{7},\quad \cos\frac{4\pi}{7},\quad \cos\frac{6\pi}{7}$$

is

$$8x^3+4x^2-4x-1=0;$$

whence by substituting $x=2y-1$, it follows that

$$\cos^2\frac{\pi}{7},\quad \cos^2\frac{2\pi}{7},\quad \cos^2\frac{3\pi}{7}$$

are the roots of the equation

$$8(2y-1)^3+4(2y-1)^2-4(2y-1)-1=0,$$

or

$$64y^3-80y^2+24y-1=0.$$

$$\therefore\ \cos^2\frac{\pi}{7}+\cos^2\frac{2\pi}{7}+\cos^2\frac{3\pi}{7}=\frac{80}{64}=\frac{5}{4};$$

and

$$\Sigma\cos^2\frac{\pi}{7}\cos^2\frac{2\pi}{7}=\frac{24}{64}=\frac{3}{8}.$$

By squaring the first of these equations and subtracting twice the second equation, we have

$$\cos^4\frac{\pi}{7}+\cos^4\frac{2\pi}{7}+\cos^4\frac{3\pi}{7}=\frac{13}{16}.$$

By putting $z=\dfrac{1}{y}$, we see that

$$\sec^2\frac{\pi}{7},\quad \sec^2\frac{2\pi}{7},\quad \sec^2\frac{3\pi}{7}$$

are the roots of the equation

$$z^3-24z^2+80z-64=0;$$

$$\therefore\ \sec^4\frac{\pi}{7}+\sec^4\frac{2\pi}{7}+\sec^4\frac{3\pi}{7}=(24)^2-(2\times 80)=416.$$

332. To find $\cos 5\theta$ and $\sin 5\theta$, we may proceed as follows:

$$\cos 5\theta+\cos\theta=2\cos 3\theta\cos 2\theta$$
$$=(4\cos^3\theta-3\cos\theta)(4\cos^2\theta-2);$$
$$\therefore\ \cos 5\theta=16\cos^5\theta-20\cos^3\theta+5\cos\theta.$$

$$\sin 5\theta+\sin\theta=2\sin 3\theta\cos 2\theta$$
$$=(3\sin\theta-4\sin^3\theta)(2-4\sin^2\theta);$$
$$\therefore\ \sin 5\theta=16\sin^5\theta-20\sin^3\theta+5\sin\theta.$$

It is easy to prove that

$$\cos 6\theta=32\cos^6\theta-48\cos^4\theta+18\cos^2\theta-1,$$

and

$$\sin 6\theta=\cos\theta\,(32\sin^5\theta-32\sin^3\theta+6\sin\theta).$$

EXAMPLES. XXV. c.

Solve the following equations :

1. $x^3-3x-1=0.$
2. $x^3-3x+1=0.$
3. $x^3-3x-\sqrt{3}=0.$
4. $8x^3-6x+\sqrt{2}=0.$
5. $8a^3x^3-6ax+2\sin 3A=0.$
6. $x^3-3a^2x-2a^3\cos 3A=0.$

7. If $\sin\alpha$ and $\sin\beta$ are the roots of the equation

$$a\sin^2\theta+b\sin\theta+c=0,$$

shew that (1) if $\sin\alpha+2\sin\beta=1$, then $a^2+2b^2+3ab+ac=0$,

(2) if $c\sin\alpha=a\sin\beta$, then $a+c=\pm b$.

8. If $\tan\alpha$ and $\tan\beta$ are the roots of the equation

$a\tan^2\theta-b\tan\theta+c=0$, and if $a\tan\alpha+b\tan\beta=2b$,

shew that $$b^2(2a-b)+c(a-b)^2=0.$$

9. If $\tan\alpha$, $\tan\beta$, $\tan\gamma$ are the roots of the equation

$$a\tan^3\theta+(2a-x)\tan\theta+y=0,$$

and if $a(\tan^2\alpha+\tan^2\beta)=2x-5a$, shew that $x\pm y=3a$.

10. If $\cos\alpha$, $\cos\beta$, $\cos\gamma$ are the roots of the equation

$$\cos^3\theta+a\cos^2\theta+b\cos\theta+c=0,$$

and if $\cos\alpha(\cos\beta+\cos\gamma)=2b$, prove that $abc+2b^3+c^2=0$.

Prove the following identities:

11. $\sec\alpha+\sec\left(\frac{2\pi}{3}+\alpha\right)+\sec\left(\frac{2\pi}{3}-\alpha\right)=-3\sec 3\alpha.$

12. $\sin^2\alpha+\sin^2\left(\frac{2\pi}{3}+\alpha\right)+\sin^2\left(\frac{4\pi}{3}+\alpha\right)=\frac{3}{2}.$

13. $\operatorname{cosec}\alpha+\operatorname{cosec}\left(\frac{2\pi}{3}+\alpha\right)+\operatorname{cosec}\left(\frac{4\pi}{3}+\alpha\right)=3\operatorname{cosec}3\alpha.$

14. $\operatorname{cosec}^2\frac{\pi}{5}+\operatorname{cosec}^2\frac{2\pi}{5}=4.$

15. $\cos\frac{2\pi}{5}+\cos\frac{4\pi}{5}=-\frac{1}{2}$, and $\cos\frac{2\pi}{5}\cos\frac{4\pi}{5}=-\frac{1}{4}.$

16. Form the equation whose roots are

(1) $\cos\frac{\pi}{7}$, $\cos\frac{3\pi}{7}$, $\cos\frac{5\pi}{7}$;

(2) $\sin^2\frac{\pi}{14}$, $\sin^2\frac{3\pi}{14}$, $\sin^2\frac{5\pi}{14}$.

17. Form the equation whose roots are

$$\sin^2\frac{\pi}{7},\quad \sin^2\frac{2\pi}{7},\quad \sin^2\frac{3\pi}{7};$$

and shew that $\sum_{n=1}^{n=3}\sin^4\frac{n\pi}{7}=\frac{21}{16}$ and $\sum_{n=1}^{n=3}\operatorname{cosec}^4\frac{n\pi}{7}=32$.

18. Form the equation whose roots are

(1) $\cos\frac{2\pi}{9}$, $\cos\frac{4\pi}{9}$, $\cos\frac{6\pi}{9}$, $\cos\frac{8\pi}{9}$;

(2) $\cos\frac{\pi}{9}$, $\cos\frac{3\pi}{9}$, $\cos\frac{5\pi}{9}$, $\cos\frac{7\pi}{9}$.

19. Form the equation whose roots are

$$\cos^2\frac{\pi}{9},\quad \cos^2\frac{2\pi}{9},\quad \cos^2\frac{3\pi}{9},\quad \cos^2\frac{4\pi}{9},$$

and shew that $\sum_{n=1}^{n=4}\cos^4\frac{n\pi}{9}=\frac{19}{16}$, and $\sum_{n=1}^{n=4}\sec^4\frac{n\pi}{9}=1120$.

20. Form the equation whose roots are

$$\tan^2\frac{\pi}{9},\quad \tan^2\frac{2\pi}{9},\quad \tan^2\frac{3\pi}{9},\quad \tan^2\frac{4\pi}{9},$$

and shew that $\cot^2\frac{\pi}{9}+\cot^2\frac{2\pi}{9}+\cot^2\frac{4\pi}{9}=9$.

21. Prove that

(1) $\operatorname{cosec}^2\frac{\pi}{7}+\operatorname{cosec}^2\frac{2\pi}{7}+\operatorname{cosec}^2\frac{3\pi}{7}=8$;

(2) $\cos\frac{\pi}{11}\cos\frac{2\pi}{11}\cos\frac{3\pi}{11}\cos\frac{4\pi}{11}\cos\frac{5\pi}{11}=\frac{1}{32}$.

MISCELLANEOUS EXAMPLES. I.

1. If $$a\tan\alpha+b\tan\beta=(a+b)\tan\frac{\alpha+\beta}{2},$$
prove that $$a\cos\beta=b\cos\alpha.$$

2. If $$\frac{\sin^4\alpha}{a}+\frac{\cos^4\alpha}{b}=\frac{1}{a+b},$$
prove that $$\frac{\sin^8\alpha}{a^3}+\frac{\cos^8\alpha}{b^3}=\frac{1}{(a+b)^3}.$$

3. Shew that
$$2\tan^{-1}\left\{\tan\frac{\alpha}{2}\tan\left(\frac{\pi}{4}-\frac{\beta}{2}\right)\right\}=\tan^{-1}\left(\frac{\sin\alpha\cos\beta}{\cos\alpha+\sin\beta}\right).$$

4. If the equation
$$\frac{\sin^{2n+2}\theta}{\sin^{2n}\alpha}+\frac{\cos^{2n+2}\theta}{\cos^{2n}\alpha}=1$$
is true when $n=1$, prove that it will be true when n is any positive integer.

5. If $a\cos\theta+b\sin\theta=c$ and $a\cos^2\theta+b\sin^2\theta=c$,
prove that $$4a^2b^2+(b-c)(a-c)(a-b)^2=0.$$

6. Prove the following identities:

(i) $\Sigma\sin(\beta-\gamma)\cos(\alpha-\beta)\cos(\alpha-\gamma)=-\Pi\sin(\beta-\gamma)$;

(ii) $\Sigma\sin\alpha\sin(\beta-\gamma)\cos(\beta+\gamma-\alpha)=0$;

(iii) $\Sigma\sin\alpha\sin(\beta-\gamma)\sin(\beta+\gamma-\alpha)=2\Pi\sin(\beta-\gamma)$.

7. If P be a point within a triangle ABC, such that
$$\angle PAB=\angle PBC=\angle PCA=\omega,$$
prove that (1) $\cot\omega=\cot A+\cot B+\cot C$;

(2) $\operatorname{cosec}^2\omega=\operatorname{cosec}^2A+\operatorname{cosec}^2B+\operatorname{cosec}^2C$.

8. A hill of inclination 1 in 169 faces West. Shew that a railway on it which runs S.E. has an inclination of 1 in 239.

9. Two vertical walls of equal height a are inclined to one another at an angle α. At noon the breadth of their shadows are b and c: shew that the altitude θ of the sun is given by the equation
$$a^2\sin^2\gamma\cot^2\theta=b^2+c^2+2bc\cos\gamma.$$

MISCELLANEOUS EXAMPLES. K.

I. (*Including Chapters I—VII.*)

1. Express in degrees and minutes and also in grades the vertical angle of an isosceles triangle in which each of the angles at the base is twelve times the vertical angle.

2. The angles of a triangle are as 4 : 5 : 6. Express them in radians.

3. Prove that $\dfrac{\cot A - \tan A}{\cot A + \tan A} = 1 - 2\sin^2 A$.

4. If A is an acute angle and $\sin A = \dfrac{5}{13}$, find the value of $\tan A + \sec A$.

5. The adjacent sides of a parallelogram are 15 ft. and 30 ft., and the included angle is 60°, find the length of the shorter diagonal to two places of decimals.

6. A tower 50 ft. high stands on the edge of a cliff. From a point in a horizontal plane through the foot of the cliff, the angular elevations of the top and bottom of the tower are observed to be α and β, where $\tan\alpha = 1\cdot26$ and $\tan\beta = 1\cdot185$. Find the height of the cliff.

---

7. Find the length of 10 degrees of a meridian upon a globe 60 ft. in diameter.

8. The sine of an angle is to its cosine as 8 to 15, find their actual values.

9. Find the values of θ from the equation

$$4\sin^2\theta+\sqrt{3}=2(1+\sqrt{3})\sin\theta.$$

10. If $\tan\alpha=\frac{4}{15}$, find the value of $\frac{5\sin\alpha+7\cos\alpha}{6\cos\alpha-3\sin\alpha}$.

11. Prove that

$$(1+\sin A+\cos A)^2=2(1+\sin A)(1+\cos A).$$

12. Simplify the expression

$$2\sec^2 A-\sec^4 A-2\operatorname{cosec}^2 A+\operatorname{cosec}^4 A,$$

giving the result in terms of $\tan A$.

---

13. If $\tan\theta=\frac{\sin\alpha-\cos\alpha}{\sin\alpha+\cos\alpha}$, prove that

$$\sqrt{2}\sin\theta=\sin\alpha-\cos\alpha.$$

14. Shew that the values of

$$\frac{\sin 45^\circ-\sin 30^\circ}{\cos 45^\circ+\cos 60^\circ} \text{ and } \frac{\sec 45^\circ-\tan 45^\circ}{\operatorname{cosec} 45^\circ+\cot 45^\circ}$$

are the same.

15. Prove that the multiplier which will convert any number of centesimal seconds into English minutes is ·0054.

16. Prove the identities:

(1) $\frac{\tan A-\tan B}{\cot B-\cot A}=\frac{\tan B}{\cot A}$;

(2) $\frac{1+\tan^2\theta}{1+\cot^2\theta}=\left(\frac{1-\tan\theta}{1-\cot\theta}\right)^2$.

17. Solve the equations:

(1) $\sin\theta+\operatorname{cosec}\theta=\frac{3}{\sqrt{2}}$; (2) $\cos\theta+\sec\theta=2\frac{1}{2}$.

18. A man running on a circular track at the rate of 10 miles an hour traverses an arc which subtends 56° at the centre in 36 seconds. Find the diameter of the circle. Take $\pi = \frac{22}{7}$.

---

19. If AD is drawn perpendicular to BC, the base of an equilateral triangle, and $BC = 2m$, find AD. Thence, from the figure, shew that

$$\cos^2 60° + \cot^2 30° = \frac{13}{4}.$$

20. Prove the identities:

(1) $(\sin^2 A + \cos^2 A)(\tan^2 A - 1) = (\tan^2 A + 1)(\sin^2 A - \cos^2 A)$.

(2) $\sin^2\alpha \cos^2\beta - \cos^2\alpha \sin^2\beta = \sin^2\alpha \sin^2\beta (\operatorname{cosec}^2\beta - \operatorname{cosec}^2\alpha)$.

21. In a triangle, right-angled at C, find c and b, given that

$$a + c = 281, \quad \cos B = \cdot 405.$$

22. On a globe of 6 miles diameter an arc of 2 fur. 55 yds. is measured: find the radian measure of the angle subtended at the centre of the globe.

If this was taken as the unit of measurement, how would a right angle be represented?

23. Shew that

(1) $$\sin\theta\cos\theta\left\{\sin\left(\frac{\pi}{2} - \theta\right)\operatorname{cosec}\theta + \cos\left(\frac{\pi}{2} - \theta\right)\sec\theta\right\} = 1;$$

(2) $$\frac{\tan\left(\frac{\pi}{2} - \theta\right)}{\sec\theta} \cdot \frac{\cot^2\theta}{\sec\left(\frac{\pi}{2} - \theta\right)} \cdot \frac{\sin\left(\frac{\pi}{2} - \theta\right)}{\sin^3\theta} = \cot^5\theta.$$

24. From a station two lighthouses A, B, are seen in directions N. and N.E. respectively; but if A were half as far off as it really is, it would appear due W. from B. Compare the distances of A and B from the station.

---

25. Find the numerical value of

$$3\tan^2 45° - \sin^2 60° - \frac{1}{2}\cot^2 30° + \frac{1}{8}\sec^2 45°;$$

and find x from the equation

$$\operatorname{cosec}(90° - A) - x\cos A \cot(90° - A) = \sin(90° - A).$$

26. Prove the identities:

(1) $(\sin A - \operatorname{cosec} A)^2 + (\cos A - \sec A)^2 = \cot^2 A + \tan^2 A - 1$;

(2) $(\cot\theta - 3)(3\cot\theta - 1) = (3\operatorname{cosec}^2\theta - 10\cot\theta)$.

27. If $\cot A = 4{\cdot}5$, find the value of $\dfrac{2\sin A - \cos A}{2\sin A + 3\cos A}$.

28. Find two values of θ which satisfy

$$2\cos\theta\cot\theta + 1 = \cot\theta + 2\cos\theta.$$

29. If an arc subtends $20°\ 17'$ at the centre of a circle whose radius is 6 inches, find in sexagesimal measure the angle it will subtend in a circle whose radius is 8 inches.

30. Looking due South from the top of a cliff the angles of depression of a rock and a life-buoy are found to be 45° and 60°. If these objects are known to be 110 yards apart, find the height of the cliff.

---

31. Prove that

$$\frac{1+\cos A}{1-\cos A} - \frac{\sec A - 1}{1+\sec A} - 4\cot^2 A = \frac{4}{1+\sec A}.$$

32. Solve the equations:

(1) $8\sin^2\theta - 2\cos\theta = 5$; (2) $5\tan^2 x - \sec^2 x = 11$.

33. What is the difference in latitude of two places on the same meridian whose distance apart is 11 inches on a globe whose radius is 5 feet? Take $\pi = \dfrac{22}{7}$.

34. Given that $\sec A = \dfrac{25}{7}$, find all the other Trigonometrical ratios of A.

35. Which of the following statements are possible, and which impossible?

(1) $4\sin^2\theta = 5$; (2) $(a^2+b^2)\cos\theta = 2ab$;

(3) $(m^2+n^2)\operatorname{cosec}\theta = m^2 - n^2$; (4) $\sec\theta = 2{\cdot}375$.

36. Walking down a hill inclined to the horizon at an angle θ a man observes an object in the horizontal plane whose angle of depression is a. Half way down the hill the angle of depression is β. Prove that $\cot\theta = 2\cot a - \cot\beta$.

II. (*After Chapter XII.*)

37. In a triangle $a = 25\sqrt{2}$, $c = 50$, $C = 90°$: find B, b and the perpendicular from C on c.

38. Prove that

$$(2\sec A + 3\sin A)(3\operatorname{cosec} A - 2\cos A) = (2\operatorname{cosec} A + 3\cos A)(3\sec A - 2\sin A).$$

39. Find the values of $\sin 960°$, $\operatorname{cosec}(-510°)$, $\tan 570°$.

40. Find all the angles between 0° and 500° which satisfy the equation $\tan^2\theta = 1$.

41. The angle of elevation of the top of a steeple is 58° from a point in the same level as its base, and is 44° from a point 42 feet directly above the former point. Given that $\tan 58° = 1{\cdot}6$ and $\tan 44° = {\cdot}965$, shew that the height of the steeple is 105 ft. approximately.

42. From the formula $\tan A = \dfrac{\sin 2A}{1 + \cos 2A}$ find $\tan 15°$ and $\tan 75°$, and solve the equation $\sec^2\theta = 4\tan\theta$.

---

43. Shew that

$$(1 + \sec\theta + \tan\theta)(1 + \operatorname{cosec}\theta + \cot\theta) = 2(1 + \tan\theta + \cot\theta + \sec\theta + \operatorname{cosec}\theta).$$

44. In a triangle ABC right-angled at C shew that

$$\frac{\sin^2 A}{\sin^2 B} - \frac{\cos^2 A}{\cos^2 B} = \frac{a^4 - b^4}{a^2 b^2}.$$

45. Find all the angles less than four right angles which satisfy the equation

$$2\cos^2\theta = 1 + \sin\theta.$$

46. Determine the value of $\sin(270°+A)$ when $\sin A = \cdot 6$.

47. Given $\sin\alpha = \frac{5}{13}$, $\sin\beta = \frac{4}{5}$, find the value of $\cos(\alpha+\beta)$, and deduce $\sin(45°+\alpha+\beta) = \frac{79\sqrt{2}}{130}$.

48. Reduce $\frac{\cos A - \cos 3A}{\sin 3A - \sin A}$ to a single term and trace the changes of the expression in sign and magnitude as A increases from 0° to 180°.

---

49. If $\cos A = -\frac{\sqrt{3}}{2}$, find $\tan A$, drawing a diagram to explain the two values.

50. From a balloon vertically over a straight road, the angles of depression of two consecutive milestones are observed to be 45° and 60°; find the height of the balloon.

51. Find the value of

(1) $\cot^2\frac{\pi}{6} - 2\cos^2\frac{\pi}{3} - \frac{3}{4}\sec^2\frac{\pi}{4} - 4\sin^2\frac{\pi}{6}$;

(2) $2\sec^2 180° \sin 0° - \cos 2\pi + \operatorname{cosec}\frac{3\pi}{2}$.

52. Prove the following identities:

(1) $\sin^4\alpha + 2\sin^2\alpha\left(1 - \frac{1}{\operatorname{cosec}^2\alpha}\right) = 1 - \cos^4\alpha$;

(2) $\dfrac{1+\tan^2\left(\frac{\pi}{4}-\theta\right)}{1-\tan^2\left(\frac{\pi}{4}-\theta\right)} = \operatorname{cosec} 2\theta$;

(3) $\cos 10° + \sin 40° = \sqrt{3}\sin 70°$.

53. If $b\tan\theta = a$, find the value of

$$\frac{a\sin\theta - b\cos\theta}{a\sin\theta + b\cos\theta}.$$

54. Prove that

$$4\cos 18° - 3\sec 18° = 2\tan 18°.$$

---

55. Find the values of

$$\tan(-240°),\ \cos 3360°,\ \cot(-840°).$$

Prove also that

$$\sin\frac{3\pi}{2} - \cos\frac{\pi}{2} + \cos\pi = \sec\frac{2\pi}{3}.$$

56. A railway train is travelling on a curve of half-a-mile radius at the rate of 20 miles an hour: through what angle has it turned in 10 seconds? Take $\pi = \frac{22}{7}$.

57. If $\sec\alpha = \frac{13}{5}$, find the value of

$$\frac{2 - 3\cot\alpha}{4 - 9\sqrt{\sec^2\alpha - 1}}.$$

58. Prove

(1) $2 - 2\tan A\cot 2A = \sec^2 A$;

(2) $$\frac{\cos\left(\frac{\pi}{4} - \theta\right) - \cos\left(\frac{\pi}{4} + \theta\right)}{\sin\left(\frac{2\pi}{3} + \theta\right) - \sin\left(\frac{2\pi}{3} - \theta\right)} + \sqrt{2} = 0.$$

59. When $A + B + C = 180°$, simplify

(1) $$\frac{\cos A\cos C + \cos(A+B)\cos(C+B)}{\cos A\sin C - \sin(A+B)\cos(C+B)};$$

(2) $$\frac{\cos A}{\sin B\sin C} + \frac{\cos B}{\sin C\sin A} + \frac{\cos C}{\sin A\sin B}.$$

60. A flagstaff 100 feet high stands vertically at the centre of a horizontal equilateral triangle: if each side of the triangle subtends an angle of 60° at the top of the flagstaff, find the side of the triangle.

---

61. Prove that the product of

$$\sin\theta(1 + \sin\theta) + \cos\theta(1 + \cos\theta)$$

and

$$\sin\theta(1 - \sin\theta) + \cos\theta(1 - \cos\theta)$$

is equal to $\sin 2\theta$.

62. Shew that

$$(1-\sin\theta)(1-\sin\phi)=\left\{\sin\frac{\theta+\phi}{2}-\cos\frac{\theta-\phi}{2}\right\}^2.$$

63. Prove that the value of

$$\frac{\sin(\alpha+\theta)-\sin(\alpha-\theta)}{\cos(\beta-\theta)-\cos(\beta+\theta)}$$

is the same for all values of θ.

64. If $A+B+C=180°$, prove that

$$\cos\frac{A}{2}\cos\frac{B-C}{2}+\cos\frac{B}{2}\cos\frac{C-A}{2}+\cos\frac{C}{2}\cos\frac{A-B}{2}$$
$$=\sin A+\sin B+\sin C.$$

65. If $\tan\frac{\theta}{2}=\operatorname{cosec}\theta-\sin\theta$, shew that

$$\cos^2\frac{\theta}{2}=\cos 36° \text{ or } \cos 108°.$$

66. A man stands at a point X on the bank XY of a river with straight and parallel sides, and observes that the line joining X to a point Z on the opposite bank makes with XY an angle of 30°. He then goes 200 yards along the bank to Y and finds that YZ makes with YX an angle of 60°. Find the breadth of the river.

---

67. It is found that the driving wheel of a bicycle, 32 inches in diameter, makes very nearly 1000 revolutions in travelling $2792\frac{1}{2}$ yards. Use this observation to calculate (to three places of decimals) the ratio of the circumference of a circle to its diameter.

68. If $\alpha+\beta+\gamma=\frac{\pi}{2}$, prove that

$$\sin^2\alpha+\sin^2\beta+\sin^2\gamma+2\sin\alpha\sin\beta\sin\gamma=1.$$

69. Prove that

(1) $(\tan A+\tan 2A)(\cos A+\cos 3A)=2\sin 3A$;

(2) $\sin^2 A\cos^4 A=\frac{1}{16}+\frac{1}{32}\cos 2A-\frac{1}{16}\cos 4A-\frac{1}{32}\cos 6A.$

70. If $a=\frac{\pi}{19}$, find the value of $\frac{\sin 23a-\sin 3a}{\sin 16a+\sin 4a}$.

71. If $A+B=225°$, prove that

$$\frac{\cot A}{1+\cot A}\cdot\frac{\cot B}{1+\cot B}=\frac{1}{2}.$$

72. Prove that $\cot\theta-\tan\theta=2\cot 2\theta$: and hence shew that

$$\tan\theta+2\tan 2\theta+4\tan 4\theta=\cot\theta-8\cot 8\theta.$$

---

73. Simplify $1-\frac{\sin^2\theta}{1+\cot\theta}-\frac{\cos^2\theta}{1+\tan\theta}$.

74. Eliminate A between the equations

$$x=3\sin A-\sin 3A,\quad y=\cos 3A+3\cos A.$$

75. Two flagstaffs stand on a horizontal plane. A, B are two points on the line joining the bases of the flagstaffs and between them. The angles of elevation of the tops of the flagstaffs as seen from A are 30° and 60°, and as seen from B, 60° and 45°. If the length of AB is 30 ft., find the heights of the flagstaffs and the distance between them.

76. Prove the identities:

(1) $\cos^2 A+\cos^2 B-2\cos A\cos B\cos(A+B)=\sin^2(A+B)$;

(2) $2\sin 5A-\sin 3A-3\sin A=4\sin A\cos^2 A(1-8\sin^2 A)$.

77. A square is inscribed in a circle the circumference of which is 3 feet. Find the number of inches in the length of a side, correct to two places of decimals. Given

$$\frac{1}{\pi}=\cdot 3183,\quad \sqrt{2}=1\cdot 4142.$$

78. Points A, B, C, D are taken on the circumference of a circle so that the arcs AB, BC, and CD subtend respectively at the centre angles of 108°, 60°, and 36°. Shew that

$$AB=BC+CD.$$

79. Prove that $\cot 15^\circ + \cot 75^\circ + \cot 135^\circ - \operatorname{cosec} 30^\circ = 1$.

80. From the equations

$$\cot\theta\,(1+\sin\theta)=4m,$$
$$\cot\theta\,(1-\sin\theta)=4n,$$

derive the relation $(m^2-n^2)^2=mn$.

81. Prove the identities:

(1) $\sin(\alpha+\beta)\cos\beta-\sin(\gamma+\alpha)\cos\gamma=\sin(\beta-\gamma)\cos(\alpha+\beta+\gamma)$;

(2) $(\tan 2A-\tan A)(\sec A+\sec 3A)=2\sin A\sec A\sec 3A$.

82. Prove that $\cos 6^\circ \cos 66^\circ \cos 42^\circ \cos 78^\circ = \dfrac{1}{16}$.

83. From the formula $\cot A=\dfrac{1+\cos 2A}{\sin 2A}$, prove that

$$\cot 22^\circ\,30' = \sqrt{2}+1.$$

84. An observer on board a ship sailing due North at the rate of ten miles an hour, sees a lighthouse in the East, and an hour later notices that the same lighthouse bears S.S.E.; find in miles, to two places of decimals, the distance of the ship from the lighthouse at the first observation.

III. (*After Chapter XVI.*)

85. Prove that

(1) $\sin A\sin(B-C)+\sin B\sin(C-A)+\sin C\sin(A-B)=0$;

(2) $\tan\theta=\dfrac{\sin\theta+\sin 2\theta}{1+\cos\theta+\cos 2\theta}$.

86. If $\alpha+\beta+\gamma=0$, prove that

$$\cos\alpha+\cos\beta+\cos\gamma=4\cos\frac{\alpha}{2}\cos\frac{\beta}{2}\cos\frac{\gamma}{2}-1.$$

87. In any triangle prove that

$$\frac{b^2-c^2}{a}\cos A+\frac{c^2-a^2}{b}\cos B+\frac{a^2-b^2}{c}\cos C=0.$$

88. If
$$\frac{\cos\theta}{a}=\frac{\sin\theta}{b},$$
prove that
$$\frac{a}{\sec 2\theta}+\frac{b}{\operatorname{cosec} 2\theta}=a.$$

89. Prove that $\log_a b \log_b c \log_c a=1$.

Given $\log_{10} 3=\cdot47712$, $\log_{10} 8=\cdot90309$, find the values of
$$\log_{10} 2\cdot4,\quad \log_{10} 5400,\quad L\tan 30^\circ.$$

90. If $A+B+C=90^\circ$, prove that
$$\cot A+\cot B+\cot C=\cot A\cot B\cot C;$$
and if A, B, C are in Arithmetical Progression, shew that this equation gives the value of $\cot 15^\circ$.

---

91. Shew that
$$(1+\sin 2A+\cos 2A)^2=4\cos^2 A\,(1+\sin 2A).$$

92. In a triangle where a, b, A are given, shew that c is one of the roots of the equation
$$x^2-2bx\cos A+b^2-a^2=0.$$

93. Prove that $\dfrac{\sin 9^\circ}{\sin 48^\circ}=\dfrac{\sin 12^\circ}{\sin 81^\circ}$.

94. If $A+B+C=180^\circ$, prove that
$$\cos\frac{A}{2}+\cos\frac{B}{2}+\cos\frac{C}{2}$$
$$=4\cos\left(45^\circ-\frac{A}{4}\right)\cos\left(45^\circ-\frac{B}{4}\right)\cos\left(45^\circ-\frac{C}{4}\right).$$

95. Given
$$L\sin 27^\circ 45'=9\cdot6680265,$$
$$L\sin 27^\circ 46'=9\cdot6682665,$$
$$L\sin\theta=9\cdot6682007,$$
find θ.

96. Prove that if A, B, C are three angles such that the sum of their cosines is zero, the product of their cosines is one-twelfth of the sum of the cosines of $3A$, $3B$, $3C$.

---

97. If A be between $270°$ and $360°$, and $\sin A = -\frac{7}{25}$, find the values of $\sin 2A$ and $\tan \frac{A}{2}$.

98. Solve the equation

$$2 \cot \frac{\theta}{2} = (1 + \cot \theta)^2.$$

Hence find the value of $\tan 15°$.

99. Given $\log_{10} 2 = \cdot 3010300$, $\log_{10} 360 = 2\cdot 5563025$, find the logarithms of $\cdot 04$, 24, $\dot{\cdot 6}$, and shew that $\log_2 30 = 4\cdot 90689$.

100. Prove that

$$\cos(x - y - z) + \cos(y - z - x) + \cos(z - x - y) - 4 \cos x \cos y \cos z$$

vanishes when $x + y + z$ is an odd multiple of a right angle.

101. If $\cot \alpha = (x^3 + x^2 + x)^{\frac{1}{2}}$, $\cot \beta = (x + x^{-1} + 1)^{\frac{1}{2}}$,

$$\tan \gamma = (x^{-3} + x^{-2} + x^{-1})^{\frac{1}{2}},$$

shew that $\alpha + \beta = \gamma$.

102. Shew how to solve a right-angled triangle of which one acute angle and the opposite side are given.

Apply this to the triangle in which the side is 28 and the angle $31° \, 53' \, 26\cdot 8''$, given $\log 2\cdot 8 = \cdot 4471580$, $\log 4\cdot 5 = \cdot 6532127$,

$$L \cot 31° \, 53' = 10\cdot 2061805, \quad \text{diff. for } 1' = 2816.$$

---

103. If $\tan A = \frac{\sqrt{3}}{4 - \sqrt{3}}$ and $\tan B = \frac{\sqrt{3}}{4 + \sqrt{3}}$,

prove that $$\tan(A - B) = \cdot 375.$$

104. The sides of a triangle are x, y, and $\sqrt{x^2 + xy + y^2}$, find its greatest angle.

105. Prove that $\cos A - \sin A$ is a factor of $\cos 3A + \sin 3A$; and that

$$\cos^2 A + \cos^2\left(A + \frac{2\pi}{3}\right) + \cos^2\left(A - \frac{2\pi}{3}\right) = \frac{3}{2}.$$

106. In any triangle, if $\tan\frac{A}{2}=\frac{5}{6}$, and $\tan\frac{B}{2}=\frac{20}{37}$, find $\tan C$.

Shew also that, in such a triangle, $a+c=2b$.

107. Simplify

$$\left\{\cot\theta+\cot\left(\theta-\frac{\pi}{2}\right)\right\}\left\{\tan\left(\frac{\pi}{4}-\theta\right)+\tan\left(\frac{\pi}{4}+\theta\right)\right\}.$$

108. If $a=40$, $b=51$, $c=43$, find the value of A, given

$$\log 1{\cdot}28=\cdot107210,\quad \log 6{\cdot}03=\cdot780317,$$
$$L\tan 24°\,44'\,16''=9{\cdot}6634465.$$

---

109. If $$\tan B=\frac{n\sin A\cos A}{1-n\sin^2 A},$$

prove that $$\tan(A-B)=(1-n)\tan A.$$

110. Given $\log 5=\cdot69897$, find $\log 200$, $\log\cdot025$, $\log\sqrt[3]{62{\cdot}5}$, and also $L\sin 30°$ and $L\cos 45°$.

111. Prove the identities:

(1) $(\sec 2A-2)\cot(A-30°)=(\sec 2A+2)\tan(A+30°)$;

(2) $1+\cos 2\alpha\cos 2\beta=2(\cos^2\alpha\cos^2\beta+\sin^2\alpha\sin^2\beta)$.

112. In a triangle, $B=60°$, $C=30°$, $BC=132$ yards. BC is produced to D and the angle $ADB=15°$; find CD and the perpendicular from A on BC, given that $\sqrt{3}=1\frac{8}{11}$ approximately.

113. In any triangle prove that

$$(a+b+c)\tan\frac{C}{2}=a\cot\frac{A}{2}+b\cot\frac{B}{2}-c\cot\frac{C}{2}.$$

114. If the sides of a triangle are 68 ft., 75 ft., 77 ft. respectively, find the least angle of the triangle, given

$$\log 2=\cdot30103,\ L\cos 26°\,34'=9{\cdot}9515389,\ \text{diff. for }1'=632.$$

---

115. If $\sin A = \cdot 6$ and A lies between 90° and 180°, find the values of $\sin(A-90°)$, $\operatorname{cosec}(270°-A)$.

116. Prove that

$$\log_a d = \log_a b \times \log_b c \times \log_c d.$$

Given $\log_{10} 5 = \cdot 69897$, find $\log_{10} 8$, $\log_8 10$, $\log_{10}(\cdot 032)^5$.

117. Prove that

$$\cos(420° + A) + \cos(60° - A) = \cos A.$$

Deduce the value of $\cos 105° + \cos 15°$.

118. Find the values of $\tan\frac{x}{2}$ from the equation

$$\cos x - \sin\alpha \cot\beta \sin x = \cos\alpha.$$

119. If $\sin A : \sin(2A+B) = n : m$, prove that

$$\cot(A+B) = \frac{m-n}{m+n}\cot A.$$

120. A tower AB stands on a horizontal plane, and AC, AD are the shadows at noon and 6 P.M. If AD is 17 ft. longer than AC, and BC is 53 ft., find the height of the tower and the altitude of the sun at noon, when the altitude at 6 P.M. is 45°; given $\tan 31° 48' = \cdot 6\dot{2}$.

---

121. Prove that

(1) $\sin 8\theta + \sin 2\theta = 4 \sin\frac{5\theta}{2}\cos\frac{5\theta}{2}\cos 3\theta$;

(2) $\sin 18° + \cos 18° = \sqrt{2}\cos 27°$.

122. Given $\log 36 = 1\cdot 556302$, $\log 48 = 1\cdot 681241$, find $\log 40$ and $\log\sqrt{\frac{2}{15}}$.

123. Given $b = 9\cdot 5$, $c = \cdot 5$, $A = 144°$, find the remaining angles; given $\log 3 = \cdot 4771213$, $L\cot 72° = 9\cdot 5117760$,

$$L\tan 16° 19' = 9\cdot 4664765, \quad L\tan 16° 18' = 9\cdot 4660078.$$

124. In any triangle prove that

(1) $bc \sin^2 A = a^2 (\cos A + \cos B \cos C)$;

(2) $bc \cos A + ca \cos B + 2ab \cos C = a^2 + b^2$.

125. If $\tan \frac{\beta}{2} = 4 \tan \frac{\alpha}{2}$, prove that

$$\tan \frac{\beta - \alpha}{2} = \frac{3 \sin \alpha}{5 - 3 \cos \alpha}.$$

126. Shew that

$$\sin (36° + A) - \sin (36° - A) + \sin (72° - A) - \sin (72° + A) = \sin A.$$

---

127. If $\sin \theta = -\frac{2}{3}$, find $\tan \theta$, and explain by means of a figure why there are two values.

128. Prove that

(1) $\sin 2A + \cos 2B = 2 \sin \left(\frac{\pi}{4} + A - B\right) \cos \left(\frac{\pi}{4} - A - B\right)$;

(2) $(\sin \theta - \sin \phi)(\cos \phi + \cos \theta) = 2 \sin (\theta - \phi) \cos^2 \frac{\theta + \phi}{2}$.

129. In any triangle, if

$$(\sin A + \sin B + \sin C)(\sin A + \sin B - \sin C) = 3 \sin A \sin B,$$

prove that $C = 60°$.

130. Prove that $\log_a b \times \log_c d = \log_a d \times \log_c b$.

131. If $\log 2001 = 3{\cdot}3012471$, $\log 2 = {\cdot}30103$, find $\log 20{\cdot}0075$.

132. If $a = 7$, $b = 8$, $c = 9$, shew that the length of line joining B to the middle point of AC is 7.

---

133. If $\tan A + \sec A = 2$, prove that $\sin A = \frac{3}{5}$ when A is less than 90°.

134. Prove that

$$\frac{3 - 4 \cos 2A + \cos 4A}{3 + 4 \cos 2A + \cos 4A} = \tan^4 A.$$

135. Shew that

$$\frac{\sin 3A+\cos 3A}{\sin 3A-\cos 3A}=\frac{1+2\sin 2A}{1-2\sin 2A}\tan(A-45°).$$

136. If
$$\frac{x}{y}=\frac{\cos A}{\cos B},$$
prove that
$$x\tan A+y\tan B=(x+y)\tan\frac{A+B}{2}.$$

137. Given

$$\log 3{\cdot}5={\cdot}544068,\ \log 3{\cdot}25={\cdot}511883,\ \log 2{\cdot}45={\cdot}389166,$$

find $\log 5$, $\log 7$, and $\log 13$.

138. In a triangle, $a=384$, $b=330$, $C=90°$; find the other angles; given

$$\log 11=1{\cdot}0413927,\quad L\tan 49°\,19'=10{\cdot}0656886;$$
$$\log 20=1{\cdot}3010300,\quad L\tan 49°\,20'=10{\cdot}0659441.$$

---

139. If $\cos\theta=\cos a\cos\beta$, prove that

$$\tan\frac{\theta+a}{2}\tan\frac{\theta-a}{2}=\tan^2\frac{\beta}{2}.$$

140. Prove that

$$\frac{\sin\theta}{\cos\theta+\sin\phi}+\frac{\sin\phi}{\cos\phi-\sin\theta}=\frac{\sin\theta}{\cos\theta-\sin\phi}+\frac{\sin\phi}{\cos\phi+\sin\theta}.$$

141. If in a triangle $c(a+b)\cos\frac{B}{2}=b(a+c)\cos\frac{C}{2}$, prove that $b=c$.

142. Prove the identities:

(1) $$\frac{\cot A+\operatorname{cosec} A}{\tan A+\sec A}=\cot\left(\frac{\pi}{4}+\frac{A}{2}\right)\cot\frac{A}{2};$$

(2) $$\sin^3 A+\sin^3(120°+A)+\sin^3(240°+A)=-\frac{3}{4}\sin 3A.$$

143. Calculate the value of $\sqrt[5]{18\times{\cdot}0015}$, having given

$$\log 3={\cdot}4771213,\ \log 48559=4{\cdot}6862697,\ \log 48560=4{\cdot}6862787.$$

144. Find the other angles of a triangle when one angle is $112^\circ 4'$, the side opposite to it 573 yards long, and another side 394 yards long; given

$$\log 573 = 2\cdot 7581546, \quad \log 394 = 2\cdot 5954962,$$
$$L\cos 22^\circ 4' = 9\cdot 9669614, \quad L\sin 39^\circ 35' = 9\cdot 8042757,$$
$$L\sin 39^\circ 36' = 9\cdot 8044284.$$

IV. (*After Chapter XVIII.*)

145. In any triangle prove

$$\frac{\cos A}{c\cos B + b\cos C} + \frac{\cos B}{a\cos C + c\cos A} + \frac{\cos C}{b\cos A + a\cos B} = \frac{a^2+b^2+c^2}{2abc}.$$

146. Given $\log 7 = \cdot 8450980$, and $\log 17 = 1\cdot 2304489$, find $\log 119$, $\log \frac{17}{7}$, and $\log \frac{289}{343}$.

147. If A, B, C are the angles of a triangle, and if

$$\cos\theta(\sin B + \sin C) = \sin A,$$

prove that
$$\tan^2\frac{\theta}{2} = \tan\frac{B}{2}\tan\frac{C}{2}.$$

148. Prove that the diameter of a circle is a mean proportional between the lengths of the sides of the equilateral triangle and the regular hexagon that circumscribe it.

149. Given that the sides a and b of a triangle are respectively $50\sqrt{5}$ feet and 150 ft., and that the angle opposite the side a is 45°, find (*without logarithms*) the two values of c. Also having given

$$\log 3 = \cdot 4771213, \quad L\sin 71^\circ 33' = 9\cdot 9770832,$$
$$L\sin 71^\circ 34' = 9\cdot 9771253,$$

find the two values of the angle B.

150. Prove that

$$2\cos 2x \operatorname{cosec} 3x = \operatorname{cosec} x - \operatorname{cosec} 3x.$$

Thence find the sum to n terms of the series

$$\cos 2x \operatorname{cosec} 3x + \cos(2\,.\,3x)\operatorname{cosec} 3^2x + \cos(2\,.\,3^2x)\operatorname{cosec} 3^3x + \ldots.$$

---

151. Prove the identities:

(1) $\cos^2 A + \sin^2 A \cos 2B = \cos^2 B + \sin^2 B \cos 2A$;

(2) $\sin 33° + \cos 63° = \cos 3°$.

152. Find all the positive angles less than two right angles which satisfy the equation

$$\tan^4 A - 4\tan^2 A + 3 = 0.$$

153. Prove that

$$\cot\frac{\theta}{2} - 3\cot\frac{3\theta}{2} = \frac{4\sin\theta}{1+2\cos\theta}.$$

154. The tangents of two angles of a triangle are $\frac{3}{4}$, and $\frac{5}{12}$ respectively. Find the tangent of the third angle, and the cosine of each angle of the triangle. Also find the third angle to the nearest second, having given

$$\log 33 = 1{\cdot}5185139, \quad \log 56 = 1{\cdot}7481880,$$
$$L\tan 59°\,29' = 10{\cdot}2295627; \quad \text{Diff. for } 1' = 2888.$$

155. If in a triangle

$$(a^2+b^2)\sin(A-B) = (a^2-b^2)\sin(A+B),$$

shew that the triangle is either isosceles or right-angled.

156. If r and R are the radii of the in-circle and circum-circle of a triangle, prove that

$$8rR\left\{\cos^2\frac{A}{2} + \cos^2\frac{B}{2} + \cos^2\frac{C}{2}\right\} = 2bc + 2ca + 2ab - a^2 - b^2 - c^2.$$

---

157. In any triangle prove that

$$\cot B + \frac{\cos C}{\sin B\cos A} = \cot C + \frac{\cos B}{\sin C\cos A}.$$

158. Given $\log 6 = {\cdot}778151$, $\log 4{\cdot}4 = {\cdot}643453$, $\log 1{\cdot}8 = {\cdot}255273$, find $\log 2$, $\log 3$, $\log 11$.

159. Prove the identities:

(1) $\sin 3A = \sin A\,(2\cos 2A - 1)\tan(60° + A)\tan(60° - A)$;

(2) $(\sin 2A - \sin 2B)\tan(A+B) = 2(\sin^2 A - \sin^2 B)$.

160. Find the greatest angle of a triangle whose sides are 183, 195, and 214 feet respectively; given

$$\log 82 = 1{\cdot}9138139, \qquad \log 296 = 2{\cdot}4712917,$$
$$\log 101 = 2{\cdot}0043214, \qquad L\tan 34^\circ\,26' = 9{\cdot}8360513,$$
$$\log 113 = 2{\cdot}0530784, \qquad L\tan 34^\circ\,27' = 9{\cdot}8363221.$$

161. A circle and a regular octagon have the same perimeter; compare their areas, given $\sqrt{2} = 1{\cdot}414$, $\pi = 3{\cdot}1416$.

162. If the sides of a triangle be in arithmetical progression, and if a be the least side, then

$$\cos A = \frac{4c - 3b}{2c}.$$

---

163. If $a\sin(\theta + \alpha) = b\sin(\theta + \beta)$, prove that

$$\cot\theta = \frac{a\cos\alpha - b\cos\beta}{b\sin\beta - a\sin\alpha}.$$

164. In the ambiguous case shew that the circum-circles of the two triangles are equal.

165. From a point A on a level plain the angle of elevation of a kite is α, and its direction South; and from a place B, which is c yards South of A on the plain, the kite is seen due North at an angle of elevation β. Find the distance of the kite from A and its height above the ground.

166. If $\alpha + \beta + \gamma = 2\pi$, express $\cos\alpha + \cos\beta + \cos\gamma + 1$ in the form of a product.

167. Prove that

$$\cos 10A + \cos 8A + 3\cos 4A + 3\cos 2A = 8\cos A\cos^3 3A.$$

168. In any triangle shew that

$$R = \frac{(r_2 + r_3)(r_3 + r_1)(r_1 + r_2)}{4(r_2 r_3 + r_3 r_1 + r_1 r_2)}.$$

---

169. If $\tan^2\theta = 2\tan^2\phi + 1$, then

$$\cos 2\theta + \sin^2\phi = 0.$$

170. Prove that

$$\tan A\tan(60^\circ + A)\tan(120^\circ + A) = -\tan 3A.$$

171. If in a triangle $A=2B$, then $a^2=b(c+b)$.

172. Shew that the length of a side of an equilateral triangle inscribed in a circle is to that of a square inscribed in the same circle as $\sqrt{3} : \sqrt{2}$.

173. In any triangle prove that $\tan\left(\frac{A}{2}+B\right)=\frac{c+b}{c-b}\tan\frac{A}{2}$.

If $3c=7b$, and $A=6° 37' 24''$, find the other angles; given

$L \tan 3° 18' 42''=8·7624069$, $L \tan 8° 13' 50''=9·1603083$,

$\log 2 = ·30103$, diff. for $10''=1486$.

174. If D be the middle point of the side BC of a triangle ABC, and if Δ be the area of the triangle, prove that

$$\cot ADB=\frac{AC^2-AB^2}{4\Delta}.$$

---

175. Prove that $\tan 20° \tan 40° \tan 80°=\tan 60°$.

176. If, in a triangle, $b=\sqrt{3}+1$, $c=2$, and $A=30°$, find B, C, and a.

177. Prove that the rectangle contained by the diameters of the circumscribed and inscribed circles of a triangle is equal to

$$\frac{2abc}{a+b+c}.$$

178. Solve the triangle when $a=7$, $b=8\sqrt{3}$, $A=30°$; given

$\log 2=·30103$, $L \sin 81° 47'=9·9955188$,

$\log 3=·4771213$, diff. for $1'=183$.

$\log 7=·8450980$,

179. If $$\sin 2\beta=\frac{\sin 2\alpha+\sin 2\alpha'}{1+\sin 2\alpha \sin 2\alpha'},$$

prove that $\tan\left(\frac{\pi}{4}+\beta\right)=\pm\tan\left(\frac{\pi}{4}+\alpha\right)\tan\left(\frac{\pi}{4}+\alpha'\right)$.

180. On a plain at some distance from its base, a mountain is found to have an elevation of 28°. At a station lying 3 miles 77 yards further away from the mountain the angle is reduced to 16°. Find the height of the mountain in feet.

$$\log 1{\cdot}6071 = {\cdot}2060, \quad L\sin 16° = 9{\cdot}4403,$$
$$L\sin 28° = 9{\cdot}6716, \quad L\sin 12° = 9{\cdot}3179.$$

---

181. Prove that

(1) $\tan\dfrac{A+B}{2} - \tan\dfrac{A-B}{2} = \dfrac{2\sin B}{\cos A + \cos B}$;

(2) $4\cos^8 A - 4\sin^8 A = 4\cos 2A - \sin 2A \sin 4A$.

182. If $A+B+C = \dfrac{\pi}{2}$, and $\cos A + \cos C = 2\cos B$,

shew that $\quad 1 + \tan\dfrac{A}{2}\tan\dfrac{C}{2} = 2\left(\tan\dfrac{A}{2} + \tan\dfrac{C}{2}\right)$,

or else $A+C$ is an odd multiple of π.

183. Shew that in any triangle

$$\cos A + \cos B - \sin C = 4\sin\frac{C}{2}\sin\left(45° - \frac{A}{2}\right)\sin\left(45° - \frac{B}{2}\right).$$

184. With the usual notation in any triangle, prove that

$$\frac{bc}{r_1} + \frac{ca}{r_2} + \frac{ab}{r_3} = 2R\left\{\frac{b}{a} + \frac{c}{a} + \frac{c}{b} + \frac{a}{b} + \frac{a}{c} + \frac{b}{c} - 3\right\}.$$

185. The bisector of the angle A meets the side BC in D and the circumscribed circle in E, shew that

$$DE = \frac{a^2\sec\frac{A}{2}}{2(b+c)}.$$

186. If $a=4090$, $b=3850$, $c=3811$, find A, given

$$\log 5{\cdot}8755 = {\cdot}7690448, \qquad \log 3{\cdot}85 = {\cdot}5854607,$$
$$\log 1{\cdot}7855 = {\cdot}2517599, \qquad \log 3{\cdot}811 = {\cdot}5810389,$$
$$L\cos 32°\,15' = 9{\cdot}9272306, \qquad L\cos 32°\,16' = 9{\cdot}9271509.$$

---

187. Prove that

(1) $\operatorname{cosec}^6\theta - \cot^6\theta = 1 + 3\operatorname{cosec}^2\theta\cot^2\theta$;

(2) $\cos(15° - \alpha)\sec 15° - \sin(15° - \alpha)\operatorname{cosec} 15° = 4\sin\alpha$.

188. Prove that

$$\frac{\sin(A+B+C)}{\cos A\cos B\cos C}=\tan A+\tan B+\tan C-\tan A\tan B\tan C.$$

189. If $\log\frac{1025}{1024}=p$, and $\log 2=q$,

prove that $$\log 4100=p+12q.$$

190. In any triangle prove that

(1) $(a^2-b^2-c^2)\tan A+(a^2-b^2+c^2)\tan B=0$;

(2) $\frac{\cos 2A}{a^2}-\frac{\cos 2B}{b^2}=\frac{1}{a^2}-\frac{1}{b^2}$.

191. Find the area of the triangle, whose sides are 68 ft., 75 ft., 77 ft., respectively; and also find the radii of the three escribed circles.

192. If the bisector of the angle A of the triangle ABC meet the opposite side in D, prove that

$$AD=\frac{2bc}{b+c}\cos\frac{A}{2}.$$

V. (*After Chapter XIX.*)

193. Solve the equations:

(1) $\sin 5\theta-\sin 3\theta=\sin\theta\sec 45°$;

(2) $\cot\theta+\cot\left(\frac{\pi}{4}+\theta\right)=2$.

194. If $2\sec 2\alpha=\tan\beta+\cot\beta$, shew that one value of $\alpha+\beta$ is $\frac{\pi}{4}$.

195. If $$\cos^2\beta\tan(\alpha+\theta)=\sin^2\beta\cot(\alpha-\theta),$$

then $$\tan^2\theta=\tan(\alpha+\beta)\tan(\alpha-\beta).$$

196. If p_1, p_2, p_3 are the perpendiculars from the angular points on the sides of a triangle, prove that

(1) $8R^3 = \dfrac{a^2b^2c^2}{p_1p_2p_3}$;

(2) $\dfrac{1}{p_3^2} = \dfrac{1}{p_1^2} + \dfrac{1}{p_2^2} - \dfrac{2}{p_1p_2}\cos C.$

197. Find the perimeter of a regular quindecagon circumscribed about a circle whose area is 1386 sq. ft.; given

$$\tan 12^\circ = \cdot 213.$$

198. The top of a pole, placed against a wall at an angle α with the horizon, just touches the coping, and when its foot is moved a feet further from the wall, and its angle of inclination is β, it rests on the sill of a window: prove that the perpendicular distance from the coping to the sill $= a \cot \dfrac{\alpha+\beta}{2}$.

---

199. In any triangle prove that

$$\frac{ab - r_1r_2}{r_3} = \frac{bc - r_2r_3}{r_1} = \frac{ca - r_3r_1}{r_2}.$$

200. Prove that

(1) $\cos^{-1}\dfrac{41}{49} = 2\sin^{-1}\dfrac{2}{7}$; (2) $3\tan^{-1}\dfrac{1}{4} = \tan^{-1}\dfrac{47}{52}$.

201. Prove the identities:

(1) $(\tan A + \sec A)\cot\dfrac{A}{2} = (\cot A + \operatorname{cosec} A)\tan\left(45^\circ + \dfrac{A}{2}\right)$;

(2) $\cos 2A + \cos 2B - 4\sin(45^\circ - A)\sin(45^\circ - B)\cos(A+B)$
$= \sin 2(A+B)$.

202. Given $\log 3 = \cdot 4771213$, $\log 7 = \cdot 8450980$,

$$L \sin 25\tfrac{5}{7}^\circ = 9\cdot 6373733;$$

shew that the perimeter of a regular figure of seven sides is greater than 3 times the diameter of the circle circumscribing the figure.

203. If $\tan\phi = \dfrac{a-b}{a+b}\cot\dfrac{C}{2}$, in any triangle, prove that

$$c = (a+b)\frac{\sin\dfrac{C}{2}}{\cos\phi}.$$

204. The sides of a triangle are 237 and 158, and the contained angle is 66° 40′; use the formulæ in the last question to find the base.

$$\log 2 = \cdot 30103, \quad L\cot 33°\,20' = 10\cdot 18197,$$
$$\log 79 = 1\cdot 89763, \quad L\sin 33°\,20' = 9\cdot 73998,$$
$$\log 22687 = 4\cdot 35578,$$
$$L\tan 16°\,54' = 9\cdot 48262, \quad L\sec 16°\,54' = 10\cdot 01917,$$
$$L\tan 16°\,55' = 9\cdot 48308, \quad L\sec 16°\,55' = 10\cdot 01921.$$

---

205. Shew that $\sec\theta = \dfrac{2}{\sqrt{2+\sqrt{2+2\cos 4\theta}}}$.

206. Prove that

$$\sin^{-1}\frac{3}{\sqrt{73}} + \cos^{-1}\frac{11}{\sqrt{146}} + \sin^{-1}\frac{1}{2} = \frac{5\pi}{12},$$

and solve the equation

$$\tan^{-1}\frac{x-1}{x+1} + \tan^{-1}\frac{2x-1}{2x+1} = \tan^{-1}\frac{23}{36}.$$

207. If x, y, z are the perpendiculars from the angular points of a triangle upon the opposite sides a, b, c, shew that

$$\frac{bx}{c} + \frac{cy}{a} + \frac{az}{b} = \frac{a^2+b^2+c^2}{2R}.$$

208. If $\sin(a-\theta) = \cos(a+\theta)$, shew that either

$$\theta = m\pi - \frac{\pi}{4} \text{ or } a = m\pi + \frac{\pi}{4},$$

where m is zero or any integer.

209. The vertical angle of an isosceles triangle is 120°; shew that the distance between the centres of the inscribed and circumscribed circles is to the base of the triangle in the ratio

$$\sqrt{3} - 1 : \sqrt{3}.$$

210. If in a triangle $3R = 4r$, shew that

$$4(\cos A + \cos B + \cos C) = 7.$$

---

211. If $\frac{\sin(\theta+a)}{\cos(\theta-a)}=\frac{1-m}{1+m}$, prove that

$$\tan\left(\frac{\pi}{4}-\theta\right)=m\cot\left(\frac{\pi}{4}-a\right).$$

212. Solve the equations:

(1) $\sin 5\theta-\sin 3\theta=\sqrt{2}\cos 4\theta$;

(2) $(1-\tan\theta)(1+\sin 2\theta)=1+\tan\theta$.

213. If $\cos A+\cos B=4\sin^2\frac{C}{2}$ in any triangle, prove that $a+b=2c$.

214. A flagstaff standing on the top of a tower 80 feet high subtends an angle $\tan^{-1}\frac{1}{9}$ at a point 100 feet from the foot of the tower: find the height of the flagstaff.

215. Prove that

(1) $\cot^{-1}7+\cot^{-1}8+\cot^{-1}18=\cot^{-1}3$;

(2) $4\tan^{-1}\frac{1}{5}-\tan^{-1}\frac{1}{239}=\frac{\pi}{4}$.

216. If $2\sin\frac{A}{2}=-\sqrt{1+\sin A}+\sqrt{1-\sin A}$, shew that A lies between $(8n+3)\frac{\pi}{2}$ and $(8n+5)\frac{\pi}{2}$.

---

217. Prove that

$$\sin^2\left(\frac{\pi}{8}+\frac{\theta}{2}\right)-\sin^2\left(\frac{\pi}{8}-\frac{\theta}{2}\right)=\frac{1}{\sqrt{2}}\sin\theta.$$

218. If $\tan\theta=\frac{x\sin\phi}{1-x\cos\phi}$, $\tan\phi=\frac{y\sin\theta}{1-y\cos\theta}$,

prove that
$$\frac{\sin\theta}{\sin\phi}=\frac{x}{y}.$$

219. Solve the equation

$$\tan^{-1}(x+1)+\tan^{-1}(x-1)=\tan^{-1}\frac{8}{31};$$

and prove that

$$\sec^2(\tan^{-1}2)+\operatorname{cosec}^2(\cot^{-1}3)=15.$$

220. Prove that in any triangle

$$\sin 10A+\sin 10B+\sin 10C=4\sin 5A\sin 5B\sin 5C;$$

also that the sum of the cotangents of $\frac{5\pi+A}{2^5}$, $\frac{5\pi+B}{2^5}$, $\frac{5\pi+C}{2^5}$ is equal to their product.

221. If d_1, d_2, d_3 are the diameters of the three escribed circles, shew that

$$d_1d_2+d_2d_3+d_3d_1=(a+b+c)^2.$$

222. To determine the breadth AB of a ravine an observer places himself at C in the straight line AB produced through B, and then walks 100 yards at right angles to this line. He then finds that AB and BC subtend angles of 15° and 25° at his eye. Find the breadth of the ravine, given

$$L\cos 25°=9{\cdot}9572757,\quad L\cos 40°=9{\cdot}8842540,$$
$$L\cos 75°=9{\cdot}4129962,$$
$$\log 37279=4{\cdot}5714643,\quad \log 3728=3{\cdot}5714759.$$

---

223. Prove that

$$(1-\cos\theta)\{\sec\theta+\operatorname{cosec}\theta(1+\sec\theta)\}^2=2\sec^2\theta(1+\sin\theta).$$

224. If in a triangle $C=60°$, prove that

$$\frac{1}{a+c}+\frac{1}{b+c}=\frac{3}{a+b+c}.$$

225. Prove that

$$2\cos\frac{A}{2^n}=\sqrt{2+\sqrt{2+\sqrt{\ldots\ldots\sqrt{2+2\cos A}}}};$$

the symbol indicating the extraction of the square root being repeated n times.

226. If $$\frac{m\tan(a-\theta)}{\cos^2\theta}=\frac{n\tan\theta}{\cos^2(a-\theta)},$$

then $$\theta=\frac{1}{2}\left\{a-\tan^{-1}\left(\frac{n-m}{n+m}\tan a\right)\right\}.$$

227. The sides of a triangle are such that
$$\frac{a}{1+m^2n^2}=\frac{b}{m^2+n^2}=\frac{c}{(1-m^2)(1+n^2)};$$

prove that $A=2\tan^{-1}\frac{m}{n}$, $B=2\tan^{-1}mn$, and the area of the triangle $=\frac{mnbc}{m^2+n^2}$.

228. A flagstaff h feet high placed on the top of a tower l feet high subtends the same angle β at two points a feet apart in a horizontal line through the foot of the tower. If θ be the angle subtended by the line a at the top of the flagstaff, shew that
$$h=a\sin\beta\operatorname{cosec}\theta,\ \text{and}\ 2l=a\operatorname{cosec}\theta(\cos\theta-\sin\beta).$$

---

229. Prove that
$$\frac{1}{2}\tan\frac{\theta}{2}+\frac{1}{4}\tan\frac{\theta}{4}=\frac{1}{4}\cot\frac{\theta}{4}-\cot\theta.$$

230. A regular polygon is inscribed in a circle such that each side is $\frac{1}{m}$th of the radius; shew that the angle at the centre subtended by each side is equal to $\sec^{-1}\frac{2m^2}{2m^2-1}$.

231. At what distance will an inch subtend an angle of one second?

232. If $\tan^{-1}y=4\tan^{-1}x$, find y as an algebraical function of x.

Hence prove that $\tan 22°\,30'$ is a root of the equation
$$x^4-6x^2+1=0.$$

233. If $\cos 2\alpha = \frac{240}{289}$, find $\tan \alpha$ and explain the double answer.

234. If θ, ϕ be the greatest and least angles of a triangle, the sides of which are in Arithmetical Progression, shew that

$$4(1-\cos\theta)(1-\cos\phi)=\cos\theta+\cos\phi.$$

---

235. Solve the equations:

(1) $\sin 7\theta = \sin 4\theta - \sin\theta$;

(2) $\tan x - \sqrt{3}\cot x + 1 = \sqrt{3}$.

236. In any triangle prove that

(1) $\sin 3A \sin(B-C) + \sin 3B \sin(C-A) + \sin 3C \sin(A-B) = 0$;

(2) $a^3 \sin(B-C) + b^3 \sin(C-A) + c^3 \sin(A-B) = 0$.

237. ABC is a triangle and a point P is taken on AB so that $AP : BP = m : n$. If the angle CPB is θ, shew that

$$(m+n)\cot\theta = n\cot A - m\cot B.$$

238. If α, β are unequal values of θ satisfying the equation

$$a\tan\theta + b\sec\theta = 1,$$

find a and b in terms of α and β, and prove that

$$\sin\alpha + \cos\alpha + \sin\beta + \cos\beta = \frac{2b(1-a)}{1+a^2}.$$

239. If $u_n = \sin^n\theta + \cos^n\theta$, prove that

$$\frac{u_3 - u_5}{u_1} = \frac{u_5 - u_7}{u_3}.$$

240. A building on a square base $ABCD$ has the sides of the base AB and CD, parallel to the banks of a river. An observer standing on the bank of the river furthest from the building in the same straight line as DA finds that the side AB subtends at his eye an angle of 45°, and after walking a yards along the bank he finds that DA subtends the angle whose sine is $\frac{1}{3}$. Prove that the length of each side of the base in yards is $\frac{a\sqrt{2}}{2}$.

---

241. Prove the identities:

(1) $(\operatorname{cosec} A - \sin A)(\sec A - \cos A) = (\tan A + \cot A)^{-1}$;

(2) $\dfrac{\tan\theta}{(1+\tan^2\theta)^2} + \dfrac{\cot\theta}{(1+\cot^2\theta)^2} = \dfrac{1}{2}\sin 2\theta.$

242. If $\sin 4\theta \cos\theta = \frac{1}{4} + \sin\frac{5\theta}{2}\cos\frac{5\theta}{2}$, find *one* value of θ.

243. Prove that

$$\tan^{-1}\frac{2mn}{m^2-n^2} + \tan^{-1}\frac{2pq}{p^2-q^2} = \tan^{-1}\frac{2MN}{M^2-N^2},$$

where $$M = mp - nq,\ N = np + mq.$$

244. In any triangle, prove that

$$\frac{\tan\frac{A}{2}}{(a-b)(a-c)} + \frac{\tan\frac{B}{2}}{(b-c)(b-a)} + \frac{\tan\frac{C}{2}}{(c-a)(c-b)} = \frac{1}{\Delta}.$$

245. If r_1, r_2, r_3 be the radii of the three escribed circles, and

$$\left(1-\frac{r_1}{r_2}\right)\left(1-\frac{r_1}{r_3}\right) = 2,$$

shew that the triangle must be right-angled.

246. The sides of a triangle are 237 and 158, and the contained angle is 58° 40′ 3·9″. Find by the aid of Tables the value of the base, without previously determining the other angles.

---

247. If $\tan(A+B) = 3\tan A$, shew that

$$\sin(2A+2B) + \sin 2A = 2\sin 2B.$$

248. Prove that

$$4\sin(\theta-a)\sin(m\theta-a)\cos(\theta-m\theta)$$
$$= 1 + \cos(2\theta-2m\theta) - \cos(2\theta-2a) - \cos(2m\theta-2a).$$

249. Perpendiculars are drawn from the angles A, B, C of an acute-angled triangle on the opposite sides, and produced to meet the circumscribing circle: if those produced parts be α, β, γ respectively, shew that

$$\frac{a}{\alpha}+\frac{b}{\beta}+\frac{c}{\gamma}=2(\tan A+\tan B+\tan C).$$

250. If A and B are two angles, each positive, and less than 90° and such that

$$3\sin^2 A+2\sin^2 B=1,$$
$$3\sin 2A-2\sin 2B=0,$$

prove that $$A+2B=90°.$$

251. Prove that

(1) $\cot^{-1}(\tan 2x)+\cot^{-1}(-\tan 3x)=x$;

(2) $\tan^{-1}\dfrac{1-x}{1+x}-\tan^{-1}\dfrac{1-y}{1+y}=\sin^{-1}\dfrac{y-x}{\sqrt{1+x^2}\sqrt{1+y^2}}$.

252. In any triangle prove that

$$a\sec\theta=b+c,$$

where $$(b+c)\sin\theta=2\sqrt{bc}\cos\frac{A}{2}.$$

Compass observations are taken from a station to two points distant respectively 1250 yards and 1575 yards. The bearing of one point is 7° 30′ West of North, and that of the other is 42° 15′ East of North. Find the distance between the points by the aid of Tables.

---

253. Prove the identities:

(1) $\cos\beta\cos(2\alpha-\beta)=\cos^2\alpha-\sin^2(\alpha-\beta)$;

(2) $(x\tan\alpha+y\cot\alpha)(x\cot\alpha+y\tan\alpha)$
$$=(x+y)^2+4xy\cot^2 2\alpha.$$

254. If $2S=A+B+C$, shew that

$$\cos^2 S+\cos^2(S-A)+\cos^2(S-B)+\cos^2(S-C)$$
$$=2+2\cos A\cos B\cos C.$$

255. If $\sin^{-1}\alpha+\sin^{-1}\beta+\sin^{-1}\gamma=\pi$, prove that

$$\alpha\sqrt{1-\alpha^2}+\beta\sqrt{1-\beta^2}+\gamma\sqrt{1-\gamma^2}=2\alpha\beta\gamma.$$

256. In a triangle $a=36$, $B=73°\ 15'$, $C=45°\ 30'$; find R and r by the aid of Tables.

257. If ρ be the radius of the circle inscribed in the pedal triangle, prove that

$$\rho=R(1-\cos^2 A-\cos^2 B-\cos^2 C).$$

258. A, B, C are the tops of posts at equal intervals by the side of a road; t and t' are the tangents of the angles which AB and BC subtend at any point P; T is the tangent of the angle which the road makes with PB: shew that

$$\frac{2}{T}=\frac{1}{t'}-\frac{1}{t}.$$

---

259. In any triangle prove that

$$\frac{(\cos B+\cos C)(1+2\cos A)}{1+\cos A-2\cos^2 A}=\frac{b+c}{a}.$$

260. With the notation of Art. 219, prove that

$$\frac{AI}{AI_1}+\frac{BI}{BI_2}+\frac{CI}{CI_3}=1.$$

261. Prove that

$$\sin^{-1}\frac{1}{3}+\sin^{-1}\frac{1}{3\sqrt{11}}+\sin^{-1}\frac{3}{\sqrt{11}}=\frac{\pi}{2}.$$

262. Find the relation between α, β, and γ in order that

$$\cot\alpha\cot\beta\cot\gamma-\cot\alpha-\cot\beta-\cot\gamma$$

should vanish.

263. If $A+B+C=\pi$, prove that

$$\frac{\tan A}{\tan B\tan C}+\frac{\tan B}{\tan C\tan A}+\frac{\tan C}{\tan A\tan B}$$
$$=\tan A+\tan B+\tan C-2(\cot A+\cot B+\cot C).$$

264. A man travelling due North along a straight road observes that at a certain milestone two objects lie due N.E. and S.W. respectively, and that when he reaches the next milestone their directions have become S.S.E. and S.S.W. respectively. Find the distance between the two objects, and prove that the sum of their shortest distances from the road is exactly a mile.

---

265. Prove that

$$\tan 20^\circ + \tan 40^\circ + \sqrt{3} \tan 20^\circ \tan 40^\circ = \sqrt{3}.$$

266. Solve the equation

$$\cot^3 \theta + 6 \operatorname{cosec} 2\theta - 8 \operatorname{cosec}^3 2\theta = 0.$$

267. If $A+B+C=180^\circ$, prove that

$$1 - 2 \sin B \cos C \cos A + \cos^2 A = \cos^2 B + \cos^2 C,$$

and if $A+B+C=0$, prove that

$$1 + 2 \sin B \sin C \cos A + \cos^2 A = \cos^2 B + \cos^2 C.$$

268. If in a triangle $\cot A$, $\cot B$, $\cot C$ are in A.P., shew that a^2, b^2, c^2 are also in A.P.

269. If α, β, γ are the angles of a triangle, prove that

$$\cos\left(\frac{3\beta}{2}+\gamma-2\alpha\right)+\cos\left(\frac{3\gamma}{2}+\alpha-2\beta\right)+\cos\left(\frac{3\alpha}{2}+\beta-2\gamma\right)$$

$$=4\cos\frac{5\alpha-2\beta-\gamma}{4}\cos\frac{5\beta-2\gamma-\alpha}{4}\cos\frac{5\gamma-2\alpha-\beta}{4}.$$

270. If the sum of the pairs of radii of the escribed circles of a triangle taken in order round the triangle be denoted by s_1, s_2, s_3, and the corresponding differences by d_1, d_2, d_3, prove that

$$d_1 d_2 d_3 + d_1 s_2 s_3 + d_2 s_3 s_1 + d_3 s_1 s_2 = 0.$$

---

271. If $\cos A = \frac{3}{4}$, shew that

$$32 \sin\frac{A}{2} \sin\frac{5A}{2} = 11.$$

272. Prove that all angles which satisfy the equation

$$\tan^2\theta + 2\tan\theta = 1,$$

are included in the formula $(8n-1)\frac{\pi}{8} \pm \frac{\pi}{4}$ where n is zero or any integer.

273. Prove

(1) $\cos\left(2\tan^{-1}\frac{1}{7}\right) = \sin\left(4\tan^{-1}\frac{1}{3}\right)$;

(2) $\tan^{-1}\frac{3\sin 2a}{5+3\cos 2a} + \tan^{-1}\left(\frac{\tan a}{4}\right) = a.$

274. If in any triangle

$$\cos\frac{A}{2} = \frac{1}{2}\sqrt{\frac{b}{c}+\frac{c}{b}},$$

shew that the square described with one side of the triangle as diagonal is equal to the rectangle contained by the other two sides.

275. Find B and C, having given $A = 50°$, $b = 119$, $a = 97$.

$\log 1{\cdot}19 = {\cdot}0755470$, $\quad L\sin 70° = 9{\cdot}9729858$,

$\log 9{\cdot}7 = {\cdot}9867717$, $\quad L\sin 70°\ 1' = 9{\cdot}9730318$,

$L\sin 50° = 9{\cdot}8842540$.

276. Circles are inscribed in the triangles $D_1E_1F_1$, $D_2E_2F_2$, $D_3E_3F_3$, where D_1, E_1, F_1 are the points of contact of the circle escribed to the side BC. Shew that if r_a, r_b, r_c be the radii of these circles

$$\frac{1}{r_a} : \frac{1}{r_b} : \frac{1}{r_c} = 1-\tan\frac{A}{4} : 1-\tan\frac{B}{4} : 1-\tan\frac{C}{4}.$$

---

277. Reduce to its simplest form

$$\tan^{-1}\left(\frac{x\cos\theta}{1-x\sin\theta}\right) - \cot^{-1}\left(\frac{\cos\theta}{x-\sin\theta}\right).$$

278. If $\cos A + \cos B = 4\sin^2\frac{C}{2}$ in any triangle, shew that the sides are in A.P.

279. Express $4\cos\alpha\cos\beta\cos\gamma\cos\delta+4\sin\alpha\sin\beta\sin\gamma\sin\delta$ as the sum of four cosines.

280. If I be the in-centre of a triangle and ρ_1, ρ_2, ρ_3 are the circum-radii of the triangles BIC, CIA, AIB, prove that

$$\rho_1\rho_2\rho_3=2rR^2.$$

281. A monument $ABCDE$ stands on level ground. At a point P on the ground the portions AB, AC, AD subtend angles α, β, γ respectively. Supposing that $AB=2$, $AC=16$, $AD=18$, and $\alpha+\beta+\gamma=180^\circ$, find AP.

282. If α and β be two angles both satisfying the equation

$$a\cos 2\theta+b\sin 2\theta=c,$$

prove that

$$\cos^2\alpha+\cos^2\beta=\frac{a^2+ac+b^2}{a^2+b^2}.$$

---

283. If $C=22\tfrac{1}{2}^\circ$, $a=\sqrt{2}$, $b=\sqrt{2+\sqrt{2}}$, solve the triangle.

284. If $A+B+C=180^\circ$, prove that

$$\sin^3 A+\sin^3 B+\sin^3 C$$
$$=3\cos\frac{A}{2}\cos\frac{B}{2}\cos\frac{C}{2}+\cos\frac{3A}{2}\cos\frac{3B}{2}\cos\frac{3C}{2}.$$

285. In the ambiguous case in which a, b, A are given, if one angle of one triangle be twice the corresponding angle of the other triangle, shew that

$$a\sqrt{3}=2b\sin A, \text{ or } 4b^3\sin^2 A=a^2(a+3b).$$

286. If the roots of $x^3-px^2-r=0$ are $\tan\alpha$, $\tan\beta$, $\tan\gamma$, find the value of $\sec^2\alpha\sec^2\beta\sec^2\gamma$.

287. If $\alpha+\beta+\gamma=\pi$, and

$$\tan\frac{1}{4}(\beta+\gamma-\alpha)\tan\frac{1}{4}(\gamma+\alpha-\beta)\tan\frac{1}{4}(\alpha+\beta-\gamma)=1,$$

prove that $$1+\cos\alpha+\cos\beta+\cos\gamma=0.$$

288. Prove that the side of a regular heptagon inscribed in a circle of radius unity is given by one of the roots of the equation

$$x^6-7x^4+14x^2-7=0,$$

and give the geometrical signification of the other roots.

---

289. If in a triangle the angle B is 45°, prove that

$$(1+\cot A)(1+\cot C)=2.$$

290. If twice the square on the diameter of a circle is equal to the sum of the squares on the sides of the inscribed triangle ABC, prove that

$$\sin^2 A+\sin^2 B+\sin^2 C=2,$$

and that the triangle is right-angled.

291. If $\cos A=\tan B$, $\cos B=\tan C$, $\cos C=\tan A$,

prove that $\sin A=\sin B=\sin C=2\sin 18°$.

292. In any triangle shew that a, b, c are the roots of the equation

$$x^3-2sx^2+(r^2+s^2+4Rr)\,x-4Rrs=0.$$

293. Shew that $\sin\frac{\pi}{14}$ is a root of the equation

$$8x^3-4x^2-4x+1=0.$$

294. The stones from a circular field (radius r) are collected into n heaps at regular intervals along the hedge. Prove that the distance a labourer will have to travel with a wheelbarrow, which justs holds one heap, in bringing them together to one of the heaps (supposing him to start from this heap) is $4r\cot\frac{\pi}{2n}$.

---

295. Shew that

$$\cos\frac{\pi}{15}\cos\frac{2\pi}{15}\cos\frac{3\pi}{15}\cos\frac{4\pi}{15}\cos\frac{5\pi}{15}\cos\frac{6\pi}{15}\cos\frac{7\pi}{15}=\left(\frac{1}{2}\right)^7.$$

296. If x, y, z are the perpendiculars drawn to the sides from any point within a triangle, shew that $x^2+y^2+z^2$ is a minimum when

$$\frac{x}{a}=\frac{y}{b}=\frac{z}{c}=\frac{2\Delta}{a^2+b^2+c^2}.$$

297. If r_a, r_b, r_c, r_d be the radii of the circles which touch each side and the adjacent two sides produced of a quadrilateral, prove that

$$\frac{a}{r_a}+\frac{c}{r_c}=\frac{b}{r_b}+\frac{d}{r_d}.$$

298. If the diameters AA', BB', CC' of the circum-circle cut the sides BC, CA, AB in P, Q, R respectively, prove that

$$\frac{1}{AP}+\frac{1}{BQ}+\frac{1}{CR}=\frac{2}{R},$$

$$\frac{1}{A'P}+\frac{1}{B'Q}+\frac{1}{C'R}=\frac{1}{2R}(4+\sec A\sec B\sec C).$$

299. If α, β, γ are angles, unequal and less than 2π, which satisfy the equation $\dfrac{a}{\cos\theta}+\dfrac{b}{\sin\theta}+c=0$, prove that

$$\sin(\alpha+\beta)+\sin(\beta+\gamma)+\sin(\gamma+\alpha)=0.$$

300. Shew that

$$\left(\sec^2\frac{\pi}{7}+\sec^2\frac{2\pi}{7}+\sec^2\frac{3\pi}{7}\right)\left(\operatorname{cosec}^2\frac{\pi}{7}+\operatorname{cosec}^2\frac{2\pi}{7}+\operatorname{cosec}^2\frac{3\pi}{7}\right)=192.$$

TABLES OF LOGARITHMS OF NUMBERS, ANTILOGARITHMS, NATURAL AND LOGARITHMIC SINES, COSINES, AND TANGENTS.

Logarithms.

No.	0	1	2	3	4	5	6	7	8	9	1	2	3	4	5	6	7	8	9
10	0000	0043	0086	0128	0170	0212	0253	0294	0334	0374	4	8	12	17	21	25	29	33	37
11	0414	0453	0492	0531	0569	0607	0645	0682	0719	0755	4	8	11	15	19	23	26	30	34
12	0792	0828	0864	0899	0934	0969	1004	1038	1072	1106	3	7	10	14	17	21	24	28	31
13	1139	1173	1206	1239	1271	1303	1335	1367	1399	1430	3	6	10	13	16	19	23	26	29
14	1461	1492	1523	1553	1584	1614	1644	1673	1703	1732	3	6	9	12	15	18	21	24	27
15	1761	1790	1818	1847	1875	1903	1931	1959	1987	2014	3	6	8	11	14	17	20	22	25
16	2041	2068	2095	2122	2148	2175	2201	2227	2253	2279	3	5	8	11	13	16	18	21	24
17	2304	2330	2355	2380	2405	2430	2455	2480	2504	2529	2	5	7	10	12	15	17	20	22
18	2553	2577	2601	2625	2648	2672	2695	2718	2742	2765	2	5	7	9	12	14	16	19	21
19	2788	2810	2833	2856	2878	2900	2923	2945	2967	2989	2	4	7	9	11	13	16	18	20
20	3010	3032	3054	3075	3096	3118	3139	3160	3181	3201	2	4	6	8	11	13	15	17	19
21	3222	3243	3263	3284	3304	3324	3345	3365	3385	3404	2	4	6	8	10	12	14	16	18
22	3424	3444	3464	3483	3502	3522	3541	3560	3579	3598	2	4	6	8	10	12	14	15	17
23	3617	3636	3655	3674	3692	3711	3729	3747	3766	3784	2	4	6	7	9	11	13	15	17
24	3802	3820	3838	3856	3874	3892	3909	3927	3945	3962	2	4	5	7	9	11	12	14	16
25	3979	3997	4014	4031	4048	4065	4082	4099	4116	4133	2	3	5	7	9	10	12	14	15
26	4150	4166	4183	4200	4216	4232	4249	4265	4281	4298	2	3	5	7	8	10	11	13	15
27	4314	4330	4346	4362	4378	4393	4409	4425	4440	4456	2	3	5	6	8	9	11	13	14
28	4472	4487	4502	4518	4533	4548	4564	4579	4594	4609	2	3	5	6	8	9	11	12	14
29	4624	4639	4654	4669	4683	4698	4713	4728	4742	4757	1	3	4	6	7	9	10	12	13
30	4771	4786	4800	4814	4829	4843	4857	4871	4886	4900	1	3	4	6	7	9	10	11	13
31	4914	4928	4942	4955	4969	4983	4997	5011	5024	5038	1	3	4	6	7	8	10	11	12
32	5051	5065	5079	5092	5105	5119	5132	5145	5159	5172	1	3	4	5	7	8	9	11	12
33	5185	5198	5211	5224	5237	5250	5263	5276	5289	5302	1	3	4	5	6	8	9	10	12
34	5315	5328	5340	5353	5366	5378	5391	5403	5416	5428	1	3	4	5	6	8	9	10	11
35	5441	5453	5465	5478	5490	5502	5514	5527	5539	5551	1	2	4	5	6	7	9	10	11
36	5563	5575	5587	5599	5611	5623	5635	5647	5658	5670	1	2	4	5	6	7	8	10	11
37	5682	5694	5705	5717	5729	5740	5752	5763	5775	5786	1	2	3	5	6	7	8	9	10
38	5798	5809	5821	5832	5843	5855	5866	5877	5888	5899	1	2	3	5	6	7	8	9	10
39	5911	5922	5933	5944	5955	5966	5977	5988	5999	6010	1	2	3	4	5	7	8	9	10
40	6021	6031	6042	6053	6064	6075	6085	6096	6107	6117	1	2	3	4	5	6	8	9	10
41	6128	6138	6149	6160	6170	6180	6191	6201	6212	6222	1	2	3	4	5	6	7	8	9
42	6232	6243	6253	6263	6274	6284	6294	6304	6314	6325	1	2	3	4	5	6	7	8	9
43	6335	6345	6355	6365	6375	6385	6395	6405	6415	6425	1	2	3	4	5	6	7	8	9
44	6435	6444	6454	6464	6474	6484	6493	6503	6513	6522	1	2	3	4	5	6	7	8	9
45	6532	6542	6551	6561	6571	6580	6590	6599	6609	6618	1	2	3	4	5	6	7	8	9
46	6628	6637	6646	6656	6665	6675	6684	6693	6702	6712	1	2	3	4	5	6	7	7	8
47	6721	6730	6739	6749	6758	6767	6776	6785	6794	6803	1	2	3	4	5	5	6	7	8
48	6812	6821	6830	6839	6848	6857	6866	6875	6884	6893	1	2	3	4	4	5	6	7	8
49	6902	6911	6920	6928	6937	6946	6955	6964	6972	6981	1	2	3	4	4	5	6	7	8
50	6990	6998	7007	7016	7024	7033	7042	7050	7059	7067	1	2	3	3	4	5	6	7	8
51	7076	7084	7093	7101	7110	7118	7126	7135	7143	7152	1	2	3	3	4	5	6	7	8
52	7160	7168	7177	7185	7193	7202	7210	7218	7226	7235	1	2	2	3	4	5	6	7	7
53	7243	7251	7259	7267	7275	7284	7292	7300	7308	7316	1	2	2	3	4	5	6	6	7
54	7324	7332	7340	7348	7356	7364	7372	7380	7388	7396	1	2	2	3	4	5	6	6	7

Logarithms.

No.	0	1	2	3	4	5	6	7	8	9	1	2	3	4	5	6	7	8	9
55	7404	7412	7419	7427	7435	7443	7451	7459	7466	7474	1	2	2	3	4	5	5	6	7
56	7482	7490	7497	7505	7513	7520	7528	7536	7543	7551	1	2	2	3	4	5	5	6	7
57	7559	7566	7574	7582	7589	7597	7604	7612	7619	7627	1	2	2	3	4	5	5	6	7
58	7634	7642	7649	7657	7664	7672	7679	7686	7694	7701	1	1	2	3	4	4	5	6	7
59	7709	7716	7723	7731	7738	7745	7752	7760	7767	7774	1	1	2	3	4	4	5	6	7
60	7782	7789	7796	7803	7810	7818	7825	7832	7839	7846	1	1	2	3	4	4	5	6	6
61	7853	7860	7868	7875	7882	7889	7896	7903	7910	7917	1	1	2	3	4	4	5	6	6
62	7924	7931	7938	7945	7952	7959	7966	7973	7980	7987	1	1	2	3	3	4	5	6	6
63	7993	8000	8007	8014	8021	8028	8035	8041	8048	8055	1	1	2	3	3	4	5	5	6
64	8062	8069	8075	8082	8089	8096	8102	8109	8116	8122	1	1	2	3	3	4	5	5	6
65	8129	8136	8142	8149	8156	8162	8169	8176	8182	8189	1	1	2	3	3	4	5	5	6
66	8195	8202	8209	8215	8222	8228	8235	8241	8248	8254	1	1	2	3	3	4	5	5	6
67	8261	8267	8274	8280	8287	8293	8299	8306	8312	8319	1	1	2	3	3	4	5	5	6
68	8325	8331	8338	8344	8351	8357	8363	8370	8376	8382	1	1	2	3	3	4	4	5	6
69	8388	8395	8401	8407	8414	8420	8426	8432	8439	8445	1	1	2	2	3	4	4	5	6
70	8451	8457	8463	8470	8476	8482	8488	8494	8500	8506	1	1	2	2	3	4	4	5	6
71	8513	8519	8525	8531	8537	8543	8549	8555	8561	8567	1	1	2	2	3	4	4	5	5
72	8573	8579	8585	8591	8597	8603	8609	8615	8621	8627	1	1	2	2	3	4	4	5	5
73	8633	8639	8645	8651	8657	8663	8669	8675	8681	8686	1	1	2	2	3	4	4	5	5
74	8692	8698	8704	8710	8716	8722	8727	8733	8739	8745	1	1	2	2	3	4	4	5	5
75	8751	8756	8762	8768	8774	8779	8785	8791	8797	8802	1	1	2	2	3	3	4	5	5
76	8808	8814	8820	8825	8831	8837	8842	8848	8854	8859	1	1	2	2	3	3	4	5	5
77	8865	8871	8876	8882	8887	8893	8899	8904	8910	8915	1	1	2	2	3	3	4	4	5
78	8921	8927	8932	8938	8943	8949	8954	8960	8965	8971	1	1	2	2	3	3	4	4	5
79	8976	8982	8987	8993	8998	9004	9009	9015	9020	9025	1	1	2	2	3	3	4	4	5
80	9031	9036	9042	9047	9053	9058	9063	9069	9074	9079	1	1	2	2	3	3	4	4	5
81	9085	9090	9096	9101	9106	9112	9117	9122	9128	9133	1	1	2	2	3	3	4	4	5
82	9138	9143	9149	9154	9159	9165	9170	9175	9180	9186	1	1	2	2	3	3	4	4	5
83	9191	9196	9201	9206	9212	9217	9222	9227	9232	9238	1	1	2	2	3	3	4	4	5
84	9243	9248	9253	9258	9263	9269	9274	9279	9284	9289	1	1	2	2	3	3	4	4	5
85	9294	9299	9304	9309	9315	9320	9325	9330	9335	9340	1	1	2	2	3	3	4	4	5
86	9345	9350	9355	9360	9365	9370	9375	9380	9385	9390	1	1	2	2	3	3	4	4	5
87	9395	9400	9405	9410	9415	9420	9425	9430	9435	9440	0	1	1	2	2	3	3	4	4
88	9445	9450	9455	9460	9465	9469	9474	9479	9484	9489	0	1	1	2	2	3	3	4	4
89	9494	9499	9504	9509	9513	9518	9523	9528	9533	9538	0	1	1	2	2	3	3	4	4
90	9542	9547	9552	9557	9562	9566	9571	9576	9581	9586	0	1	1	2	2	3	3	4	4
91	9590	9595	9600	9605	9609	9614	9619	9624	9628	9633	0	1	1	2	2	3	3	4	4
92	9638	9643	9647	9652	9657	9661	9666	9671	9675	9680	0	1	1	2	2	3	3	4	4
93	9685	9689	9694	9699	9703	9708	9713	9717	9722	9727	0	1	1	2	2	3	3	4	4
94	9731	9736	9741	9745	9750	9754	9759	9763	9768	9773	0	1	1	2	2	3	3	4	4
95	9777	9782	9786	9791	9795	9800	9805	9809	9814	9818	0	1	1	2	2	3	3	4	4
96	9823	9827	9832	9836	9841	9845	9850	9854	9859	9863	0	1	1	2	2	3	3	4	4
97	9868	9872	9877	9881	9886	9890	9894	9899	9903	9908	0	1	1	2	2	3	3	4	4
98	9912	9917	9921	9926	9930	9934	9939	9943	9948	9952	0	1	1	2	2	3	3	4	4
99	9956	9961	9965	9969	9974	9978	9983	9987	9991	9996	0	1	1	2	2	3	3	3	4

Antilogarithms.

Log.	0	1	2	3	4	5	6	7	8	9	1	2	3	4	5	6	7	8	9
·00	1000	1002	1005	1007	1009	1012	1014	1016	1019	1021	0	0	1	1	1	1	2	2	2
·01	1023	1026	1028	1030	1033	1035	1038	1040	1042	1045	0	0	1	1	1	1	2	2	2
·02	1047	1050	1052	1054	1057	1059	1062	1064	1067	1069	0	0	1	1	1	1	2	2	2
·03	1072	1074	1076	1079	1081	1084	1086	1089	1091	1094	0	0	1	1	1	1	2	2	2
·04	1096	1099	1102	1104	1107	1109	1112	1114	1117	1119	0	1	1	1	1	2	2	2	2
·05	1122	1125	1127	1130	1132	1135	1138	1140	1143	1146	0	1	1	1	1	2	2	2	2
·06	1148	1151	1153	1156	1159	1161	1164	1167	1169	1172	0	1	1	1	1	2	2	2	2
·07	1175	1178	1180	1183	1186	1189	1191	1194	1197	1199	0	1	1	1	1	2	2	2	2
·08	1202	1205	1208	1211	1213	1216	1219	1222	1225	1227	0	1	1	1	1	2	2	2	2
·09	1230	1233	1236	1239	1242	1245	1247	1250	1253	1256	0	1	1	1	1	2	2	2	3
·10	1259	1262	1265	1268	1271	1274	1276	1279	1282	1285	0	1	1	1	1	2	2	2	3
·11	1288	1291	1294	1297	1300	1303	1306	1309	1312	1315	0	1	1	1	2	2	2	2	3
·12	1318	1321	1324	1327	1330	1334	1337	1340	1343	1346	0	1	1	1	2	2	2	2	3
·13	1349	1352	1355	1358	1361	1365	1368	1371	1374	1377	0	1	1	1	2	2	2	3	3
·14	1380	1384	1387	1390	1393	1396	1400	1403	1406	1409	0	1	1	1	2	2	2	3	3
·15	1413	1416	1419	1422	1426	1429	1432	1435	1439	1442	0	1	1	1	2	2	2	3	3
·16	1445	1449	1452	1455	1459	1462	1466	1469	1472	1476	0	1	1	1	2	2	2	3	3
·17	1479	1483	1486	1489	1493	1496	1500	1503	1507	1510	0	1	1	1	2	2	2	3	3
·18	1514	1517	1521	1524	1528	1531	1535	1538	1542	1545	0	1	1	1	2	2	2	3	3
·19	1549	1552	1556	1560	1563	1567	1570	1574	1578	1581	0	1	1	1	2	2	3	3	3
·20	1585	1589	1592	1596	1600	1603	1607	1611	1614	1618	0	1	1	1	2	2	3	3	3
·21	1622	1626	1629	1633	1637	1641	1644	1648	1652	1656	0	1	1	2	2	2	3	3	3
·22	1660	1663	1667	1671	1675	1679	1683	1687	1690	1694	0	1	1	2	2	2	3	3	3
·23	1698	1702	1706	1710	1714	1718	1722	1726	1730	1734	0	1	1	2	2	2	3	3	4
·24	1738	1742	1746	1750	1754	1758	1762	1766	1770	1774	0	1	1	2	2	2	3	3	4
·25	1778	1782	1786	1791	1795	1799	1803	1807	1811	1816	0	1	1	2	2	2	3	3	4
·26	1820	1824	1828	1832	1837	1841	1845	1849	1854	1858	0	1	1	2	2	3	3	3	4
·27	1862	1866	1871	1875	1879	1884	1888	1892	1897	1901	0	1	1	2	2	3	3	3	4
·28	1905	1910	1914	1919	1923	1928	1932	1936	1941	1945	0	1	1	2	2	3	3	4	4
·29	1950	1954	1959	1963	1968	1972	1977	1982	1986	1991	0	1	1	2	2	3	3	4	4
·30	1995	2000	2004	2009	2014	2018	2023	2028	2032	2037	0	1	1	2	2	3	3	4	4
·31	2042	2046	2051	2056	2061	2065	2070	2075	2080	2084	0	1	1	2	2	3	3	4	4
·32	2089	2094	2099	2104	2109	2113	2118	2123	2128	2133	0	1	1	2	2	3	3	4	4
·33	2138	2143	2148	2153	2158	2163	2168	2173	2178	2183	0	1	1	2	2	3	3	4	4
·34	2188	2193	2198	2203	2208	2213	2218	2223	2228	2234	1	1	2	2	3	3	4	4	5
·35	2239	2244	2249	2254	2259	2265	2270	2275	2280	2286	1	1	2	2	3	3	4	4	5
·36	2291	2296	2301	2307	2312	2317	2323	2328	2333	2339	1	1	2	2	3	3	4	4	5
·37	2344	2350	2355	2360	2366	2371	2377	2382	2388	2393	1	1	2	2	3	3	4	4	5
·38	2399	2404	2410	2415	2421	2427	2432	2438	2443	2449	1	1	2	2	3	3	4	4	5
·39	2455	2460	2466	2472	2477	2483	2489	2495	2500	2506	1	1	2	2	3	3	4	5	5
·40	2512	2518	2523	2529	2535	2541	2547	2553	2559	2564	1	1	2	2	3	4	4	5	5
·41	2570	2576	2582	2588	2594	2600	2606	2612	2618	2624	1	1	2	2	3	4	4	5	5
·42	2630	2636	2642	2649	2655	2661	2667	2673	2679	2685	1	1	2	2	3	4	4	5	6
·43	2692	2698	2704	2710	2716	2723	2729	2735	2742	2748	1	1	2	3	3	4	4	5	6
·44	2754	2761	2767	2773	2780	2786	2793	2799	2805	2812	1	1	2	3	3	4	4	5	6
·45	2818	2825	2831	2838	2844	2851	2858	2864	2871	2877	1	1	2	3	3	4	5	5	6
·46	2884	2891	2897	2904	2911	2917	2924	2931	2938	2944	1	1	2	3	3	4	5	5	6
·47	2951	2958	2965	2972	2979	2985	2992	2999	3006	3013	1	1	2	3	3	4	5	5	6
·48	3020	3027	3034	3041	3048	3055	3062	3069	3076	3083	1	1	2	3	4	4	5	6	6
·49	3090	3097	3105	3112	3119	3126	3133	3141	3148	3155	1	1	2	3	4	4	5	6	6

Antilogarithms.

Log.	0	1	2	3	4	5	6	7	8	9	1	2	3	4	5	6	7	8	9
·50	3162	3170	3177	3184	3192	3199	3206	3214	3221	3228	1	1	2	3	4	4	5	6	7
·51	3236	3243	3251	3258	3266	3273	3281	3289	3296	3304	1	2	2	3	4	5	5	6	7
·52	3311	3319	3327	3334	3342	3350	3357	3365	3373	3381	1	2	2	3	4	5	5	6	7
·53	3388	3396	3404	3412	3420	3428	3436	3443	3451	3459	1	2	2	3	4	5	6	6	7
·54	3467	3475	3483	3491	3499	3508	3516	3524	3532	3540	1	2	2	3	4	5	6	6	7
·55	3548	3556	3565	3573	3581	3589	3597	3606	3614	3622	1	2	2	3	4	5	6	7	7
·56	3631	3639	3648	3656	3664	3673	3681	3690	3698	3707	1	2	3	3	4	5	6	7	8
·57	3715	3724	3733	3741	3750	3758	3767	3776	3784	3793	1	2	3	3	4	5	6	7	8
·58	3802	3811	3819	3828	3837	3846	3855	3864	3873	3882	1	2	3	4	4	5	6	7	8
·59	3890	3899	3908	3917	3926	3936	3945	3954	3963	3972	1	2	3	4	5	5	6	7	8
·60	3981	3990	3999	4009	4018	4027	4036	4046	4055	4064	1	2	3	4	5	6	6	7	8
·61	4074	4083	4093	4102	4111	4121	4130	4140	4150	4159	1	2	3	4	5	6	7	8	9
·62	4169	4178	4188	4198	4207	4217	4227	4236	4246	4256	1	2	3	4	5	6	7	8	9
·63	4266	4276	4285	4295	4305	4315	4325	4335	4345	4355	1	2	3	4	5	6	7	8	9
·64	4365	4375	4385	4395	4406	4416	4426	4436	4446	4457	1	2	3	4	5	6	7	8	9
·65	4467	4477	4487	4498	4508	4519	4529	4539	4550	4560	1	2	3	4	5	6	7	8	9
·66	4571	4581	4592	4603	4613	4624	4634	4645	4656	4667	1	2	3	4	5	6	7	9	10
·67	4677	4688	4699	4710	4721	4732	4742	4753	4764	4775	1	2	3	4	5	7	8	9	10
·68	4786	4797	4808	4819	4831	4842	4853	4864	4875	4887	1	2	3	4	6	7	8	9	10
·69	4898	4909	4920	4932	4943	4955	4966	4977	4989	5000	1	2	3	5	6	7	8	9	10
·70	5012	5023	5035	5047	5058	5070	5082	5093	5105	5117	1	2	4	5	6	7	8	9	11
·71	5129	5140	5152	5164	5176	5188	5200	5212	5224	5236	1	2	4	5	6	7	8	10	11
·72	5248	5260	5272	5284	5297	5309	5321	5333	5346	5358	1	2	4	5	6	7	9	10	11
·73	5370	5383	5395	5408	5420	5433	5445	5458	5470	5483	1	3	4	5	6	8	9	10	11
·74	5495	5508	5521	5534	5546	5559	5572	5585	5598	5610	1	3	4	5	6	8	9	10	12
·75	5623	5636	5649	5662	5675	5689	5702	5715	5728	5741	1	3	4	5	7	8	9	10	12
·76	5754	5768	5781	5794	5808	5821	5834	5848	5861	5875	1	3	4	5	7	8	9	11	12
·77	5888	5902	5916	5929	5943	5957	5970	5984	5998	6012	1	3	4	5	7	8	10	11	12
·78	6026	6039	6053	6067	6081	6095	6109	6124	6138	6152	1	3	4	6	7	8	10	11	13
·79	6166	6180	6194	6209	6223	6237	6252	6266	6281	6295	1	3	4	6	7	9	10	11	13
·80	6310	6324	6339	6353	6368	6383	6397	6412	6427	6442	1	3	4	6	7	9	10	12	13
·81	6457	6471	6486	6501	6516	6531	6546	6561	6577	6592	2	3	5	6	8	9	11	12	14
·82	6607	6622	6637	6653	6668	6683	6699	6714	6730	6745	2	3	5	6	8	9	11	12	14
·83	6761	6776	6792	6808	6823	6839	6855	6871	6887	6902	2	3	5	6	8	9	11	13	14
·84	6918	6934	6950	6966	6982	6998	7015	7031	7047	7063	2	3	5	6	8	10	11	13	15
·85	7079	7096	7112	7129	7145	7161	7178	7194	7211	7228	2	3	5	7	8	10	12	13	15
·86	7244	7261	7278	7295	7311	7328	7345	7362	7379	7396	2	3	5	7	8	10	12	13	15
·87	7413	7430	7447	7464	7482	7499	7516	7534	7551	7568	2	3	5	7	9	10	12	14	16
·88	7586	7603	7621	7638	7656	7674	7691	7709	7727	7745	2	4	5	7	9	11	12	14	16
·89	7762	7780	7798	7816	7834	7852	7870	7889	7907	7925	2	4	5	7	9	11	13	14	16
·90	7943	7962	7980	7998	8017	8035	8054	8072	8091	8110	2	4	6	7	9	11	13	15	17
·91	8128	8147	8166	8185	8204	8222	8241	8260	8279	8299	2	4	6	8	9	11	13	15	17
·92	8318	8337	8356	8375	8395	8414	8433	8453	8472	8492	2	4	6	8	10	12	14	15	17
·93	8511	8531	8551	8570	8590	8610	8630	8650	8670	8690	2	4	6	8	10	12	14	16	18
·94	8710	8730	8750	8770	8790	8810	8831	8851	8872	8892	2	4	6	8	10	12	14	16	18
·95	8913	8933	8954	8974	8995	9016	9036	9057	9078	9099	2	4	6	8	10	12	15	17	19
·96	9120	9141	9162	9183	9204	9226	9247	9268	9290	9311	2	4	6	8	11	13	15	17	19
·97	9333	9354	9376	9397	9419	9441	9462	9484	9506	9528	2	4	7	9	11	13	15	17	20
·98	9550	9572	9594	9616	9638	9661	9683	9705	9727	9750	2	4	7	9	11	13	16	18	20
·99	9772	9795	9817	9840	9863	9886	9908	9931	9954	9977	2	5	7	9	11	14	16	18	20

Natural Sines.

Deg.	0′	6′ 0°·1	12′ 0°·2	18′ 0°·3	24′ 0°·4	30′ 0°·5	36′ 0°·6	42′ 0°·7	48′ 0°·8	54′ 0°·9	Mean Differences 1′	2′	3′	4′	5′
0	0000	0017	0035	0052	0070	0087	0105	0122	0140	0157	3	6	9	12	15
1	0175	0192	0209	0227	0244	0262	0279	0297	0314	0332	3	6	9	12	15
2	0349	0366	0384	0401	0419	0436	0454	0471	0488	0506	3	6	9	12	15
3	0523	0541	0558	0576	0593	0610	0628	0645	0663	0680	3	6	9	12	15
4	0698	0715	0732	0750	0767	0785	0802	0819	0837	0854	3	6	9	12	14
5	0872	0889	0906	0924	0941	0958	0976	0993	1011	1028	3	6	9	12	14
6	1045	1063	1080	1097	1115	1132	1149	1167	1184	1201	3	6	9	12	14
7	1219	1236	1253	1271	1288	1305	1323	1340	1357	1374	3	6	9	12	14
8	1392	1409	1426	1444	1461	1478	1495	1513	1530	1547	3	6	9	12	14
9	1564	1582	1599	1616	1633	1650	1668	1685	1702	1719	3	6	9	12	14
10	1736	1754	1771	1788	1805	1822	1840	1857	1874	1891	3	6	9	11	14
11	1908	1925	1942	1959	1977	1994	2011	2028	2045	2062	3	6	9	11	14
12	2079	2096	2113	2130	2147	2164	2181	2198	2215	2233	3	6	9	11	14
13	2250	2267	2284	2300	2317	2334	2351	2368	2385	2402	3	6	8	11	14
14	2419	2436	2453	2470	2487	2504	2521	2538	2554	2571	3	6	8	11	14
15	2588	2605	2622	2639	2656	2672	2689	2706	2723	2740	3	6	8	11	14
16	2756	2773	2790	2807	2823	2840	2857	2874	2890	2907	3	6	8	11	14
17	2924	2940	2957	2974	2990	3007	3024	3040	3057	3074	3	6	8	11	14
18	3090	3107	3123	3140	3156	3173	3190	3206	3223	3239	3	6	8	11	14
19	3256	3272	3289	3305	3322	3338	3355	3371	3387	3404	3	5	8	11	14
20	3420	3437	3453	3469	3486	3502	3518	3535	3551	3567	3	5	8	11	14
21	3584	3600	3616	3633	3649	3665	3681	3697	3714	3730	3	5	8	11	14
22	3746	3762	3778	3795	3811	3827	3843	3859	3875	3891	3	5	8	11	14
23	3907	3923	3939	3955	3971	3987	4003	4019	4035	4051	3	5	8	11	14
24	4067	4083	4099	4115	4131	4147	4163	4179	4195	4210	3	5	8	11	13
25	4226	4242	4258	4274	4289	4305	4321	4337	4352	4368	3	5	8	11	13
26	4384	4399	4415	4431	4446	4462	4478	4493	4509	4524	3	5	8	10	13
27	4540	4555	4571	4586	4602	4617	4633	4648	4664	4679	3	5	8	10	13
28	4695	4710	4726	4741	4756	4772	4787	4802	4818	4833	3	5	8	10	13
29	4848	4863	4879	4894	4909	4924	4939	4955	4970	4985	3	5	8	10	13
30	5000	5015	5030	5045	5060	5075	5090	5105	5120	5135	3	5	8	10	13
31	5150	5165	5180	5195	5210	5225	5240	5255	5270	5284	3	5	7	10	12
32	5299	5314	5329	5344	5358	5373	5388	5402	5417	5432	2	5	7	10	12
33	5446	5461	5476	5490	5505	5519	5534	5548	5563	5577	2	5	7	10	12
34	5592	5606	5621	5635	5650	5664	5678	5693	5707	5721	2	5	7	10	12
35	5736	5750	5764	5779	5793	5807	5821	5835	5850	5864	2	5	7	9	12
36	5878	5892	5906	5920	5934	5948	5962	5976	5990	6004	2	5	7	9	12
37	6018	6032	6046	6060	6074	6088	6101	6115	6129	6143	2	5	7	9	12
38	6157	6170	6184	6198	6211	6225	6239	6252	6266	6280	2	5	7	9	11
39	6293	6307	6320	6334	6347	6361	6374	6388	6401	6414	2	4	7	9	11
40	6428	6441	6455	6468	6481	6494	6508	6521	6534	6547	2	4	7	9	11
41	6561	6574	6587	6600	6613	6626	6639	6652	6665	6678	2	4	7	9	11
42	6691	6704	6717	6730	6743	6756	6769	6782	6794	6807	2	4	6	9	11
43	6820	6833	6845	6858	6871	6884	6896	6909	6921	6934	2	4	6	8	11
44	6947	6959	6972	6984	6997	7009	7022	7034	7046	7059	2	4	6	8	10

Natural Sines.

Deg.	0′	6′ 0°·1	12′ 0°·2	18′ 0°·3	24′ 0°·4	30′ 0°·5	36′ 0°·6	42′ 0°·7	48′ 0°·8	54′ 0°·9	Mean Differences 1′	2′	3′	4′	5′
45	7071	7083	7096	7108	7120	7133	7145	7157	7169	7181	2	4	6	8	10
46	7193	7206	7218	7230	7242	7254	7266	7278	7290	7302	2	4	6	8	10
47	7314	7325	7337	7349	7361	7373	7385	7396	7408	7420	2	4	6	8	10
48	7431	7443	7455	7466	7478	7490	7501	7513	7524	7536	2	4	6	8	10
49	7547	7558	7570	7581	7593	7604	7615	7627	7638	7649	2	4	6	8	9
50	7660	7672	7683	7694	7705	7716	7727	7738	7749	7760	2	4	6	7	9
51	7771	7782	7793	7804	7815	7826	7837	7848	7859	7869	2	4	5	7	9
52	7880	7891	7902	7912	7923	7934	7944	7955	7965	7976	2	4	5	7	9
53	7986	7997	8007	8018	8028	8039	8049	8059	8070	8080	2	3	5	7	9
54	8090	8100	8111	8121	8131	8141	8151	8161	8171	8181	2	3	5	7	8
55	8192	8202	8211	8221	8231	8241	8251	8261	8271	8281	2	3	5	7	8
56	8290	8300	8310	8320	8329	8339	8348	8358	8368	8377	2	3	5	6	8
57	8387	8396	8406	8415	8425	8434	8443	8453	8462	8471	2	3	5	6	8
58	8480	8490	8499	8508	8517	8526	8536	8545	8554	8563	2	3	5	6	8
59	8572	8581	8590	8599	8607	8616	8625	8634	8643	8652	1	3	4	6	7
60	8660	8669	8678	8686	8695	8704	8712	8721	8729	8738	1	3	4	6	7
61	8746	8755	8763	8771	8780	8788	8796	8805	8813	8821	1	3	4	6	7
62	8829	8838	8846	8854	8862	8870	8878	8886	8894	8902	1	3	4	5	7
63	8910	8918	8926	8934	8942	8949	8957	8965	8973	8980	1	3	4	5	6
64	8988	8996	9003	9011	9018	9026	9033	9041	9048	9056	1	3	4	5	6
65	9063	9070	9078	9085	9092	9100	9107	9114	9121	9128	1	2	4	5	6
66	9135	9143	9150	9157	9164	9171	9178	9184	9191	9198	1	2	3	5	6
67	9205	9212	9219	9225	9232	9239	9245	9252	9259	9265	1	2	3	4	6
68	9272	9278	9285	9291	9298	9304	9311	9317	9323	9330	1	2	3	4	5
69	9336	9342	9348	9354	9361	9367	9373	9379	9385	9391	1	2	3	4	5
70	9397	9403	9409	9415	9421	9426	9432	9438	9444	9449	1	2	3	4	5
71	9455	9461	9466	9472	9478	9483	9489	9494	9500	9505	1	2	3	4	5
72	9511	9516	9521	9527	9532	9537	9542	9548	9553	9558	1	2	3	4	4
73	9563	9568	9573	9578	9583	9588	9593	9598	9603	9608	1	2	2	3	4
74	9613	9617	9622	9627	9632	9636	9641	9646	9650	9655	1	2	2	3	4
75	9659	9664	9668	9673	9677	9681	9686	9690	9694	9699	1	1	2	3	4
76	9703	9707	9711	9715	9720	9724	9728	9732	9736	9740	1	1	2	3	3
77	9744	9748	9751	9755	9759	9763	9767	9770	9774	9778	1	1	2	3	3
78	9781	9785	9789	9792	9796	9799	9803	9806	9810	9813	1	1	2	2	3
79	9816	9820	9823	9826	9829	9833	9836	9839	9842	9845	1	1	2	2	3
80	9848	9851	9854	9857	9860	9863	9866	9869	9871	9874	0	1	1	2	2
81	9877	9880	9882	9885	9888	9890	9893	9895	9898	9900	0	1	1	2	2
82	9903	9905	9907	9910	9912	9914	9917	9919	9921	9923	0	1	1	2	2
83	9925	9928	9930	9932	9934	9936	9938	9940	9942	9943	0	1	1	1	2
84	9945	9947	9949	9951	9952	9954	9956	9957	9959	9960	0	1	1	1	2
85	9962	9963	9965	9966	9968	9969	9971	9972	9973	9974	0	0	1	1	1
86	9976	9977	9978	9979	9980	9981	9982	9983	9984	9985	0	0	1	1	1
87	9986	9987	9988	9989	9990	9990	9991	9992	9993	9993	0	0	0	1	1
88	9994	9995	9995	9996	9996	9997	9997	9997	9998	9998	0	0	0	0	0
89	9998	9999	9999	9999	9999	9999	9999	9999	9999	9999	0	0	0	0	0

Natural Cosines.

Deg.	0′	6′ 0°·1	12′ 0°·2	18′ 0°·3	24′ 0°·4	30′ 0°·5	36′ 0°·6	42′ 0°·7	48′ 0°·8	54′ 0°·9	Mean Differences 1′	2′	3′	4′	5′
0	1·000	9999	9999	9999	9999	9999	9999	9999	9999	9999	0	0	0	0	0
1	9998	9998	9998	9997	9997	9997	9996	9996	9995	9995	0	0	0	0	0
2	9994	9993	9993	9992	9991	9990	9990	9989	9988	9987	0	0	0	1	1
3	9986	9985	9984	9983	9982	9981	9980	9979	9978	9977	0	0	1	1	1
4	9976	9974	9973	3972	9971	9969	9968	9966	9965	9963	0	0	1	1	1
5	9962	9960	9959	9957	9956	9954	9952	9951	9949	9947	0	1	1	1	2
6	9945	9943	9942	9940	9938	9936	9934	9932	9930	9928	0	1	1	1	2
7	9925	9923	9921	9919	9917	9914	9912	9910	9907	9905	0	1	1	2	2
8	9903	9900	9898	9895	9893	9890	9888	9885	9882	9880	0	1	1	2	2
9	9877	9874	9871	9869	9866	9863	9860	9857	9854	9851	0	1	1	2	2
10	9848	9845	9842	9839	9836	9833	9829	9826	9823	9820	1	1	2	2	3
11	9816	9813	9810	9806	9803	9799	9796	9792	9789	9785	1	1	2	2	3
12	9781	9778	9774	9770	9767	9763	9759	9755	9751	9748	1	1	2	3	3
13	9744	9740	9736	9732	9728	9724	9720	9715	9711	9707	1	1	2	3	3
14	9703	9699	9694	9690	9686	9681	9677	9673	9668	9664	1	1	2	3	4
15	9659	9655	9650	9646	9641	9636	9632	9627	9622	9617	1	2	2	3	4
16	9613	9608	9603	9598	9593	9588	9583	9578	9573	9568	1	2	2	3	4
17	9563	9558	9553	9548	9542	9537	9532	9527	9521	9516	1	2	3	3	5
18	9511	9505	9500	9494	9489	9483	9478	9472	9466	9461	1	2	3	4	5
19	9455	9449	9444	9438	9432	9426	9421	9415	9409	9403	1	2	3	4	5
20	9397	9391	9385	9379	9373	9367	9361	9354	9348	9342	1	2	3	4	5
21	9336	9330	9323	9317	9311	9304	9298	9291	9285	9278	1	2	3	4	5
22	9272	9265	9259	9252	9245	9239	9232	9225	9219	9212	1	2	3	4	6
23	9205	9198	9191	9184	9178	9179	9164	9157	9150	9143	1	2	3	5	6
24	9135	9128	9121	9114	9107	9100	9092	9085	9078	9070	1	2	4	5	6
25	9063	9056	9048	9041	9033	9026	9018	9011	9003	8996	1	3	4	5	6
26	8988	8980	8973	8965	8957	8949	8942	8934	8926	8918	1	3	4	5	6
27	8910	8902	8894	8886	8878	8870	8862	8854	8846	8838	1	3	4	5	7
28	8829	8821	8813	8805	8796	8788	8780	8771	8763	8755	1	3	4	6	7
29	8746	8738	8729	8721	8712	8704	8695	8686	8678	8669	1	3	4	6	7
30	8660	8652	8643	8634	8625	8616	8607	8599	8590	8581	1	3	4	6	7
31	8572	8563	8554	8545	8536	8526	8517	8508	8499	8490	2	3	5	6	8
32	8480	8471	8462	8453	8443	8434	8425	8415	8406	8396	2	3	5	6	8
33	8387	8377	8368	8358	8348	8339	8329	8320	8310	8300	2	3	5	6	8
34	8290	8281	8271	8261	8251	8241	8231	8221	8211	8202	2	3	5	7	8
35	8192	8181	8171	8161	8151	8141	8131	8121	8111	8100	2	3	5	7	8
36	8090	8080	8070	8059	8049	8039	8028	8018	8007	7997	2	3	5	7	9
37	7986	7976	7965	7955	7944	7934	7923	7912	7902	7891	2	4	5	7	9
38	7880	7869	7859	7848	7837	7826	7815	7804	7793	7782	2	4	5	7	9
39	7771	7760	7749	7738	7727	7716	7705	7694	7683	7672	2	4	6	7	9
40	7660	7649	7638	7627	7615	7604	7593	7581	7570	7559	2	4	6	8	9
41	7547	7536	7524	7513	7501	7490	7478	7466	7455	7443	2	4	6	8	10
42	7431	7420	7408	7396	7385	7373	7361	7349	7337	7325	2	4	6	8	10
43	7314	7302	7290	7278	7266	7254	7242	7230	7218	7206	2	4	6	8	10
44	7193	7181	7169	7157	7145	7133	7120	7108	7096	7083	2	4	6	8	10

Natural Cosines.

Deg.	0′	6′ 0°·1	12′ 0°·2	18′ 0°·3	24′ 0°·4	30′ 0°·5	36′ 0°·6	42′ 0°·7	48′ 0°·8	54′ 0°·9	Mean Differences 1′	2′	3′	4′	5′
45	7071	7059	7046	7034	7022	7009	6997	6984	6972	6959	2	4	6	8	10
46	6947	6934	6921	6909	6896	6884	6871	6858	6845	6833	2	4	6	8	11
47	6820	6807	6794	6782	6769	6756	6743	6730	6717	6704	2	4	6	9	11
48	6691	6678	6665	6652	6639	6626	6613	6600	6587	6574	2	4	7	9	11
49	6561	6547	6534	6521	6508	6494	6481	6468	6455	6441	2	4	7	9	11
50	6428	6414	6401	6388	6374	6361	6347	6334	6320	6307	2	4	7	9	11
51	6293	6280	6266	6252	6239	6225	6211	6198	6184	6170	2	5	7	9	11
52	6157	6143	6129	6115	6101	6088	6074	6060	6046	6032	2	5	7	9	12
53	6018	6004	5990	5976	5962	5948	5934	5920	5906	5892	2	5	7	9	12
54	5878	5864	5850	5835	5821	5807	5793	5779	5764	5750	2	5	7	9	12
55	5736	5721	5707	5693	5678	5664	5650	5635	5621	5606	2	5	7	10	12
56	5592	5577	5563	5548	5534	5519	5505	5490	5476	5461	2	5	7	10	12
57	5446	5432	5417	5402	5388	5373	5358	5344	5329	5314	2	5	7	10	12
58	5299	5284	5270	5255	5240	5225	5210	5195	5180	5165	2	5	7	10	12
59	5150	5135	5120	5105	5090	5075	5060	5045	5030	5015	3	5	8	10	13
60	5000	4985	4970	4955	4939	4924	4909	4894	4879	4863	3	5	8	10	13
61	4848	4833	4818	4802	4787	4772	4756	4741	4726	4710	3	5	8	10	13
62	4695	4679	4664	4648	4633	4617	4602	4586	4571	4555	3	5	8	10	13
63	4540	4524	4509	4493	4478	4462	4446	4431	4415	4399	3	5	8	10	13
64	4384	4368	4352	4337	4321	4305	4289	4274	4258	4242	3	5	8	11	13
65	4226	4210	4195	4179	4163	4147	4131	4115	4099	4083	3	5	8	11	13
66	4067	4051	4035	4019	4003	3987	3971	3955	3939	3923	3	5	8	11	14
67	3907	3891	3875	3859	3843	3827	3811	3795	3778	3762	3	5	8	11	14
68	3746	3730	3714	3697	3681	3665	3649	3633	3616	3600	3	5	8	11	14
69	3584	3567	3551	3535	3518	3502	3486	3469	3453	3437	3	5	8	11	14
70	3420	3404	3387	3371	3355	3338	3322	3305	3289	3272	3	5	8	11	14
71	3256	3239	3223	3206	3190	3173	3156	3140	3123	3107	3	6	8	11	14
72	3090	3074	3057	3040	3024	3007	2990	2974	2957	2940	3	6	8	11	14
73	2924	2907	2890	2874	2857	2840	2823	2807	2790	2773	3	6	8	11	14
74	2756	2740	2723	2706	2689	2672	2656	2639	2622	2605	3	6	8	11	14
75	2588	2571	2554	2538	2521	2504	2487	2470	2453	2436	3	6	8	11	14
76	2419	2402	2385	2368	2351	2334	2317	2300	2284	2267	3	6	8	11	14
77	2250	2233	2215	2198	2181	2164	2147	2130	2113	2096	3	6	9	11	14
78	2079	2062	2045	2028	2011	1994	1977	1959	1942	1925	3	6	9	11	14
79	1908	1891	1874	1857	1840	1822	1805	1788	1771	1754	3	6	9	11	14
80	1736	1719	1702	1685	1668	1650	1633	1616	1599	1582	3	6	9	12	14
81	1564	1547	1530	1513	1495	1478	1461	1444	1426	1409	3	6	9	12	14
82	1392	1374	1357	1340	1323	1305	1288	1271	1253	1236	3	6	9	12	14
83	1219	1201	1184	1167	1149	1132	1115	1097	1080	1063	3	6	9	12	14
84	1045	1028	1011	0993	0976	0958	0941	0924	0906	0889	3	6	9	12	14
85	0872	0854	0837	0819	0802	0785	0767	0750	0732	0715	3	6	9	12	15
86	0698	0680	0663	0645	0628	0610	0593	0576	0558	0541	3	6	9	12	15
87	0523	0506	0488	0471	0454	0436	0419	0401	0384	0366	3	6	9	12	15
88	0349	0332	0314	0297	0279	0262	0244	0227	0209	0192	3	6	9	12	15
89	0175	0157	0140	0122	0105	0087	0070	0052	0035	0017	3	6	9	12	15

Natural Tangents.

Deg.	0′	6′ 0°·1	12′ 0°·2	18′ 0°·3	24′ 0°·4	30′ 0°·5	36′ 0°·6	42′ 0°·7	48′ 0°·8	54′ 0°·9	Mean Differences 1′	2′	3′	4′	5′
0	0000	0017	0035	0052	0070	0087	0105	0122	0140	0157	3	6	9	12	15
1	0175	0192	0209	0227	0244	0262	0279	0297	0314	0332	3	6	9	12	15
2	0349	0367	0384	0402	0419	0437	0454	0472	0489	0507	3	6	9	12	15
3	0524	0542	0559	0577	0594	0612	0629	0647	0664	0682	3	6	9	12	15
4	0699	0717	0734	0752	0769	0787	0805	0822	0840	0857	3	6	9	12	15
5	0875	0892	0910	0928	0945	0963	0981	0998	1016	1033	3	6	9	12	15
6	1051	1069	1086	1104	1122	1139	1157	1175	1192	1210	3	6	9	12	15
7	1228	1246	1263	1281	1299	1317	1334	1352	1370	1388	3	6	9	12	15
8	1405	1423	1441	1459	1477	1495	1512	1530	1548	1566	3	6	9	12	15
9	1584	1602	1620	1638	1655	1673	1691	1709	1727	1745	3	6	9	12	15
10	1763	1781	1799	1817	1835	1853	1871	1890	1908	1926	3	6	9	12	15
11	1944	1962	1980	1998	2016	2035	2053	2071	2089	2107	3	6	9	12	15
12	2126	2144	2162	2180	2199	2217	2235	2254	2272	2290	3	6	9	12	15
13	2309	2327	2345	2364	2382	2401	2419	2438	2456	2475	3	6	9	12	15
14	2493	2512	2530	2549	2568	2586	2605	2623	2642	2661	3	6	9	12	16
15	2679	2698	2717	2736	2754	2773	2792	2811	2830	2849	3	6	9	13	16
16	2867	2886	2905	2924	2943	2962	2981	3000	3019	3038	3	6	9	13	16
17	3057	3076	3096	3115	3134	3153	3172	3191	3211	3230	3	6	10	13	16
18	3249	3269	3288	3307	3327	3346	3365	3385	3404	3424	3	6	10	13	16
19	3443	3463	3482	3502	3522	3541	3561	3581	3600	3620	3	6	10	13	16
20	3640	3659	3679	3699	3719	3739	3759	3779	3799	3819	3	7	10	13	17
21	3839	3859	3879	3899	3919	3939	3959	3979	4000	4020	3	7	10	13	17
22	4040	4061	4081	4101	4122	4142	4163	4183	4204	4224	3	7	10	14	17
23	4245	4265	4286	4307	4327	4348	4369	4390	4411	4431	3	7	10	14	17
24	4452	4473	4494	4515	4536	4557	4578	4599	4621	4642	4	7	11	14	18
25	4663	4684	4706	4727	4748	4770	4791	4813	4834	4856	4	7	11	14	18
26	4877	4899	4921	4942	4964	4986	5008	5029	5051	5073	4	7	11	15	18
27	5095	5117	5139	5161	5184	5206	5228	5250	5272	5295	4	7	11	15	18
28	5317	5340	5362	5384	5407	5430	5452	5475	5498	5520	4	8	11	15	19
29	5543	5566	5589	5612	5635	5658	5681	5704	5727	5750	4	8	12	15	19
30	5774	5797	5820	5844	5867	5890	5914	5938	5961	5985	4	8	12	16	20
31	6009	6032	6056	6080	6104	6128	6152	6176	6200	6224	4	8	12	16	20
32	6249	6273	6297	6322	6346	6371	6395	6420	6445	6469	4	8	12	16	20
33	6494	6519	6544	6569	6594	6619	6644	6669	6694	6720	4	8	13	17	21
34	6745	6771	6796	6822	6847	6873	6899	6924	6950	6976	4	9	13	17	21
35	7002	7028	7054	7080	7107	7133	7159	7186	7212	7239	4	9	13	18	22
36	7265	7292	7319	7346	7373	7400	7427	7454	7481	7508	5	9	14	18	23
37	7536	7563	7590	7618	7646	7673	7701	7729	7757	7785	5	9	14	18	23
38	7813	7841	7869	7898	7926	7954	7983	8012	8040	8069	5	9	14	19	24
39	8098	8127	8156	8185	8214	8243	8273	8302	8332	8361	5	10	15	20	24
40	8391	8421	8451	8481	8511	8541	8571	8601	8632	8662	5	10	15	20	25
41	8693	8724	8754	8785	8816	8847	8878	8910	8941	8972	5	10	16	21	26
42	9004	9036	9067	9099	9131	9163	9195	9228	9260	9293	5	11	16	21	27
43	9325	9358	9391	9424	9457	9490	9523	9556	9590	9623	6	11	17	22	28
44	9657	9691	9725	9759	9793	9827	9861	9896	9930	9965	6	11	17	23	29

Natural Tangents.

Deg.	0′	6′ 0°·1	12′ 0°·2	18′ 0°·3	24′ 0°·4	30′ 0°·5	36′ 0°·6	42′ 0°·7	48′ 0°·8	54′ 0°·9	Mean Differences 1′	2′	3′	4′	5′
45	1·0000	0035	0070	0105	0141	0176	0212	0247	0283	0319	6	12	18	24	30
46	1·0355	0392	0428	0464	0501	0538	0575	0612	0649	0686	6	12	18	25	31
47	1·0724	0761	0799	0837	0875	0913	0951	0990	1028	1067	6	13	19	25	32
48	1·1106	1145	1184	1224	1263	1303	1343	1383	1423	1463	7	13	20	27	33
49	1·1504	1544	1585	1626	1667	1708	1750	1792	1833	1875	7	14	21	28	34
50	1·1918	1960	2002	2045	2088	2131	2174	2218	2261	2305	7	14	22	29	36
51	1·2349	2393	2437	2482	2527	2572	2617	2662	2708	2753	8	15	23	30	38
52	1·2799	2846	2892	2938	2985	3032	3079	3127	3175	3222	8	16	24	31	39
53	1·3270	3319	3367	3416	3465	3514	3564	3613	3663	3713	8	16	25	33	41
54	1·3764	3814	3865	3916	3968	4019	4071	4124	4176	4229	9	17	26	34	43
55	1·4281	4335	4388	4442	4496	4550	4605	4659	4715	4770	9	18	27	36	45
56	1·4826	4882	4938	4994	5051	5108	5166	5224	5282	5340	10	19	29	38	48
57	1·5399	5458	5517	5577	5637	5697	5757	5818	5880	5941	10	20	30	40	50
58	1·6003	6066	6128	6191	6255	6319	6383	6447	6512	6577	11	21	32	43	53
59	1·6643	6709	6775	6842	6909	6977	7045	7113	7182	7251	11	23	34	45	56
60	1·7321	7391	7461	7532	7603	7675	7747	7820	7893	7966	12	24	36	48	60
61	1·8040	8115	8190	8265	8341	8418	8495	8572	8650	8728	13	26	38	51	64
62	1·8807	8887	8967	9047	9128	9210	9292	9375	9458	9542	14	27	41	55	68
63	1·9626	9711	9797	9883	9970	$\overline{0057}$	$\overline{0145}$	$\overline{0233}$	$\overline{0323}$	$\overline{0413}$	15	29	44	58	73
64	2·0503	0594	0686	0778	0872	0965	1060	1155	1251	1348	16	31	47	63	78
65	2·1445	1543	1642	1742	1842	1943	2045	2148	2251	2355	17	34	51	68	85
66	2·2460	2566	2673	2781	2889	2998	3109	3220	3332	3445					
67	2·3559	3673	3789	3906	4023	4142	4262	4383	4504	4627					
68	2·4751	4876	5002	5129	5257	5386	5517	5649	5782	5916					
69	2·6051	6187	6325	6464	6605	6746	6889	7034	7179	7326					
70	2·7475	7625	7776	7929	8083	8239	8397	8556	8716	8878					
71	2·9042	9208	9375	9544	9714	9887	$\overline{0061}$	$\overline{0237}$	$\overline{0415}$	$\overline{0595}$					
72	3·0777	0961	1146	1334	1524	1716	1910	2106	2305	2506					
73	3·2709	2914	3122	3332	3544	3759	3977	4197	4420	4646					
74	3·4874	5105	5339	5576	5816	6059	6305	6554	6806	7062					
75	3·7321	7583	7848	8118	8391	8667	8947	9232	9520	9812					
76	4·0108	0408	0713	1022	1335	1653	1976	2303	2635	2972					
77	4·3315	3662	4015	4374	4737	5107	5483	5864	6252	6646					
78	4·7046	7453	7867	8288	8716	9152	9594	$\overline{0045}$	$\overline{0504}$	$\overline{0970}$					
79	5·1446	1929	2422	2924	3435	3955	4486	5026	5578	6140					
80	5·6713	7297	7894	8502	9124	9758	$\overline{0405}$	$\overline{1066}$	$\overline{1742}$	$\overline{2432}$					
81	6·3138	3859	4596	5350	6122	6912	7720	8548	9395	$\overline{0264}$					
82	7·1154	2066	3002	3962	4947	5958	6996	8062	9158	$\overline{0285}$					
83	8·1443	2636	3863	5126	6427	7769	9152	$\overline{0579}$	$\overline{2052}$	$\overline{3572}$					
84	9·5144	9·677	9·845	10·02	10·20	10·39	10·58	10·78	10·99	11·20					
85	11·43	11·66	11·91	12·16	12·43	12·71	13·00	13·30	13·62	13·95					
86	14·30	14·67	15·06	15·46	15·89	16·35	16·83	17·34	17·89	18·46					
87	19·08	19·74	20·45	21·20	22·02	22·90	23·86	24·90	26·03	27·27					
88	28·64	30·14	31·82	33·69	35·80	38·19	40·92	44·07	47·74	52·08					
89	57·29	63·66	71·62	81·85	95·49	114·6	143·2	191·0	286·5	573·0					

Mean Differences are not given here because they cannot be used so as to secure sufficient accuracy, owing to the rapidity of change in the tangent as the angle increases. It is safer to use Proportional Parts. Compare Note to Art. 197A.

Logarithmic Sines.

Deg.	0′	6′ 0°·1	12′ 0°·2	18′ 0°·3	24′ 0°·4	30′ 0°·5	36′ 0°·6	42′ 0°·7	48′ 0°·8	54′ 0°·9	Mean Differences 1′	2′	3′	4′	5′
0		7·2419	5429	7190	8439	9408	$\overline{0200}$	$\overline{0870}$	$\overline{1450}$	$\overline{1961}$					
1	8·2419	2832	3210	3558	3880	4179	4459	4723	4971	5206					
2	8·5428	5640	5842	6035	6220	6397	6567	6731	6889	7041					
3	8·7188	7330	7468	7602	7731	7857	7979	8098	8213	8326	21	42	62	84	104
4	8·8436	8543	8647	8749	8849	8946	9042	9135	9226	9315	16	32	48	64	80
5	8·9403	9489	9573	9655	9736	9816	9894	9970	$\overline{0046}$	$\overline{0120}$	13	26	39	52	65
6	9·0192	0264	0334	0403	0472	0539	0605	0670	0734	0797	11	22	33	44	55
7	9·0859	0920	0981	1040	1099	1157	1214	1271	1326	1381	10	19	29	38	48
8	9·1436	1489	1542	1594	1646	1697	1747	1797	1847	1895	8	17	25	34	42
9	9·1943	1991	2038	2085	2131	2176	2221	2266	2310	2353	8	15	23	30	38
10	9·2397	2439	2482	2524	2565	2606	2647	2687	2727	2767	7	14	20	27	34
11	9·2806	2845	2883	2921	2959	2997	3034	3070	3107	3143	6	12	19	25	31
12	9·3179	3214	3250	3284	3319	3353	3387	3421	3455	3488	6	11	17	23	28
13	9·3521	3554	3586	3618	3650	3682	3713	3745	3775	3806	5	11	16	21	26
14	9·3837	3867	3897	3927	3957	3986	4015	4044	4073	4102	5	10	15	20	24
15	9·4130	4158	4186	4214	4242	4269	4296	4323	4350	4377	5	9	14	18	23
16	9·4403	4430	4456	4482	4508	4533	4559	4584	4609	4634	4	9	13	17	21
17	9·4659	4684	4709	4733	4757	4781	4805	4829	4853	4876	4	8	12	16	20
18	9·4900	4923	4946	4969	4992	5015	5037	5060	5082	5104	4	8	11	15	19
19	9·5126	5148	5170	5192	5213	5235	5256	5278	5299	5320	4	7	11	14	18
20	9·5341	5361	5382	5402	5423	5443	5463	5484	5504	5523	3	7	10	14	17
21	9·5543	5563	5583	5602	5621	5641	5660	5679	5698	5717	3	6	10	13	16
22	9·5736	5754	5773	5792	5810	5828	5847	5865	5883	5901	3	6	9	12	15
23	9·5919	5937	5954	5972	5990	6007	6024	6042	6059	6076	3	6	9	12	15
24	9·6093	6110	6127	6144	6161	6177	6194	6210	6227	6243	3	6	8	11	14
25	9·6259	6276	6292	6308	6324	6340	6356	6371	6387	6403	3	5	8	11	13
26	9·6418	6434	6449	6465	6480	6495	6510	6526	6541	6556	3	5	8	10	13
27	9·6570	6585	6600	6615	6629	6644	6659	6673	6687	6702	2	5	7	10	12
28	9·6716	6730	6744	6759	6773	6787	6801	6814	6828	6842	2	5	7	9	12
29	9·6856	6869	6883	6896	6910	6923	6937	6950	6963	6977	2	4	7	9	11
30	9·6990	7003	7016	7029	7042	7055	7068	7080	7093	7106	2	4	6	9	11
31	9·7118	7131	7144	7156	7168	7181	7193	7205	7218	7230	2	4	6	8	10
32	9·7242	7254	7266	7278	7290	7302	7314	7326	7338	7349	2	4	6	8	10
33	9·7361	7373	7384	7396	7407	7419	7430	7442	7453	7454	2	4	6	8	10
34	9·7476	7487	7498	7509	7520	7531	7542	7553	7564	7575	2	4	6	7	9
35	9·7586	7597	7607	7618	7629	7640	7650	7661	7671	7682	2	4	5	7	9
36	9·7692	7703	7713	7723	7734	7744	7754	7764	7774	7785	2	3	5	7	9
37	9·7795	7805	7815	7825	7835	7844	7854	7864	7874	7884	2	3	5	7	8
38	9·7893	7903	7913	7922	7932	7941	7951	7960	7970	7979	2	3	5	6	8
39	9·7989	7998	8007	8017	8026	8035	8044	8053	8063	8072	2	3	5	6	8
40	9·8081	8090	8099	8108	8117	8125	8134	8143	8152	8161	1	3	4	6	7
41	9·8169	8178	8187	8195	8204	8213	8221	8230	8238	8247	1	3	4	6	7
42	9·8255	8264	8272	8280	8289	8297	8305	8313	8322	8330	1	3	4	6	7
43	9·8338	8346	8354	8362	8370	8378	8386	8394	8402	8410	1	3	4	5	7
44	9·8418	8426	8433	8441	8449	8457	8464	8472	8480	8487	1	3	4	5	6

Logarithmic Sines.

Deg.	0′	6′ 0°·1	12′ 0°·2	18′ 0°·3	24′ 0°·4	30′ 0°·5	36′ 0°·6	42′ 0°·7	48′ 0°·8	54′ 0°·9	Mean Differences 1′	2′	3′	4′	5′
45	9·8495	8502	8510	8517	8525	8532	8540	8547	8555	8562	1	2	4	5	6
46	9·8569	8577	8584	8591	8598	8606	8613	8620	8627	8634	1	2	4	5	6
47	9·8641	8648	8655	8662	8669	8676	8683	8690	8697	8704	1	2	3	5	6
48	9·8711	8718	8724	8731	8738	8745	8751	8758	8765	8771	1	2	3	4	6
49	9·8778	8784	8791	8797	8804	8810	8817	8823	8830	8836	1	2	3	4	5
50	9·8843	8849	8855	8862	8868	8874	8880	8887	8893	8899	1	2	3	4	5
51	9·8905	8911	8917	8923	8929	8935	8941	8947	8953	8959	1	2	3	4	5
52	9·8965	8971	8977	8983	8989	8995	9000	9006	9012	9018	1	2	3	4	5
53	9·9023	9029	9035	9041	9046	9052	9057	9063	9069	9074	1	2	3	4	5
54	9·9080	9085	9091	9096	9101	9107	9112	9118	9123	9128	1	2	3	4	5
55	9·9134	9139	9144	9149	9155	9160	9165	9170	9175	9181	1	2	3	3	4
56	9·9186	9191	9196	9201	9206	9211	9216	9221	9226	9231	1	2	3	3	4
57	9·9236	9241	9246	9251	9255	9260	9265	9270	9275	9279	1	2	2	3	4
58	9·9284	9289	9294	9298	9303	9308	9312	9317	9322	9326	1	2	2	3	4
59	9·9331	9335	9340	9344	9349	9353	9358	9362	9367	9371	1	1	2	3	4
60	9·9375	9380	9384	9388	9393	9397	9401	9406	9410	9414	1	1	2	3	4
61	9·9418	9422	9427	9431	9435	9439	9443	9447	9451	9455	1	1	2	3	3
62	9·9459	9463	9467	9471	9475	9479	9483	9487	9491	9495	1	1	2	3	3
63	9·9499	9503	9507	9510	9514	9518	9522	9525	9529	9533	1	1	2	3	3
64	9·9537	9540	9544	9548	9551	9555	9558	9562	9566	9569	1	1	2	2	3
65	9·9573	9576	9580	9583	9587	9590	9594	9597	9601	9604	1	1	2	2	3
66	9·9607	9611	9614	9617	9621	9624	9627	9631	9634	9637	1	1	2	2	3
67	9·9640	9643	9647	9650	9653	9656	9659	9662	9666	9669	1	1	2	2	3
68	9·9672	9675	9678	9681	9684	9687	9690	9693	9696	9699	0	1	1	2	2
69	9·9702	9704	9707	9710	9713	9716	9719	9722	9724	9727	0	1	1	2	2
70	9·9730	9733	9735	9738	9741	9743	9746	9749	9751	9754	0	1	1	2	2
71	9·9757	9759	9762	9764	9767	9770	9772	9775	9777	9780	0	1	1	2	2
72	9·9782	9785	9787	9789	9792	9794	9797	9799	9801	9804	0	1	1	2	2
73	9·9806	9808	9811	9813	9815	9817	9820	9822	9824	9826	0	1	1	2	2
74	9·9828	9831	9833	9835	9837	9839	9841	9843	9845	9847	0	1	1	1	2
75	9·9849	9851	9853	9855	9857	9859	9861	9863	9865	9867	0	1	1	1	2
76	9·9869	9871	9873	9875	9876	9878	9880	9882	9884	9885	0	1	1	1	2
77	9·9887	9889	9891	9892	9894	9896	9897	9899	9901	9902	0	1	1	1	1
78	9·9904	9906	9907	9909	9910	9912	9913	9915	9916	9918	0	1	1	1	1
79	9·9919	9921	9922	9924	9925	9927	9928	9929	9931	9932	0	0	1	1	1
80	9·9934	9935	9936	9937	9939	9940	9941	9943	9944	9945	0	0	1	1	1
81	9·9946	9947	9949	9950	9951	9952	9953	9954	9955	9956	0	0	1	1	1
82	9·9958	9959	9960	9961	9962	9963	9964	9965	9966	9967	0	0	1	1	1
83	9·9968	9968	9969	9970	9971	9972	9973	9974	9975	9975	0	0	0	1	1
84	9·9976	9977	9978	9978	9979	9980	9981	9981	9982	9983	0	0	0	0	1
85	9·9983	9984	9985	9985	9986	9987	9987	9988	9988	9989	0	0	0	0	0
86	9·9989	9990	9990	9991	9991	9992	9992	9993	9993	9994	0	0	0	0	0
87	9·9994	9994	9995	9995	9996	9996	9996	9996	9997	9997	0	0	0	0	0
88	9·9997	9998	9998	9998	9998	9999	9999	9999	9999	9999	0	0	0	0	0
89	9·9999	9999	$\bar{0}000$	$\bar{0}000$	0000	$\bar{0}000$	$\bar{0}000$	0000	$\bar{0}000$	$\bar{0}000$	0	0	0	0	0

Logarithmic Cosines.

Deg.	0′	6′ 0°·1	12′ 0°·2	18′ 0°·3	24′ 0°·4	30′ 0°·5	36′ 0°·6	42′ 0°·7	48′ 0°·8	54′ 0°·9	Mean Differences 1′	2′	3′	4′	5′
0	10·0000	0000	0000	0000	0000	0000	0000	0000	0000	9·9999	0	0	0	0	0
1	9·9999	9999	9999	9999	9999	9999	9998	9998	9998	9998	0	0	0	0	0
2	9·9997	9997	9997	9396	9996	9996	9996	9995	9995	9994	0	0	0	0	0
3	9·9994	9994	9993	9993	9992	9992	9991	9991	9990	9990	0	0	0	0	0
4	9·9989	9989	9988	9988	9987	9987	9986	9985	9985	9984	0	0	0	0	0
5	9·9983	9983	9982	9981	9981	9980	9979	9978	9978	9977	0	0	0	0	1
6	9·9976	9975	9975	9974	9973	9972	9971	9970	9969	9968	0	0	0	1	1
7	9·9968	9967	9966	9965	9964	9963	9962	9961	9960	9959	0	0	1	1	1
8	9·9958	9956	9955	9954	9953	9952	9951	9950	9949	9047	0	0	1	1	1
9	9·9946	9945	9944	9943	9941	9940	9939	9937	9936	9935	0	0	1	1	1
10	9·9934	9932	9931	9929	9928	9927	9925	9924	9922	9921	0	0	1	1	1
11	9·9919	9918	9916	9915	9913	9912	9910	9909	9907	9906	0	1	1	1	1
12	9·9904	9902	9901	9899	9897	9896	9894	9892	9891	9889	0	1	1	1	1
13	9·9887	9885	9884	9882	9880	9878	9876	9875	9873	9871	0	1	1	1	2
14	9·9869	9867	9865	9863	9861	9859	9857	9855	9853	9851	0	1	1	1	2
15	9·9849	9847	9845	9843	9841	9839	9837	9835	9833	9831	0	1	1	1	2
16	9·9828	9826	9824	9822	9820	9817	9815	9813	9811	9808	0	1	1	2	2
17	9·9806	9804	9801	9799	9797	9794	9792	9789	9787	9785	0	1	1	2	2
18	9·9782	9780	9777	9775	9772	9770	9767	9764	9762	9759	0	1	1	2	2
19	9·9757	9754	9751	9749	9746	9743	9741	9738	9735	9733	0	1	1	2	2
20	9·9730	9727	9724	9722	9719	9716	9713	9710	9707	9704	0	1	1	2	2
21	9·9702	9699	9696	9693	9690	9687	9684	9681	9678	9675	0	1	1	2	2
22	9·9672	9669	9666	9662	9659	9656	9653	9650	9647	9643	1	1	2	2	3
23	9·9640	9637	9634	9631	9627	9624	9621	9617	9614	9611	1	1	2	2	3
24	9·9607	9604	9601	9597	9594	9590	9587	9583	9580	9576	1	1	2	2	3
25	9·9573	9569	9566	9562	9558	9555	9551	9548	9544	9540	1	1	2	2	3
26	9·9537	9533	9529	9525	9522	9518	9514	9510	9507	9503	1	1	2	3	3
27	9·9499	9495	9491	9487	9483	9479	9475	9471	9467	9463	1	1	2	3	3
28	9·9459	9455	9451	9447	9443	9439	9435	9431	9427	9422	1	1	2	3	3
29	9·9418	9414	9410	9406	9401	9397	9393	9388	9384	9380	1	1	2	3	4
30	9·9375	9371	9367	9362	9358	9353	9349	9344	9340	9335	1	1	2	3	4
31	9·9331	9326	9322	9317	9312	9308	9303	9298	9294	9289	1	2	2	3	4
32	9·9284	9279	9275	9270	9265	9260	9255	9251	9246	9241	1	2	2	3	4
33	9·9236	9231	9226	9221	9216	9211	9206	9201	9196	9191	1	2	3	3	4
34	9·9186	9181	9175	9170	9165	9160	9155	9149	9144	9139	1	2	3	3	4
35	9·9134	9128	9123	9118	9112	9107	9101	9096	9091	9085	1	2	3	4	5
36	9·9080	9074	9069	9063	9057	9052	9046	9041	9035	9029	1	2	3	4	5
37	9·9023	9018	9012	9006	9000	8995	8989	8983	8977	8971	1	2	3	4	5
38	9·8965	8959	8953	8947	8941	8935	8929	8923	8917	8911	1	2	3	4	5
39	9·8905	8899	8893	8887	8880	8874	8868	8862	8855	8849	1	2	3	4	5
40	9·8843	8836	8830	8823	8817	8810	8804	8797	8791	8784	1	2	3	4	5
41	9·8778	8771	8765	8758	8751	8745	8738	8731	8724	8718	1	2	3	5	6
42	9·8711	8704	8697	8690	8683	8676	8669	8662	8655	8648	1	2	3	5	6
43	9·8641	8634	8627	8620	8613	8606	8598	8591	8584	8577	1	2	4	5	6
44	9·8569	8562	8555	8547	8540	8532	8525	8517	8510	8502	1	2	4	5	6

Logarithmic Cosines.

Deg.	0′	6′ 0°·1	12′ 0°·2	18′ 0°·3	24′ 0°·4	30′ 0°·5	36′ 0°·6	42′ 0°·7	48′ 0°·8	54′ 0°·9	Mean Differences 1′	2′	3′	4′	5′
45	9·8495	8487	8480	8472	8464	8457	8449	8441	8433	8426	1	3	4	5	6
46	9·8418	8410	8402	8394	8386	8378	8370	8362	8354	8346	1	3	4	5	7
47	9·8338	8330	8322	8313	8305	8297	8289	8280	8272	8264	1	3	4	6	7
48	9·8255	8247	8238	8230	8221	8213	8204	8195	8187	8178	1	3	4	6	7
49	9·8169	8161	8152	8143	8134	8125	8117	8108	8099	8090	1	3	4	6	7
50	9·8081	8072	8063	8053	8044	8035	8026	8017	8007	7998	2	3	5	6	8
51	9·7989	7979	7970	7960	7951	7941	7932	7922	7913	7903	2	3	5	6	8
52	9·7893	7884	7874	7864	7854	7844	7835	7825	7815	7805	2	3	5	7	8
53	9·7795	7785	7774	7764	7754	7744	7734	7723	7713	7703	2	3	5	7	9
54	9·7692	7682	7671	7661	7650	7640	7629	7618	7607	7597	2	4	5	7	9
55	9·7586	7575	7564	7553	7542	7531	7520	7509	7498	7487	2	4	6	7	9
56	9·7476	7464	7453	7442	7430	7419	7407	7396	7384	7373	2	4	6	8	10
57	9·7361	7349	7338	7326	7314	7302	7290	7278	7266	7254	2	4	6	8	10
58	9·7242	7230	7218	7205	7193	7181	7168	7156	7144	7131	2	4	6	8	10
59	9·7118	7106	7093	7080	7068	7055	7042	7029	7016	7003	2	4	6	9	11
60	9·6990	6977	6963	6950	6937	6923	6910	6896	6883	6869	2	4	7	9	11
61	9·6856	6842	6828	6814	6801	6787	6773	6759	6744	6730	2	5	7	9	12
62	9·6716	6702	6687	6673	6659	6644	6629	6615	6600	6585	2	5	7	10	12
63	9·6570	6556	6541	6526	6510	6495	6480	6465	6449	6434	3	5	8	10	13
64	9·6418	6403	6387	6371	6356	6340	6324	6308	6292	6276	3	5	8	11	13
65	9·6259	6243	6227	6210	6194	6177	6161	6144	6127	6110	3	6	8	11	14
66	9·6093	6076	6059	6042	6024	6007	5990	5972	5954	5937	3	6	9	12	15
67	9·5919	5901	5883	5865	5847	5828	5810	5792	5773	5754	3	6	9	12	15
68	9·5736	5717	5698	5679	5660	5641	5621	5602	5583	5563	3	6	10	13	16
69	9·5543	5523	5504	5484	5463	5443	5423	5402	5382	5361	3	7	10	14	17
70	9·5341	5320	5299	5278	5256	5235	5213	5192	5170	5148	4	7	11	14	18
71	9·5126	5104	5082	5060	5037	5015	4992	4969	4946	4923	4	8	11	15	19
72	9·4900	4876	4853	4829	4805	4781	4757	4733	4709	4684	4	8	12	16	20
73	9·4659	4634	4609	4584	4559	4533	4508	4482	4456	4430	4	9	13	17	21
74	9·4403	4377	4350	4323	4296	4269	4242	4214	4186	4158	5	9	14	18	23
75	9·4130	4102	4073	4044	4015	3986	3957	3927	3897	3867	5	10	15	20	24
76	9·3837	3806	3775	3745	3713	3682	3650	3618	3586	3554	5	11	16	21	26
77	9·3521	3488	3455	3421	3387	3353	3319	3284	3250	3214	6	11	17	23	28
78	9·3179	3143	3107	3070	3034	2997	2959	2921	2883	2845	6	12	19	25	31
79	9·2806	2767	2727	2687	2647	2606	2565	2524	2482	2439	7	14	20	27	34
80	9·2397	2353	2310	2266	2221	2176	2131	2085	2038	1991	8	15	23	30	38
81	9·1943	1895	1847	1797	1747	1697	1646	1594	1542	1489	8	17	25	34	42
82	9·1436	1381	1326	1271	1214	1157	1099	1040	0981	0920	10	19	29	38	48
83	9·0859	0797	0734	0670	0605	0539	0472	0403	0334	0264	11	22	33	44	55
84	9·0192	0120	0046	8·9970	9894	9816	9736	9655	9573	9489	13	26	39	52	65
85	8·9408	9315	9226	9135	9042	8946	8849	8749	8647	8543	16	32	48	64	80
86	8·8436	8326	8213	8098	7979	7857	7731	7602	7468	7330	21	42	62	84	104
87	7·7188	7041	6889	6731	6567	6397	6220	6035	5842	5640					
88	8·5428	5206	4971	4723	4459	4179	3880	3558	3210	2832					
89	8·2419	1961	1450	0870	0200	7·9408	8439	7190	5429	2419					

Logarithmic Tangents.

Deg.	0′	6′ 0°·1	12′ 0°·2	18′ 0°·3	24′ 0°·4	30′ 0°·5	36′ 0°·6	42′ 0°·7	48′ 0°·8	54′ 0°·9	Mean Differences 1′	2′	3′	4′	5′
0		7·2419	5429	7190	8439	9409	$\overline{0200}$	$\overline{0870}$	$\overline{1450}$	$\overline{1962}$					
1	8·2419	2833	3211	3559	3881	4181	4461	4725	4973	5208					
2	8·5431	5643	5845	6038	6223	6401	6571	6736	6894	7046					
3	8·7194	7337	7475	7609	7739	7865	7988	8107	8223	8336	21	42	62	84	104
4	8·8446	8554	8659	8762	8862	8960	9056	9150	9241	9331	16	32	48	64	81
5	8·9420	9506	9591	9674	9756	9836	9915	9992	$\overline{0068}$	$\overline{0143}$	13	26	40	53	66
6	9·0216	0289	0360	0430	0499	0567	0633	0699	0764	0828	11	22	34	45	56
7	9·0891	0954	1015	1076	1135	1194	1252	1310	1367	1423	10	20	29	39	49
8	9·1478	1533	1587	1640	1693	1745	1797	1848	1898	1948	9	17	26	35	43
9	9·1997	2046	2094	2142	2189	2236	2282	2328	2374	2419	8	16	23	31	39
10	9·2463	2507	2551	2594	2637	2680	2722	2764	2805	2846	7	14	21	28	35
11	9·2887	2927	2967	3006	3046	3085	3123	3162	3200	3237	6	13	19	26	32
12	9·3275	3312	3349	3385	3422	3458	3493	3529	3564	3599	6	12	18	24	30
13	9·3634	3668	3702	3736	3770	3804	3837	3870	3903	3935	6	11	17	22	28
14	9·3968	4000	4032	4064	4095	4127	4158	4189	4220	4250	5	10	16	21	26
15	9·4281	4311	4341	4371	4400	4430	4459	4488	4517	4546	5	10	15	20	25
16	9·4575	4603	4632	4660	4688	4716	4744	4771	4799	4826	5	9	14	19	23
17	9·4853	4880	4907	4934	4961	4987	5014	5040	5066	5092	4	9	13	18	22
18	9·5118	5143	5169	5195	5220	5245	5270	5295	5320	5345	4	8	13	17	21
19	9·5370	5394	5419	5443	5467	5491	5516	5539	5563	5587	4	8	12	16	20
20	9·5611	5634	5658	5681	5704	5727	5750	5773	5796	5819	4	8	12	15	19
21	9·5842	5864	5887	5909	5932	5954	5976	5998	6020	6042	4	7	11	15	19
22	9·6064	6086	6108	6129	6151	6172	6194	6215	6236	6257	4	7	11	14	18
23	9·6279	6300	6321	6341	6362	6383	6404	6424	6445	6465	3	7	10	14	17
24	9·6486	6506	6527	6547	6567	6587	6607	6627	6647	6667	3	7	10	13	17
25	9·6687	6706	6726	6746	6765	6785	6804	6824	6843	6863	3	7	10	13	16
26	9·6882	6901	6920	6939	6958	6977	6996	7015	7034	7053	3	6	9	13	16
27	9·7072	7090	7109	7128	7146	7165	7183	7202	7220	7238	3	6	9	12	15
28	9·7257	7275	7293	7311	7330	7348	7366	7384	7402	7420	3	6	9	12	15
29	9·7438	7455	7473	7491	7509	7526	7544	7562	7579	7597	3	6	9	12	15
30	9·7614	7632	7649	7667	7684	7701	7719	7736	7753	7771	3	6	9	12	14
31	9·7788	7805	7822	7839	7856	7873	7890	7907	7924	7941	3	6	9	11	14
32	9·7958	7975	7992	8008	8025	8042	8059	8075	8092	8109	3	6	8	11	14
33	9·8125	8142	8158	8175	8191	8208	8224	8241	8257	8274	3	5	8	11	14
34	9·8290	8306	8323	8339	8355	8371	8388	8404	8420	8436	3	5	8	11	14
35	9·8452	8468	8484	8501	8517	8533	8549	8565	8581	8597	3	5	8	11	13
36	9·8613	8629	8644	8660	8676	8692	8708	8724	8740	8755	3	5	8	11	13
37	9·8771	8787	8803	8818	8834	8850	8865	8881	8897	8912	3	5	8	10	13
38	9·8928	8944	8959	8975	8990	9006	9022	9037	9053	9068	3	5	8	10	13
39	9·9084	9099	9115	9130	9146	9161	9176	9192	9207	9223	3	5	8	10	13
40	9·9238	9254	9269	9284	9300	9315	9330	9346	9361	9376	3	5	8	10	13
41	9·9392	9407	9422	9438	9453	9468	9483	9499	9514	9529	3	5	8	10	13
42	9·9544	9560	9575	9590	9605	9621	9636	9651	9666	9681	3	5	8	10	13
43	9·9697	9712	9727	9742	9757	9773	9788	9803	9818	9833	3	5	8	10	13
44	9·9848	9864	9879	9894	9909	9924	9939	9955	9970	9985	3	5	8	10	13

Logarithmic Tangents.

Deg.	0′	6′ 0°·1	12′ 0°·2	18′ 0°·3	24′ 0°·4	30′ 0°·5	36′ 0°·6	42′ 0°·7	48′ 0°·8	54′ 0°·9	Mean Differences 1′	2′	3′	4′	5′
45	10·0000	0015	0030	0045	0061	0076	0091	0106	0121	0136	3	5	8	10	13
46	10·0152	0167	0182	0197	0212	0228	0243	0258	0273	0288	3	5	8	10	13
47	10·0303	0319	0334	0349	0364	0379	0395	0410	0425	0440	3	5	8	10	13
48	10·0456	0471	0486	0501	0517	0532	0547	0562	0578	0593	3	5	8	10	13
49	10·0608	0624	0639	0654	0670	0685	0700	0716	0731	0746	3	5	8	10	13
50	10·0762	0777	0793	0808	0824	0839	0854	0870	0885	0901	3	5	8	10	13
51	10·0916	0932	0947	0963	0978	0994	1010	1025	1041	1056	3	5	8	10	13
52	10·1072	1088	1103	1119	1135	1150	1166	1182	1197	1213	3	5	8	10	13
53	10·1229	1245	1260	1276	1292	1308	1324	1340	1356	1371	3	5	8	11	13
54	10·1387	1403	1419	1435	1451	1467	1483	1499	1516	1532	3	5	8	11	13
55	10·1548	1564	1580	1596	1612	1629	1645	1661	1677	1694	3	5	8	11	14
56	10·1710	1726	1743	1759	1776	1792	1809	1825	1842	1858	3	5	8	11	14
57	10·1875	1891	1908	1925	1941	1958	1975	1992	2008	2025	3	6	8	11	14
58	10·2042	2059	2076	2093	2110	2127	2144	2161	2178	2195	3	6	9	11	14
59	10·2212	2229	2247	2264	2281	2299	2316	2333	2351	2368	3	6	9	12	14
60	10·2386	2403	2421	2438	2456	2474	2491	2509	2527	2545	3	6	9	12	15
61	10·2562	2580	2598	2616	2634	2652	2670	2689	2707	2725	3	6	9	12	15
62	10·2743	2762	2780	2798	2817	2835	2854	2872	2891	2910	3	6	9	12	15
63	10·2928	2947	2966	2985	3004	3023	3042	3061	3080	3099	3	6	9	13	16
64	10·3118	3137	3157	3176	3196	3215	3235	3254	3274	3294	3	6	10	13	16
65	10·3313	3333	3353	3373	3393	3413	3433	3453	3473	3494	3	7	10	13	17
66	10·3514	3535	3555	3576	3596	3617	3638	3659	3679	3700	3	7	10	14	17
67	10·3721	3743	3764	3785	3806	3828	3849	3871	3892	3914	4	7	11	14	18
68	10·3936	3958	3980	4002	4024	4046	4068	4091	4113	4136	4	7	11	15	19
69	10·4158	4181	4204	4227	4250	4273	4296	4319	4342	4366	4	8	12	15	19
70	10·4389	4413	4437	4461	4484	4509	4533	4557	4581	4606	4	8	12	16	20
71	10·4630	4655	4680	4705	4730	4755	4780	4805	4831	4857	4	8	13	17	21
72	10·4882	4908	4934	4960	4986	5013	5039	5066	5093	5120	4	9	13	18	22
73	10·5147	5174	5201	5229	5256	5284	5312	5340	5368	5397	5	9	14	19	23
74	10·5425	5454	5483	5512	5541	5570	5600	5629	5659	5689	5	10	15	20	25
75	10·5719	5750	5780	5811	5842	5873	5905	5936	5968	6000	5	10	16	21	26
76	10·6032	6065	6097	6130	6163	6196	6230	6264	6298	6332	6	11	17	22	28
77	10·6366	6401	6436	6471	6507	6542	6578	6615	6651	6688	6	12	18	24	30
78	10·6725	6763	6800	6838	6877	6915	6954	6994	7033	7073	6	13	19	26	32
79	10·7113	7154	7195	7236	7278	7320	7363	7406	7449	7493	7	14	21	28	35
80	10·7537	7581	7626	7672	7718	7764	7811	7858	7906	7954	8	16	23	31	39
81	10·8003	8052	8102	8152	8203	8255	8307	8360	8413	8467	9	17	26	35	43
82	10·8522	8577	8633	8690	8748	8806	8865	8924	8985	9046	10	20	29	39	49
83	10·9109	9172	9236	9301	9367	9433	9501	9570	9640	9711	11	22	34	45	56
84	10·9784	9857	9932	0008	0085	0164	0244	0326	0409	0494	13	26	40	53	66
85	11·0580	0669	0759	0850	0944	1040	1138	1238	1341	1446	16	32	48	64	81
86	11·1554	1664	1777	1893	2012	2135	2261	2391	2525	2663	20	42	62	84	104
87	11·2806	2954	3106	3264	3429	3599	3777	3962	4155	4357					
88	11·4569	4792	5027	5275	5539	5819	6119	6441	6789	7167					
89	11·7581	8038	8550	9130	9800	0591	1561	2810	4571	7581					

ANSWERS.

I. PAGE 4.

1. $\cdot 75$. 2. $\cdot 125$. 3. $\cdot 375$. 4. $\cdot 024\dot{1}$.

5. $\cdot 089$. 6. $\cdot 0204045$. 7. $76^g\,91'\,66\cdot 7''$. 8. $21^g\,12'\,50''$.

9. $56^g\,24'\,25''$. 10. $48^g\,75'\,25''$. 11. $12^g\,23'\,40\cdot 7''$.

12. $158^g\,6'\,94\cdot 4''$. 13. $22'\,50''$. 14. $6'\,36\cdot 7''$.

15. $51^\circ\,11'\,15''$. 16. $35^\circ\,9'\,22\cdot 5''$. 17. $36^\circ\,0'\,40\cdot 6''$.

18. $55'\,5\cdot 8''$. 19. $2^\circ\,43'\,6\cdot 4''$. 20. $7^\circ\,17'\,26\cdot 1''$.

21. $3'\,22\cdot 5''$. 22. $20'\,0\cdot 4''$. 23. $45^\circ, 27^\circ$. 24. 72°.

II b. PAGE 11.

1. $\frac{15}{17}, \frac{17}{8}, \frac{8}{15}, \frac{17}{15}$. 2. $\frac{12}{5}, \frac{13}{12}, \frac{5}{13}, \frac{12}{13}$.

3. $25, \frac{4}{5}, \frac{4}{5}, \frac{3}{4}, \frac{5}{3}$. 4. $7, \frac{7}{25}, \frac{24}{7}, \frac{25}{24}$.

5. $\frac{37}{35}, \frac{37}{12}, \frac{35}{12}, \frac{35}{37}$. 6. 12 inches, $\frac{4}{5}, \frac{3}{5}, \frac{4}{3}$.

7. $25, \frac{24}{25}, \frac{7}{25}$. 8. 40 ft., $\frac{40}{41}, \frac{9}{40}$.

9. 20 ft., sine $=\frac{20}{29}$, cosine $=\frac{21}{29}$, tangent $=\frac{20}{21}$.

10. sine $=\frac{1}{\sqrt{5}}$, cosine $=\frac{2}{\sqrt{5}}$, tangent $=\frac{1}{2}$.

11. $\frac{12}{13}, \frac{13}{12}, \frac{77}{85}, \frac{85}{77}$. 12. $\frac{3}{5}, \frac{4}{3}, \frac{20}{29}, \frac{29}{20}$.

III. c. PAGE 23.

1. $\frac{2}{\sqrt{3}}, \sqrt{3}$. 2. $\frac{4}{5}, \frac{3}{5}$. 3. $\frac{1}{\sqrt{15}}, \frac{\sqrt{15}}{4}$. 4. $\frac{\sqrt{5}}{2}, \sqrt{5}$.

5. $\frac{\sqrt{48}}{7}, \frac{1}{\sqrt{48}}$. 6. $\frac{7}{24}, \frac{25}{24}$. 7. $\sqrt{1-\cos^2 A}, \frac{\sqrt{1-\cos^2 A}}{\cos A}$.

8. $\sqrt{1+\cot^2 \alpha}, \frac{\cot \alpha}{\sqrt{1+\cot^2 \alpha}}$. 9. $\frac{\sqrt{\sec^2\theta-1}}{\sec\theta}, \frac{1}{\sqrt{\sec^2\theta-1}}$.

10. $\operatorname{cosec} A=\frac{1}{\sin A}$, $\cos A=\sqrt{1-\sin^2 A}$, $\sec A=\frac{1}{\sqrt{1-\sin^2 A}}$,

$\tan A=\frac{\sin A}{\sqrt{1-\sin^2 A}}$, $\cot A=\frac{\sqrt{1-\sin^2 A}}{\sin A}$. 11. $\sqrt{2}$.

13. $\frac{p}{q}$. 14. $\frac{m^2-1}{2m}, \frac{m^2-1}{m^2+1}$. 15. $\frac{p^2-q^2}{p^2+q^2}, \frac{p^2+q^2}{2pq}$.

16. 3. 17. $\frac{p^2-q^2}{p^2+q^2}$.

IV. a. Page 26.

1. 5. 2. $1\frac{1}{2}$. 3. 0. 4. $2\frac{1}{3}$. 5. $\frac{1}{2}$.

6. $1\frac{1}{2}$. 7. 9. 8. 2. 9. $2\frac{1}{12}$. 10. $\frac{1}{2}$.

11. $\frac{3}{4}$. 12. 0. 13. $1\frac{11}{12}$. 14. $\frac{\sqrt{3}}{2}$. 15. 6.

IV. b. Page 28.

1. 22° 30′. 2. 64° 59′ 30″. 3. 79° 58′ 57″. 4. 45° + A.

5. 45° − B. 6. 60° + B. 7. 50°. 8. 60°.

9. 18°. 10. 9°. 11. 22° 30′. 12. 45°.

13. 30°. 14. 15°. 30. 1. 31. $\tan A$.

IV. c. Page 29_B.

1. ·4179. 2. ·4192. 3. ·5922. 4. ·5840.

5. ·4874. 6. ·6648. 7. 1·1015. 8. 1·0881.

9. 1·8078. 10. 62° 7′. 11. 26° 19′. 12. 71° 13′.

13. 36° 48′. 14. 51° 14′. 15. 29° 44′.

IV. d. Page 31.

1. 45°.	**2.** 60°.	**3.** 60°.	**4.** 45°.	**5.** 60°.
6. 30°.	**7.** 45°.	**8.** 60°.	**9.** 45°.	**10.** 60°.
11. 45°.	**12.** 60°.	**13.** 45°.	**14.** 30°.	**15.** 45°.
16. 60°.	**17.** 30°.	**18.** 60°.	**19.** 45°, 71° 34′.	
20. 60°.	**21.** 30°.	**22.** 30°.	**23.** 60°.	**24.** 45°.

25. 45° or 30°. **26.** 60°. **27.** 45°, 50° 12′. **28.** 45°, 26° 34′.

MISCELLANEOUS EXAMPLES. A. Page 32.

1. (1) ·2537064; (2) ·704. **3.** $\frac{20}{29}, \frac{29}{21}$. **4.** $\frac{15}{8}, \frac{17}{8}$.

6. (1) 15° 28′ 7·5″; (2) 1′ 37·2″. **7.** 41, $\frac{9}{40}, \frac{41}{9}, \frac{41}{40}$.

8. (1) possible; (2) impossible; (3) possible.

10. $\frac{\sqrt{1+\cot^2 a}}{\cot a}$, $\sqrt{1+\cot^2 a}$. **11.** 6.

12. $\frac{m}{\sqrt{m^2+n^2}}, \frac{\sqrt{m^2+n^2}}{n}$. **16.** $\frac{20}{21}, \frac{29}{20}$.

18. 10°. **20.** (1) 30°; (2) 45°, or 75° 58′.

22. 30°. **25.** 19° 28′.

V. a. Page 37.

1. $c=2$, $B=60°$, $C=30°$. **2.** $a=6\sqrt{3}$, $A=60°$, $C=30°$.

3. $c=8\sqrt{3}$, $A=30°$, $B=60°$. **4.** $c=30\sqrt{3}$, $B=30°$, $C=60°$.

5. $b=20\sqrt{2}$, $A=C=45°$. **6.** $c=10\sqrt{3}$, $A=30°$, $B=60°$.

7. $a=2\sqrt{2}$, $B=C=45°$. **8.** $a=9$, $A=60°$, $C=30°$.

9. $B=60°$, $b=27$, $c=18\sqrt{3}$. **10.** $C=65°$, $b=1·69$, $c=2\sqrt{3}$.

11. $B=36°$, $a=3·236$, $b=4·702$. **12.** $B=90°$, $a=2·724$, $c=5·346$.

13. $A=53°$, $a=60·18$, $c=79·86$. **14.** $C=90°$, $a=20$, $c=40$.

15. $A=90°$, $a=4\sqrt{2}$, $b=4$. **16.** $A=90°$, $b=4$, $c=4\sqrt{3}$.

17. 700. **18.** 31. **19.** 86·47. **20.** 978.

21. $C=54°$, $a=73$, $b=124$. **22.** $B=68° 17′$, $C=21° 43′$, $b=93$.

23. $C=50° 36′$, $a=39·3875$, $c=30·435$.

24. $c=353$, $A=39° 36′$, $B=50° 24′$.

25. $A=24° 30′$, $B=65° 30′$, $a=10·37$.

V. b. PAGE 39.

1. $10\sqrt{3}=17\cdot32$.
2. $a=10\sqrt{2}=14\cdot14$, $c=20$.
3. $AB=17\cdot32$ ft., $AC=10$ ft., $AD=8\cdot66$ ft.
4. 12, 4.
5. 22·56.
6. 22·89.
7. $20(3+\sqrt{3})=94\cdot64$.
8. $DC=BD=100$.
9. 13·382, 36° 25′.

VI. a. PAGE 42.

1. 173·2 ft.
2. 277·12 ft.
3. 60°.
4. 50 ft.; 100 ft.
5. 22·5 ft.; 38·97 ft.
6. 30 ft.
7. 200 yds.
8. 51 yds., 81 yds.
9. 86·6 yds.
10. 46·19 ft.
11. 273·2 ft.
12. Each = 70·98 ft.
13. 5 miles.
14. 73·2 ft.
15. 64 ft.
16. 300 ft.
17. 1193 yds.
18. 277·12 yds.

VI. b. PAGE 47.

1. 565·6 yds.; 1131·2 yds.
2. 3·464 miles; 6 miles.
3. 29 miles.
4. 10 miles per hour.
5. 10 miles; 24·14 miles.
6. 16 miles; S. 25° W.
7. 9·656 miles.
8. 5·77 miles; 11·54 miles.
9. 295·1 knots.
10. 5·196 miles per hour; 18 miles.
11. 31 minutes past midnight.
12. 38·97 miles per hour.

VI. c. PAGE 48_A.

1. 36 yds. 1 ft.
2. 340 ft.
3. 161·8 m.
4. 586 ft.
5. 24 yds.
6. 26° 34′, 63° 26′.
7. 80° 26′, 80° 26′, 19° 8′.
8. 107 ft.
9. 244 ft.
10. 668 yds.
11. 28° 15′.
12. 467·9 m., 784·7 m.
13. 118·35 ft.
14. 271 m.
15. 7·9 mi.
16. 970 m.
17. 441·5.
18. N. 38° 23′ E.
19. 13·49 mi., 24·12 mi.

VII. a. PAGE 54.

1. $\frac{\pi}{4}$.
2. $\frac{\pi}{6}$.
3. $\frac{7\pi}{12}$.
4. $\frac{\pi}{8}$.
5. $\frac{\pi}{10}$.
6. $\frac{23\pi}{72}$.
7. $\frac{2\pi}{25}$.
8. $\frac{7\pi}{16}$.
9. ·4509.
10. ·6545.

11. 1·4399. **12.** 1·1999. **13.** 2·7489. **14.** ·9163.
15. 135°. **16.** 28°. **17.** 33° 20′. **18.** 37° 30′.
19. 22° 30′. **20.** 38° 30′. **21.** 29° 48′. **22.** 165°.
23. ·638. **24.** 1·232. **25.** 2·0262. **26.** 2·9979.

VII. b. Page 56.

1. $\frac{3}{4}$. **2.** $\frac{1}{3\sqrt{2}}$. **3.** $4\frac{1}{4}$. **4.** $\frac{3}{\sqrt{2}}$. **5.** 9.
6. $\frac{3}{4}$. **7.** 1. **13.** $\frac{\pi}{4}$, $\frac{2\pi}{15}$. **14.** $\frac{5\pi}{6}$, $\frac{5\pi}{7}$.

VII. c. Page 60.

1. $\frac{1}{5}$. **2.** 300 ft. **3.** A radian.
4. 5·85 yards. **5.** 330. **6.** $\frac{1}{44}$ of a second.
7. $58\frac{2}{3}$. **8.** 40 yds. **9.** 1·15192 miles.
10. 17·904. **11.** 2° 6′. **12.** 45 feet.

MISCELLANEOUS EXAMPLES. B. Page 61.

1. 9°. **2.** 95·26. **3.** 54°. **4.** 3438 inches.
6. 30°. **8.** $22\frac{1}{2}°$, $\frac{\pi}{8}$. **9.** $67\frac{1}{2}°$.
10. $a=6\sqrt{3}$, $c=12$, perp. $=3\sqrt{3}$. **12.** 17·32 ft.
14. 120°, 36°, 24°. **15.** $-\frac{35}{8}$.
17. (1) possible; (2) impossible. **18.** 8·66 miles.
19. $\frac{\pi}{5}$, $\frac{\pi}{3}$, $\frac{7\pi}{15}$. **21.** 90. **24.** 4 miles per hour, 1·732 miles.
25. $\frac{\pi}{8}$. **26.** (1) 30°; (2) 30°. **27.** $\frac{5}{14}$.
29. 200 yards. **30.** 33 feet.

VIII. a. Page 69.

1.	Second.	2.	Third.	3.	First.	4.	Third.
5.	Second.	6.	Second.	7.	Third.	8.	Third.
9.	Sine.	10.	Cosine.	11.	Tangent.	12.	Sine.
13.	Sine.	14.	Tangent.	15.	Sine.	16.	All.
17.	Cosine.	18.	$60°, \frac{\sqrt{3}}{2}$.	19.	$30°, \frac{\sqrt{3}}{2}$.	20.	$45°$, 1.
21.	$45°, \sqrt{2}$.	22.	$30°$, 2.	23.	$60°, \frac{2}{\sqrt{3}}$.	24.	$45°$, 1.
25.	$60°$, 2.	26.	$60°, \sqrt{3}$.				

VIII. b. Page 72.

1.	$-\sqrt{3}$.	2.	$\frac{1}{\sqrt{2}}$.	3.	$-\frac{1}{2}$.
4.	$-\frac{12}{13}, \frac{12}{5}$.	5.	$-\frac{5}{4}, -\frac{3}{4}$.	6.	$\frac{4}{3}, -\frac{3}{5}$.
7.	$\frac{\sqrt{3}}{2}, -\frac{1}{\sqrt{3}}$.	8.	$1, -\sqrt{2}$.	9.	$\pm\frac{5}{13}, \pm\frac{5}{12}$.

IX. Page 79.

1. $\cot A$ decreases from ∞ to 0, then increases numerically from 0 to $-\infty$, then decreases from ∞ to 0, then increases numerically from 0 to $-\infty$. 2. $\operatorname{cosec}\theta$ decreases from ∞ to 1, then increases from 1 to ∞. 3. $\cos\theta$ decreases numerically from -1 to 0, then increases from 0 to 1. 4. $\tan A$ decreases from ∞ to 0, then increases numerically from 0 to $-\infty$. 5. $\sec\theta$ decreases numerically from $-\infty$ to -1, then increases numerically from -1 to $-\infty$. 6. 3. 7. 1. 8. -2. 9. 2.

MISCELLANEOUS EXAMPLES. C. Page 80.

1. $\pm\frac{4}{5}$. 3. $A=60°$, $B=30°$, $a=\frac{21\sqrt{3}}{2}$. 4. $\frac{7}{24}$.

5. 1313 miles, nearly. 6. 301 feet. 7. $3\frac{7}{10}$.

8. 12·003 inches. 10. 200 feet. 11. 45°.

12. 36° 52′, 126° 52′. 13. 10 mi., 1480 mi.

14. 6 km. per hour; 3464 m. 15. $\alpha=49° 19'$, $\beta=2° 14'$.

16. 1. 17. 27·35 mi.; 18·65 mi.

X. a. PAGE 87.

1. $-\frac{1}{\sqrt{2}}$. 2. $\frac{1}{2}$. 3. $\sqrt{3}$. 4. $-\sqrt{2}$.

5. $-\frac{\sqrt{3}}{2}$. 6. 1. 7. -1. 8. $-\frac{1}{2}$.

9. 2. 10. -1. 11. $-\frac{\sqrt{3}}{2}$. 12. -2.

13. -2. 14. $-\frac{1}{\sqrt{2}}$. 15. $\sqrt{3}$. 16. $\sin A$.

17. $\tan A$. 18. $-\cos A$. 19. $-\sec A$. 20. $-\cos A$.

21. $-\tan A$. 22. $-\cos \theta$. 23. $\tan \theta$. 24. $-\text{cosec}\, \theta$.

25. 1. 26. $2 \sin A$. 27. 1.

X. b. PAGE 91.

1. $-\frac{1}{2}$. 2. $-\frac{\sqrt{3}}{2}$. 3. $\frac{1}{2}$. 4. $-\frac{1}{2}$. 5. -1.

6. 1. 7. $\frac{2}{\sqrt{3}}$. 8. $-\frac{1}{\sqrt{3}}$. 9. $-\sqrt{2}$. 10. $\frac{1}{\sqrt{2}}$.

11. 0. 12. $\frac{1}{\sqrt{2}}$. 13. $-\sqrt{2}$. 14. $-\frac{1}{2}$. 15. $-\sqrt{2}$.

16. $-\frac{1}{\sqrt{2}}$. 17. -1. 18. 2. 19. $\frac{1}{\sqrt{3}}$. 20. $\frac{2}{\sqrt{3}}$.

21. $\pm 30°$, $\pm 330°$. 22. $210°$, $330°$, $-30°$, $-150°$.

23. $120°$, $300°$, $-60°$, $-240°$. 24. $135°$, $315°$, $-45°$, $-225°$.

30. 3. 31. $\cot^2 A$. 32. -1. 34. -4.

XI. a. PAGE 97.

4. $1; \frac{24}{25}$. 5. $\frac{33}{65}; \frac{16}{65}$. 6. $-\frac{85}{36}$.

XI. b. PAGE 100.

1. 1. 2. $\frac{1}{7}$. 3. $0; \frac{12}{35}$. 4. $-\frac{278}{29}; \frac{1}{2}$.

11. $\cos A \cos B \cos C - \cos A \sin B \sin C - \sin A \cos B \sin C - \sin A \sin B \cos C$;

$\sin A \cos B \cos C - \cos A \sin B \cos C + \cos A \cos B \sin C + \sin A \sin B \sin C$.

12. $\dfrac{\tan A - \tan B - \tan C - \tan A \tan B \tan C}{1 - \tan B \tan C + \tan C \tan A + \tan A \tan B}$.

13. $\dfrac{\cot A \cot B \cot C - \cot A - \cot B - \cot C}{\cot B \cot C + \cot C \cot A + \cot A \cot B - 1}$.

XI. d. PAGE 104.

1. $-\dfrac{7}{9}$. 2. $\dfrac{17}{25}$. 3. $\dfrac{24}{25}$. 4. $\dfrac{3}{4}$.

5. $\dfrac{7}{25}$; $\dfrac{24}{25}$. 6. $\dfrac{1}{3}$. 7. $\dfrac{1}{7}$.

XI. e. PAGE 106.

1. $-\dfrac{23}{27}$. 2. $\dfrac{117}{125}$. 3. $\dfrac{9}{13}$.

XII. a. PAGE 112.

1. $\sin 4\theta + \sin 2\theta$. 2. $\sin 9\theta - \sin 3\theta$. 3. $\cos 12A + \cos 2A$.

4. $\cos A - \cos 5A$. 5. $\sin 9\theta - \sin \theta$. 6. $\sin 12\theta - \sin 4\theta$.

7. $\cos 6\theta - \cos 12\theta$. 8. $\sin 16\theta - \sin 2\theta$. 9. $\cos 13\alpha + \cos 9\alpha$.

10. $\cos 5\alpha - \cos 15\alpha$. 11. $\dfrac{1}{2}(\sin 11\alpha - \sin 3\alpha)$.

12. $\dfrac{1}{2}(\cos 2\alpha - \cos 4\alpha)$. 13. $\dfrac{1}{2}(\sin 2A + \sin A)$.

14. $\dfrac{1}{2}(\sin 6A - \sin A)$. 15. $\cos\dfrac{7\theta}{3} + \cos\theta$.

16. $\dfrac{1}{2}\left(\cos\dfrac{\theta}{2} - \cos\theta\right)$. 17. $\cos(\alpha+\beta) + \cos(\alpha - 3\beta)$.

18. $\cos(2\alpha - \beta) - \cos(4\alpha + \beta)$. 19. $\sin(3\theta - \phi) + \sin(\theta + 3\phi)$.

20. $\sin(4\theta - \phi) - \sin(2\theta + 3\phi)$. 21. $\dfrac{1}{2}\left(\dfrac{\sqrt{3}}{2} - \sin 2\alpha\right)$.

XII. b. PAGE 114.

1. $2\sin 6\theta \cos 2\theta$. 2. $2\cos 3\theta \sin 2\theta$. 3. $2\cos 5\theta \cos 2\theta$.

4. $2\sin 10\theta \sin\theta$. 5. $2\cos 6\alpha \sin\alpha$. 6. $2\cos\dfrac{11\alpha}{2}\cos\dfrac{5\alpha}{2}$.

7. $2\sin 8\alpha \cos 5\alpha$. 8. $-2\sin 3\alpha \sin 2\alpha$. 9. $2\cos\dfrac{11A}{2}\cos\dfrac{7A}{2}$.

10. $-2\cos 7A \sin 4A$. 11. $\sin 20°$. 12. $\sqrt{3}\cos 10°$.

XII. f. Page 122_A.

10. $1+2\cos A$. 11. $\cdot352$; $69^\circ 23'$.
14. $\sin 10\alpha+\sin 6\alpha+\sin 4\alpha$. 15. $2\cdot4936$. 16. 10°.
18. $\sin 2\theta=\cdot96$; $\cos 2\theta=\cdot28$.
26. $\sin 2\theta=\cdot8$; $\cos 2\theta=-\cdot6$; $\theta=63^\circ 26'$. 27. $\theta=50^\circ 12'$, or 90°.

XIII. a. Page 128.

1. 60°. 2. 120°. 3. $A=30^\circ$, $B=120^\circ$, $C=30^\circ$. 4. 45°.
5. 90°. 6. $A=75^\circ$, $B=45^\circ$, $C=60^\circ$.
7. $A=30^\circ$, $B=135^\circ$, $C=15^\circ$. 8. $28^\circ 56'$. 9. $101^\circ 32'$.
10. $\sqrt{6}$. 11. 7. 12. 8. 13. 14. 14. 9.
15. $b=2\sqrt{6}$, $A=75^\circ$, $C=30^\circ$. 16. $a=\sqrt{5}+1$, $B=36^\circ$, $C=72^\circ$.
17. $C=75^\circ$, $a=c=2\sqrt{3}+2$. 18. $A=105^\circ$, $a=\sqrt{3}+1$, $c=\sqrt{3}-1$.
19. $C=30^\circ$, $a=2$, $b=\sqrt{3}+1$. 21. 2. 22. 6. 23. 60°.
24. 105°, 45°, 30°. 25. 105°, 15°, 60°. 26. $\frac{\sqrt{3}}{2}$, 105°, 15°.

XIII. b. Page 132.

1. $B=60^\circ, 120^\circ$; $C=90^\circ, 30^\circ$; $c=2, 1$.
2. $B=60^\circ, 120^\circ$; $A=75^\circ, 15^\circ$; $a=3+\sqrt{3}, 3-\sqrt{3}$.
3. $A=45^\circ$, $B=75^\circ$, $b=\sqrt{3}+1$; no ambiguity. 4. Impossible.
5. $C=45^\circ, 135^\circ$; $A=105^\circ, 15^\circ$; $a=3+\sqrt{3}, 3-\sqrt{3}$.
6. $C=75^\circ, 105^\circ$; $A=45^\circ, 15^\circ$; $a=2\sqrt{3}, 3-\sqrt{3}$.
7. $A=75^\circ, 105^\circ$; $B=90^\circ, 60^\circ$; $b=2\sqrt{6}, 3\sqrt{2}$.
8. $B=90^\circ$, $C=72^\circ$, $c=4\sqrt{5+2\sqrt{5}}$; no ambiguity. 9. Impossible.

XIII. d. Page 136.

1. 72°, 72°, 36°; each side $=\sqrt{5}+1$.
2. $A=60^\circ$, $a=9-3\sqrt{3}$, $b=3(\sqrt{6}-\sqrt{2})$, $c=3\sqrt{2}$.
3. $A=105^\circ$, $B=15^\circ$, $C=60^\circ$. 4. $B=54^\circ, 126^\circ$; $C=108^\circ, 36^\circ$.
5. $C=60^\circ, 120^\circ$; $A=90^\circ, 30^\circ$; $a=100\sqrt{3}$. No, for $C=90^\circ$.
6. 18°, 126°. 8. $A=90^\circ$, $B=30^\circ$, $C=60^\circ$; $2c=a\sqrt{3}$.

MISCELLANEOUS EXAMPLES. D. Page 138.

2. 43. 3. ∞, 1. 4. -1. 6. $a=2$, $B=30°$, $C=105°$.
9. $A=30°$, $B=75°$, $C=75°$.

XIV. a. Page 145.

1. 10, 8, $-\frac{3}{2}$, $\frac{2}{3}$, $\frac{1}{2}$, -1 2. $\frac{4}{3}$, $\frac{5}{4}$, $-\frac{1}{2}$, $\frac{7}{4}$.
3. 2401, ·5, $\frac{10000}{9}$, 1, $\frac{5}{4}$, 1000, 10000.
4. 5, 3, 3, 4, 0. 5. 0, 2, $\bar{2}$, 0, $\bar{4}$, 3, $\bar{1}$.
6. $\bar{1}$·8091488, 6·8091488, $\bar{4}$·8091488. 7. 3·25, 325, ·000325.
8. 2·8853613. 9. 3·3714373. 10. 1·5475286.
11. 1·9163822. 12. $\bar{1}$·4419030. 13. $\bar{2}$·3380134.
14. $\bar{1}$·6989700. 15. $\bar{1}$·8125919. 16. ·0501716.
17. $\log 2 = ·3010300$. 18. $1 - \log 2 = ·6989700$. 19. 1·320469
20. ·0260315. 21. ·2898431. 22. $\bar{7}$·2621538.
23. 7; 4. 24. 2058.

XIV. b. Page 149.

1. 9·076226. 2. 3·01824. 3. 2467·266.
4. 2·23. 5. 3·54. 6. 1·72. 7. 32, 79.
8. 22·2398. 9. 3·32. 10. 5·77. 11. 2·05.
12. $x = 2\log 2 = ·60206$, $y = -2\log 5 = -1·39794$.
13. $x = \dfrac{\log 3}{\log 3 - \log 2} = 2·71$; $y = \dfrac{\log 2}{\log 3 - \log 2} = x - 1 = 1·71$.
14. $3(b - a - c + 2)$, $\frac{1}{2}(2a - 3c + 6)$.
15. $b + c - 2$, $\frac{1}{6}(3a + 2b + 3c - 5)$.

MISCELLANEOUS EXAMPLES. E. Page 150.

3. $b = \sqrt{3} - 1$, $A = 135°$, $C = 30°$. 8. $A = 105°$, $B = 45°$.

XV. a. Page 155.

1. 6·6947486. 2. ·5404924. 3. 6·4547860.
4. 1·7606731. 5. 6·7840083. 6. 55740·83.
7. 673·5466. 8. ·0106867. 9. ·008287771.
10. ·2531925. 11. 2·031324. 12. 1·389495.
13. 2·424463. 14. 2·069138.

XV. b. PAGE 159.

1. ·6164825. 2. ·7928863. 3. 1·2154838. 4. 62° 42′ 31″.
5. 30° 40′ 23″. 6. 48° 45′ 44″. 7. 9·8440554.
8. 10·1317778. 9. 9·7530545. 10. 44° 17′ 8″.
11. 55° 30′ 39″. 12. 9·6656561. 13. 10·1912872.

XV. c. PAGE 161.

1. 2·36952. 2. 84336. 3. 33·27475.
4. ·03803142. 5. 112184. 6. 1225·508.
7. 27·90209. 8. ·580303. 9. 6·84829.
10. 3·288754, 1·236122. 11. 2273·54.
12. ·5095328. 13. 7·29889. 14. ·045800373.
15. ·1972945. 16. ·0001706363. 17. ·644065.
18. 9·52912. 19. ·3175271. 20. ·335859.
21. ·4221836. 22. 124272·2. 23. 250·2357.
24. (1) 36° 45′ 22″; (2) 19° 28′ 16″. 25. ·441785. 26. 68° 25′ 6″.

XV. d. PAGE 163_C.

1. 49940. 2. 15210. 3. ·0001685. 4. ·7573.
5. 467·3. 6. 13·60. 7. ·05868. 8. 002374.
9. 2·429. 10. ·07612. 11. 1·923. 12. 1·444.
13. 19·97. 14. ·2008. 15. 61·86. 16. 2·258.
17. 2·224. 18. 4·354. 19. 1·784. 20. ·00008855.
21. 64·49. 22. 13·81. 23. ·2510. 24. ·0006814.
25. ·9811. 26. 16. 27. 9·29; 2560. 28. 4·616.
29. $x=1·151$, $y=1·353$. 30. 1·874. 31. 11410.
32. (i) 105·5; (ii) 849·4. 33. 316·9. 34. 5·044.
35. 38·53. 36. 4·015. 37. 2·007 cm. 38. 45·16 cm.
39. ·2905.

XV. e. PAGE 163_H.

1. ·4944. 2. ·7931. 3. ·9651. 4. 1·5171.
5. 1·0932. 6. 1·2153. 7. 18° 13′. 8. 79° 44′.

5. $A=28°20'49''$; $C=39°35'11''$. 6. $A=81°45'2''$, or $23°2'58''$.

7. (1) Not ambiguous, for $C=90°$;
(2) ambiguous, $b=60·3893$ ft.;
(3) not ambiguous.

XVI. f. Page 180.

1. $A=58°24'43''$, $B=48°11'23''$, $C=73°23'54''$.

2. $112°12'54''$, $45°53'33''$, $21°53'33''$. 3. $75°48'54''$.

4. 4227·4815. 5. $B=108°12'26''$, $C=49°27'34''$.

6. $A=105°38'57''$, $B=15°38'57''$. 7. 17·1 or 3·68.

8. $108°26'6''$, $53°7'48''$, $18°26'6''$. 9. $126°22'$; $96°27'$, or $19°3'$.

10. $B=80°46'26·5''$, $C=63°48'33·5''$. 11. $70°0'56''$, or $109°59'4''$.

12. 4·0249. 13. $41°45'14''$.

14. $A=42°0'14''$, $B=55°56'46''$, $C=82°3'$.

15. $41°24'35''$. 16. $A=60°5'34''$, $C=29°54'26''$.

17. 889·2554 ft. 18. $72°12'59''$, $47°47'1''$.

19. 44·4878 ft. 20. $A=102°56'38''$, $B=42°3'22''$.

21. $B=99°54'23''$, $C=32°50'37''$, $a=18·7254$. 22. $72°26'26''$.

23. $A=27°29'56''$, $B=98°55'$, $C=53°35'4''$.

24. $B=32°15'49''$, $C=44°31'17''$, $a=1180·525$.

25. $a=20·9059$, $c=33·5917$. 26. $a=2934·124$, $b=3232·846$.

27. $B=1°1'23''$, $C=147°28'37''$, $a=4389·8$.

28. $A=26°24'23''$, $B=118°18'25''$, $b=642·756$.

29. $53°17'55''$, or $126°42'5''$.

30. $A=31°39'33''$, $C=96°1'27''$, $a=878·753$.

31. $b=4028·38$, $c=2831·67$.

32. $B=75°53'29''$, or $104°6'31''$; $A=60°54'19''$, or $32°41'17''$.

33. Base$=2·44845$ ft., altitude$=·713321$ ft.

34. 90°, nearly. 35. (1) impossible; (2) ambiguous; (3) 63·996.

36. $\theta=72°31'53''$, $c=12·8255$. 37. $\theta=60°13'52''$, $c=19·523977$.

XVI. g. Page 183_D.

1. $108° 38'$. 2. $90°$. 3. $41° 8'$.
4. $A=30° 50'$, $B=131° 15'$, $C=17° 55'$.
5. $A=28° 24'$, $B=44° 30'$, $C=107° 6'$.
6. $A=27° 40'$, $B=95° 27'$, $C=56° 53'$.
7. $116° 28'$. 8. $25° 31'$.
9. $A=38° 10'$, $B=60°$, $C=81° 50'$.
10. $A=53° 8'$, $B=59° 30'$, $C=67° 22'$.
11. $B=51° 35'$, $C=20° 59'$. 12. $A=23° 3'$, $B=33° 15'$.
13. $A=1° 3'$, $C=118° 27'$.
14. $A=97° 15{\cdot}5'$, $B=37° 37{\cdot}5'$, $c=19{\cdot}49$.
15. $B=129° 29'$, $C=13° 31'$, $a=102{\cdot}6$.
16. $B=49° 29'$, $C=70° 31'$. 17. $A=71° 30'$, $B=26° 16'$.
18. $A=97° 31{\cdot}5'$, $B=29° 41{\cdot}5'$. 19. $B=132° 20'$, $C=29° 24'$.
20. $A=24° 15'$, $B=34° 7'$, $c=36{\cdot}49$.
21. $68{\cdot}41$. 22. $65{\cdot}42$. 23. $4{\cdot}200$. 24. 1215.
25. $79{\cdot}75$. 26. $a=4{\cdot}328$ in., $b=5{\cdot}499$ in. 27. $130{\cdot}3$.
28. $a=214{\cdot}2$, $b=223{\cdot}4$. 29. $b=3{\cdot}841$, $c=4{\cdot}762$.
30. $a=26{\cdot}71$, $c=99{\cdot}68$. 31. $57° 18'$.
32. $44° 17'$ or $135° 43'$. 33. $A=36° 18'$, $c=29{\cdot}18$.
34. $B=74° 36'$ or $105° 24'$, $C=65° 24'$ or $34° 36'$, $c=133{\cdot}2$ or $82{\cdot}22$.
35. $B=24° 53'$ or $155° 7'$, $C=134° 26'$ or $4° 12'$, $c=232{\cdot}5$ or $23{\cdot}84$.
36. $A=31° 40'$, $C=96° 2'$, $a=878{\cdot}8$.
37. $A=26° 12'$, $B=118° 48'$, $b=850{\cdot}7$.
38. $A=102° 57'$, $B=42° 3'$. 39. $41° 45'$.
40. $B=75° 12'$ or $104° 7'$, $A=60° 56'$ or $32° 40'$.
41. $A=42°$, $B=55° 58'$, $C=82° 2'$.
42. $41° 24'$. 43. $B=99° 54{\cdot}5'$, $C=32° 50{\cdot}5'$, $a=18{\cdot}72$.
44. $B=1° 1'$, $C=147° 29'$, $a=4391$. 45. $53° 17'$ or $126° 43'$.
46. $12{\cdot}81$. 47. $19{\cdot}53$.

XVII. a. Page 185.

1. $146{\cdot}4$ ft. 2. $880\sqrt{3}=1524$ ft. 5. $ab/(a-b)$ ft.
6. $\frac{1}{3}\sqrt{6}={\cdot}816$ miles. 7. $10(\sqrt{10}+\sqrt{2})=45{\cdot}76$ ft.

9. 1 or $\frac{1}{3}$. **10.** $9\frac{3}{7}$ ft. **12.** $48\sqrt{6}=117\cdot6$ ft.

14. $750\sqrt{6}=1837$ ft. **15.** $2640(3+\sqrt{3})=12492$ ft.

XVII. b. PAGE 190.

1. 30 ft. **2.** $a\sqrt{2}$ ft. **5.** 100 ft.

12. $\sqrt{500-200\sqrt{3}}=12\cdot4$ ft.

XVII. c. PAGE 195.

1. $1060\cdot5$ ft. **2.** $\frac{500\sqrt{6}}{3}=408$ ft.

3. $120\sqrt{6}=294$ ft. **5.** 106 ft.

10. Height $=40\sqrt{6}=98$ ft.; distance $=40(\sqrt{14}+\sqrt{2})=206$ ft.

11. $50\sqrt{120+30\sqrt{6}}=696$ yds.

XVII. d. PAGE 197.

1. 5 miles nearly. **2.** Height $=19\cdot4$ yds.; distance $=102\cdot9$ yds.

3. $200\cdot1$ ft. **4.** Height $=394\cdot4$ ft.; distance $=406\cdot4$ ft.

5. Height $=916\cdot8$ ft.; distance $=\cdot9848$ mile.

6. Height $=45\cdot91$ ft.; distance $=99\cdot17$ ft.

7. $11\cdot55$ or $25\cdot97$ miles per hour.

8. Height $=159\cdot2$ ft.; distance $=215\cdot5$ ft.

XVIII. a. PAGE 206.

1. 9000 sq. ft. **2.** 15390. **3.** $\frac{84}{85}$.

4. $24, \frac{117}{5}, \frac{936}{25}$. **5.** 225 sq. ft. **6.** 672 sq. ft.

7. 36 yds. **8.** $r=4$, $R=8\frac{1}{8}$. **9.** 12, 6, 28.

10. 12, 16, 20.

XVIII. b. Page 210.

1. 26·46 sq. ft.
2. 9·585 yds., 7·18875 sq. yds.
4. 216·23 sq. ft.
5. 128·352 in.
6. 101·78 ft.
7. 57·232 ft.
8. 63·09 sq. ft.

XVIII. c. Page 218.

17. $\frac{\pi}{3}+(-1)^n\frac{1}{2^n}\left(A-\frac{\pi}{3}\right)$,

$\frac{\pi}{3}+(-1)^n\frac{1}{2^n}\left(B-\frac{\pi}{3}\right)$,

$\frac{\pi}{3}+(-1)^n\frac{1}{2^n}\left(C-\frac{\pi}{3}\right)$.

XVIII. d. Page 223.

1. $1\frac{5}{7}$, $2\frac{1}{3}$.
4. Diagonals 65, 63; area 1764.
5. $2\sqrt{77}+6\sqrt{11}$.

XVIII. e. Page 225.

2. 7071 sq. yds.
5. $\sqrt{\frac{x}{y}+\frac{y}{z}+\frac{z}{x}}$.
13. 20, 21, 29.

MISCELLANEOUS EXAMPLES. F. Page 228.

3. Expression $=\cot A+\cot B+\cot C$.
4. $B=45°$, $135°$; $C=105°$, $15°$; $c=\sqrt{6}+\sqrt{2}$, $\sqrt{6}-\sqrt{2}$.
6. 126.
7. 68·3 yds., 35·35 yds.
11. $C=45°$, $135°$; $A=105°$, $15°$; $a=2\sqrt{3}$, $4\sqrt{3}-6$.
12. 10 miles; $10\sqrt{2-\sqrt{2}}$ miles.
24. (1) $90°-\frac{A}{2}$, $90°-\frac{B}{2}$, $90°-\frac{C}{2}$;
(2) $180°-2A$, $180°-2B$, $180°-2C$.
25. Expression $=\sin^2(\alpha-\beta)$.
28. 21·3 miles per hour.
29. 1 hr. 30′; 2 hrs. 16′.

XIX. a. PAGE 235.

1. $n\pi+(-1)^n\frac{\pi}{6}$. 2. $n\pi+(-1)^n\frac{\pi}{4}$. 3. $2n\pi\pm\frac{\pi}{3}$.

4. $n\pi+\frac{\pi}{3}$. 5. $n\pi-\frac{\pi}{6}$. 6. $2n\pi\pm\frac{3\pi}{4}$.

7. $n\pi\pm\frac{\pi}{4}$. 8. $n\pi\pm\frac{\pi}{6}$. 9. $n\pi\pm\frac{\pi}{3}$.

10. $2n\pi\pm\alpha$. 11. $n\pi\pm\alpha$. 12. $n\pi\pm\alpha$.

13. $n\pi$. 14. $\frac{n\pi}{3}+(-1)^n\alpha$. 15. $2n\pi$, or $\frac{2n\pi}{5}$.

16. $\frac{n\pi}{3}$, or $n\pi\pm\frac{\pi}{6}$. 17. $\frac{n\pi}{4}$, or $\frac{n\pi}{3}+(-1)^n\frac{\pi}{18}$.

18. $\frac{(2n+1)\pi}{2}$, or $2n\pi$, or $\frac{(2n+1)\pi}{5}$.

19. $\frac{(2n+1)\pi}{2}$, or $\frac{(2n+1)\pi}{4}$, or $\frac{(2n+1)\pi}{8}$.

20. $n\pi$, or $\frac{(2n+1)\pi}{14}$. 21. $\frac{n\pi}{6}$, or $\frac{n\pi}{9}$.

22. $\frac{(2n+1)\pi}{6}$, or $n\pi\pm\frac{\pi}{8}$. 23. $\frac{(2n+1)\pi}{8}$, or $\frac{n\pi}{3}+(-1)^n\frac{\pi}{9}$.

24. $(2n+1)\pi$, or $2n\pi\pm\frac{\pi}{3}$. 25. $2n\pi$, or $2n\pi\pm\frac{2\pi}{3}$.

26. $n\pi+(-1)^n\frac{\pi}{6}$, or $n\pi+(-1)^n\frac{3\pi}{2}$. 27. $\frac{n\pi}{2}+\frac{\pi}{8}$.

28. $2n\pi+\frac{2\pi}{3}$. 29. $2n\pi-\frac{\pi}{4}$.

XIX. b. PAGE 237.

1. $\frac{(2n+1)\pi}{2(p+q)}$. 2. $\frac{(4k+1)\pi}{2(n-m)}$, or $\frac{(4k-1)\pi}{2(n+m)}$. 3. $2n\pi$, or $2n\pi-\frac{2\pi}{3}$.

4. $2n\pi+\frac{\pi}{2}$, or $(2n+1)\pi+\frac{\pi}{6}$. 5. $2n\pi+\frac{\pi}{2}$, or $2n\pi+\frac{\pi}{6}$.

6. $2n\pi+\frac{5\pi}{12}$, or $2n\pi-\frac{\pi}{12}$. 7. $2n\pi+\frac{\pi}{12}$, or $2n\pi-\frac{7\pi}{12}$.

8. $2n\pi+\frac{5\pi}{4}$, or $2n\pi-\frac{3\pi}{4}$. 9. $2n\pi+\frac{\pi}{3}$, or $(2n+1)\pi$.

10. $\frac{n\pi}{2}+(-1)^n\frac{\pi}{12}$. 11. $\frac{(2n+1)\pi}{4}$, or $n\pi\pm\frac{\pi}{6}$.

12. $n\pi$, or $n\pi \pm \frac{\pi}{6}$.

13. $\frac{n\pi}{2}$.

14. $n\pi + \frac{\pi}{4}$, or $2n\pi$, or $2n\pi + \frac{\pi}{2}$.

[*In some of the following examples, the equations have to be squared, so that extraneous solutions are introduced.*]

15. $\frac{2n\pi}{3} + \frac{\pi}{4}$, or $2n\pi + \frac{\pi}{4}$.

16. $n\pi - \frac{\pi}{4}$, or $\frac{n\pi}{2} + (-1)^n \frac{\pi}{12}$.

17. $\frac{(2n+1)\pi}{10}$, or $\frac{(2n+1)\pi}{2}$.

18. $\frac{(2n+1)\pi}{2}$, or $n\pi \pm \frac{\pi}{3}$.

19. $n\pi + \frac{\pi}{4}$, or $n\pi + \frac{\pi}{6}$.

20. $\frac{n\pi}{2} + (-1)^{n+1} \frac{\pi}{12}$, or $\frac{n\pi}{2}$.

21. $\theta = n\pi \pm \frac{\pi}{4}$, $\phi = n\pi \pm \frac{\pi}{6}$.

22. $\theta = n\pi \pm \frac{\pi}{6}$, $\phi = n\pi \pm \frac{\pi}{3}$.

23. $\theta = n\pi \pm \frac{\pi}{4}$, $\phi = n\pi \pm \frac{\pi}{3}$.

XIX. d. PAGE 244.

1. $\pm \frac{1}{\sqrt{2}}$.

2. ± 1.

3. ± 2.

4. $\frac{-3 \pm \sqrt{17}}{4}$.

5. 1, or $\frac{1}{2}$.

6. 0, or $\pm \frac{1}{2}$.

7. $\pm \frac{1}{\sqrt{2}}$.

8. $\pm \frac{25}{24}$.

9. $\frac{1}{2}$.

10. $\frac{a-b}{1+ab}$.

11. $\frac{b-a}{1+ab}$.

12. $\sqrt{3}$.

13. $x = ac - bd$, $y = bc + ad$.

14. ± 1, or $\pm (1 \pm \sqrt{2})$.

15. $n\pi + \frac{\pi}{4}$.

19. $x=1$, $y=2$; $x=2$, $y=7$.

MISCELLANEOUS EXAMPLES. G. PAGE 246.

2. (1) $\frac{(2n+1)\pi}{2}$, $2n\pi \pm \frac{\pi}{3}$; (2) $2n\pi \pm \frac{\pi}{3}$.

6. 78° 27′ 4″.

9. 6.

10. 800 yds, 146·4 yds, 546·4 yds.

XX. a. PAGE 255.

2. $\sin \frac{A}{2} = \frac{5}{13}$, $\cos \frac{A}{2} = -\frac{12}{13}$.

3. $\sin \frac{A}{2} = -\frac{15}{17}$, $\cos \frac{A}{2} = \frac{8}{17}$.

4. $2\sin\frac{A}{2}=-\sqrt{1+\sin A}+\sqrt{1-\sin A}$;

$2\cos\frac{A}{2}=-\sqrt{1+\sin A}-\sqrt{1-\sin A}$.

5. $2\sin\frac{A}{2}=-\sqrt{1+\sin A}-\sqrt{1-\sin A}$;

$2\cos\frac{A}{2}=-\sqrt{1+\sin A}+\sqrt{1-\sin A}$.

6. $2\sin\frac{A}{2}=+\sqrt{1+\sin A}+\sqrt{1-\sin A}$;

$2\cos\frac{A}{2}=+\sqrt{1+\sin A}-\sqrt{1-\sin A}$.

7. $\sin\frac{A}{2}=\frac{4}{5}$, $\cos\frac{A}{2}=\frac{3}{5}$.

8. $\sin\frac{A}{2}=\frac{8}{17}$, $\cos\frac{A}{2}=-\frac{15}{17}$.

9. (1) $2n\pi-\frac{\pi}{4}$ and $2n\pi+\frac{\pi}{4}$; (2) $2n\pi+\frac{5\pi}{4}$ and $2n\pi+\frac{7\pi}{4}$;

(3) $2n\pi+\frac{3\pi}{4}$ and $2n\pi+\frac{5\pi}{4}$.

10. No; $2\sin\frac{A}{2}=\sqrt{1+\sin A}+\sqrt{1-\sin A}$.

14. (1) $=\sqrt{2}\cos\left(\theta-\frac{\pi}{4}\right)$; (2) $=2\sin\left(\theta-\frac{\pi}{3}\right)$.

15. (1) $=-\sec 2\theta$; (2) $=\tan^2\frac{\theta}{2}$.

XX. b. Page 260.

3. $\frac{1}{5}$.

4. $-\frac{1}{3}$.

XXI. a. Page 267.

1. 1440 yards.
2. $342\frac{6}{7}$ yards.
3. 22 yards.
4. 6′ 34″.
5. 4′35″.
6. 11 ft. 11 in.
7. 210 yards.
8. 9′ 33″.
10. 50 ft.
11. (1) $\frac{\pi}{10800}$; (2) $\frac{\pi}{648000}$.
12. πr^2.
13. $\frac{1}{2}$.
14. $m-n$.
15. $\frac{1}{2}-\frac{\sqrt{3}}{200}=\cdot 491$.
16. $\frac{1}{2}+\frac{11\sqrt{3}}{7200}=\cdot 503$.
17. $\frac{21\sqrt{3}}{55}=39\cdot 7'$.

XXI. b. PAGE 271.

1. 12 miles. **2.** 150 ft. **3.** 15 miles. **4.** 80 ft. 8 in.

5. 204 ft. 2 in. **6.** 54′ 33″. **7.** 104 ft. 2 in.

8. 10560 ft. **9.** 610 ft., $\frac{5}{2}\sqrt{110}$ minutes = 26′ 13″.

11. 8. **12.** -1. **13.** (1) $\cos\alpha$; (2) $-\sin\alpha$.

14. 45° 54′ 33″, 44° 5′ 27″.

MISCELLANEOUS EXAMPLES. H. PAGE 283.

3. 18° 26′ 6″. **6.** 35 miles or 13 miles per hour.

XXIII. a. PAGE 291.

1. $\dfrac{\sin^2 n\alpha}{\sin\alpha}$. **2.** $\sin\dfrac{n\beta}{2}\cos\left(\alpha-\dfrac{n-1}{2}\beta\right)\Big/\sin\dfrac{\beta}{2}$.

3. $-\cos\left(\alpha+\dfrac{\pi}{2n}\right)\Big/\sin\dfrac{\pi}{2n}$. **4.** $\sin\dfrac{n\pi}{2k}\cos\dfrac{(n+1)\pi}{2k}\Big/\sin\dfrac{\pi}{2k}$.

5. $\dfrac{1}{2}$. **6.** $-\dfrac{1}{2}$. **7.** $\cot\dfrac{\pi}{2n}$. **8.** $-\cos\dfrac{\pi}{n}$.

9. $\sin n\alpha$. **10.** $\sin\dfrac{n(\theta+\pi)}{2}\sin\left(\dfrac{n+1}{2}\theta+\dfrac{n-1}{2}\pi\right)\Big/\sin\dfrac{\theta+\pi}{2}$.

11. $\sin\dfrac{n(\pi-\beta)}{2}\cos\left\{\alpha+\dfrac{(n-1)(\pi-\beta)}{2}\right\}\Big/\sin\dfrac{\pi-\beta}{2}$.

12. $\sin\dfrac{n(\pi-2\beta)}{4}\cos\left\{\alpha+\dfrac{(n-1)(\pi-2\beta)}{4}\right\}\Big/\sin\dfrac{\pi-2\beta}{4}$.

13. $\dfrac{n\cos\theta}{2}-\dfrac{\sin n\theta\cos(n+2)\theta}{2\sin\theta}$.

14. $\dfrac{\sin 2n\alpha\sin 2(n+1)\alpha}{2\sin 2\alpha}-\dfrac{n\sin 2\alpha}{2}$.

15. $\operatorname{cosec}\alpha\,\{\tan(n+1)\alpha-\tan\alpha\}$.

16. $\operatorname{cosec}2\theta\,\{\cot\theta-\cot(2n+1)\theta\}$. **17.** $\tan\alpha-\tan\dfrac{\alpha}{2^n}$.

18. $\dfrac{1}{2}(\operatorname{cosec}\alpha-\operatorname{cosec}3^n\alpha)$. **19.** $\dfrac{1}{2}(\tan 3^n\alpha-\tan\alpha)$.

9. 35° 32′. **10.** 51° 35′. **11.** 33° 8′. **12.** 48° 12′.
13. 9·8439. **14.** 10·2823. **15.** 9·4841. **16.** ·3161.
17. ·3033. **18.** 4·159. **19.** ·6992. **20.** 44° 19′.
21. 459·5. **22.** (i) 77·25; (ii) 32·00. **23.** 33° 33′.
24. 19·92. **25.** 166·0.

XVI. a. Page 166.

6. $\frac{1}{2}$. **7.** $\frac{3}{2}$.

XVI. b. Page 169.

1. 113° 34′ 41″. **2.** 49° 28′ 26″. **3.** 55° 46′ 16″.
4. 78° 27′ 47″. **5.** 64° 37′ 23″. **6.** 35° 5′ 49″. **7.** 93° 35′.
8. A = 67° 22′ 49″, B = 53° 7′ 48″, C = 59° 29′ 23″.
9. A = 46° 34′ 3″, B = 104° 28′ 39″, C = 28° 57′ 18″.

XVI. c. Page 173.

1. A = 79° 6′ 24″, B = 40° 53′ 36″. **2.** A = 6° 1′ 54″, C = 108° 58′ 6″.
3. A = 24° 10′ 59″, B = 95° 49′ 1″. **4.** B = 78° 48′ 52″, C = 56° 41′ 8″.
5. A = 27° 38′ 45″, C = 117° 38′ 45″. **6.** 82° 57′ 15″, 36° 32′ 45″.
7. A = 74° 32′ 44″, C = 48° 59′ 16″.
8. B = 100° 47′ 1″, C = 14° 12′ 59″.
9. A = 136° 35′ 21·8″, B = 13° 14′ 33·2″.

XVI. d. Page 174.

1. 89·646162. **2.** 255·3864. **3.** 92·788. **4.** b = 185, c = 192.
5. 321·0793. **6.** a = 765·4321, c = 1035·43.
7. b = 767·792, c = 1263·58.

XVI. e. Page 176.

1. 32° 25′ 35″. **2.** 41° 41′ 28″ or 138° 18′ 32″.
3. A = 100° 34′, B = 34° 26′. **4.** 51° 18′ 21″ or 128° 41′ 39″.

XXIII. b. PAGE 294.

1. $\dfrac{n}{2}+\dfrac{\sin 4n\theta}{4\sin 2\theta}$. 2. $\dfrac{n}{2}$. 3. $\dfrac{n}{2}$.

4. $\dfrac{3\sin\dfrac{n\theta}{2}\sin\dfrac{(n+1)\theta}{2}}{4\sin\dfrac{\theta}{2}}-\dfrac{\sin\dfrac{3n\theta}{2}\sin\dfrac{3(n+1)\theta}{2}}{4\sin\dfrac{3\theta}{2}}$.

5. 0. 6. 0. 7. $\cot\theta-2^n\cot 2^n\theta$.

8. $\dfrac{1}{2}\operatorname{cosec}\alpha\{\tan(n+1)\alpha-\tan\alpha\}$. 9. $\dfrac{\sin 2\theta}{2}-\dfrac{\sin 2^{n+1}\theta}{2^{n+1}}$.

10. $\sin^2\theta-2^n\sin^2\dfrac{\theta}{2^n}$. 11. $\tan^{-1}x-\tan^{-1}\dfrac{x}{n+1}$.

12. $\tan^{-1}(n+1)-\dfrac{\pi}{4}$. 13. $\tan^{-1}\{1+n(n+1)\}-\dfrac{\pi}{4}$.

14. $\tan^{-1}n(n+1)$.

XXIV. a. PAGE 301.

2. $x=a\cos\dfrac{\alpha+\beta}{2}\Big/\cos\dfrac{\alpha-\beta}{2},\quad y=b\sin\dfrac{\alpha+\beta}{2}\Big/\cos\dfrac{\alpha-\beta}{2}$.

3. $x=a(\cos\alpha+\sin\alpha),\ y=b(\sin\alpha-\cos\alpha)$.

13. $4\sin\dfrac{\alpha+\beta+\gamma}{2}\sin\dfrac{\beta+\gamma-\alpha}{2}\sin\dfrac{\gamma+\alpha-\beta}{2}\sin\dfrac{\alpha+\beta-\gamma}{2}$.

14. $4\sin\dfrac{\alpha+\beta+\gamma}{2}\sin\dfrac{\alpha+\beta-\gamma}{2}\cos\dfrac{\beta+\gamma-\alpha}{2}\cos\dfrac{\gamma+\alpha-\beta}{2}$.

15. $-4\cos\left(\dfrac{\alpha+\beta+\gamma}{2}-\dfrac{\pi}{4}\right)\Pi\cos\left(\dfrac{\beta+\gamma-\alpha}{2}+\dfrac{\pi}{4}\right)$.

22. (1) $(a^2+b^2)x^2-2bcx+c^2-a^2=0$;

(2) $(a^2+b^2)^2x^2-2(a^2-b^2)(2c^2-a^2-b^2)x+a^4+b^4+4c^4-2a^2b^2$
$-4a^2c^2-4b^2c^2=0$.

[*Use* $\cos 2\alpha\cos 2\beta=\cos^2(\alpha-\beta)-\sin^2(\alpha+\beta)$.]

XXV. a. Page 318.

1. $2\sqrt{pq}$. 2. 4. 3. 24. 4. 2. 7. $\sqrt{2}$. 8. 2.

9. $\sqrt{a^2-2ab\sin\alpha+b^2}$. 10. $\sqrt{p^2+2pq\sin\alpha+q^2}$.

11. Maximum $=2\sin\frac{\sigma}{2}$. 12. Maximum $=\sin^2\frac{\sigma}{2}$.

13. Minimum $=2\tan\frac{\sigma}{2}$. 14. Minimum $=2\operatorname{cosec}\frac{\sigma}{2}$.

15. Maximum $=\frac{1}{8}$. 16. Minimum $=\sqrt{3}$.

17. Minimum $=\frac{3}{4}$. 18. Minimum $=6$.

19. Minimum $=1$. 20. Minimum $=1$.

21. $\frac{1}{2}(a+c)\pm\frac{1}{2}\sqrt{b^2+(a-c)^2}$. 25. $\frac{5}{3}$.

26. $k^2/(a^2+b^2+c^2)$; $k^2/(a+b+c)$.

XXV. b. Page 324.

1. $\frac{x^2}{a^2}+\frac{y^2}{b^2}=2$. 2. $x^2+y^2=a^2+b^2$. 3. $b^2=a^2(2-a^2)$.

4. $y(x^2-1)=2$. 5. $(a^2-b^2)^2=16ab$. 6. $x^{\frac{4}{3}}y^{\frac{2}{3}}-x^{\frac{2}{3}}y^{\frac{4}{3}}=1$.

7. $a^2b^2(a^2+b^2)=1$. 8. $x^{\frac{2}{3}}+y^{\frac{2}{3}}=a^{\frac{2}{3}}$. 9. $x^{\frac{4}{5}}y^{\frac{6}{5}}-x^{\frac{6}{5}}y^{\frac{4}{5}}=a^2$.

10. $\frac{x^2}{a^2}+\frac{y^2}{b^2}=1$. 12. $x^{\frac{1}{2}}+y^{\frac{1}{2}}=2$. 13. $(x+y)^{\frac{2}{3}}+(x-y)^{\frac{2}{3}}=2$.

16. $a^2+b^2=2c^2$. 20. $(x+y)^{\frac{2}{3}}+(x-y)^{\frac{2}{3}}=2a^{\frac{2}{3}}$. 21. $\frac{x^2}{b^2}+\frac{y^2}{a^2}=1$.

22. $\frac{x^2}{a}+\frac{y^2}{b}=a+b$, or $\{a(y^2-b^2)-b(x^2-a^2)\}^2=-4abx^2y^2$.

24. $xy=(y-x)\tan\alpha$. 25. $a^2+b^2-2\cos\alpha=2$. 26. $a+b=2ab$.

29. $(a+b)(m+n)=2mn$. 30. $x^2+y^2=16a^2$.

31. $(a-b)\{c^2-(a+b)^2\}=4abcm$.

XXV. c. Page 334.

1. $2\cos 20^\circ,\ -2\cos 40^\circ,\ -2\cos 80^\circ$.

2. $2\sin 10^\circ,\ 2\sin 50^\circ,\ -2\sin 70^\circ$.

3. $2\cos 10^\circ,\ -2\cos 50^\circ,\ -2\cos 70^\circ$.

4. $\sin 15^\circ,\ \sin 45^\circ,\ -\sin 75^\circ$.

5. $\frac{1}{a}\sin A,\ \frac{1}{a}\sin(60^\circ - A),\ -\frac{1}{a}\sin(60^\circ + A)$.

6. $2a\cos A,\ 2a\cos(120^\circ \pm A)$.

16. (1) $8x^3 - 4x^2 - 4x + 1 = 0$; (2) $64y^3 - 80y^2 + 24y - 1 = 0$.

17. $64y^3 - 112y^2 + 56y - 7 = 0$.

18. (1) $16x^4 + 8x^3 - 12x^2 - 4x + 1 = 0$; (2) $16x^4 - 8x^3 - 12x^2 + 4x + 1 = 0$.

19. $256y^4 - 448y^3 + 240y^2 - 40y + 1 = 0$.

20. $t^8 - 36t^6 + 126t^4 - 84t^2 + 9 = 0$.

MISCELLANEOUS EXAMPLES. K. Page 337.

1. $7^\circ 12'$: 8 grades. 2. $\frac{4\pi}{15},\ \frac{\pi}{3},\ \frac{2\pi}{5}$. 4. $\frac{3}{2}$.

5. $15\sqrt{3} = 25{\cdot}98$ ft. 6. 790 ft. 7. 5·236 ft.

8. $\frac{8}{17},\ \frac{15}{17}$. 9. $30^\circ,\ 60^\circ$. 10. $\frac{125}{78}$. 12. $\frac{1-\tan^8 A}{\tan^4 A}$.

17. (1) 45°; (2) 60°. 18. 360 yards. 21. 200, 183 nearly.

22. ·09375; 16·7552. 24. $\sqrt{2} : 1$. 25. 1; $\tan A$.

27. $-\frac{5}{31}$. 28. 45° or 60°. 29. $15^\circ 12' 45''$.

30. 260·26 yards. 32. (1) 60°; (2) 60°. 33. $10\frac{1}{2}^\circ$.

34. sine $= \frac{24}{25}$, cosine $= \frac{7}{25}$, tangent $= \frac{24}{7}$.

35. (1) and (3) are impossible, (2) and (4) possible.

37. $B = 45^\circ,\ b = 25\sqrt{2},\ p = 25$. 39. $-\frac{\sqrt{3}}{2},\ -2,\ \frac{1}{\sqrt{3}}$.

40. $45^\circ,\ 135^\circ,\ 225^\circ,\ 315^\circ,\ 405^\circ,\ 495^\circ$. 42. $15^\circ,\ 75^\circ$.

45. $30^\circ,\ 150^\circ,\ 270^\circ$. 46. $\pm{\cdot}8$. 47. $\frac{16}{65}$.

48. $\tan 2A$. **49.** $\pm\frac{1}{\sqrt{3}}$. **50.** $880(3+\sqrt{3})=4164\cdot16$ yds.

51. (1) 0; (2) -2. **53.** $\frac{a^2-b^2}{a^2+b^2}$. **55.** $-\sqrt{3}$, $-\frac{1}{2}$, $\frac{1}{\sqrt{3}}$.

56. $6\frac{4}{11}°$. **57.** $-\frac{15}{352}$. **59.** (1) $\cot C$; (2) 2.

60. $50\sqrt{6}$ ft. **66.** $50\sqrt{3}=86\cdot6$ yds. **67.** 3·141.

70. -1. **73.** $\frac{1}{2}\sin 2\theta$. **74.** $x^{\frac{2}{3}}+y^{\frac{2}{3}}=4^{\frac{2}{3}}$.

75. $15\sqrt{3}$ ft., $15(3+\sqrt{3})$ ft., $60+15\sqrt{3}$ ft. **77.** 8·10.

84. 4·14. **89.** ·38021, 3·73239, 9·76143

95. 27° 45′ 44″. **97.** $\sin 2A=-\frac{336}{625}$; $\tan\frac{A}{2}=-\frac{1}{7}$.

98. $2-\sqrt{3}$. **99.** $\bar{2}\cdot60206$, 1·3802113, $\bar{1}\cdot8239087$.

102. 45, 53; 58° 6′ 33·2″. **104.** 120°. **106.** $\frac{20}{21}$.

107. $4\operatorname{cosec} 2\theta$. **108.** 49° 28′ 32″.

110. 2·30103, $\bar{2}\cdot39794$, ·598626, 9·69897, 9·849485.

112. 114 yds., 57 yds. **114.** 53° 7′ 48″. **115.** ·8, 1·25.

116. ·90309, 1·10739, $\bar{8}\cdot52575$. **117.** $\frac{1}{\sqrt{2}}$.

118. $-\tan\frac{\alpha}{2}\cot\frac{\beta}{2}$, $\tan\frac{\alpha}{2}\tan\frac{\beta}{2}$. **120.** 45 ft.; 58° 12′.

122. 1·60206, $\bar{1}\cdot562469$. **123.** 34° 18′ 1″, 1° 41′ 59″.

127. $\pm\frac{2\sqrt{5}}{5}$. **131.** 1·3011928.

137. ·69897, ·845098, 1·113943.

138. 49° 19′ 30″, 40° 40′ 30″. **143.** ·4855934.

144. 39° 35′ 11″, 28° 20′ 49″.

146. 2·0755469, ·3853509, $\bar{1}\cdot9256038$.

149. $100\sqrt{2}$, $50\sqrt{2}$; 71° 33′ 54″, 108° 26′ 6″.

150. $\operatorname{cosec} x-\operatorname{cosec} 3^n x$. **152.** 45°, 60°, 120°, 135°.

154. $-\frac{56}{33}$; $\frac{4}{5}$, $\frac{12}{13}$, $-\frac{33}{65}$; 120° 30′ 37″.

158. ·30103, ·477121, 1·041393. **160.** 68° 52′ 42″.

161. $\frac{\text{area of circle}}{\text{area of octagon}}=\frac{1380}{1309}$. **165.** $\frac{c\sin\beta}{\sin(\alpha+\beta)}$, $\frac{c\sin\alpha\sin\beta}{\sin(\alpha+\beta)}$.

166. $-4\cos\frac{\alpha}{2}\cos\frac{\beta}{2}\cos\frac{\gamma}{2}$. **173.** $B=4°\,55'\,11''$, $C=168°\,27'\,25''$.

176. $B=105°$, $C=45°$, $a=\sqrt{2}$.

178. $B=81° 47' 12''$ or $98° 12' 48''$; $c=13$ or 11;
$C=68° 12' 48''$ or $51° 47' 12''$.

180. 10000 ft. **186.** $64° 31' 58''$.

191. 2310 sq. ft.; 55 ft., 66 ft., 70 ft.

193. (1) $n\pi$, $\frac{n\pi}{2} \pm \frac{\pi}{16}$; (2) $n\pi \pm \frac{\pi}{6}$.

197. 134·19 ft. **204.** 226·87. **206.** $\frac{4}{3}$, $-\frac{3}{8}$.

212. (1) $(2n+1)\frac{\pi}{8}$, $n\pi + (-1)^n \frac{\pi}{4}$; (2) $n\pi$, $n\pi + \frac{3\pi}{4}$.

214. 20 ft. **219.** $\frac{1}{4}$ or -8. **222.** 37·27919.

231. $\frac{1800}{176\pi} = 3\frac{1}{4}$ miles nearly. **233.** $\pm \frac{7}{23}$.

235. (1) $\frac{n\pi}{4}$, $\frac{2n\pi}{3} \pm \frac{\pi}{9}$; (2) $n\pi + \frac{\pi}{3}$, $n\pi + \frac{3\pi}{4}$.

242. 10°. **246.** 205·4. **252.** 1224·35 yards.

256. 9·65146, 20·5309. **262.** $\alpha + \beta + \gamma = (2n+1)\frac{\pi}{2}$.

264. $\sqrt{2}$ miles. **266.** $\theta = n\pi$.

275. $B=70° 0' 57''$ or $109° 59' 3''$;
$C=59° 59' 3''$ or $20° 0' 57''$. **277.** θ.

279. $\cos(\alpha+\beta+\gamma+\delta) + \cos(\alpha+\beta-\gamma-\delta) + \cos(\alpha+\gamma-\beta-\delta) + \cos(\alpha+\delta-\beta-\gamma)$.

281. 4. **283.** $A=45°$, $B=112\frac{1}{2}°$, $c=\sqrt{2-\sqrt{2}}$.

CAMBRIDGE: PRINTED BY JOHN CLAY, M.A. AT THE UNIVERSITY PRESS.